Microbiologically Influenced Corrosion in the Upstream Oil and Gas Industry

Microbiologically Influenced Corrosion in the Upstream Oil and Gas Industry

Edited by
Torben Lund Skovhus, Dennis Enning,
and Jason S. Lee

CRC Press
Taylor & Francis Group
Boca Raton London New York

CRC Press is an imprint of the
Taylor & Francis Group, an **informa** business

CRC Press
Taylor & Francis Group
6000 Broken Sound Parkway NW, Suite 300
Boca Raton, FL 33487-2742

Printed on acid-free paper

International Standard Book Number-13: 978-1-4987-2656-6 (Hardback)

Library of Congress Cataloging-in-Publication Data

Names: Skovhus, Torben Lund, editor. | Enning, Dennis, editor. | Lee, Jason S., editor.
Title: Microbiologically influenced corrosion in the upstream oil and gas industry / edited by Dr. Torben Lund Skovhus, Dr. Dennis Enning, Dr. Jason S. Lee.
Description: Boca Raton : Taylor & Francis, CRC Press, 2017.
Identifiers: LCCN 2016038559 | ISBN 9781498726566 (hardback) | ISBN 9781498726603 (ebook)
Subjects: LCSH: Petroleum pipelines--Corrosion. | Oilfields--Equipment and supplies--Protection. | Microbiologically influenced corrosion.
Classification: LCC TN879.55 .M529 2017 | DDC 665.5/44--dc23
LC record available at https://lccn.loc.gov/2016038559

Visit the Taylor & Francis Web site at
http://www.taylorandfrancis.com

and the CRC Press Web site at
http://www.crcpress.com

Contents

SECTION III Microbiologically Influenced Corrosion Management

SECTION IV Field Cases and Laboratory Studies

Foreword

MICROBIOLOGICALLY INFLUENCED CORROSION: AS SEEN FROM THE INDUSTRIAL PERSPECTIVE

Microbiologically influenced corrosion (MIC) is a phenomenon that has undoubtedly always occurred in the oil and gas industry; however, it has been and is indeed still poorly understood and recognized. Reluctance among corrosion professionals in all fields to recognize the possibility for MIC to occur has often led to it being considered as the least likely cause of a corrosion failure.

The reviews and case histories in this book demonstrate how the understanding of MIC has progressed in order to enable us to differentiate between electrochemical (abiotic) and microbial (biotic) corrosion. In the past, it was accepted that there was a degree of risk due to microbially generated hydrogen sulfide giving potential for stress corrosion cracking; however, that the microorganisms themselves are an intrinsic part of the corrosion mechanism has taken longer to be acknowledged.

While we have, historically, always acknowledged the need for some form of "good housekeeping" in order to maintain a degree of microbial cleanliness in injection and production systems, this unfortunately has not always translated into practical action—sometimes through ignorance or misunderstanding or, at other times, deliberately to save expenditure on chemicals or equipment. As either industry professionals or academics, it is beholden on us to ensure that those responsible for decision making are informed—in an understandable way—of the risks and uncertainties of MIC.

Over the past few years, as our ability to identify and quantify causative microorganisms has improved, MIC has begun to be more readily accepted as the potential cause of corrosion events. There is, however, no single tool available to specifically identify and confirm MIC or predict the likelihood or severity of its occurrence. As our knowledge increases, it has become apparent that the complexity and interaction of the chemical and physical conditions with microbial communities in oil and gas systems is far beyond that which was initially imagined. That we are able to collect so much practical information from both academia and industry in one volume serves to demonstrate how the underlying science and understanding of MIC has progressed in the past decade.

Reliable detection and quantification of implicated microorganisms using qPCR analysis of functional genes has proven to be challenging because a primer set often fails to detect many species in its target group. This is due to the fact that the design of a single primer set to detect all species in a target community is not feasible due to the variance in genetic sequences that code for the target gene. Significantly higher taxonomic coverage can be realized using primer sets that target the 16S rRNA gene enabling quantification of bacteria, archaea, or total prokaryotes to give less uncertainty. Metagenomic (presence or frequency of the pertinent functional genes) and, more recently, metabolomic (expression of the metabolic pathways) analyses have also been gaining acceptance in the industry although this has often led to little more than a list of microbial families or genera. How best to translate these scientific observations into practical applications and decision making is an ongoing challenge in the industry.

As we move forward, the need for standardized, cost-effective, and practical methods to give input to risk assessment tools is becoming more apparent to allow us to capitalize on the enormous increase in understanding and technical capability. One must, however, not lose sight of the need for sampling methods and data interpretation tools that are fit for purpose in an industry that, at the time of writing is undergoing, yet again, a refocusing of resources and manpower.

Anthony F. Mitchell
Statoil ASA
Norway

Foreword

WHAT'S NEW IN MIC IN THE UPSTREAM OIL AND GAS INDUSTRY: A SCIENTIFIC PERSPECTIVE

Understanding the root causes of corrosion, including microbially influenced corrosion (MIC), has a new urgency for two reasons. First, pipeline spills are increasingly viewed as unacceptable, forcing an increase in efforts and expenditures to prevent these. Second, prices for oil and gas are low, decreasing the margins within which increased corrosion safety must be realized. In view of this, the appearance of a book focused on MIC in the upstream oil and gas industry is timely and potentially important.

The book's chapters point to several common trends. Because corrosion is a multidisciplinary problem with chemical, physical, and microbial causes, a lot of data need to be integrated to come to a more complete understanding, which must ultimately be translated into more effective and economic corrosion control. This includes knowledge of salinity, temperature, pH, water chemistry, and microbiology as well as of the nature and rates of the corrosion process. For instance, knowing the concentration of sulfate is important to evaluate the potential for growth of sulfate-reducing microbes (SRMs).

However, in addition to SRMs, hydrocarbon resource environments harbor a wide variety of microbes. The steel surfaces of upstream oil and gas facilities are also colonized by fermentative, acid-producing bacteria (APB) and by methanogenic *Archaea*. The sulfide produced by SRMs can be oxidized to corrosive sulfur and polysulfide (S-PS) under conditions of oxygen ingress or nitrate amendment by sulfide-oxidizing, nitrate-reducing bacteria (soNRB). Next-generation sequencing technology of 16S rRNA genes can capture the entire diversity of microbes present on corroding surfaces, allowing potential recognition of novel corrosion-causing microbes. Quantitative PCR (qPCR) assays using primers targeting functional genes of SRMs or methanogens can provide numbers, which may indicate success of treatment. Measurement of universal indicators of microbial life, such as ATP, can also quantitate microbial populations. These methods are more sophisticated than the trusted bottle tests to quantitate numbers of lactate-utilizing SRB and APB. Different from bottle tests, some can give readings even when no live bacteria are present. Hence, understanding these methods and acting only on firmly established thresholds is vital. Interestingly, a multipronged approach, in which some or all of these analysis methods are used, is considered the best approach by several authors.

MIC is often evaluated by introducing carbon steel coupons in produced waters in pipelines or storage tanks and by evaluating how these are colonized and corroded. This is a slow process, requiring weeks or months of exposure. A new technique is to use coupons that are precolonized by SRMs and determine how immersion in field waters changes the corrosion rate of these SRM-colonized coupons by rapid electrochemical measurements. This approach has been used for estimating external pipeline corrosion, which is slowed down by the use of coatings or cathodic protection. The latter technology provides soil microbial communities with low potential electrons or hydrogen, allowing growth of hydrogenotrophic microbes (methanogens, acetogens, and SRMs). The effects of this on corrosion protection are currently unknown.

Several chapters describe the corrosive effects of nitrate when this is added as an SRM-control agent. When nitrate is added to sulfide-containing waters, soNRB will couple oxidation of sulfide to reduction of nitrate, forming S-PS, sulfate, nitrite, and nitrogen. If nitrate is added to a sulfide-containing flowing system, S-PS may be widely distributed, causing increased corrosion throughout. The correct way to do this for an aboveground system would be to limit the soNRB reaction to a bioreactor from which elemental sulfur is removed, preventing its dispersal. Nitrate is usually injected to limit or prevent reservoir souring. S-PS formation is prevented in this application by adding excess nitrate, which will oxidize sulfide (if present) to sulfate. Mixing of sulfide-containing produced water and incompletely deaerated seawater in offshore produced water reinjection systems can give rise to S-PS with the associated increased corrosion threat. Increased fractions of sulfide-oxidizing *Epsilonproteobacteria*, as determined by sequencing-based microbial community analyses of mixed waters, may indicate that this threat is real.

Managing all this information to achieve decreased corrosion while limiting costs is key. The late Jan Larsen suggested a two-step approach in which qPCR assays were used initially to determine maximum levels of SRMs and methanogens, present (first-level modeling). This was followed by RT-qPCR assays to identify active cells and evaluate the corrosion threat more precisely (second-level monitoring). Others have described ways to integrate information from more diverse monitoring methods, including the old workhorse, bug bottles.

Complex models based on integrating data from advanced microbial monitoring methods with other corrosion-relevant information are used to predict MIC threats and are described in several chapters. Automated monitoring in which detection of microbial activity over a defined threshold triggers a protective response (e.g., a biocide injection) is also being actively researched. Material choices, that is, carbon steel or corrosion-resistant alloys, in the design phase of a field are crucial to prevent corrosion failure. In the example of an offshore field injected with a mixture of seawater and produced water, the pipeline transporting the mixed water could be made from more corrosion-resistant material than those transporting either of the component waters.

In summary, the study of MIC has progressed to include more advanced monitoring, modeling, and knowledge through case studies, which, when interpreted appropriately, can save money by channeling expensive chemical treatments and material choices to units of the upstream oil and gas industry, where these are most needed. Fundamentals of how microbes can withdraw electrons from metal to stimulate their metabolism are also being pursued and may give further insights, which may eventually translate into better management of corrosion.

Gerrit Voordouw
Petroleum Microbiology Research Group
University of Calgary
Canada

In Memory of Jan Larsen (1955–2015)

Mr. Jan Larsen (born September 19, 1955) passed away on October 9, 2015, after a period of sickness leaving behind his wife Birthe Steinmetz and his family.

Jan had been with Maersk Oil for more than 30 years, working in the production technology discipline, primarily focusing on production chemistry.

Jan was a dedicated, productive, and highly acknowledged scholar, and he leaves behind a strong heritage of technical landmarks. Jan had a deep interest in oilfield microbiology and devoted a lot of time and energy into developing tools and strategies to mitigate the impact of bacteria and other microorganisms in the oil industry.

Jan engineered and introduced the nitrate injection scheme on the Halfdan oilfield (Denmark) and the in-house Maersk MIC model. A model that can predict the threat of microbiologically influenced corrosion in oilfield pipelines and production systems.

In March 2015, Jan received NACE's Award for Technical Achievements, recognizing his contribution toward solving tricky technical challenges in wells. It was the first time that a Maersk Oil employee received this award. In connection with receiving the award, Jan said the following:

> *I am proud that others value the results of work done by Maersk Oil. Preparing papers is partly a way to collect and structure all our experience in one document and partly to share knowledge with others. Presenting papers at conferences is also a test of whether the solutions are sustainable and whether others can use them.*

This shows us how committed Jan was to developing and sharing knowledge, something that will be truly missed in our community—especially by many young industry professionals and scientists to whom Jan was a true mentor and role model.

We will all miss your warm personality, professionalism, dry sense of humor, and sharp paper-writing pen.

Torben Lund Skovhus

In Memory of Hector A. Videla (1938–2012)

Throughout a career that spanned from the 1970s to 2012, Dr. Hector Videla contributed to our understanding of microbiologically influenced corrosion (MIC). Dr. Videla received his PhD in pharmacy and biochemistry from the National University of La Plata, Argentina. Classical training in biochemistry is evident in Hector's early approach to MIC. He was among the first to acknowledge that the impact of microorganisms on corrosion depended on the medium in which reactions were monitored. Over his career, Hector published 260 peer-reviewed papers in international journals, 14 book chapters, and 7 books. He worked with metals exposed to hydrocarbon fuels, milk, seawater, and artificial media. He expanded his research to include biodeterioration of materials other than metals, including cultural artifacts exposed in atmospheric conditions. In 1985, Hector organized a joint congress between Argentina and the United States funded by the U.S. National Science Foundation and the Consejo Nacional de Investigaciones Cientificas y Tecnicas. The congress, one of the first dedicated solely to the topic of MIC, was held in La Plata, Argentina, and included some of the pioneering researchers on that subject. Hector served on the Board of Directors for NACE International from 2003 to 2006, representing the Latin America area. He was a regular contributor to the *Journal of Corrosion* and annual conference proceedings. NACE International acknowledged his contributions in 2001 with the NACE Fellow Honor. During his career, Hector trained nine PhD students who are now contributing in industry and academia. He was passionate about his country, his family, his Siamese cats, and his science. Dr. Hector A. Videla was a gentleman and a scholar, generous with ideas and gracious with his praise of others.

Brenda J. Little

In Memory of Hector A. Videla (1936–2015)

Preface

This book contains the latest technical and scientific contributions to the field of microbiologically influenced corrosion (MIC) in the upstream oil and gas industry. In 25 chapters, a total of 59 authors have contributed by sharing their industry perspective on MIC through profound scientific treatise of the underlying principles or with insightful case studies.

As researchers and industry professionals, we have come a long way in the past decade with our understanding of and approach toward the management of MIC. A large body of knowledge has been generated on the interactions in biofilms, electrochemical reactions, improved surface deposit characterization, inhibitor development, modeling, and the integration of MIC as a relevant corrosion mechanism in the process of corrosion management. These efforts have been accomplished in close and fruitful collaboration between academic researchers and industry professionals.

In a farsighted review on MIC by Videla and Herrera in 2005, the authors envisioned a future where MIC research and development was progressing at an unprecedented pace. Videla and Herrera predicted that DNA-based technology would revolutionize MIC management in the oil and gas industry. As it turns out, they were right in this prediction. However, the authors underestimated the vast amounts of data that modern sequencing technology could produce at a low cost. Finally, Videla and Herrera could not foresee the challenges that this vast volume of new data would present to managers and corrosion engineers who are attempting to manage MIC in everyday oil and gas operations where MIC is still not well understood.

Both the amount and type of data available, and thus the level of complexity, have increased over the past decade with the increased utilization of molecular microbiological methods (MMM). Along with this new type of data, which, in many cases, has replaced the traditional cultivation-based microbiological testing in oil and gas fields, came new challenges. The interpretation of data and development of models describing microbial ecology in the context of biofilm–metal interactions have become an ever more important area of research.

The late Jan Larsen from Maersk Oil, Denmark, was a pioneer in transforming such academic research into applicable industry work practices. As an example, his visionary contribution led to the development of a MIC model that uses molecular biology data for the assessment of MIC in pipelines and topside facilities (Skovhus et al. 2012).

Documentation of research, development, and practical experience in the field of MIC in the oil and gas industry has been fragmented into a multitude of technical forums, journals, and standards. To our knowledge, no single, practical book on MIC is available to industry professionals working as corrosion engineers in the oil and gas industry or engineering students and scientists who desire an in-depth introduction to the topic.

As editors, we therefore took on the task to prepare this book as a comprehensive resource for students, researchers, early career professionals, and integrity managers alike.

Similar to other corrosion mechanisms, MIC can be managed as part of a corrosion management framework (Skovhus and Eckert 2014). Corrosion management is focused on reducing the likelihood of corrosion threats resulting in an unintended outcome (e.g., a leak or failure) rather than addressing consequence or overall risk. At the core of corrosion management is the process of *detecting corrosion threats, identifying preventive and mitigative barriers, and monitoring the effectiveness of those barriers.*

These three fundamental corrosion management processes (detection, mitigation, and monitoring) are at the core of how we have structured this book.

Section I provides an overview of materials and corrosion in oil and gas production in two chapters. Chapter 1 (Heidersbach) reviews the nature of various corrosive environments in upstream oil and gas operations. Chapter 2 (Eckert and Amend) describes the materials used in these corrosive environments, including their susceptibilities to MIC and the process of material selection used when designing for corrosion control.

Section II presents the fundamentals of MIC, including the associated microorganisms and biotic corrosion mechanisms. Chapter 3 (Head) explores the microorganisms found in petroleum systems, both in the deep biosphere (i.e., the reservoir) as well as in the produced fluids, which enter topside facilities. An overview of biofilm formation and some of the common mechanisms associated with MIC, including biotic sulfide and methane production, is provided in Chapter 4 (Skovhus et al.). Chapter 5 (Lahme and Hubert) examines the consequences of shifts between oxidizing and reducing conditions on the microbiology, biogeochemistry, and the resultant corrosivity of oilfield fluids. This topic is particularly relevant where oxidant nitrate is injected into oil reservoirs for the mitigation of reservoir souring. Chapter 6 (Fischer et al.) examines such reservoir souring with a particular focus on the impact of biogenic sulfide on corrosion and materials performance.

Section III is devoted to the ever-increasing field of MIC management. Chapter 7 (Skovhus and Eckert) overviews the steps and practices in MIC management, followed by three parts devoted to the detection (Chapters 8, 9, and 10), mitigation (Chapters 11 and 12), and monitoring (Chapters 13 and 14) of MIC. Starting with MIC detection, Chapter 8 (Lee and Little) examines the techniques used in diagnosing MIC, including the latest molecular methods used in microbial ecology. Chapter 9 (Sharma and Voordouw) explains the importance of using a holistic, multipronged approach to the detection and evaluation of MIC. Chapter 10 (de Rezende) discusses the use of quantitative polymerase chain reaction (qPCR) in the study of sulfate-reducing prokaryotes and critically reflects on the advantages and disadvantages of this "workhorse" technique in petroleum molecular microbiology. For MIC mitigation, Chapter 11 (Morris and van der Kraan) examines the biocides and chemical treatments used to combat microorganisms associated with internal corrosion of pipelines while Chapter 12 (Wilson and Jack) examines the strategies to control external corrosion, which is usually achieved through coating and cathodic protection. Section III concludes with MIC monitoring, where Chapter 13 (De Paula and Keasler) depicts the general approach to monitoring the microbiological threat, including nomenclature, decision loops, and recently developed tools. Chapter 14 (Andersen et al.) reviews the current models available for MIC threat assessment in the upstream oil and gas industry.

Section IV is a collection of MIC case studies comprising field investigations and laboratory experiments. The consequences of oxygen and nitrate ingress into anaerobic systems are described in Chapter 15 (Mand et al.). Chapter 16 (Marks et al.) details the multidisciplinary methodology used to characterize microbial populations and their metabolite profiles in an oil and gas production facility. Chapter 17 (Kilbane) illustrates the specific mitigation and monitoring strategies pertinent to sour systems. The material challenges due to lay-up periods during shutdowns of subsea pipelines, referred to as wet parking, are presented in Chapter 18 (Machuca). The misapplication of biocides and the impact on mitigation of MIC is explored in Chapter 19 (Tidwell et al.). Chapter 20 (Roalkvam and Drønen) presents two field studies involving seawater injection pipelines in the North Sea and the potential threats of nitrate injection to mitigate MIC. Laboratory and field studies of electrical resistance probes used in the monitoring of buried pipelines are detailed in Chapter 21 (Jansen et al.). Chapter 22 (Fichter and Summer) illustrates four case histories in different operating environments, including in a highly saline and sour gas shale, a natural gas sandstone, a light oil shale, and a tight shale gas field. Chapter 23 (de Oliveira et al.) evaluates two stainless steels in a hypersaline laboratory study specifically designed to simulate the operating conditions of a pipeline. Oil and gas production via seawater injection results in a waterflooded reservoir with drastically different chemistries and microbiological populations. Chapter 24 (Wrangham) examines the challenges of identifying, treating, and monitoring MIC in a waterflooded environment. Finally, Chapter 25 (Alabbas) presents four oil and gas corrosion case histories, including their backgrounds, diagnoses, and the remedial actions taken.

The editors wish to thank all of the authors and reviewers who contributed their time, knowledge, and expertise to this book, making it an invaluable resource for many years to come. Thank you!

Torben Lund Skovhus
Dennis Enning
Jason S. Lee

REFERENCES

Videla, HA and Herrera, LK. 2005. Microbiologically influenced corrosion: Looking to the future. *International Microbiology*. 8:169–180.

Skovhus, TL, Holmkvist, L, Andersen, K, Pedersen, H and Larsen, J. 2012. MIC risk assessment of the Halfdan oil export spool. *SPE International Conference and Exhibition on Oilfield Corrosion.* SPE-155080. Society of Petroleum Engineers.

Skovhus, TL and Eckert, RB. 2014. Practical aspects of MIC detection, monitoring and management in the oil and gas industry. *Corrosion 2014.* Paper C2014-3920, Houston, TX: NACE International.

Editors

Torben Lund Skovhus, PhD, is a researcher and project manager at VIA University College in the Centre of Applied Research and Development in Building, Energy & Environment (Horsens, Denmark). He graduated from Aarhus University, Denmark, in 2002 with a master's degree (Cand. Scient.) in biology. In 2005, he earned his PhD from the Department of Microbiology, Aarhus University.

In 2005, Dr. Skovhus was employed at the Danish Technological Institute (DTI) in the Centre for Chemistry and Water Technology, where he was responsible for the consultancy activities for the oil and gas industry around the North Sea. He was heading the DTI Microbiology Laboratory while he was developing several consultancy and business activities with the oil and gas industry. He founded DTI Oil & Gas in both Denmark and Norway, where he was the team and business development leader for 5 years. Thereafter, Dr. Skovhus worked as a project manager at DNV GL (Det Norske Veritas) in the field of corrosion management in both Bergen and Esbjerg.

Dr. Skovhus is currently the chair of NACE TEG286X and ISMOS TSC, an organization he cofounded in 2006. He is an international scientific reviewer and the author of more than 50 technical and scientific papers and book chapters related to industrial microbiology, applied biotechnology, corrosion management, oilfield microbiology, water treatment and safety, reservoir souring, and biocorrosion.

Dennis Enning, PhD, is an engineering associate at ExxonMobil Upstream Research Company (EMURC) in Houston, Texas. He earned his MSc and PhD at the Max Planck Institute for Marine Microbiology (Bremen, Germany) where he investigated the fundamental mechanisms of microbiologically influenced corrosion by lithotrophic, sulfate-reducing bacteria. He earned his undergraduate education in microbiology, biochemistry, and cell biology at Ludwig-Maximilians University in Munich (Germany) and at the University of Edinburgh (UK). In his current role at EMURC, Dr. Enning leads a research program on the assessment, mitigation, and monitoring of MIC with the aim of improving MIC management in oil and gas operations. He further serves as the global subject matter expert on microbial corrosion within ExxonMobil Corporation.

Jason S. Lee, PhD, is a materials engineer at the U.S. Naval Research Laboratory, Stennis Space Center (Mississippi). He earned his MS and PhD in materials science and engineering from the University of Virginia (Charlottesville) and his BA in chemistry and cellular/molecular biology from the University of Michigan (Ann Arbor). His research focuses on improved fundamental understanding of the mechanisms for microbiologically influenced corrosion, biodeterioration, and biodegradation of metals, coatings, and fabrics. His has expertise in electrochemical techniques, computational modeling, and environmental electron microscopy for biological analysis. He is an active member of the National Association of Corrosion Engineers (NACE) and The Electrochemical Society and has coauthored more than 80 peer-reviewed publications in the fields of MIC and localized corrosion.

Contributors

Deniz F. Aktas
OU Biocorrosion Center
Department of Microbiology and Plant Biology
University of Oklahoma
Norman, Oklahoma

Faisal M. Alabbas
Saudi Aramco
Dhahran, Saudi Arabia

Bill Amend
DNV GL
Dublin, Ohio

Erlend Stokstad Andersen
Øst-Riv AS
Oslo, Norway

Vincent Bonifay
OU Biocorrosion Center
Department of Microbiology and Plant Biology
University of Oklahoma
Norman, Oklahoma

Kathryn R. Brown
OU Biocorrosion Center
Department of Microbiology and Plant Biology
University of Oklahoma
Norman, Oklahoma

Monica Canalizo-Hernandez
ExxonMobil Upstream Research Company
Houston, Texas

Joshua T. Cooper
OU Biocorrosion Center
Department of Microbiology and Plant Biology
University of Oklahoma
Norman, Oklahoma

Diogo Coutinho
National Institute of Technology
Rio de Janeiro, Brazil

Annette De Capite
OU Biocorrosion Center
Department of Microbiology and Plant Biology
University of Oklahoma
Norman, Oklahoma

Viviane de Oliveira
National Institute of Technology
Rio de Janeiro, Brazil

Renato De Paula
Nalco Champion—An Ecolab Company
RD&E—Microbiology
Sugar Land, Texas

Júlia R. de Rezende
School of Civil Engineering and Geosciences
Newcastle University
Newcastle upon Tyne, United Kingdom

and

The Lyell Centre for Earth and Marine Science and Technology
Heriot-Watt University
Edinburgh, United Kingdom

Karine Drønen
University of Bergen
Bergen, Norway

Kathleen E. Duncan
OU Biocorrosion Center
Department of Microbiology and Plant Biology
University of Oklahoma
Norman, Oklahoma

Richard B. Eckert
DNV GL
Dublin, Ohio

Jennifer Fichter
Encana Services Company Ltd.
Denver, Colorado

David Fischer
ExxonMobil Upstream Research Company
Houston, Texas

Mariana Galvão
National Institute of Technology
Rio de Janeiro, Brazil

Jan Gerritse
Deltares
Utrecht, the Netherlands

Yves Gunaltun
TOTAL EP
Paris, France

Brian H. Harriman
OU Biocorrosion Center
Department of Microbiology and Plant Biology
University of Oklahoma
Norman, Oklahoma

Ian M. Head
School of Civil Engineering and Geosciences
Newcastle University
Newcastle upon Tyne, United Kingdom

Robert Heidersbach
Dr. Rust, Inc.
Cape Canaveral, Florida

Elizabeth Hillier
DNV GL
London, United Kingdom

Heike Hoffmann
Intertek Production and Integrity Assurance
Aberdeen, United Kingdom

Casey Hubert
School of Civil Engineering and Geosciences
Newcastle University
Newcastle upon Tyne, United Kingdom

and

Department of Biological Sciences
University of Calgary
Calgary, Alberta, Canada

Thomas R. Jack
Biological Sciences
University of Calgary
Calgary, Alberta, Canada

Stefan Jansen
Deltares
Utrecht, the Netherlands

Victor Keasler
Nalco Champion—An Ecolab Company
RD&E—Microbiology
Sugar Land, Texas

John J. Kilbane II
Intertek
Houston, Texas

Amit Kumar
ExxonMobil Upstream Research Company
Houston, Texas

Sven Lahme
School of Civil Engineering and Geosciences
Newcastle University
Newcastle upon Tyne, United Kingdom

Huynh M. Le
OU Biocorrosion Center
Department of Microbiology and Plant Biology
University of Oklahoma
Norman, Oklahoma

Jason S. Lee
Naval Research Laboratory
Stennis Space Center
Hancock County, Mississippi

Brenda J. Little
Naval Research Laboratory
Stennis Space Center
Hancock County, Mississippi

Márcia Lutterbach
National Institute of Technology
Rio de Janeiro, Brazil

Laura L. Machuca
Curtin Corrosion Engineering Industry Centre (CCEIC)
Department of Chemical Engineering
Curtin University
Perth, Australia

Jaspreet Mand
Petroleum Microbiology Research Group
Department of Biological Sciences
University of Calgary
Calgary, Alberta, Canada

Christopher R. Marks
OU Biocorrosion Center
Department of Microbiology and Plant Biology
University of Oklahoma
Norman, Oklahoma

Michael J. McInerney
OU Biocorrosion Center
Department of Microbiology and Plant Biology
University of Oklahoma
Norman, Oklahoma

Brandon E. L. Morris
Dow Europe GmbH
Horgen, Switzerland

Irene Roalkvam
Center for Geobiology and Department for Biology
University of Bergen
Bergen, Norway

Mohita Sharma
Petroleum Microbiology Research Group
Department of Biological Sciences
University of Calgary
Calgary, Alberta, Canada

Yin Shen
Petroleum Microbiology Research Group
Department of Biological Sciences
University of Calgary
Calgary, Alberta, Canada

Torben Lund Skovhus
Centre of Applied Research and Development in Building, Energy and Environment
VIA University College
Horsens, Denmark

Berthil Slim
DNV GL Oil and Gas
Groningen, the Netherlands

Pierre Souquet
Technology Division
TOTAL SA
Pau, France

Blake W. Stamps
OU Biocorrosion Center
Department of Microbiology and Plant Biology
University of Oklahoma
Norman, Oklahoma

Bradley S. Stevenson
OU Biocorrosion Center
Department of Microbiology and Plant Biology
University of Oklahoma
Norman, Oklahoma

Joseph M. Suflita
OU Biocorrosion Center
Department of Microbiology and Plant Biology
University of Oklahoma
Norman, Oklahoma

Elizabeth J. Summer
Ecolyse, Inc.
College Station, Texas

Jan Sunner
OU Biocorrosion Center
Department of Microbiology and Plant Biology
University of Oklahoma
Norman, Oklahoma

Timothy J. Tidwell
Nalco Champion—An Ecolab Company
RD&E—Microbiology
Sugar Land, Texas

Geert M. van der Kraan
Dow Europe GmbH
Horgen, Switzerland

Mirjam van Burgel
DNV GL Oil and Gas
Groningen, the Netherlands

Gerrit Voordouw
Petroleum Microbiology Research Group
Department of Biological Sciences
University of Calgary
Calgary, Alberta, Canada

Sandra L. Wilson
Biological Sciences
University of Calgary
Calgary, Alberta, Canada

Jodi B. Wrangham
Ecolyse, Inc.
College Station, Texas

Section I

Materials and Corrosion in Oil and Gas Production

1 Corrosion in Oil and Gas Production

Robert Heidersbach

CONTENTS

ABSTRACT

Oil and gas production operations are involved with exploring, drilling, and bringing to the surface oil and gas resources that are typically buried deep below the surface of land or the ocean bottom. Once hydrocarbon resources are brought to the surface, separation processes are necessary to segregate liquid crude oil from natural gases and from the formation water that invariably comes to the surface with the oil and gas. The produced formation water is combined with other chemicals and leads to corrosion.

This chapter briefly describes the methods of oil and gas production, the forms of corrosion associated with this production, and some of the means employed to control corrosion. The chemistry of corrosion is discussed with emphasis on those environmental factors most likely to affect corrosion rates. Particular emphasis is placed on pitting corrosion and on environmental cracking, often due to the presences of hydrogen sulfide (H_2S), because these are the two forms of corrosion that are most likely to cause problems in oil and gas production. The differences between corrosion-related inspection and monitoring are explained, as these procedures are frequently misunderstood by many managers and professionals.

Microbiologically influenced corrosion, the subject of the rest of this book, is not discussed in detail in this introductory chapter.

1.1 INTRODUCTION

The purpose of this chapter is to introduce the twin subjects of oil and gas production and corrosion control in oil and gas production.

The American Petroleum Institute (API) divides the petroleum industry into the following categories:

- Upstream
- Downstream
- Pipelines

Many aspects of pipeline corrosion are closely associated with upstream operations, and gathering lines (lines transporting fluids from the wellhead to processing facilities and then to pipelines) have almost the same corrosion problems as pipelines, which generally transport fluids for longer distances after some preliminary processing. The processing often reduces the corrosivity of the fluids being transported. This chapter will concentrate on discussing corrosion problems found in pipelines, storage and process equipment, and wells.

1.2 OIL AND GAS PRODUCTION

Upstream oil and gas operations are those associated with exploring, drilling, and bringing to the surface oil and gas resources that are typically below ground. Figure 1.1

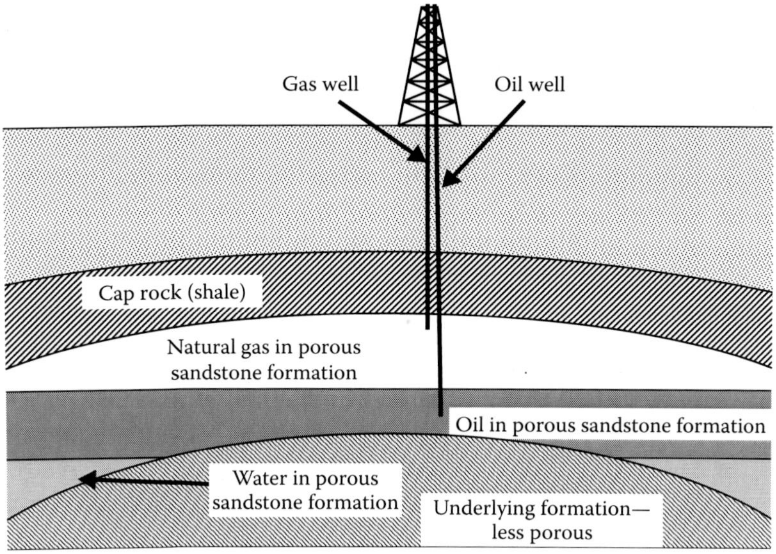

FIGURE 1.1 A typical oil and gas field.

shows a simplified diagram of a conventional oil and gas field along with the associated geological formations. Many fields will produce both oil and natural gas, and as the field ages, the amount of water found in both oil and natural gas will normally increase. This water, when combined with other chemicals found in both oil and natural gas leads to corrosion, and corrosion is a major concern in many oil and gas operations.

Figure 1.1 shows a conventional oil and gas field. Recent developments in hydraulic fracturing ("fracking") have allowed recovery of natural gas from shale formations that previously served as "cap rock" preventing natural gas from rising from the shale formations. The shale, which has very fine-grain structure, serves as a permeation barrier to the upward migration of both oil and natural gas. By creating fractures or fissures in the shale formation, previously nonproducible fluids can be released and then recovered from these formations. It has been estimated that U.S. corrosion costs in oil and gas production are more than $1.3 billion per year with most of these costs associated with pipelines, downhole tubing, and capital expenditures (Ruschau and Al-Anezi 2001). In October 2016, the pipelines running from the largest new oilfield in the world (the Kashagan field in the Caspian Sea near Kazakhstan) returned to production after a delay of over 3 years due to stress-corrosion cracking problems on the original pipelines (Mannheimer, 2016). The original pipelines, which cracked due to hydrogen sulfide in the produced oil and gas, were replaced with more corrosion-resistant pipelines.

Monetary costs are not the only reason for corrosion control. Figure 1.2 shows a fatal natural gas pipeline rupture caused by corrosion that was, at least in part, associated with microbial activity, and Figure 1.3 shows an oil boom used to prevent the spread of crude oil leaking from a submerged pipeline in a small river.

Although corrosion problems with new systems, for example, the Kashagan field in Kazakhstan, sometimes occur, it is more common for corrosion problems to occur in mature oilfields. As oilfields age, the amount of water in produced fluids (water

FIGURE 1.2 Corrosion-induced pipeline explosion.

FIGURE 1.3 Pollution control of an oil slick from a leaking submerged pipeline.

cut) increases, and this causes the produced fluids to become more corrosive. At the same time, production and profits decrease and the need for increased inspection, and maintenance funding becomes harder to obtain. The widely publicized leaks of an Alaska pipeline in 2006 were due, at least in part, to postponement of inspections that might have shown problems developing before they produced leaks and environmental damage (Bailey 2006).

1.3 CORROSION CHEMISTRY

Corrosion can be defined as degradation of a material, usually metallic, due to a reaction with its environment. Most, but not all, corrosion is electrochemical in nature. This means that most corrosion involves the oxidation of metal, which must necessarily be accompanied by reduction of some other chemical species.

A typical oxidation reaction for carbon steel would be

$$Fe \rightarrow Fe^{+2} + 2e^- \quad (1.1)$$

Common reduction reactions associated with corrosion include

$$\text{Hydrogen evolution } 2H^+ + 2e^- \rightarrow H_2 \quad (1.2)$$

Oxygen reduction in acid solutions

$$O_2 + 4H^+ + 4e^- \rightarrow 2H_2O \quad (1.3)$$

In neutral or basic solutions

$$H_2 + 2H_2O + 4e^- \rightarrow 4OH^- \quad (1.4)$$

The reduction reaction is usually corrosion rate–controlling because of the low concentrations of the reducible species in most environments compared with the higher concentrations (essentially 100%) of the metal. As one example, the dissolved oxygen concentration in most air-exposed surface waters is slightly lower than 10 ppm (parts per million). This relatively low dissolved oxygen concentration is usually much higher than the concentration of any other reducible species, and the control of air leakage into surface facilities is a primary means of controlling internal corrosion in topside oilfield equipment and piping.

Figure 1.4 shows the galvanic series of several metals in surface seawater. If two metals are in electrical contact, the anodic metal (closer to the top in Figure 1.4) will corrode at an accelerated rate, and the more cathodic metal will corrode at a slower rate. This is the principle behind cathodic protection, a widely used method of controlling corrosion of buried or immersed metal systems, such as pipelines or offshore platforms.

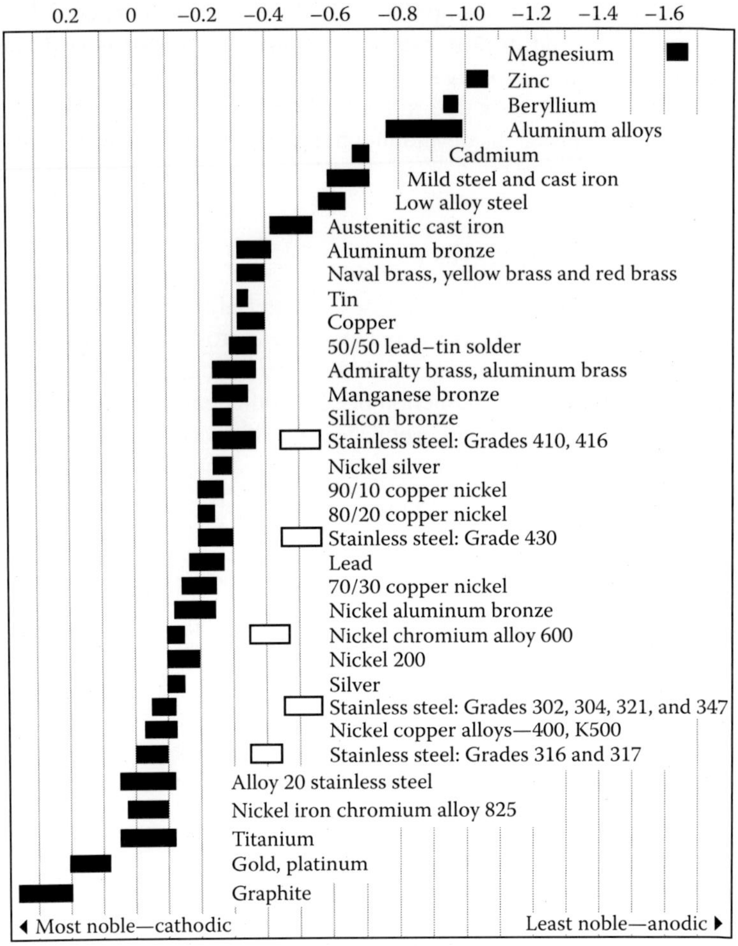

FIGURE 1.4 Galvanic series of common metals in surface seawater.

1.3.1 Effects of pH on Corrosion

The pHs of liquid or soil environments have a strong effect on corrosion. Figure 1.5 shows the effect of pH on the corrosion of iron and carbon steel. At low pHs (acids) the bare metal surface is exposed to the environment, and corrosion proceeds rapidly. Intermediate environments have lower corrosion rates due to the partially protective properties of rust or other iron minerals (passive films) forming on the surface (Uhlig 1948; Revie and Uhlig 2008). At higher pHs, most minerals are relatively insoluble, and mineral scales from the environment (often calcium carbonates) deposit on the surface and lower the corrosion rates even further. The use of protective scales on metal surfaces has been a common means of controlling corrosion in water since the 1920s, and most domestic water suppliers provide slightly hard water with excess dissolved calcium or magnesium for corrosion control purposes (Revie and Uhlig 2008).

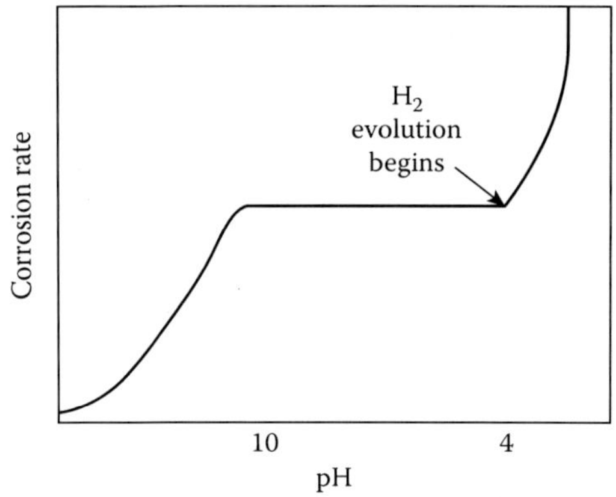

FIGURE 1.5 The effect of pH on the corrosion rate of iron in water at room temperature.

1.3.2 Passivity

Passivity in metals means that the metal is covered with reaction products, usually oxides, of the metal with the environment. Passive films often slow the corrosion rates as shown in Figure 1.5 for iron in water, but carbon steels and many other metals require some other means of lowering corrosion rates to acceptable levels. Passive films on aluminum, stainless steel, or titanium alloys often lower corrosion rates enough for use in many environments. On carbon steel, the most common metal used in oil and gas production, it is usually necessary to use protective coatings, corrosion inhibitors, or cathodic protection to achieve acceptably low corrosion rates.

1.4 CORROSIVE ENVIRONMENTS

1.4.1 Oilfield Corrosion

The majority of oilfield corrosion requires liquid water. For this reason, as an oilfield ages and the water cut increases, corrosion also increases. This is shown in Figure 1.6, which shows the effect of water cut on corrosion rates. For some oilfields, it may take several years before corrosion becomes a problem. Unfortunately, this means that corrosion and other maintenance problems become more important at a time when maintenance funds, often related to production rates, decrease.

Water has very limited solubility in hydrocarbons, and the presence of a separated water phase is necessary for corrosion. The low-corrosion region in Figure 1.6 is where most of the metal surface is in contact with a water-in-oil emulsion. Figure 1.7 shows corrosion at the 6 o'clock position of a gathering line. The field had aged to the point where production rates were relatively low, and even though the oil had been processed in an oil–water separator, there was enough residual water in the oil to form a water layer at the bottom of the line. When production rates were higher, the faster liquid flow kept the water in an emulsified suspension, but as the field declined in production rates, the small amount of water in the oil was able to separate out and cause corrosion.

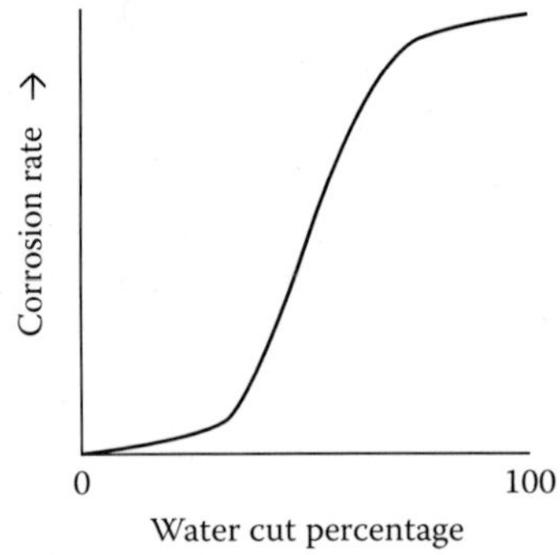

FIGURE 1.6 The effects of water cut on the corrosion rates of oil well tubing. (From R. Heidersbach, *Metallurgy and Corrosion Control in Oil and Gas Production*, John Wiley & Sons, Hoboken, NJ, 2011.)

FIGURE 1.7 Corrosion at the bottom (6 o'clock position) of a gathering line in an old oilfield with reduced production rates.

1.4.2 Natural Gas Corrosion

While most crude oil is not corrosive, and oilfields may take years before they become corrosive, natural gas production must deal with corrosion from the beginning of production. This is because all natural gas reservoirs will produce some water, and minor components of the natural gas will dissolve in this water and make it corrosive. Corrosion occurs when condensation of water with these corrosive components, usually small organic acid molecules, accumulates on metal surfaces.

1.4.3 Environmental Factors Influencing Corrosion

Most corrosion rates are controlled by the concentrations of the chemically reducible impurities in the water phases of hydrocarbon liquids (Shankardass 2004). Oxygen is the most common problem in oilfield fluids, and it is several times more corrosive than carbon dioxide (CO_2) and hydrogen sulfide (H_2S). While oxygen is chemically reducible when dissolved in water, the other two gases form weak mineral acids, and the reducible species are the hydrogen ions in the acids.

1.4.4 H_2S Embrittlement

The discussion so far has concentrated on corrosion where the metal oxidizes due to the presence of a chemically reducible species, such as oxygen or the ions in an acid, which produces what some authorities call weight-loss corrosion. H_2S can also cause embrittlement of many metals, including high-strength steels. ANSI/NACE MR 0175/ISO 15156 is the international standard defining conditions where H_2S cracking may occur in upstream oil and gas environments (ANSI/NACE/ISO 2015). This three-part standard is hundreds of pages long, and a discussion of it is beyond the scope of this brief report.

1.5 METALS USED IN OIL AND GAS PRODUCTION

Until the 1980s, metals used in upstream oil and gas production operations were primarily carbon steels (Kane 2006). The development of deep, hot gas wells in the 1980s led to the use of corrosion-resistant alloys (CRAs), and this trend continues as the industry becomes involved in more aggressive environments. The consequences of corrosion, for example, in remote offshore locations, mean that material and equipment reliability are becoming increasingly important.

Most pipelines and gathering lines continue to be made from carbon steel although some operators are now using stainless steel for offshore pipelines where the consequences of potential leaks justify the added cost. Because material costs are only about 25% of the cost of installing offshore pipelines, the use of CRAs such as 13-chrome stainless steel only increases the cost by about 25%.

1.5.1 Carbon Steels

Carbon steels are usually specified by their strength levels. Figure 1.8 shows the stress–strain curve for low-strength carbon steels. Most materials textbooks have detailed discussions on the equipment and procedures used to develop stress–strain curves as well as on procedures to determine other mechanical properties of metals and nonmetallic materials (Fischer 2008; Smith and Hashemi 2009; Mittemeijer 2011; Callister and Rethwisch 2013). Hardness (resistance to penetration), ductility (ability to stretch or elongate without fracture), and toughness (resistance to impact loading, often at reduced temperatures) are also included in many specifications. API standards for oil country tubular goods (OCTGs; tubing, casing, drill pipe, and pipeline steel) are usually based on mechanical properties. CRA specifications emphasize alloy chemistry in addition to the mechanical properties listed above.

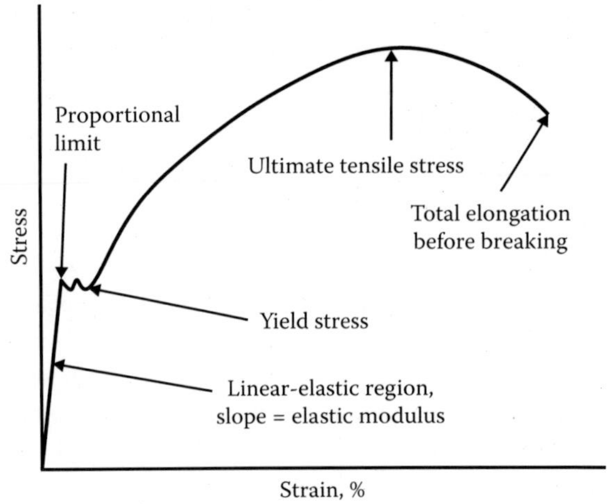

FIGURE 1.8 Stress–strain curve for low-strength carbon steel showing upper and lower yield stresses.

1.5.2 Corrosion-Resistant Alloys (CRAs)

The most common CRAs used in oilfields are stainless steels although titanium, aluminum, nickel-based alloys, and other chemistries also find uses in processing equipment and other specialized applications. When CRAs are specified, it is usually because the chemical environment or elevated temperatures are too aggressive for carbon steels and the use of corrosion inhibitors, protective coatings, or cathodic protection. Detailed discussions on these alloys are available, and they are unlikely to be used in situations where microbially influenced corrosion (MIC) is likely (Kane 2006; Heidersbach 2011).

1.5.3 Welds

Welding is used to join large metal structures, and welded locations produce problems for oilfield operations. One of the problems with welds is that they often have rougher surfaces, which can make them harder to paint and more likely to have biofilms or other deposits. The structure of welds also means that they are more likely to have defects than the parent metal, and this means they are more likely to fracture. For this reason, it is common to specify that longitudinally welded pipe joints be placed in pipelines with their longitudinal welds alternating at the 10 o'clock and 2 o'clock positions. This puts the welds in the upper quadrant of the pipeline, where corrosion is least likely to occur, and it decreases the chance of a rupture that may be associated with one longitudinal weld running beyond the girth weld and into the next joint. The seam (longitudinal) welds on older pipelines frequently had poor toughness properties, but the practice continues for newer construction with supposedly better weld quality and better steel-making practices. This construction practice is shown in Figure 1.9.

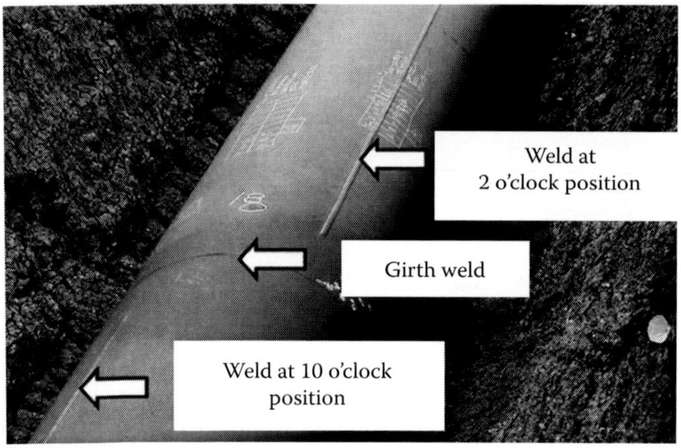

FIGURE 1.9 Pipeline with alternating longitudinal welds at 10 o'clock and 2 o'clock.

1.6 FORMS OF CORROSION

There are a number of ways of describing corrosion, and many authors classify the forms of corrosion using the terms first popularized by M. Fontana, who was a longtime corrosion authority at Ohio State University. This discussion follows the general format developed by Dr. Fontana and then lists special forms of corrosion or terms to describe them that are unique to the oilfields.

1.6.1 General Corrosion

This is the most common form of corrosion, but it receives relatively little technical attention because the means of controlling it seldom require the attention of management or engineers.

Control of general corrosion is usually amenable to the use of protective coatings, corrosion inhibitors, or cathodic protection. Figures 1.10 and 1.11 show instances where improper control of the environmental chemistry caused general corrosion. Figure 1.12 shows an example where cathodic protection did not work because the electrical current could not pass beyond the impermeable pipeline coating that

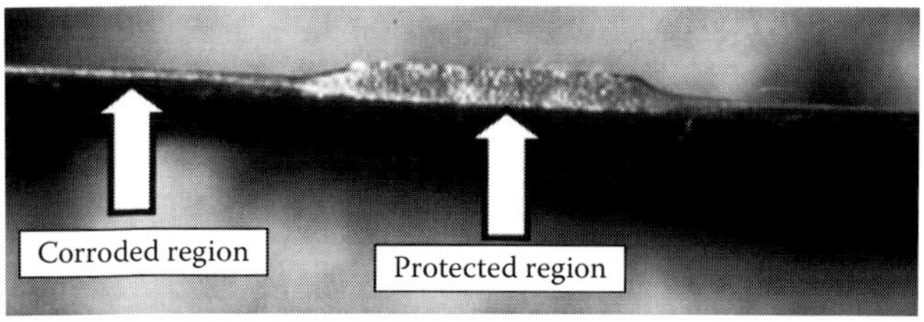

FIGURE 1.10 Cross-section of carbon steel with uniform corrosion.

FIGURE 1.11 General corrosion along the bottom of a gas well flowline. Arrows indicate water level variation causing differences in total corrosion.

FIGURE 1.12 General attack beneath a pipeline debonded coating.

had debonded and allowed corrosive water to accumulate on the pipeline surface. Better coatings that do not shield cathodic protection currents are commercially available.

1.6.2 Galvanic Corrosion

Figure 1.4 shows the relative galvanic series for metals in seawater, and most designers assume that similar relationships pertain to other environments. Galvanic corrosion is controlled by electrical isolation between different metals (Figure 1.13) or by using galvanically compatible metals in the same system. Cathodic protection can also be used.

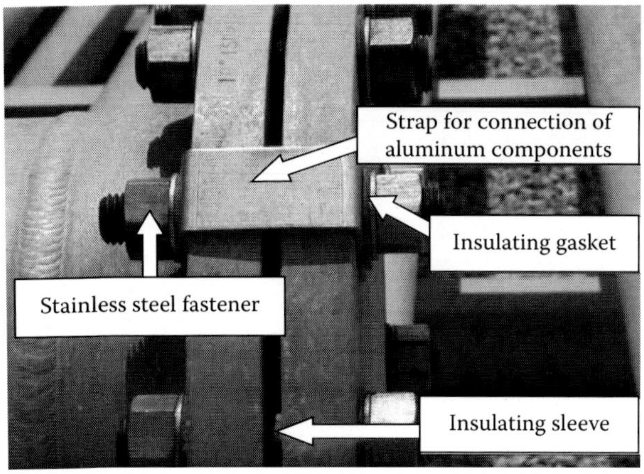

FIGURE 1.13 Electrical isolation of aluminum piping from stainless steel fasteners.

1.6.3 Pitting Corrosion

Pitting corrosion causes the most leaks in oilfield equipment. Pits are randomly located at microscopic defects in the metal or the protective films on their surface. Figure 1.14 shows pitting on a carbon steel potable water line. The arrow indicates the deepest pit, but notice that the "rust tubercules" over the pits are much higher than the underlying pits. The chemistry of the water within a pit is invariably more acidic than the overall environment (Figure 1.15) (Brown 1970; NASA 2014).

The rough surfaces of pitted metal serve as attachment points for biofilms and can lead to MIC. The environments behind the tubercules are protected from biocides. This is one reason why effective MIC control requires a combination of mechanical cleaning and chemical biocides.

FIGURE 1.14 Pitting corrosion in a potable water line.

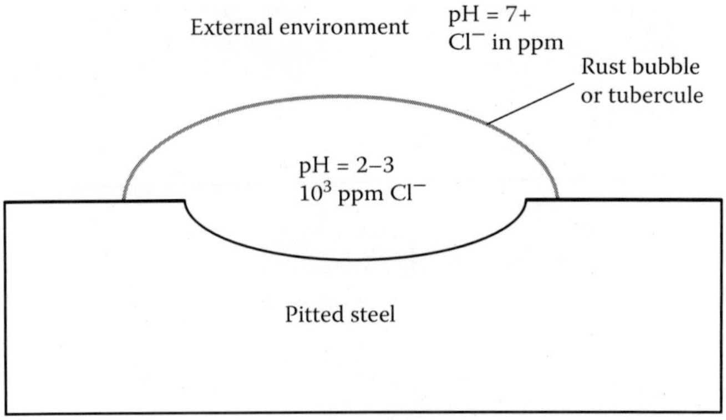

FIGURE 1.15 Pit with pH and chloride concentration changes indicated.

1.6.4 Crevice Corrosion

Crevice corrosion has the same mechanisms as pitting corrosion, and the two are sometimes discussed under the umbrella term "occluded cell corrosion." Designs that minimize the presence of tight crevices limit the amount of crevice corrosion that will occur. Inspectors should also pay special attention to these areas. Figures 1.16 and 1.17 show examples of crevice corrosion.

The titanium flange in Figure 1.17 is an example of improper alloy selection. A titanium aerospace alloy with high strength-to-weight properties was installed instead of a more corrosion-resistant alloy. The oil company in question now routinely specifies titanium alloys with palladium additions to minimize crevice corrosion problems.

FIGURE 1.16 Crevice corrosion underneath the restraining strap on a pipeline.

FIGURE 1.17 Crevice corrosion on a flanged fitting in a seawater chlorinator.

1.6.5 Intergranular Corrosion

Metal structures are aggregates of crystals, and impurities in these crystals are most likely to segregate to grain (crystal) boundaries where the interatomic spaces are larger than in the bulk crystals.

Most grain boundaries are more reactive than the surrounding base metal due to the presence of impurities and other defects. The increased reactivity of grain boundaries over crystal matrices is usually negligible and is of minimal consequence. Intergranular corrosion occurs when significant differences in alloy chemistry near grain boundaries cause significant attack and the alloy disintegrates. Intergranular corrosion can be a significant problem in stainless steels and in some aluminum alloys. Other alloy systems, for example, highly alloyed austenitic iron–chromium–nickel alloys, can also have this problem.

Most upstream intergranular corrosion of stainless steels is due to improper welding practices. The temperatures involved in welding create heat-affected zones (HAZs) where carbon in the stainless steel can react with chromium, which depletes the adjacent crystal structure and sets up microscopic galvanic cells. This is shown in Figure 1.18, which is common to virtually all corrosion engineering textbooks.

This corrosion can be minimized by using austenitic stainless steels with lower carbon contents, 304L or 316L (UNS S30403 or UNS S31603) or by using stabilized grades of stainless steel, types 321 or 347 (UNS S32100 or S34700), which have titanium or niobium additions. The titanium or niobium additions are added as "carbide getters" that preferentially form titanium or niobium carbides, thus tying up the carbon and minimizing the formation of chromium carbides. Both of the above approaches are similarly effective in minimizing sensitization and the resulting intergranular weld decay in HAZs of austenitic stainless steels. North American steel producers tend to prefer the low-carbon approach, and European producers have traditionally preferred the stabilizing addition approach.

Figure 1.19 shows intergranular corrosion of a stainless steel nut. The alloying additions necessary to allow efficient cutting have aligned along the forming direction of the rod from which the nut was formed. These alloying additions are anodic to the stainless steel and led to rapid corrosion.

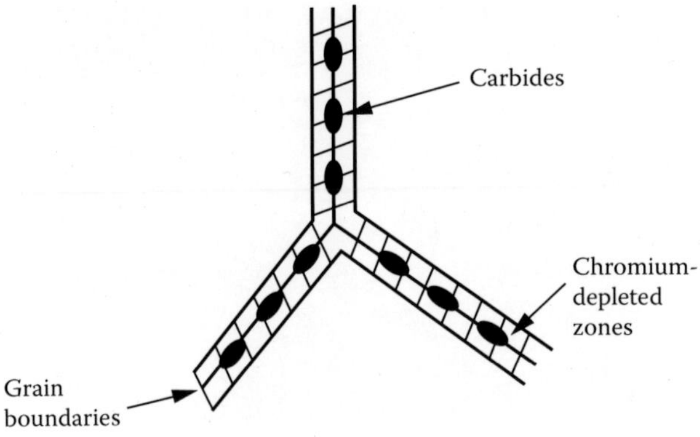

FIGURE 1.18 Sensitized regions in austenitic stainless steel.

FIGURE 1.19 Intergranular corrosion of a stainless steel nut.

1.6.6 Dealloying

Dealloying is seldom a problem in oilfield operations. It is a form of corrosion where one component of the alloy is lost, leaving a weak residual structure. Cast irons, which are seldom used in hydrocarbon service, can lose over 90% of the iron in the metal, leaving a weak residual graphite structure. This is shown in Figure 1.20. Copper-based alloys can also dealloy, and problems have been reported in cupronickel heat exchangers and in aluminum bronze pump components although these problems are rare.

1.6.7 Erosion Corrosion

Erosion corrosion is a problem in multiphase fluid flow. Examples are shown in Figures 1.21 and 1.22.

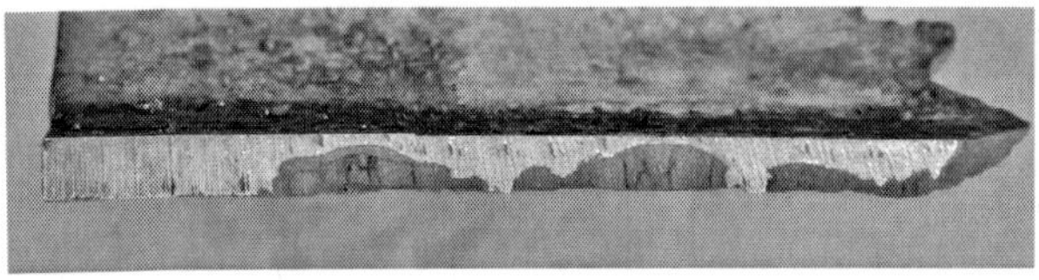

FIGURE 1.20 Dark, graphitic corrosion on the exterior of a cast iron water main. (Courtesy of Testlabs International, Ltd., Winnipeg, Canada, http://www.testlabs.ca.)

FIGURE 1.21 Eroded wellhead choke.

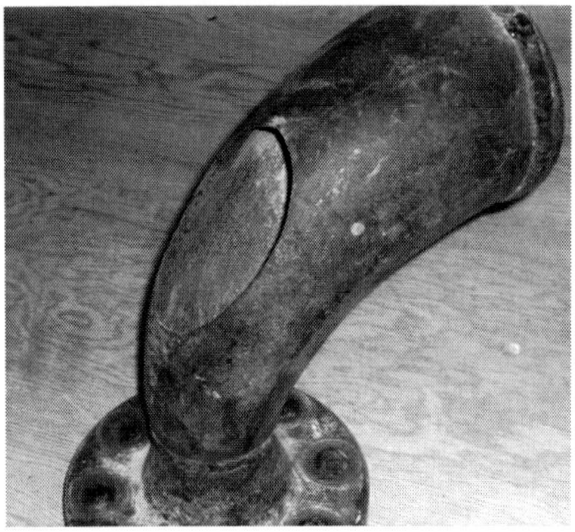

FIGURE 1.22 Erosion corrosion of an elbow in natural gas piping.

It has been estimated that up to 15% of failures in oil and gas production are due to erosion corrosion in gravel packs, nozzles, and Christmas trees before they reach a first separator (Feyerly et al. 2008). The equipment features most likely to experience erosion corrosion are shown in Table 1.1. Erosion corrosion is most likely to happen with particulates, for example, sand production, although liquid droplets and cavitation can also produce problems (Barton 2003; Feyerly et al. 2008).

TABLE 1.1
Equipment Features Susceptible to Erosion Corrosion

Chokes	Most vulnerable
Sudden constrictions	
Partially closed valves	
Standard radius elbows	
Weld intrusions	
Pipe bore mismatches at flanges	
Reducers	
Long radius elbows	
Mitre elbows	
Blind tees	
Straight pipes	Least vulnerable

Erosion corrosion can be controlled by the use of harder alloys (including flame-sprayed or welded hard facings) or by using a more corrosion-resistant alloy. Alterations in fluid velocity and changes in flow patterns can also reduce the effects of erosion corrosion.

Prediction of erosion corrosion locations and severity is limited, and there is no clear consensus on how to determine erosion thresholds. For this reason, monitoring in likely erosion locations once production has started is the primary means of controlling the effects of erosion corrosion (Kolts et al. 2006).

1.6.8 Fretting Corrosion

Fretting corrosion happens when small oscillations in metal-to-metal contact abrade the protective films on metal surfaces and produce accelerated corrosion. It is similar to, and is often considered to be, a form of erosion corrosion. Figure 1.23 shows fretting at the intermediate support baffle in a heat exchanger. The bending moment caused by the baffle restraints can produce fatigue cracking as shown in Figure 1.24.

FIGURE 1.23 Fretting at a heat exchanger intermediate support.

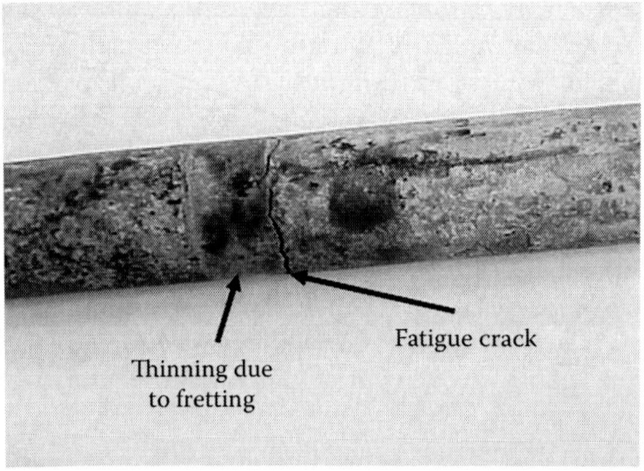

FIGURE 1.24 Thinning and fatigue cracking in a heat exchanger tube.

1.6.9 Environmentally Induced Cracking

Environmentally induced cracking is often defined as the brittle fracture of an otherwise ductile material due to the presence of tensile stresses and a specific environment. The stresses involved in environmental cracking can be from applied loads or from residual stresses caused by manufacturing and construction processes. The overall macroscopic stresses associated with this cracking are generally much below the yield stresses of the alloys in question. This cracking can result in the sudden rupture of structures, especially pressure vessels, such as pipelines, and this sudden failure can lead to significant safety concerns.

While the mechanisms of some forms of environmental cracking remain unclear, approaches to minimize the problem are available. The most widely accepted international standard on environmental cracking control is ANSI/NACE MR0175/ISO 15156. This standard emphasizes the role of H_2S in promoting environmental cracking in oilfield equipment. Many terms are commonly used for this H_2S problem, and the terms from ANSI/NACE standard should be used whenever communicating with oil and gas producers and suppliers. These terms include the following:

- Hydrogen-induced cracking (HIC)
- Hydrogen stress cracking (HSC)
- Soft zone cracking (SZC)
- Stepwise cracking (SWC)
- Stress corrosion cracking (SCC): The standard states that SCC is an *anodic* process. Many authorities disagree with this and consider SCC, especially in high-strength steels, to be environmental hydrogen embrittlement (EHE), a term not found in the ANSI/NACE/ISO standard but used in other standards, for example, those from ASTM (Heidersbach 2011; ASTM 2012, 2014)
- Stress-oriented hydrogen-induced cracking (SOHIC)
- Sulfide stress cracking (SSC)

Chloride salts, invariably present in produced water from oilfield formations, can also promote environmental cracking of carbon steel and other alloys, and CO_2-rich waters, for example, natural gas condensates, also have this problem. Environmental cracking can also be caused by hot salts (caustic embrittlement by molten salts formed inside boiler tubes), liquid metals, and other nonhydrocarbon environments.

Figures 1.25 and 1.26 show environmental cracking on pipelines. One means of controlling environmental cracking is by using metals that are resistant to brittle cracking. In general, this means lower-strength metals, and most pipelines are made from steel having yield stresses of 80 ksi or less. While this doesn't eliminate the problem, it does mean that cracks, when they form, may resist rapid propagation and fracture long enough that they can be discovered by inspection before leaks occur.

FIGURE 1.25 Catastrophic rupture of a pipeline as a result of SCC.

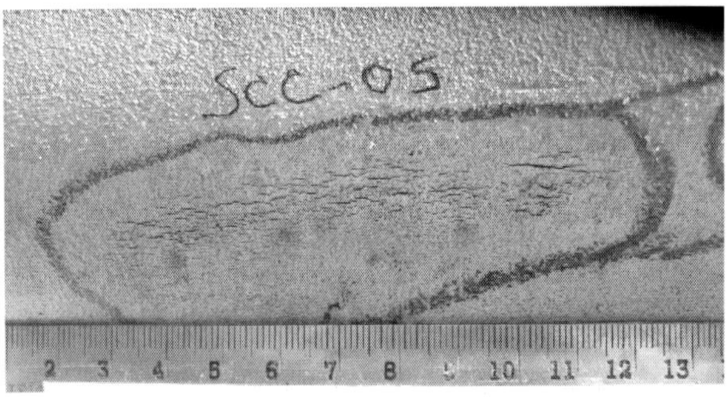

FIGURE 1.26 Clustered "colonies" of stress corrosion cracks on the outside of a carbon steel pipeline.

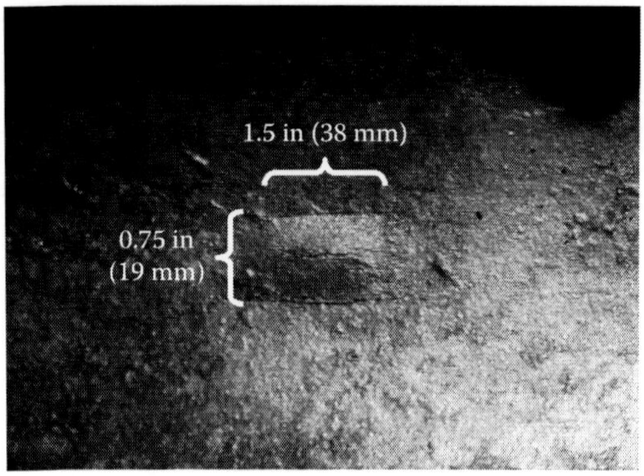

FIGURE 1.27 Hydrogen blistering on the inside of a crude oil pipeline.

Hydrogen blistering (Figure 1.27) occurs in steels when hydrogen from corrosion or other sources enters the steel and becomes trapped at imperfections in the steel. This is often a problem with low-strength steels not subject to brittle cracking, and the blisters can be identified by inspection and repaired by various methods.

1.6.10 Corrosion Fatigue

Repetitive loading can cause fracture and collapse at lower loads than would cause problems with static loading. Figure 1.28 shows the concept of a corrosive environment lowering the "endurance limit" (stress at which metal can resist fatigue cracking for an indefinite period of time), and Figure 1.29 shows a collapsed offshore platform that failed due to corrosion fatigue. Fatigue cracks can often be detected before system failure, and organizations emphasize crack inspection on susceptible equipment. It is very hard to model corrosion fatigue because environments can change during the lifetime of equipment and many assemblies, for example, offshore platforms, have complex loading histories—big waves produce more damage than many smaller waves.

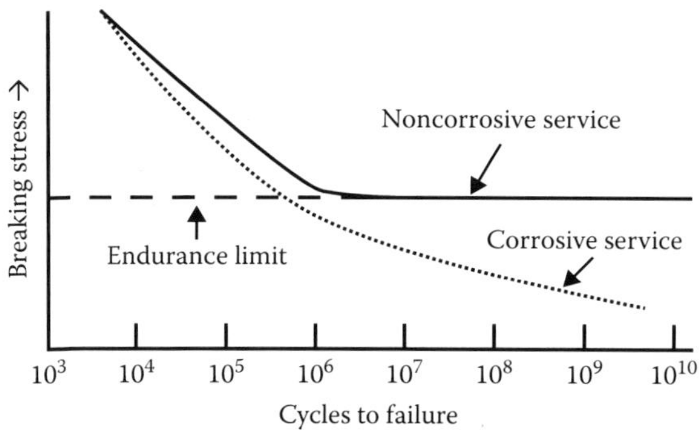

FIGURE 1.28 The reduction in fatigue resistance due to a corrosive.

FIGURE 1.29 The collapse of an offshore platform caused by a corrosion fatigue environment.

1.6.11 Other Forms of Corrosion Important to Oilfield Operations

There is no universally accepted terminology for corrosion, and the oil and gas industry has adopted a number of terms for corrosion that frequently overlap with the terms discussed in previous sections of this chapter. This list is by no means exhaustive, and more detailed discussions are available (Heidersbach 2011).

- Corrosion under insulation (CUI): This is a major problem for many insulated piping and storage systems. Moisture can collect in the annular spaces between the exterior metal jacketing (usually aluminum or galvanized metal sheeting) and the exterior of the piping or storage tank walls. These problems are hard to detect, but operating personnel can notice signs of moisture collecting and problems developing. Figure 1.30 shows pitting corrosion on offshore platform piping once the insulation has been removed. Figure 1.31 shows moss growing at seams in the insulation jacketing of a crude oil loading line. The moisture that allowed this moss to grow also led to extensive CUI at the loading platform (Heidersbach 2012).
- NACE SP0198 provides guidelines on how to limit CUI (NACE 2010).
- Oxygen attack: This often refers to pitting corrosion due to air ingress into topside equipment.
- Sweet corrosion: This is a term that is becoming outdated. It refers to corrosion in environments where the corrosion is due to the presence of dissolved CO_2. Natural gas condensates are examples of environments that produce sweet corrosion (Papavinasan 2013).
- Sour corrosion: This refers to weight-loss corrosion that happens in environments that have H_2S as the predominant chemically reducible species. Sour corrosion is different from the environmental cracking caused by the

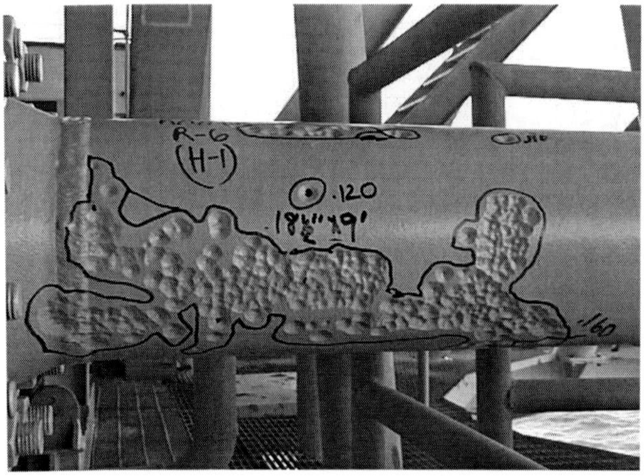

FIGURE 1.30 Pitting corrosion on offshore platform piping after thermal insulation was removed.

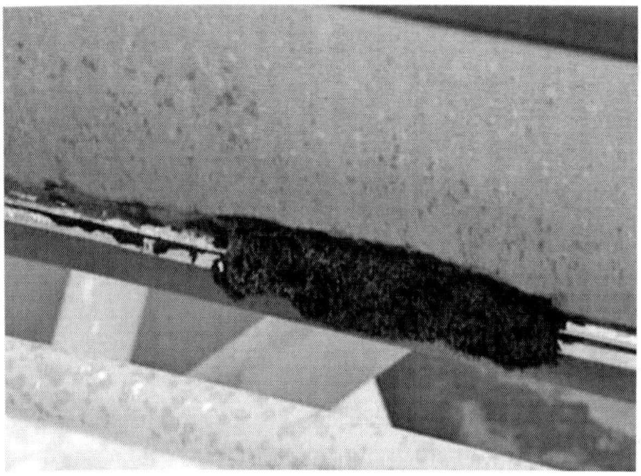

FIGURE 1.31 Moss indicating excessive moisture accumulating on insulated crude oil piping.

presence of H_2S, which is covered by the mechanisms addressed in ANSI/NACE MR 0175/ISO 15156.

- Wireline corrosion: This is an attack in drill pipes or other tubular goods caused by removal of the protective coatings or films on the inside of downhole tubular goods. The high-strength steel wires used to suspend and retrieve downhole inspection and maintenance devices can cut the protective coatings and leave bare steel exposed to corrosion. Wireline damage can be minimized by adjusting running procedures to include installing wireline guides and reducing wireline speed (<0.5 m/s or <100 ft/min). Corrosion inhibitors should be inserted immediately after wireline work (Garverick 1994).

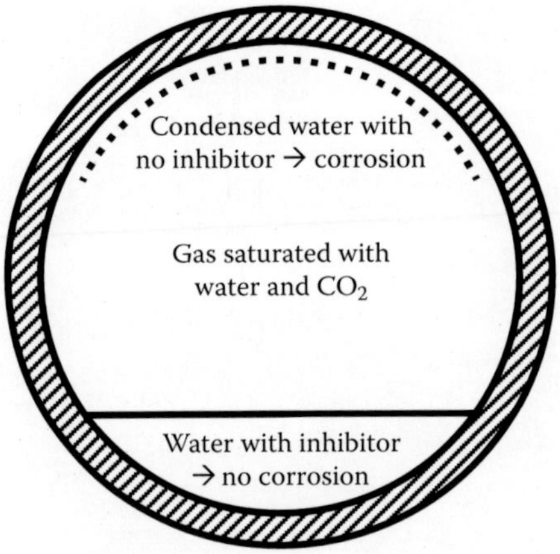

FIGURE 1.32 Top-of-line corrosion in a multiphase pipeline.

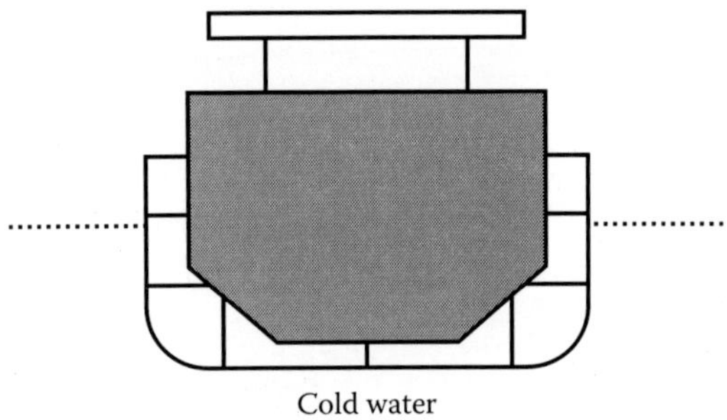

FIGURE 1.33 Thermos bottle corrosion.

- Top-of-line corrosion: This occurs in gas piping and pipelines when acidic condensate accumulates at the top of a pipeline interior and no corrosion inhibitor is present. This idea is shown in Figure 1.32.
- Thermos bottle corrosion: This happens in the annular spaces of double-walled tankers and floating production, storage, and offloading vessels (FPSOs) where condensation leads to accelerated corrosion for vessels carrying warm cargo in cold waters (Figure 1.33). The condensation on the interior air space surfaces of the exterior hulls causes water to collect in the annular spaces between the double hulls. The resulting stagnant water produces locations where MIC and other forms of corrosion can occur. Most of these annular spaces are very narrow and difficult to inspect or repair.

1.6.12 Additional Comments

Most corrosion is localized, and corrosion allowances are of limited value. Pitting corrosion, which produces the most leaks, and environmental cracking, which can lead to catastrophic failures—especially in pressurized gas systems—are the two forms of corrosion that receive the most attention in oilfield operations. Many instances of pitting corrosion are attributed to MIC, and it is not clear whether MIC is a primary cause of corrosion or whether the sheltered environments within pits merely allow detectable organisms to accumulate.

1.7 CORROSION CONTROL

Protective coatings, corrosion inhibitors, and cathodic protection are routinely used to control corrosion on oilfield equipment, which is usually made from carbon steel. Oilfield systems are often complex, but even aboveground storage tanks (ASTs) may require several different methods of corrosion control. For the typical crude oil storage tank shown in Figure 1.34, the following approaches are common:

- Exterior steel exposed to the atmosphere—protective coatings.
- Vapor spaces above the typical liquid layer—protective coatings. Note that these coatings, unlike the exterior coatings, do not require UV resistance.
- Interior walls exposed to crude oil—minimal corrosion expected due to oil-wetted surfaces.
- Interior floor and first 1 m of interior wall—water-resistant coating systems or linings.
- Exterior tank floor—often uncoated but sometimes protective coatings.
- Cathodic protection is commonly applied to the tank bottoms on both the interior (salt water exposed) and exterior (wet soil) surfaces.

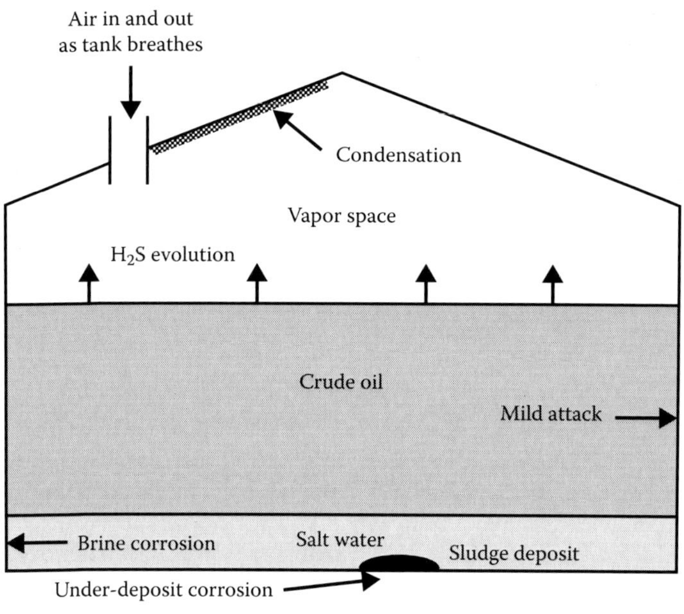

FIGURE 1.34 Different corrosive environment locations on a crude oil storage tank.

FIGURE 1.35 Ring bulging at the outer surface of AST tank bottom.

MIC is possible on the bottom of both the tank interior and exterior surfaces. Note the suggestions of sludge deposits on the interior floor.

The drawing in Figure 1.34 is an idealized representation of fairly simple equipment. Actual as-built or modified equipment can often deviate from the original designs. Figure 1.35 shows bulging caused by settlement of a storage tank's walls. The uneven surfaces shown in Figure 1.35 can serve as locations for poor drainage, sludge deposits, and other problems.

1.8 INSPECTION, MONITORING, AND TESTING

The purposes of inspection and monitoring are as follows:

- Inspection is used to determine the condition of a system at the time of the inspection.
- Monitoring is used to either periodically or continuously assess whether the corrosivity of a system is changing. It is often used to determine the effectiveness of biocides, corrosion inhibitors, and other chemical treatments.

Both inspection and monitoring are focused on existing equipment whereas testing, discussed separately below, is focused on determining the effectiveness of corrosion inhibitors or if recently fabricated/renovated equipment has been fabricated correctly.

1.8.1 MONITORING

Monitoring, either by coupon insertion or electrochemical means, cannot substitute for inspection. It is impossible to identify the most corrosion-susceptible area in a complex system, so probes or coupons cannot warn operators about corrosion where systems are most susceptible.

Monitoring is used to determine if corrosion rates or the corrosivities of environments are changing and, if so, whether the environments are becoming more or less aggressive. The techniques described below are used to evaluate the effectiveness of corrosion inhibitors, determine if corrosion rates are changing, for example, due to the inadvertent admission of air (oxygen) into piping systems and for a variety of other purposes. There are several other techniques that are available, but weight-loss coupons and electrical-resistivity probes are the most commonly used corrosion rate–monitoring techniques in oil and gas production and pipelines.

Monitoring probes must be placed in locations where corrosion is most likely to occur, for example, the bottom interior of horizontal piping. Unfortunately, it is common to place probes in convenient locations. Figure 1.36 shows locations where moisture is likely to infiltrate the jacketing system on insulated piping at a major gas processing facility. Locations A and B are jacketing seams, and location C is where the jacketing is missing for mounting on a structural support. Unfortunately the locations for ultrasonic inspection/monitoring ports are not located near these locations. Instead they were located for the convenience of the monitoring crews (Figure 1.37).

FIGURE 1.36 Locations where moisture infiltration is likely on insulated piping.

FIGURE 1.37 Ultrasonic monitoring port located for convenient access.

Weight-loss coupons like the coupon shown in Figure 1.38 are widely used to monitor the effectiveness of corrosion inhibitors. Coupons are removed at predetermined intervals ranging from 15 days to 6 months and evaluated for weight loss, pitting corrosion, and other corrosion indications. It should be noted that the coupons are often evaluated for "average rates" based on mass loss and sample surface area, usually reported in mm/y or mpy (NACE 2013).

The next most commonly used technique for oilfield corrosion monitoring is the use of electrical resistance probes like those shown in Figure 1.39. Electrical resistance probes can indicate changes in corrosion rate, depending on the corrosivity of the environment in times ranging from less than 1 day up to as much as 10 days if the environment is less corrosive (NACE 2013). This is useful in identifying if corrosion has occurred due to upsets in operating conditions, and it is a major advantage of electrical resistance probes over weight-loss coupons.

Weight-loss coupons and corrosion probes should be placed in locations where corrosion is likely to be a problem. Figure 1.40 indicates how corrosion rates will vary depending on location within a piping system.

FIGURE 1.38 Corrosion coupon used to determine weight loss and pitting corrosion.

FIGURE 1.39 Typical electrical resistance probes.

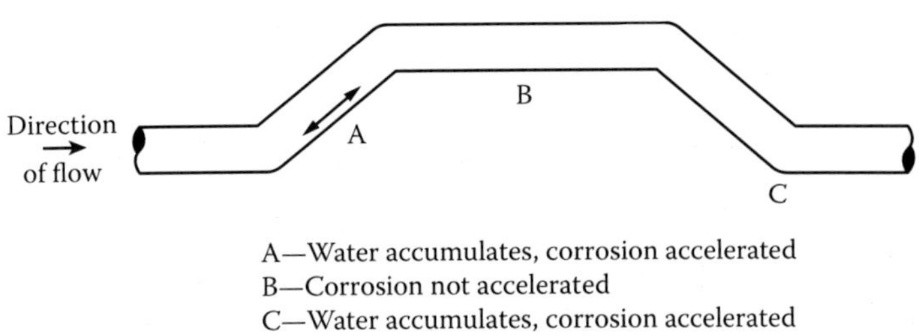

FIGURE 1.40 Locations where corrosion on coupons will detect differences in corrosion rates.

1.8.2 Inspection

Oilfield equipment is so large and complicated that it is usually impossible to inspect all locations where corrosion could occur. Thus, it becomes necessary to identify those locations where corrosion is either likely to occur or, if it does occur, will produce the most serious consequences. Risk-based inspection (RBI) procedures, which are relatively new and somewhat controversial, are intended to maximize limited inspection resources and to focus on areas having high probability of corrosion and high consequences if corrosion were to occur.

Figure 1.41 shows a pipeline cross-section. The most likely interior corrosion locations for a multiphase or natural gas pipeline are near the top, where condensate forms, and the bottom, where water, sludge, and solid deposits can create corrosive conditions. The exterior of a buried pipeline will frequently have poorly consolidated soil and air–water interfaces, which are very corrosive near the 4 o'clock and 8 o'clock positions. The likelihood of corrosion at the 12 o'clock, 4 o'clock, 6 o'clock, and 8 o'clock radial locations is the reason why longitudinal welds are located on alternate joints at 2 o'clock and 10 o'clock (see Figure 1.10). Inspections must ensure that all radial locations as well as longitudinal and girth welds are inspected.

Unfortunately, inspection and monitoring can become routine, and operating organizations can forget the purposes of inspection. The acid storage tank in Figure 1.42 was ultrasonically inspected on a regular basis with the intent of identifying when gradual wall thinning would require that the carbon steel tank be replaced. Over several years, the inspection contractor would perform appropriate inspections and submit reports and invoices to the plant operator. The reports became so routine that technical people failed to note that the wall thinning had progressed to the point where tank replacement was necessary. The corrosion moisture problems associated

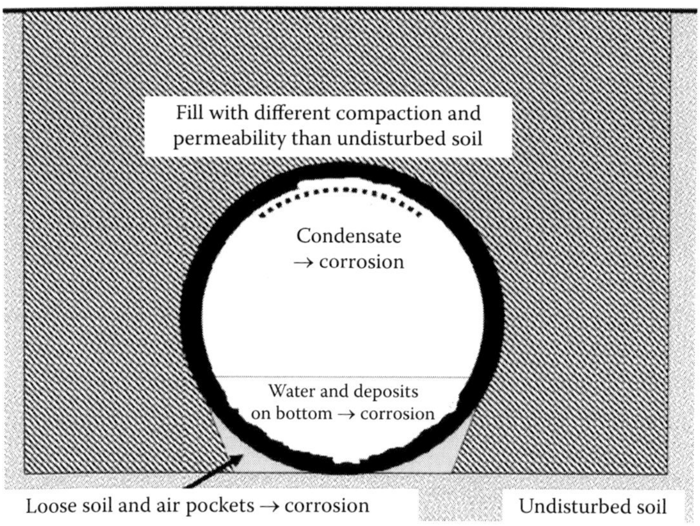

FIGURE 1.41 Radial locations where corrosion is likely on a buried pipeline.

FIGURE 1.42 Slumping storage tank due to wall thinning.

with Figure 1.31 are another example of problems noted but not acted upon in a timely fashion.

The oilfields are full of inspection reports that are not analyzed. It is far better to have a small number of inspections that receive critical evaluation than a large number of routine reports that get ignored.

1.8.3 Testing

Testing is necessary to determine the compatibility of various chemical treatments applied to oil and gas fluids. These include biocides, corrosion inhibitors, and hydrate inhibitors (important for subsea pipelines).

Hydrostatic testing is used to test the quality of construction or repairs to pipelines and other equipment. Because of the large volume of water to test pipelines, it is common to use river water or other supplies of potentially corrosive water that will require biocides and corrosion inhibitors. Draining of pipelines is seldom completely successful, and corrosion is known to occur in low spots and crevices.

1.9 CLOSING COMMENTS

This brief report was intended to introduce the reader to some of the corrosion problems associated with oilfield production and pipeline operations. More detailed reports are available that also comment in detail on corrosion control techniques, specific alloys used for different applications, and suggestions on corrosion control methods for specific types of equipment (Garverick 1994; Kane 2006; Heidersbach 2011; Papavinasan 2013; Craig 2014). Oilfield operators concentrate on pitting corrosion and environmental cracking, but the other forms of corrosion are also technically important. MIC, the subject of this book, was mentioned only briefly, but subsequent chapters deal with MIC in great detail.

REFERENCES

ANSI/NACE MR 0175/ISO 15156, 2015. *Materials for Use in H_2S Containing Environments in Oil and Gas Production*, NACE, Houston, TX, 3 parts.

ASTM F1940-07a, 2014. *Process Control Verification to Prevent Hydrogen Embrittlement in Plated or Coated Fasteners*, ASTM International, West Conshohocken, PA. http://www.astm.org.

ASTM F1624-12, 2012. *Measurement of Hydrogen Embrittlement Threshold in Steel by the Incremental Step Loading Technique*, ASTM International, West Conshohocken, PA. http://www.astm.org.

Bailey, A., 2006. BP: Learning from oil spill lessons, *Petroleum News*, 11 (20), http://www.petroleumnews.com/pntruncate/573947058.shtml.

Barton, N., 2003. *Erosion in Elbows in Hydrocarbon Production Systems: Review Document.* Health & Safety Executive (UK) Research Report, 115.

Brown, B. F., 1970. Concept of the occluded corrosion cell, *Corrosion*, 26 (8), 249–250.

Callister, W., and Rethwisch, D., 2013. *Materials Science and Engineering: An Introduction*, Wiley-Interscience, Hoboken, NJ.

Craig, B., 2014. *Oilfield Metallurgy and Corrosion*, 4th Edition, MetCorr, Denver, CO.

Feyerly, J., Mori, G., Holzleitner, S., Haberi, J., Oberndorfer, J., Havlik, M. W., Monetti, C., 2008. Erosion-corrosion of carbon steels in a laboratory: Three-phase flow, *Corrosion*, 62 (2), 175–186.

Fischer, T., 2008. *Materials Science for Engineeering Students*, Academic Press, Cambridge, MA.

Garverick, L., 1994. *Corrosion in the Petrochemical Industry*, ASM International, Metals Park, OH, p. 268.

Heidersbach, R., 2011. *Metallurgy and Corrosion Control in Oil and Gas Production*, John Wiley & Sons, Hoboken, NJ.

Heidersbach, R., 2012. *Corrosion Survey of Valdez Marine Terminal*, PWSRCAC Contract 559.12.01.

Kane, R., 2006. Corrosion in Petroleum Production Operations, *Metals Handbook, Vol. 13C—Corrosion in Specific Industries*, ASM International, Metals Park, OH, pp. 922–966.

Kolts, J., Joosten, M., and Singh, P., 2006. *An Engineering Approach to Corrosion/Erosion Prediction*, NACE 06560.

Mannheimer, A., 2016. More than 350,000 tonnes of Kashagan oil shipped for export, *Kashagan Today*, http://kashagan.today/?p=8893.

Mittemeijer, E., 2011. *Fundamentals of Materials Science*, Springer-Verlag, Berlin, Germany.

NACE SP0198-2010. *Control of Corrosion under Thermal Insulation and Fireproofing Materials.*

NACE SP0775-2013. *Preparation, Installation, Analysis and Interpretation of Corrosion Coupons in Oilfield Operations.*

NASA, 2014. *Pitting Corrosion*, http://corrosion.ksc.nasa.gov/pittcor.htm.

Papavinasan, P., 2013. *Corrosion Control in the Oil and Gas Industry*, Gulf Publishing, Houston, TX.

Revie, R. W., and Uhlig, H. H., 2008. *Corrosion and Corrosion Control*, Wiley-Interscience, Hoboken, NJ.

Ruschau, G., and Al-Anezi, M., 2001. *Oil and Gas Exploration and Production. Appendix S to Corrosion Costs and Preventive Strategies in the United States*, Report FHWA-RD-01-156.

Shankardass, A., 2004. Corrosion control in pipelines using oxygen stripping, *Oils and Water Usage Workshop*, Edmonton, Alberta, Canada.

Smith, W., and Hashemi, J., 2009. *Foundations of Materials Science and Engineeering*, McGraw-Hill, NY.

Uhlig, H. H., 1948. *The Corrosion Handbook*, The Electrochemical Society, John Wiley, New York.

2 MIC and Materials Selection

Richard B. Eckert and Bill Amend

CONTENTS

ABSTRACT

While we know that the presence and activity of microorganisms can directly or indirectly influence corrosion of metals in many different ways, the common view is often oversimplified, for example, assuming that bacteria "prefer" growth on weld metal as compared to pipe parent metal. When viewed from the scale of a microorganism, the microstructure of most engineering alloys is heterogeneous in composition, mechanical properties, residual stresses, local potentials and surface topography. Each of these characteristics can influence both abiotic and biotic corrosion initiation, propagation, and growth. Features that are visually observable (welds, crevices, pits, etc.) are enormous when considered from the microbiological scale and may have more macroscopic effects on corrosion than microscopic features.

Metallurgy is an important consideration in both the design phase and in the operations phase of oil and gas asset life. In the design phase, engineers consider corrosion threats, including microbiologically influenced corrosion (MIC), and weigh the lifecycle costs of using carbon steel with chemical treatment (with lower capital expense [CAPEX] and higher operating expense [OPEX]) or corrosion-resistant alloys (CRAs) with higher CAPEX and lower OPEX. CRAs offer many advantages in corrosion resistance as compared to carbon steels; however, they are also susceptible to localized corrosion or cracking mechanisms under specific physiochemical conditions. MIC is often a localized corrosion mechanism, and it occurs under biofilms that form occlusions on the metal surface and in crevices that challenge the pitting resistance of many CRAs, particularly in chloride-containing environments. Thus, when using CRAs, it is important to consider whether deposits and biofilms will form on the surface at some point in the operational life of the asset because alloys with good resistance to abiotic corrosion in a specific service environment may offer minimal improvement in performance over carbon steel if biofilms are present.

Metallurgy is also an important consideration when assessing potential MIC problems in an asset, such as when conducting a basic or root-cause analysis (RCA). RCA has proven to be a useful tool when MIC investigations are conducted properly and thoroughly because the end result should provide the operator a means to prevent similar incidents in the future. When assessing corrosion damage or performing a corrosion root-cause analysis, metallurgy must be considered both in terms of abiotic effects and biotic effects. Some manufacturing and fabrication processes can have a detrimental effect on biotic or abiotic corrosion resistance, resulting in performance that can be significantly different from what was predicted based upon common electrochemical or corrosion coupon testing. Understanding the effects of various manufacturing and fabrication practices on corrosion resistance can allow users to optimize those practices to reduce the likelihood of, or severity of, corrosion.

2.1 INTRODUCTION

A wide range of ferrous and nonferrous alloys are available for selection and use in oil and gas environments. The primary concern of most designers or users is typically selection of an appropriate combination of thickness and mechanical properties for the anticipated service stresses; however, in some environments, the degradation rate of the alloy is also a significant consideration. Time-dependent degradation can take many forms, including metal loss from corrosion and/or erosion, fatigue cracking, embrittlement, or stress corrosion cracking. However, this chapter primarily addresses concerns related to corrosion and, in particular, MIC.

In steel, improved metallurgical technologies have enabled the reduction of carbon and impurities such as sulfur and phosphorous. Those reductions contribute to improved mechanical properties while retaining or even improving weldability. However, they can also contribute to improved corrosion resistance by reducing the number of inclusions at which corrosion pits often initiate, and they influence diversity in the types of weld zone microstructures that form. Since site-to-site differences in composition or microstructure on either a micro or macro scale can promote corrosion, improving the uniformity and consistency of base metals and weld zones can influence the long-term performance of alloys in corrosive service.

With regard to predicting the long-term performance of alloys in corrosive environments, the predictions can be confounded by the potential inconsistency of the candidate alloys. Most of the alloys are available in multiple specifications and grades having similar and often overlapping, but slightly different, requirements for processing methods, composition, and mechanical properties. Most specifications describe the allowable *ranges* of mechanical properties and chemical compositions. Variations in the product compositions within the specified ranges not only contribute to some variations in mechanical properties, but can result in variations in corrosion resistance when supposedly replicate materials are obtained from different sources or, to a lesser degree, when multiple heats are obtained from single sources.

Further confounding the ability to predict performance in oil and gas environments is the fact that the applicable material specifications, grades, and manufacturing capabilities have evolved over time. Thus, historical observations of service performance may not be directly transferable to current versions of the same alloys. For example, API specification 5L grade X52 ERW seam steel pipe made today bears little metallurgical similarity to X52 electric resistance welded (ERW) seam pipe produced in the 1950s or 1960s. The differences are a result of significant changes in ERW seam welding practices; postweld seam heat treatment; and development and application of control-rolled, microalloyed steels. The modern pipe generally has a much lower impurity content, improved microstructure, and better toughness and resistance to selective seam corrosion.

While improved metallurgical process control during manufacturing can reduce alloy impurity content and improve mechanical properties, it can also inadvertently result in reduced corrosion resistance. For example, the corrosion resistance of modern UNS S31600 (Type 316 stainless steel) is lower than historical versions of the stainless steel made to the same specification. The reduced corrosion resistance is attributed to the improved ability of manufacturers to reproducibly limit the molybdenum content of the alloy to a value very close to the lower end of the specified range, that is, 2.0%–3.0% (NACE 2015). The reduction in molybdenum content reduced the cost of manufacturing and also reduced the resistance to pitting corrosion. As a result, UNS S31600 is no longer suitable for use in some oil and gas processing and refining operations in which the earlier version was acceptable. Unexpected failures have occurred when modern Type 316 stainless steel equipment was installed after years of successful performance by higher molybdenum versions of the same alloy (NACE 2015).

2.2 MATERIALS IN OIL AND GAS PRODUCTION

Equipment for oil and gas applications can be broadly categorized into equipment for downhole applications in wells, and surface equipment. Downhole equipment and tubulars (oil country tubular goods or OCTG) are generally selected for strength and corrosion resistance with less emphasis on toughness or weldability. (OCTG generally have threaded connections and are rarely welded except, for example, to wellhead equipment.)

2.2.1 Materials in Downhole Applications

The difficulty in accessing the downhole equipment means that it is much more difficult (or sometimes impossible) to inspect the condition of downhole tubulars or equipment unless the well is removed from service and, in some cases, the equipment is removed from the well. Premature or unexpected failure of downhole equipment can mean significant loss of revenue associated with failure to produce the oil and gas and expensive "fishing" operations to retrieve failed components from the depths of a well. Therefore, careful selection of the appropriate materials of construction is critical.

One source for information about commonly used OCTG alloys and grades is API specification 5CT "Specification for Casing and Tubing" (API 2011). The specification describes grades named in accordance with the specified minimum yield strength, ranging from 40 ksi to 125 ksi, and compositions including the following general categories, listed in order from lowest to highest corrosion resistance:

- Unalloyed and low alloy steel
- 9% chromium steel
- 13% chromium steel

In addition, manufacturers provide several additional compositions meeting a range of strength levels for downhole applications. Those more corrosion-resistant grades are generally based on the following broad categories shown in order of increasing pitting resistance (although the resistance of some categories overlaps). See Table 2.1 for examples of alloys in these categories although not all alloys in the table are used for downhole applications.

- Austenitic and super-austenitic stainless steel
- 22% chromium duplex stainless steel
- 25% chromium duplex stainless steel, that is, "superduplex stainless steel"
- More highly alloyed duplex stainless steel, that is, "hyperduplex stainless steel"
- Nickel–chromium–molybdenum alloys
- Titanium alloys

The more corrosion-resistant grades of materials generally fall into broad categories, reflecting either the range of chemical composition or the type of microstructural phases present or both. In most of these grades, high strength is generally achieved by cold working at the expense of some resistance to environmental cracking, especially in environments containing hydrogen sulfide and/or high-temperature chlorides.

The nature of the threaded connections used in OCTG creates the potential for various corrosion issues. First, the threaded connections can be susceptible to crevice corrosion. Second, some types of connection designs do not result in a smooth bore, and localized turbulence can occur, which can promote CO_2 corrosion. Third, some connections require that the ends of the tubular be "upset" to provide a local area of increased wall thickness to accommodate the machining of pipe threads. The localized heating and mechanical working of the end of the steel tubular produces local

TABLE 2.1
General Comments on Materials Used in Oil and Gas Applications

Material Group	Comments
Steel	• Covered by a range of API, ASTM, MSS, and other specifications. • Available in a wide range of strength and toughness obtained via optimized chemical compositions or heat treatment or both. • Formed by a wide range of manufacturing processes. Often coated, clad, or lined with other materials to improve resistance to corrosion or erosion.
Precipitation hardenable stainless steel	• Stainless steels alloyed to produce high strength upon heat treatment via a precipitation hardening mechanism. • Common applications include pump and valve components. • Corrosion and environmental cracking resistance is strongly related to the specific tempering temperatures used in heat treatment. Lower tempering temperatures generally produce higher strength but less resistance to corrosion and cracking.
Martensitic stainless steel	• 400 series stainless steels have somewhat better resistance to general corrosion than low-alloy steels and have better strength but may have high pitting rates in chloride-containing fluids. • 400 series stainless steels generally have inferior corrosion resistance to the 300 series stainless steels. • Strength and hardness is dependent upon the details of the heat treatment used. • Common applications include pump and valve trim and other applications requiring high strength and/or hardness with moderate corrosion resistance. • They also have better high-temperature oxidation resistance than carbon steels. • Welding practices (weld preheat, heat input, and cooling rate control) for these alloys are significantly different than for most other corrosion-resistant alloys.
Austenitic/ superaustenitic stainless steel	• 300 series stainless steels have limited resistance to pitting, crevice corrosion, and chloride stress corrosion cracking in chloride-containing environments. • The numeric designation of the austenitic stainless steel grade is not a good general indicator of the corrosion resistance. Higher nickel, chromium, and molybdenum contents improve resistance, but corrosion resistance may be reduced if not fabricated and welded with appropriate procedures. • The corrosion resistance of superaustentitic stainless steels overlaps with the corrosion resistance of duplex, superduplex, and hyperduplex stainless steels described below.
Duplex stainless steel	• 22% chromium varieties are most common, but "lean" duplex alloys of lesser alloy content have some applications. • The term "duplex" is derived from the type of microstructure, which consists of a balanced mix of austenite and ferrite to optimize corrosion resistance and stress corrosion cracking resistance. • Good combination of strength in the annealed condition, plus corrosion resistance, resistance to stress corrosion cracking, and moderate cost while retaining reasonably good weldability. • Exposure to high service temperatures or to improper welding or heat treatment practices can significantly degrade toughness and/or corrosion resistance.

(Continued)

TABLE 2.1 (CONTINUED)
General Comments on Materials Used in Oil and Gas Applications

Material Group	Comments
Superduplex, hyperduplex stainless steel	• More highly alloyed versions of the duplex stainless steels, generally containing approximately 25%–32% chromium with better corrosion resistance, but the same general advantages and limitations as described for the duplex stainless steels.
Ferritic stainless steel	• Mainly used as heat exchanger components, these stainless steels have ferritic microstructures and good resistance to chloride stress corrosion cracking.
Nickel alloys	• Trade-offs in corrosion resistance versus ease of fabrication (welding and fabrication methods must be optimized to preserve corrosion resistance). • Commonly used as liners, cladding, weld overlay because of the high cost of solid alloy components. • Nickel alloy weld metals are compatible with a wide range of steels, stainless steels, and other nickel alloys and are commonly used in joints between dissimilar metals. • Less sensitivity to corrosion at "heat tint" than most stainless steels. • Alloy compositions are optimized for specific types of corrosive environments.
Copper alloys	• Not resistant to MIC despite common beliefs. • Susceptible to corrosion by H_2S. • Velocity sensitive; Ni additions improve erosion–corrosion resistance. • Bronze alloys with small amounts of aluminum, nickel, or other elements are used in some valves, seals, and other peripheral equipment. • Admiralty brass is sometimes used as heat exchanger tubing.
Titanium alloys	• Mainly for heat exchangers or extremely corrosive conditions. • Very good MIC resistance. • Potential for hydriding (embrittlement) if galvanically coupled to steel or to other corroding, anodic alloys.

changes in the microstructure, much like those found in a weld heat-affected zone. As a result, a relatively narrow circumferentially oriented band of corrosion, often referred to as "ringworm corrosion" can occur at the transition between the original and altered microstructures unless the entire tubular is heat-treated to homogenize the microstructure after upsetting.

Other alloys can be found in limited amounts in selected applications. Beryllium copper drill string components combine high strength with nonmagnetic properties that are important in "measurement while drilling" (MWD) applications. Electroless nickel and hard chrome plating are found in some downhole pump applications. Various types of welded hardfacing, including those containing tungsten carbide or chromium carbides, are found in pump and valve components that are subject to wear and abrasion in addition to corrosion.

2.2.2 Surface Piping and Components

Surface components can use many of the same alloys as are used in downhole equipment, but there is more emphasis placed on weldability since few components in surface applications depend upon threaded connections. Common alloys used in surface piping and equipment include the groups as shown in Table 2.1.

For pipelines, steel pipe is often purchased to the API 5L specification (API 2012). The specification includes 53 unique combinations of steel strength, composition, and type of heat treatment divided into two different product specification levels (PSL 1 and PSL 2) and having separate requirements for seamless and welded seam pipe. The requirements for strength, toughness, composition, and heat treatment options have evolved over time. Note that some grades of pipe in API 5L have different requirements than grades of the same designation made to ASTM specifications; for example, ASTM A53 grade B has different composition limits and allowable heat treatment options than API 5L grade B. The differences between ASTM A53 grades and API 5L grades can result in different characteristics when welded and some differences in possible response to corrosive environments.

For corrosive environments, a large number of stainless steel and nickel alloys are available, ranging from the 300 series stainless steels, such as Type 304 (UNS S30400), to the very highly alloyed nickel–chromium–molybdenum alloys. A simple calculation that considers the amount of chromium, molybdenum, and nitrogen present in the alloy allows users to determine the relative resistance of candidate alloys to pitting corrosion in chloride environments. The pitting resistance equivalent number (PREN) is calculated as shown in Equation 2.1:

$$\text{PREN} = \%\text{Cr} + 3.3 \times \%\text{Mo} + 16 \times \%\text{N} \tag{2.1}$$

Alloys with a PREN of 32 or greater are generally considered to be resistant to pitting corrosion in ambient temperature seawater (although a higher PREN may be required for resistance to MIC in seawater). While the PREN is a general indicator of the pitting resistance of different alloys, the performance of an alloy in environments other than seawater can be influenced by specific characteristics of the environment, and the relative performance of different alloys can change if the service environment changes. Papavinasam et al. (2007) described a number of operational parameters that influence localized pitting of carbon steels, which are also likely to be involved in pitting susceptibility of CRAs.

Furthermore, the chromium, molybdenum, and nitrogen in CRAs only contribute to the corrosion resistance when they are present in solid solution. Improper welding or thermal processing can cause some alloying elements to form precipitates (e.g., chromium carbides or chromium nitrides). The precipitates decrease corrosion resistance since the matrix surrounding the precipitates is depleted in chromium or molybdenum in solution as a result of the formation of the precipitates. For example, in austenitic stainless steels, exposure to high temperatures (roughly between 480°C and 800°C, depending upon the specific alloy composition and exposure time) or improper heat control during welding can result in strings of intergranular chromium carbide precipitates (i.e., sensitization). Sensitization results in vulnerability to intergranular corrosion and cracking. When welding or exposure to high temperatures cannot be avoided, harmful precipitation is minimized through use of stainless steels having very low carbon contents (i.e., the "L" grades) or through additions of elements such as titanium or columbium that preferentially combine with the carbon to leave the chromium in solution. Also note that nitrogen does not remain in solution in ferrite, so nitrogen does not effectively contribute to the localized corrosion resistance of ferritic stainless steel or the ferrite phase in duplex stainless steel.

Another way of ranking the corrosion resistance of alloys is to measure the temperature at which pitting or crevice corrosion initiates in standard chloride-containing test solutions. The higher the critical crevice temperature (CCT) or critical pitting temperature (CPT) is, the better the relative resistance is to initiation of localized corrosion. The crevice corrosion initiation temperature tends to be lower than the pitting initiation temperature. These tests do not include the effects of biofilms on corrosion initiation, however.

The most commonly used pitting test procedure is described in ASTM standard G48 (ASTM 2011), which specifies the use a 6% ferric chloride solution adjusted to a low pH value as the test medium. Just as the PREN is best used as an indicator of relative resistance to corrosion in seawater, the pitting and corrosion initiation temperatures measured in the G48 test are best used either as indicators of performance in environments similar to the standard environments used for the testing (which are not common in upstream oil and gas service) or for illustrating the effects of welding, material processing, and fabrication on the corrosion resistance of the alloy. The G48 test can grossly underestimate the CPT and CCT in actual oil and gas service environments. Furthermore, note that CPT and CCT are not typically used to describe MIC resistance although it is not uncommon for MIC resistance to be better in alloys with high CPT and CCT.

Underestimating the CPT and CCT in the service environment can lead to overly conservative and unnecessarily expensive material selection criteria. Whenever possible, electrochemical tests such as the ASTM G61 cyclic potentiodynamic polarization test method, conducted over a range of temperatures in electrolytes approximating the actual service environment, should be used as a material selection resource (Iannuzzi 2013).

In addition to determining the conditions under which localized corrosion initiates, potentiodynamic polarization tests can also be used to determine the likelihood that a change in the environmental conditions will cause a corroding surface to repassivate. If the surface repassivates, the localized corrosion stops. In some cases, an excursion in operating conditions can initiate localized corrosion and the corrosion will continue when the environmental conditions return to normal. In comparison, other alloys will stop corroding when environmental conditions return to normal (Bäßler et al. 2015).

Table 2.2 lists a number of materials that are commonly used in corrosive service and their chemical composition.

TABLE 2.2
Examples of Common Alloys Selected for Corrosive Service

UNS, DIN Number	Common Name	General Description	Fe	Ni	Cr	Mo	Cu	N	Other	PREN
S42000	13 Cr	Martensitic (API 5CT OCTG)	Bal		11.5–14					11.5
S17400 1.4542	17-4PH	Precipitation hardenable stainless	Bal	3.0–5.0	15–17.5		3.0–5.0		Nb + Ta 0.15–0.45	15.0
S41000 1.4006	410 SS	Martensitic stainless	Bal	0.75	11.5–13.5					11.5
S30400 1.4301	304 SS	Austenitic stainless	Bal	8.0–11	17.5–20					17.5
S31600 1.4401	316 SS	Austenitic stainless	Bal	10–14	16–18	2.0–3.0				22.6
S44627	E-Brite	Ferritic stainless	Bal	0.5	25–27.5	0.75–1.5			Ni + Cu 0.50	27.5
S44735	AL29-4C	Ferritic stainless	Bal	1.0	28–30	3.6–4.2			Ti + Cb 0.2–1.014	39.9
S32304 1.4362	2304	Lean duplex stainless	Bal	2.0–5.5	21.5–24.5	0.05–0.6	0.05–0.6	0.05–0.20		22.5
S32205 1.4462	2205	Duplex stainless	Bal	4.5–6.5	22–23	3.0–3.5		0.14–0.2		34.1
S32750 1.4501	2507	Superduplex stainless	Bal	6.0–8.0	24.0–26.0	3.0–5.0		0.24–0.32	W 0.5–1.0 Si 0.2–0.7	37.7
S32760 1.4410	Zeron 100	Superduplex stainless	Bal	6.0–8.0	24.0–26.0	3.0–4.0		0.2–0.3	W 0.5–1.0	37.1
S33207	SAF 3207	Hyperduplex stainless	Bal	7	32	3.5		0.5		51.6

(*Continued*)

TABLE 2.2 (CONTINUED)
Examples of Common Alloys Selected for Corrosive Service

UNS, DIN Number	Common Name	General Description	Fe	Ni	Cr	Mo	Cu	N	Other	PREN
N08020 2.4660	Alloy 20	Superaustenitic stainless	Bal	32–38	19–21	2.0–3.0	3.0-4.0		Nb 1.0	25.6
N08367 17006	AL-6XN	Superaustenitic stainless	Bal	23.5–25.5	20–22	6.0–7.0	0.75	0.18–0.25		42.7
N08904 1.4539	904L	Superaustenitic stainless	Bal	23–28	19–23	4.0–5.0	1.0-2.0			32.2
S31254 1.4529	254 SMO	Superaustenitic stainless	Bal	17.5–18.5	19.5–20.5	6.0–6.5		0.18–0.22		42.2
N10276 1.7751	C276	NiCrMo	4.0–7.0	Bal	14.5–16.5	15–17	2.0		W 3.0–4.5	64.0
N060221.7744	C22	NiCrMo	2.0–6.0	Bal	20–22	12.5–14.5			W 2.5–3.5 V 0.35 Ti 0.6–1.2	61.3
N06625 2.4856	Alloy 625	NiCrMo	5.0	Bal	20–23	8.0–10			Nb + Ta 3.15–4.15	46.4
N04400 1.7750	Monel 400	NiCu	2.5	63.0			28–34			NA

Note: Single values represent maximum allowable values. PREN values are calculated from the minimum allowable values. Actual PREN values are generally slightly higher. Corrosion resistance may be lower than indicated by the PREN, depending upon the effects of thermal processing, especially for precipitation hardenable alloys.

2.3 METALLURGICAL AND SURFACE CONDITIONS THAT AFFECT ABIOTIC CORROSION: WHY LAB TESTS MAY NOT REPLICATE SERVICE PERFORMANCE

The previous section indicates the importance of replicating the service environment when performing corrosion tests designed to support materials selection tasks. The G48 test was used as an example of how tests might illustrate the relative performance of alloys in a standard test environment. The same test also illustrates that material processing can have an adverse effect on corrosion resistance in the test environment although the results are not useful for estimating the time to failure for a specific alloy in a specific environment. Even when standard electrochemical tests or common configurations of corrosion coupons are used in long-term tests, the results can be misleading and perhaps be only broad indictors of relative performance of candidate alloys. There are several reasons for the differences in performance in testing and performance in service. These differences described below are particularly applicable to testing for susceptibility to MIC.

First, the surface finish of standard test electrodes or corrosion coupons typically consists of a uniform surface ground to 120 grit or smoother. In comparison, a fabricated pipeline or other piece of equipment will generally have a rough mill finish. In the case of steel, the mill finish typically includes appreciable amounts of high-temperature oxides except in areas where surface imperfections have been ground off or where the surface has been ground and/or machined in preparation for welding. Corrosion-resistant alloys may be pickled with acid to remove weld heat tint, other surface oxides, and contaminants prior to being placed into service, but the surface roughness is generally not comparable to the uniformly ground surfaces of test electrodes and coupons. The surface smoothness and methods used to remove surface oxides influences pitting resistance. For example, chemical pickling has been shown to be more effective at restoring the corrosion resistance of stainless steel surfaces affected by heat tint from welding than when the surfaces are cleaned to bright metal by grinding, sand blasting, or wire brushing (Pettersson and Fager 1995).

Titanium alloy test panels measuring several square feet were used in the as-manufactured condition to test the performance of alternative materials of construction in hot, high-chloride brine produced from a geothermal reservoir (Amend 2015, Personal experience). The panels showed complete resistance to corrosion except where paint pen marks added by the alloy producer had not been removed prior to testing. At the paint marks, severe crevice corrosion was observed. The susceptibility to such localized corrosion and the importance of complete cleaning of wetted surfaces would not have been identified through standard corrosion coupon or electrochemical testing.

Other sources of inconsistency between test performance and in-service performance can be traced to material handling practices. For corrosion-resistant alloys, it is best practice to make sure that material handling equipment prevents metal-to-metal contact with other alloys. Smearing or otherwise transferring small amounts of mild steel onto the surface of a stainless steel component can cause localized formation of very corrosive by-products, for example, ferric chloride, on the alloy surface when the mild steel particles corrode in service.

In one case, a manufacturer was observed moving large-diameter stainless fittings around the shop floor with no protective coverings on the forklift tines. Scrape marks were subsequently observed on the internal surfaces of the fittings. Mitigation consisted of using abrasive discs to buff away the gouge, followed in some cases by passivation using chemicals that reestablish or reinforce the protective passive layer on the alloy surface. The selection and use of a passivating solution depends upon the alloy being processed and the severity of the service conditions. Spatter from welding or cutting operations can also result in iron contamination of stainless steel components and lead to chloride stress cracking in certain environments.

Finally, standard corrosion tests may not account for the effects of welding on the corrosion resistance of the fabricated assembly. Corrosion test coupons can be modified to contain welds so that the corrosion resistance of the weld metal, heat-affected zone and parent metal together can be examined. An example of a corrosion test using a coupon containing a weld is shown in Figure 2.1.

Welding can affect the corrosion resistance in several ways. The weld metal can consist of a different alloy composition than the base metal with the result being some galvanic difference between the weld and the base metal. Figure 2.2 illustrates selective corrosion of weld metal that occurred in a supplemental corrosion test applied to a welding procedure qualification test of a corrosion-resistant alloy.

Even if the weld metal exactly matches the composition of the base metal (which is unlikely), the cast microstructure of the weld metal deposit general results in

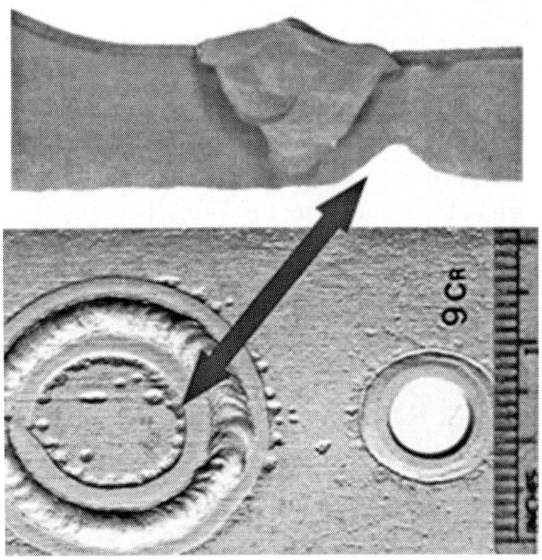

FIGURE 2.1 Localized corrosion (at arrows) at the edge of a weld heat-affected zone replicated by corrosion coupon testing that included use of circular weld beads. The top image is a longitudinal cross-section through a girth weld joining straight mild steel pipe to a steel weld neck flange. The selective corrosion in the low-temperature heat-affected zone of the pipe did not occur in the flanges, which were found to be silicon killed steel. The corrosion was not related to MIC, that is, it was abiotic.

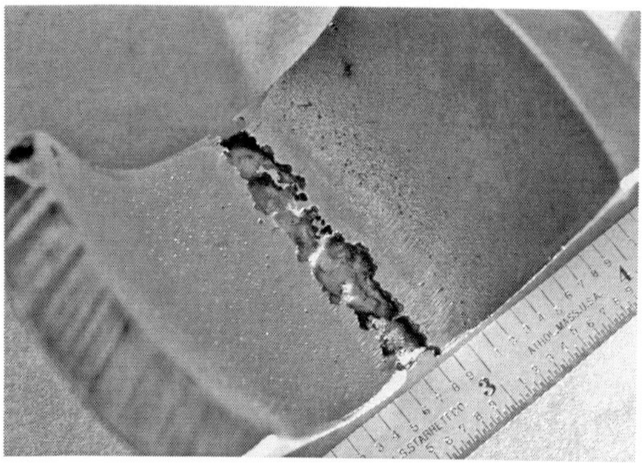

FIGURE 2.2 Selective corrosion of weld metal produced in a supplemental abiotic corrosion test applied to a welding procedure qualification test of a corrosion-resistant alloy.

microsegregation of some of the alloying elements. As a result, microscopic areas of the weld are enriched in alloying elements while adjacent areas have diminished amounts of the alloying elements responsible for corrosion resistance.

In welding, the fusion process results in exposure of localized areas of the base metal to very high temperatures regardless of whether or not weld metal is added to the joint. The high temperatures and related fast cooling rates result in formation of heat-affected zones in which large variations in microstructure are observed over short distances. Different microstructures have different responses to corrosive environments. An example of selective longitudinal pipe seam corrosion is shown in Figure 2.3. No filler metal was added to the ERW pipe seam during the manufacturing process. Therefore, the selective corrosion is unrelated to variations in composition across the joint.

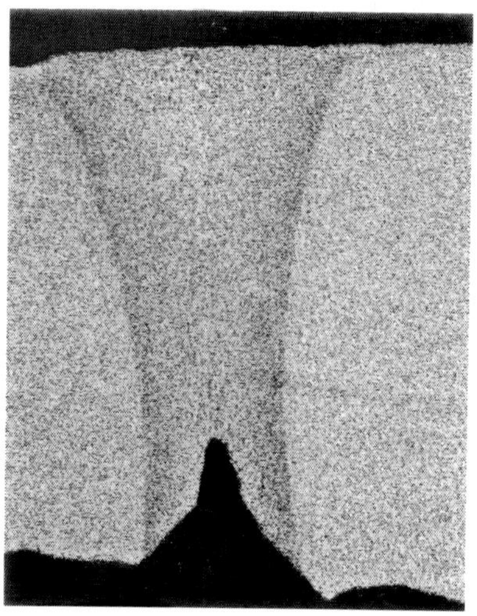

FIGURE 2.3 Metallographic cross-section of an ERW weld included here to illustrate selective seam corrosion on the internal surface of the pipe.

Even if corrosion resistance is maintained, the residual stress from welding alone can lead to stress corrosion cracking if a susceptible alloy is exposed to a crack-inducing environment or "preferential" corrosion of the more highly stressed portion of the heat-affected zone.

The surface contour of a weld also tends to be different from the surface morphology of the original base metal. On either a micro or a macro scale, the surface roughness can be different, and the oxides present can be different than oxides on adjacent surfaces. The differences can influence susceptibility to corrosion. Tall weld caps on pipe seams or field welds have been known to influence corrosion susceptibility by resulting in premature failure of field-applied external coatings. Spatter, arc burns, and rough weld ripple can result in local thin areas of liquid coatings applied by spraying or brushing, and it can be more difficult to achieve the required surface cleanliness when using manual cleaning methods. The disbondment of tape wrap coatings over large pipe seams, made by the submerged arc welding process, is well known. The disbondment is sometimes referred to as "tenting."

The presence of workmanship flaws in welds can exacerbate corrosion by creating crevices for the initiation of corrosion and irregular surfaces for establishment of biofilms. Inadequate penetration at girth welds is especially noteworthy since the resulting groove-like feature can collect sediment or water that is not easily removed by pipeline cleaning pigs. An example of internal corrosion that started at a welding defect is shown in Figure 2.4. Girth welds and seam welds that protrude significantly from the internal pipe surface may prevent adequate contact of cleaning pigs with the adjoining metal, leaving deposits and biofilm adjacent to weld.

Even in the absence of welding-related corrosion, typical corrosion tests do not evaluate the effects of coupling "new" components to "old" components of the same or similar grade and specification. It is not uncommon to observe that new

FIGURE 2.4 Internal corrosion that occurred at a girth weld defect (inadequate penetration).

attachments to existing steel pipe systems appear to have higher corrosion rates than the older, existing portions of the system. While it is often popular to blame the steel "quality," the cause of the difference in performance more likely is related to the existence of semipassive films formed over time on the older steel. For example, both sulfide corrosion products and some oxides are cathodic to bare steel. The films are often different or absent in the newly added components, resulting in the new steel being anodic to adjacent older steel.

In regard to external corrosion, it can be difficult to assess the effects of pipelines traversing different soil types in corrosion tests. An electrically continuous pipeline exposed to multiple soil types can have localized areas that have greater corrosion rates than expected from the burial of electrically isolated corrosion coupons. The different soil types promote the formation of anodic and cathodic areas in what has been termed "long cell corrosion" (Schaschl and Marsh 1963). Buried corrosion coupons that are not electrically continuous with the adjacent pipeline may not replicate the electrical potentials developed on pipelines that traverse different soil types.

2.4 MATERIALS AND MIC SUSCEPTIBILITY

MIC is typically observed to be a localized corrosion mechanism, which is logical considering the characteristics of biofilms in pipeline environments. Biofilms and surface deposits inside of pipelines are seldom completely and perfectly homogeneous, particularly at a micro- or macroscopic scale. Pipeline deposits are often a combination of inorganic and organic materials into which a biofilm is established. Heterogeneity of chemical, electrochemical, microbiological, and physical conditions at the pipe surface can promote localized concentration cells, accumulation of acidic species, and create local potential differences resulting in corrosion cells. Note this heterogeneity is not exclusively the result of biofilms and microbiological activity, which is somewhat problematic when attempting to distinguish the relative contributions of biotic versus abiotic factors for corrosion diagnosis. Deposits can also retain water in films surrounding particles, providing an aqueous niche for biofilm growth in an environment that is predominantly hydrocarbons. In some cases, microbiological activities may be the "driving force" in under-deposit corrosion (UDC) in otherwise benign surface deposits. Localized corrosion related to deposits may also be associated with velocity, the presence of water, total system pressure, O_2, CO_2 and H_2S partial pressure, temperature, and sulfate and chloride ion concentrations (Papavinasam et al. 2007).

Macroscopic MIC morphology is a persisting legend, particularly in the oil and gas industry, with biotic corrosion described as wormy, pits with striations, pits within pits, tunneling, and so on. It is mentioned here because MIC, in oil and gas at least, sometimes manifests such that subtle distinctions in metallurgy, residual stress, and so on are illustrated in the corrosion damage. The complicating factor in using metal loss morphology as an indicator of MIC is that other corrosion mechanisms can also cause these same distinctions in morphology since corrosion is a function of local chemical and electrochemical conditions. Therefore, corrosion morphology cannot be a *sole* basis for MIC diagnosis; instead, the following information is generally accepted by the industry and academia as being suitable evidence for identifying MIC (Little et al. 2006):

1. A sample of the corrosion product from the affected surface that has not been altered by collection or storage (i.e. contamination, oxidation, etc.)
2. Identification of a complete corrosion mechanism, including anodic and cathodic reactions
3. Identification of microorganisms capable of growth in the particular environment
4. Demonstration of an association of those microorganisms with the observed corrosion, both spatially and chemically/electrochemically

As in nearly all corrosion, MIC occurs as a process, not as a phenomenon that suddenly just appears. MIC occurs through a process of initiation and propagation. Assuming a new, clean surface as a starting point, upon exposure to pipeline fluids containing planktonic (free-floating) microorganisms, some initial pioneer microbes become attached to the surface, which is an ecological advantage since a biofilm provides improved access to energy sources and nutrients. The attachment point could be random or related to local flow conditions that allowed the microorganisms to attach. In some research, bacteria were shown to be attracted to anodic sites on the metal surface. Once microorganisms attach to a surface, they can express different genes that alter their behavior and attract additional microorganisms. The attached microorganisms often form exopolysaccharides that aid in adhesion to the metal (Beech et al. 1994). These polymers also cause inorganic particles to become embedded in the biofilm. As cell division occurs and additional microorganisms aggregate on the colonized surface, the biofilm begins to act as an occlusion on the metal surface, limiting ionic access to the area beneath the biofilm and also concentrating other chemical species, including corrosion products. On alloys that form passive films for protection, the species concentrated under the biofilm may damage the atomically thin passive layer, allowing localized corrosion to occur. Many reports of MIC on austenitic stainless steel demonstrate a cross-sectional morphology that is similar to chloride-induced pitting with a small pit entrance and a cavernous corroded area beneath it. The pit morphology concentrates metal chlorides, lowering the pH in the cavern and promoting sustained corrosion or metastable pitting.

As the biofilm continues to grow, it increases in physical size and may develop to include basic functional structures, such as microscopic channels that help remove metabolites and provide access to nutrients. Mature biofilms also experience cell death and may slough off parts of the film into the bulk phase. The nature of the corrosion associated with the biofilm may also change over time, particularly as the physical condition of the biofilm/deposit changes. For example, the initial biofilm may initiate microscopic "pitting" on the surface due to acid production. As the biofilm thickens and cells nearer the metal surface die off, the occlusion caused by the biofilm may simply promote a concentration cell mechanism, or the complexity of microbiological involvement may change. Iron-reducing bacteria, for example, are known to remove an electron directly from the metal for use in iron reduction even when they are not intimately in contact with the metal; that is, a conductive iron sulfide film may serve in electron transport (Videla et al. 2008). Sulfate-reducing bacteria (SRB) near a corroding surface have been shown to form iron sulfides in their cell membranes where sulfide is being produced by sulfate reduction (Eckert et al. 2003). After a point, the SRB is essentially "plugged" with iron sulfide, and the cell dies, leaving behind a shell or "fossil" that is conductive (McBeth et al. 2011).

Seeing that biofilms can form on nearly any material surface, MIC susceptibility then must be tied to a material's ability to resist localized corrosion and specific biotic corrosive species, for example, sulfide or organic acid. As described earlier in this chapter, there are different methods for testing susceptibility of a material to localized corrosion and many subtle factors that can influence the outcome of such tests. A material's PREN value may provide at least some value in estimating the relative resistance of an alloy to MIC. A general review of literature in which MIC is cited as the cause of corrosion will show that as PREN increases, the frequency of MIC case studies decreases. MIC is frequently reported for carbon steels and somewhat less frequently for stainless steels. As one moves to duplex and super-duplex or hyper-duplex alloys, the incidence of reported MIC is fairly rare. In laboratory studies using *Desulfovibrio desulfuricans* 2205, duplex stainless steel was reported (Antony et al. 2007) to experience etching, pitting, and crevice attack after 40 days exposure in a chloride-containing medium. Another study (Machuca et al. 2012) of duplex stainless steel (DSS) in natural seawater showed crevice corrosion only occurred in samples that were electrochemically polarized. Nickel–chromium–molybdenum alloys and titanium have not been reported as being susceptible to MIC (at least based on the literature search performed here); the exception being in environments containing oxygen.

In surface waters and sediments containing oxygen, a number of microorganisms can oxidize dissolved manganese to form enriched biomineralized deposits. Deposits of manganese oxides when formed on stainless steels and CRAs, such as alloy 625, are highly cathodic and result in localized potential differences that can drive severe corrosion (Lewandowski and Hamilton 2002). These deposits can be thin and brittle, resulting in fine cracks in the scale that act as crevices where corrosion is driven by the large corrosion potentials between manganese oxides and the exposed metal. Thus, although the corrosion is not directly caused by microorganisms, the mineral scales resulting from their activity end up causing localized corrosion by shifting corrosion potentials.

Copper–nickel and nickel-based alloys have been used successfully in flowing, aerated seawater service although MIC has been reported in some cases, particularly where flow is stopped for extended periods of time. One study (Little et al. 1990) discussed severe corrosion of copper–nickel (88.5% copper, 10% nickel, 1.5% iron) piping after 1 year of service and nickel alloy (66.5% nickel, 31.5% copper, 1.25% iron) after 6 months of service in stagnant estuarine water from the Gulf of Mexico. In both cases, localized corrosion was found under biofilms containing SRB.

Since MIC is a surface phenomenon, reducing the susceptibility of a material to localized corrosion is based on alloy selection and maintaining surface conditions that reduce the potential for deposit and biofilm accumulation. Clean surfaces can be promoted by using a design that prevents stratified flow and water/solids accumulation or that can achieve a re-entrainment velocity upon startup or be easily cleaned in some way. Fouling and scale can often be addressed using chemical mitigation and prevention measures. Maintenance, such as pigging of pipelines, is another way to prevent surface deposits and biofilms from remaining in the same place for extended periods of time. Increase of surface roughness or surface profile is sometimes considered as having the ability to increase susceptibility to biofilm formation as mentioned earlier; however, considering that most bacteria are only a few microns in length, a mirror finish would be required to completely avoid having niches where microorganisms could collect.

Few pipeline systems can operate continuously at velocities that generate enough shear force to prevent biofilm accumulation; however, the effect of velocity on MIC damage rates (i.e., not microbiological growth rates or numbers) has not been fully explored in oil and gas service. One laboratory study that was conducted with *D. desulfuricans* showed that at a velocity of roughly 3.5 m/s bacteria could not adhere to a metal surface to form a biofilm (Wen et al. 2007). Pre-established biofilms (e.g., during shut-down periods) could withstand such flow and remain active.

Materials susceptibility to MIC has been investigated by many researchers. One industry study (Blythe and Gauger 2000) that looked at inclusion content and surface roughness in regard to biofilm establishment and corrosion rates of welded carbon steel found the following:

- For base metals, no firm conclusion about the effects of surface finish on severity of MIC or the relationship between colonization and the inclusion content or composition of the steels tested were observed.
- Low-quality steels are likely to be more susceptible to corrosion damage related to MIC. Steels with low inclusion content and few sulfide inclusions consistently showed lower corrosion rates in the testing even though colonization was similar to other steels.
- Microorganisms did *not* preferentially attack MnS inclusions in the test.
- SRB are not required for MIC although they can increase the severity of attack.
- For welds, the combination of rough (surface) texture and high inclusion content produced high corrosion rates when tested with APB but not for other strains of bacteria. High corrosion rates were *not* observed unless the rough surface was present along with high inclusion content.

Other research has also examined the effect of surface roughness on the degree of cell attachment and extent of pitting corrosion initiation. A surface roughness of >3.3 μm was reported to affect bacterial attachment in some studies, and others reported that a smoother surface has fewer sites that can be activated into metastable pit growth or confirmed that metastable pit growth tended to occur on rougher surfaces (Chen et al. 2014). According to NACE 3T199, extended analysis coupons (NACE 2012) that are used for monitoring corrosion initiation mechanisms are prepared (Eckert et al. 2006) with a 600-grit wet-sanded finish to allow distinction of microscopic pitting initiation as shown in Figure 2.5.

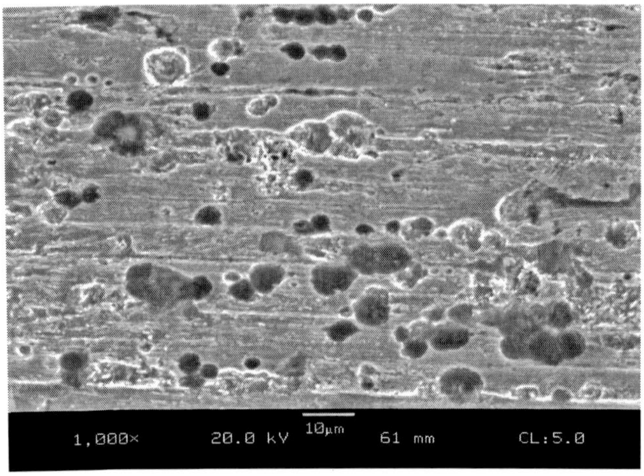

FIGURE 2.5 SEM micrograph of an extended analysis coupon surface showing biotic microscopic pit initiation at 1000× original magnification. The coupon had a 600-grit wet-sanded surface finish before being installed in the pipeline environment.

2.5 MATERIALS SELECTION AND MIC

While the use of CRAs with high resistance to localized pitting is a possible approach, it is certainly not economical in most cases. The fact is that most oil and gas operations rely on carbon steel as their primary material of construction. Some advantage can be gained, however, in selectively applying CRAs where the threat of MIC is the highest. It is fairly well established, for example, that dead-leg piping is more susceptible to MIC than pipelines that are normally flowing. A number of schemes for assessing and ranking the threat of MIC have been published (Jenneman 2012). The threat of MIC in dead legs can be managed by material selection in the design stage, retrofitting carbon steel with CRA, or eliminating the environment that promotes MIC. Produced water, seawater, and firewater systems are also highly susceptible to MIC. Nonmetallic components (i.e., epoxy composite piping, etc.) can be considered where pressures, stresses, and fire-resistance requirements allow alternatives to metals. Aboveground piping for saltwater disposal systems is sometimes constructed using fiberglass-reinforced plastic (FRP).

The application of CRAs in equipment or piping that is highly susceptible to MIC can be made more economical by the use of CRA-clad carbon steel or limiting the use of CRAs to only the most susceptible locations that cannot be temporarily isolated, cleaned, and mothballed. Limiting the extent of MIC of susceptible equipment that cannot be cleaned, flushed, or mothballed with chemical treatment is another way to reduce the need for CRAs.

There are also some cautions and considerations when implementing CRAs in an existing carbon steel system. First, if joining is done using welding, the materials must be compatible and field weldable. Qualified welding procedures may need to be developed, and welders who are experienced in welding dissimilar alloys should be used. Duplex alloys can be irreversibly damaged by improper welding, and other CRAs can experience sensitization with poor corrosion resistance or cracking as the result. Typically, a CRA-lined or -clad pipeline welding requires an all CRA girth weld that is designed to prevent initiation of corrosion and internal stress cracking while still being metallurgically compatible with the combination of pipe base metal and cladding. Second, there is the potential for galvanic corrosion at flanged connections between carbon steel and CRA; therefore, joint design and gasket selection are critical. Finally, maintenance procedures may need to be changed after implementing CRAs, and there should be awareness of any fluids that are incompatible with different process fluids or layup conditions. Well tubing made of 13Cr, for example, can resist corrosion in high-chloride well fluids; however, any exposure to oxygen under those conditions will result in severe corrosion. Chemicals, such as drilling mud, process fluids, cleaning materials, wash water with contaminants, or corrosion inhibitors and biocides, should be reviewed for compatibility with any new pipe material.

Internal coatings and linings of carbon steel equipment are another way to avoid contact between the environment and material at least for a finite period of time, that is, the life of the coating. Internal coatings are often avoided for piping although, for lines that are normally dead legs or are infrequently operated, internal coatings can be an alternative to CRA. As for any coating job, quality of materials and workmanship and adequate quality control are essential in order to have confidence in the coating system. The use of qualified and certified coating professionals is also important. Potential issues with internal coatings are damage from heat or rapid depressurization, mechanical damage during operation or maintenance, coating of tie-in welds, and lack of insight for inspection site selection. The use of high-density polyethylene (HDPE) liners in short sections of piping might also be a viable alternative. Advances are being made in coatings containing nanoparticles with antimicrobial properties and highly hydrophobic coatings that resist water attachment.

Produced water and seawater injection systems often contain an extensive amount of pipe, making the widespread use of CRAs cost-prohibitive. Oilfield water injection systems usually have the capacity for pigging main pipelines; however, well lines and smaller service piping on well pads usually cannot be cleaned or flushed. Biocides and corrosion inhibitors are usually used in seawater and produced water, and an oxygen scavenger is used in seawater.

When selecting an alloy for any type of oil and gas asset, the designer needs to consider all phases of asset life and the potential corrosion mechanisms that can develop in those phases. For example, pipelines are sometimes built in stages to accommodate increasing production needs as a field is being developed or in anticipation of new tie-ins with other systems. Pipelines that are significantly oversized for the current flow rates can result in stratified flow and sediment deposition, promoting MIC and UDC. When pipeline segments are installed and remain stagnant until further development occurs, these segments are essentially operational dead legs that are prone to liquid and solid accumulation and MIC. Designers also need to consider whether the materials selected for an asset will be compatible with corrosion mitigation chemicals and biocides as many chemicals are highly corrosive in their neat form. Undiluted biocides often have a low pH, and chemical handling systems must be designed appropriately to store and deliver these treatment chemicals. The elastomers selected for gaskets, seals, and O-rings must also be capable of withstanding exposure to neat mitigation chemicals—a major concern for umbilicals used in subsea production.

Finally, materials selection to address the threat of MIC is not limited to primary oil and gas pipelines because firewater piping, drain lines, relief lines, tanks, vessels, and other utility systems frequently experience stagnant conditions and/or sediment accumulation that promote MIC. When materials selection is limited or cost-prohibitive, the designer needs to provide capabilities for inspection, monitoring, flushing, and cleaning of equipment that will operate in a manner that promotes MIC. Quite often new facilities are designed and built in full awareness that certain parts will be stagnant most of the time, and yet no means of monitoring or cleaning are incorporated. Materials selection is certainly one means of improving asset integrity and reliability; however, when CRAs are not used, some means of MIC mitigation should still be provided.

REFERENCES

Antony, P. J., Chongdar, S., Kumarb, P., and Ramana, R., 2007, Corrosion of 2205 duplex stainless steel in chloride medium containing sulfate-reducing bacteria, *Electrochimica Acta*, 52, 3985–3994.

API, 2011, *Specification for Casing and Tubing*, 9th edition, Washington, DC, American Petroleum Institute.

API, 2012, *API 5L Specification for Line Pipe*, 45th edition, Washington, DC, American Petroleum Institute.

ASTM, 2011, *ASTM G48-11, Standard Test Methods for Pitting and Crevice Corrosion Resistance of Stainless Steels and Related Alloys by Use of Ferric Chloride Solution*, Washington, DC, ASTM.

Bäßler, R., Keserovic, A., Sobettzki, J., Klapper, H., and Dimper, M., 2015, Laboratory evaluation of metallic materials for geothermal applications, *EPRI International Conference on Corrosion in Power Plants*, Paper 006, San Diego, CA, EPRI.

Beech, I. B., Cheung, C. W. S., Chan, C. S. P., Hill, M. A., Franco, R., and Lino, A. R., 1994, Study of parameters implicated in the biodeterioration of mild steel in the presence of different species of sulphate-reducing bacteria, *International Biodeterioration and Biodegradation*, 34, 289–303.

Blythe, G., and Gauger, K., 2000, *Metallurgical Factors Affecting Microbiologically Influenced Corrosion (MIC) of Pipeline Steels*, Report #GRI-00/0241, Chicago, Gas Research Institute.

Chen, Y., Howdyshell, R., Howdyshell, S., and Ju, L., 2014, Characterizing pitting corrosion caused by a long-term starving sulfate-reducing bacterium surviving on carbon steel and effects of surface roughness, *Corrosion*, 70 (8), 767–780.

Eckert, R. B., Aldrich, H., Yang, C., and Edwards, C., 2006, Guidelines for reproducing microbiologically influenced corrosion initiation on carbon steels under controlled conditions, *Corrosion 2006*, Paper 06519, Houston, TX, NACE.

Eckert, R., Aldrich, H., Edwards, C., and Cookingham, B., 2003, Microscopic differentiation of internal corrosion initiation mechanisms in natural gas pipeline systems, *Corrosion 2003*, Paper 03544, Houston, TX, NACE.

Iannuzzi, M., 2013, Maximum allowable service temperatures and the ASTM International G48 Test, http://www.aboutcorrosion.com/2013/11/29/service-temperature-g48-test/ (accessed August 16, 2015).

Jenneman, G. E., 2012, MIC: Managing risk and uncertainties, *14th International Middle East Corrosion Conference and Exhibition*, Paper 114-MC-17, Kingdom of Bahrain, NACE.

Lewandowski, Z., and Hamilton, A., 2002, MIC of stainless steels as a model system to study metal-microbe interactions, *Corrosion 2002*, Paper 02474, Houston, TX, NACE.

Little, B. J., Lee, J. S., and Ray, R. I., 2006, Diagnosing microbiologically influenced corrosion: A state-of-the-art review, *Corrosion*, 62 (11), 1006–1017.

Little, B. J., Lee, J. S., Ray, R. I., and McNeil, M., 1990, Microbiologically influenced corrosion in copper and nickel, seawater piping systems, *MTS Journal*, 24 (3), 10–17.

Machuca, L. L., Bailey, S. I., Gubner, R., Watkin, E. L. J., Ginige, M. P., and Kaksonen, A. H., 2012, Crevice corrosion of duplex stainless steels in the presence of natural marine biofilms, *Corrosion 2012*, Paper 1486, Houston, TX, NACE.

McBeth, J. M., Little, B. J., Ray, R. I., Farrar, K. M., and Emerson, D., 2011, Neutrophilic iron-oxidizing "zetaproteobacteria" and mild steel corrosion in nearshore marine environments, *Applied and Environmental Microbiology*, 1405–1412.

NACE, 2012, *Techniques for Monitoring Corrosion—Field Applications*, Report 3T199, Houston, TX, NACE.

NACE, 2015, REFIN*COR database of petroleum refining and gas processing, Committee Information Exchange search results; entries 86C6.3-07, 86F7.1-03, 86F7.1-05, 88C7.1-51, 88C7.1-51, 89C7.1-18, and 91C9.1-10. Houston, TX, NACE.

Papavinasam, S., Doiron, A., and Revie, R. W., 2007, Model to predict pitting corrosion in oil & gas pipelines, *Corrosion 2007*, Paper 07658, Houston, TX, NACE.

Pettersson, C., and Fager, S., 1995, Welding Practice for the Sandvik duplex stainless steels SAF 2301, SAF 2205, and SAF 2507, AB Sandvik Steel, S-91-57-ENG.

Schaschl, E., and Marsh, G. A., 1963, Some new views on soil corrosion, *Materials Protection*, 2 (11), 8–17.

Videla, H., Le Borgne, S., Panter, C., and Raman, R., 2008, MIC of steels by iron reducing bacteria, *Corrosion 2008*, Paper 08505, Houston, TX, NACE.

Wen, J., Gu, T., and Nesic, S., 2007, Investigation of the effects of fluid flow on SRB biofilm, *Corrosion 2007*, Paper 07516, Houston, TX, NACE.

Section II

Microbiologically Influenced Corrosion Fundamentals: Microorganisms and Mechanisms

3 Microorganisms in the Oil and Gas Industry

Ian M. Head

CONTENTS

ABSTRACT

The deep biosphere is the largest microbial biome on the planet. Being remote, difficult to sample, and refractory to direct observation, it can often seem esoteric and detached from practical relevance. Those working in the oil and gas industry know from difficult and costly experience that this view of the deep biosphere could not be further from the truth. Petroleum systems provide the most tangible evidence that the deep biosphere should not be relegated to a curious scientific oddity. The operational and economic consequences of microbial activity in petroleum systems are immense and a testament to the power of the deep biosphere and its very direct impact on human activities.

Petroleum reservoirs harbor a diverse array of microorganisms, which, on production *and* geological time scales, influence the properties and quality of emplaced oil and affect its production and ultimately its value. Heavy oil formation over tens of millions of years, sulfide formation leading to souring, microbially influenced corrosion, and the potential to enhance oil recovery all have a central microbiological component, and in some cases, there may be microbiological solutions to detrimental processes that occur during petroleum production. Here, the range of organisms associated with petroleum systems is explored. A particular emphasis is placed on the activities they catalyze and which underline their practical importance for good or ill in the oil and gas industry.

3.1 INTRODUCTION: THE DEEP BIOSPHERE AND THE PETROLEUM RESERVOIR MICROBIOME

Subsurface sediments (defined by Whitman et al. 1998 as terrestrial habitats below 8 m and marine sediments below 10 cm) represent the largest biome on the planet with the latest estimates indicating that the deep biosphere comprises 4.7×10^{29} cells (Kallmeyer et al. 2012). Whitman's analysis considered unconsolidated sediment to a maximum depth of 4 km, which equates to a sediment volume of 1.2×10^{17} m^3. Four kilometers is considered to be the limit of the deep biosphere as this corresponds to the approximate average depth of the 122°C isotherm (Garland 1971; Kallmeyer et al. 2012), which is close to the currently known maximum temperature for life (Takai et al. 2008). However, microbial life in the deep subsurface and specifically in deep petroleum reservoirs may cease at lower temperatures (Wilhelms et al. 2001) probably due to the constraints of extreme energy limitation in deep subsurface sediments (Wilhelms et al. 2001; Jørgensen and D'Hondt 2006; Head et al. 2014).

Petroleum reservoirs occur from the Earth's surface to a depth of almost 6 km with the majority found at depths of less than 4 km (Ehrenberg and Nadeau 2005). The majority of petroleum reservoirs therefore are clearly within the bounds of the deep biosphere. The first evidence that life existed in petroleum reservoirs was reported in 1926 when Bastin reported the occurrence of sulfate-reducing bacteria in produced waters from a petroleum reservoir in the Illinois Basin (Bastin 1926), and though there have always been concerns regarding the provenance of microorganisms isolated from deep subsurface environments, and some organisms detected in these systems are almost certainly contaminants, it now seems clear that deep subsurface sediments, including petroleum reservoirs, harbor diverse indigenous microbial communities (Magot et al. 2000; Magot 2005; Wentzel et al. 2013). Irrespective of their source, the microorganisms that constitute the petroleum biosphere have significant consequences for petroleum systems, both with respects to the predrill, in situ activities of indigenous organisms and the response of either indigenous or contaminating/deliberately introduced organisms during production.

The excellent review of Wentzel and coauthors (2013) and several chapters in Ollivier and Magot (2005) provide a comprehensive survey of the microorganisms isolated from or identified in petroleum reservoirs. Readers wishing more detailed information are directed to these sources. From these review articles, it is clear that anaerobic thermophiles from the Thermotogae are common in high-temperature petroleum reservoirs (e.g., *Petrotoga*, *Thermotoga*, *Geotoga*, *Kosmotoga*, *Oceanotoga*, and *Thermosipho*), and *Petrotoga* and *Geotoga*, for example, have only been isolated from petroleum reservoirs, suggesting these are indigenous petroleum reservoir specialists. Firmicutes, such as *Thermoanaerobacter*, *Geobacillus*, *Desulfotomaculum*, *Caldanaerobacter*, *Mahella*, and *Bacillus* (Wentzel et al. 2013) are also commonly reported among petroleum reservoir isolates. *Deferribacter* spp., *Thermus* spp., *Anaerobaculum* spp., and *Thermovirga* spp. too have been isolated from a range of petroleum reservoirs. Significantly, in the context of reservoir souring and corrosion, a range of delatproteobacterial sulfate reducers have also been isolated frequently from petroleum reservoirs (Magot et al. 2000; Magot 2005), and many types of SRB have been found only in petroleum reservoirs and not other environments, strongly suggesting that they are part of an indigenous microflora (Birkeland 2005).

Many mesophilic bacteria have also been isolated from petroleum reservoirs and have been inferred to be indigenous based on their growth at in situ salinity (Magot 2005). Mesophilic, halophilic fermentative bacteria from the genus *Haloanaerobium*, for example, have been isolated from petroleum reservoirs as has *Fusibacter paucivorans* and *Spirochaeta smaragdinae* (Ollivier and Cayol 2005). A range of heterotrophic iron- and nitrate-reducing bacteria are also known from petroleum reservoirs (e.g., some members of the genera *Denitrivibrio*, *Garciella*, *Geobacillus*, *Marinobacter*, *Petrobacter*, *Shewanella*, and *Deferribacter* as mentioned above) and sulfide-oxidizing nitrate-reducers from the *Epsilonproteobacteria* (Gevertz et al. 2000).

In addition to bacteria, a range of archaea are also known to inhabit petroleum reservoirs. *Thermococcus* spp., *Pyrococcus* spp., and *Archaeoglobus* spp., all Euryarchaeota, have been isolated from high-temperature petroleum reservoirs as have thermophilic methanogens. Interestingly, it is exclusively hydrogenotrophic methanogens that have been isolated from high-temperature petroleum reservoirs, including members of the genera *Methanoculleus*, *Methermicoccus*, *Methanothermobacter*, and *Methanococcus* (Jeanthon et al. 2005; Wentzel et al. 2013). This has prompted the suggestion that syntrophic acetate oxidation to hydrogen and CO_2 may be an important process in high-temperature petroleum reservoirs (Jeanthon et al. 2005), and indeed *Thermacetogenium phaeum*, a known syntrophic acetate-oxidizing bacterium, has been identified in a high-temperature Chinese oil reservoir. Interestingly, a small number of culture-independent studies have detected members of the Methanosarcinales in high-temperature systems (Orphan et al. 2000), and Methanosarcinales sequences have been recovered from low-temperature reservoirs (Grabowski et al. 2005; Piceno et al. 2014), and in at least one low-temperature Alaskan reservoir, Methanosarcinales appeared to be the dominant group of methanogens (Pham et al. 2009).

3.2 THE PETROLEUM RESERVOIR BIOSPHERE

The organisms found in petroleum reservoirs and reservoir fluid chemistry (e.g., presence of methane, sulfide, volatile fatty acids, and gas isotope data) indicate that the petroleum reservoir biosphere is primarily anoxic. In instances where formation waters are low in sulfate and seawater injection for secondary recovery has not been practiced, the petroleum reservoir biome is primarily driven by fermentation reactions that deliver hydrogen, carbon dioxide, and acetate to methanogens. In high-temperature reservoirs with archaeal communities dominated by hydrogenotrophic methanogens, it is also likely that syntrophic acetate oxidation plays a significant role in the anaerobic food chain of petroleum reservoirs. Interestingly, many of the fermenting archaea and bacteria found in petroleum reservoirs have the capacity to reduce oxidized iron and/or elemental sulfur and thiosulfate, for example (e.g., *Thermococcus*, *Thermotoga*, *Halanaerobium*; Jeanthon et al. 2005; Ollivier and Cayol 2005), suggesting that a degree of metabolic versatility is an important adaptation even in the rather static environment of an untapped petroleum reservoir.

Where sulfate is abundant, sulfate reduction is a significant driver of the deep petroleum biosphere, and a wide range of unique sulfate-reducing bacteria, from three phyla (*Proteobacteria, Firmicutes, Thermodesulfobacteria*) are known from petroleum reservoirs as are euryarchaeotal *Archaeoglobus* spp. (Birkeland 2005). However, often *Archaeoglobus* spp. are not considered to be indigenous to petroleum reservoirs and are thought to be introduced with injected seawater (Birkeland 2005; Jeanthon et al. 2005). While the largest pool of carbon in petroleum reservoirs is clearly hydrocarbons (principally saturated and aromatic hydrocarbons), there are many other potential carbon and energy sources in petroleum reservoirs. These include organic acids, such as acetate, propionate, and butyrate, which can be present at tens of millimolar concentrations (Barth 1991). Moreover, hydrogen may act as an electron donor, driving sulfate reduction and methanogenesis in petroleum reservoirs with organic or inorganic carbon serving as carbon sources. In the case of CO_2 reduction, either for methanogenesis or biosynthesis, the carbon dioxide and hydrogen requirements can be met purely by the fermentation of organic substrates. However, it is possible that hydrogen in petroleum reservoirs may be derived from other sources, such as aromatization of organic compounds, serpentinization in deep hot rocks, or thermal maturation of organic matter in source rocks and transport of the hydrogen generated to the reservoir (Head et al. 2003).

The slow microbial utilization of hydrocarbons in petroleum reservoirs on geological time scales leads to one of the most dramatic expressions of the petroleum reservoir biosphere, if not the deep biosphere more generally: the occurrence of heavy oil. Heavy oil results from the activity of microorganisms present in the reservoir, which, over millions of years, have consumed the lighter, biodegradable components of the oil, typically components of the saturated hydrocarbon and light aromatic hydrocarbon fractions. This leads to oils that contain higher proportions of branched and cyclic saturated hydrocarbons, heavier aromatic hydrocarbons, and the more polar components of crude oils—the resin and asphaltene fractions. These heavy, biodegraded oils have physical and chemical properties that make them more difficult and costly to produce and refine. For example, heavy oils can have viscosities in the range of 1 million to 10 million centipoise and can contain as much as 5% or more sulfur by weight. By contrast, light oils typically have viscosities less than 10 centipoise at room temperature and sulfur content as low as 0.01% by weight (Carrales and Martin 1975).

Studies of heavy oil reservoirs have offered fundamental insights into the nature of the petroleum reservoir biosphere. First, gradients in oil composition often observed in biodegraded petroleum reservoirs containing heavy oil suggest that oil-degrading microorganisms are not homogeneously distributed throughout the reservoir, but are, in fact, focused at the base of the reservoir where oil-saturated rock meets underlying water-saturated rock—the oil–water transition zone (OWTZ; Head et al. 2003). This was subsequently confirmed by quantification of bacterial 16S rRNA genes through an OWTZ demonstrating that bacterial abundance in the transition zone was two to three orders of magnitude greater than in the overlying oil column (Bennett et al. 2013). Nevertheless, it is known that microorganisms can also be detected in oil (Bennett et al. 2013; Meckenstock et al. 2014; Wang et al. 2014) where they are present at lower abundance than in the OWTZ (Bennett et al. 2013) and most likely are present in highly dispersed water droplets that are typically present in crude oils (Meckenstock et al. 2014).

Produced waters from petroleum reservoirs and other deep groundwater systems contain 10^3 to 10^6 microbial cells/ml (Pedersen 1993; Nilsen et al. 1996); however, there are almost no data on microbial abundance in sediments directly recovered from petroleum reservoirs. Bacterial 16S rRNA gene abundance data from a core spanning the OWTZ of a heavy oil reservoir at a depth of approximately 600 m indicated that bacterial abundances ranged from around 10^3 (and lower) to 10^6 per cm^3 sediment with greatest abundance in the OWTZ and the lowest abundance in the oil-saturated reservoir sediments (Bennett et al. 2013). Interestingly, the abundance at the OWTZ was consistent with the "global" relationship between cell numbers and depth seen in a large sampling of deep subsurface marine sediments (Parkes et al. 1994, 2014; Jørgensen and Marshall 2016). The abundance in the oil column was two to three orders of magnitude or more lower, suggesting that lack of access to a continuous water phase or the inhibitory effect of certain components of the oil might limit microbial abundance in the bulk volume of oil reservoirs.

The undisturbed, in situ petroleum reservoir biosphere is deep and slow. It has been estimated that it can take millions to tens of millions of years to biodegrade the oil in a petroleum reservoir (Larter et al. 2003), and this equates to very low levels of microbial activity. On the basis of oil compositional gradients and other independent approaches, first-order rate constants for oil biodegradation at an OWTZ have been estimated to be on the order of 10^{-6} to 10^{-7} yr^{-1}. This is on the same order as the estimated reactivity of deeply buried sedimentary carbon that may also fuel the deep biosphere in general (Middelburg 1989; Middelburg et al. 1993; Jørgensen and Marshall 2016). Thus, in terms of broad characteristics, the petroleum reservoir deep biosphere is similar to other deep sedimentary environments, which, excluding situations where plate tectonics drive fluid circulation and mass transport, is an environment where activity is controlled by slow transport of solutes by diffusion (Jørgensen and Marshall 2016). This results in low cellular activity and low doubling times (typically on the order of hundreds to thousands of years; Jørgensen and Marshall 2016).

3.3 WAKING THE (ALMOST) DEAD

Like the majority of the deep biosphere, microbial activity in the petroleum reservoir biosphere is likely limited by low energy availability engendered by limited quantities and transport of labile organic carbon, nutrients, and electron acceptors. Regions in the deep biosphere where advection of fluids increases mass transport of limiting resource are characterized by increased cell abundance and activity (Parkes et al. 2005). This clearly illustrates the fact that undisturbed deep subsurface sediments are mass transport limited, and resident microbiota have the potential to be reawakened given favorable environmental conditions. Moreover, it is clear from the work of Parkes et al. (2005) that an increase in solute mass transport can be sufficient to revive the dormant cells of the deep biosphere. For a more complete appraisal and definition of dormancy in relation to the deep biosphere, the reader is directed to Jørgenesen and Marshall (2016).

A corollary of this observation is that any anthropogenic activity that alters fluid flow in the deep subsurface will have the likely consequence of stimulating the activity of the previously dormant indigenous microbiota. In addition, the stimulation of the native microbiota purely as a result of increased mass transport will be augmented by the delivery of exogenous organisms, electron donors, electron acceptors, and nutrients.

Potential electron donors and carbon sources are plentiful in petroleum reservoirs, and under certain circumstances, electron acceptors can also be nonlimiting (e.g., sulfate from injected seawater). The sources of inorganic macronutrients required to wake the dormant petroleum reservoir microbiota are less well understood. Ammonium concentrations of up to 15 mM have been reported from some deep (ca. 300 mbsf) sub-seafloor sediments (Morono et al. 2011), and even higher ammonium concentrations (up to 71 mM) have been detected in a limited number of deep groundwater samples from petroleum reservoirs (Manning and Hutcheon 2004).

Crude oil itself contains nitrogen (ca. 0.1% to 0.9% by weight; Ball et al. 1951). Much of this is organic nitrogen contained in the macromolecular fractions of the oil (Richter et al. 1952) and may thus be poorly available to micororganisms. The more soluble forms of organic nitrogen, mainly in the form of carbazoles, pyrolles, pyridines, and quinolones, are a potential source of nitrogen for microorganisms present, and alkylcarbazoles in particular have been shown to be degraded during heavy oil formation (Oldenburg et al. 2006).

Phosphorus in petroleum reservoirs is most likely present in mineral form, such as apatite inclusions in feldspars (Head et al. 2003), and it is known that in near-surface aquifers microorganisms preferentially colonize such mineral phases (Rogers et al. 1998). Whether this is also the case in petroleum reservoirs is not known, but again, utilization of a solid phase phosphorus source indicates that mass transfer of inorganic nutrients may have an important role to play in waking the petroleum reservoir biosphere.

Other factors that may contribute to waking the (almost) dead resident microbiota include temperature and salinity. The formation water from oil reservoirs can range from almost freshwater to near salt saturation (Ollivier and Cayol 2005), and the most halophilic hydrogenotrophic methanogen cultivated to date, *Methanocalculus halotolerans*, was isolated from a petroleum reservoir (Ollivier et al. 1998). It is apparent that there is an interaction between temperature and salinity, and halophilic thermophiles have not been found in petroleum reservoirs (Adkins et al. 1992; Bernard et al. 1992). This prompted Jeanthon et al. (2005) to suggest that high temperature and salinity together may limit microbial activity in petroleum reservoirs. This has subsequently been formalized in the palaeopickling hypothesis that attempts to explain the occurrence on nonbiodegraded oils in low-temperature saline reservoirs (Head et al. 2014).

While in situ, salinity and temperature may limit microbial activity; both may be altered considerably during oil production through practices such as seawater injection. Generally, this will lower temperatures, and depending on the formation water salinity, salinity may be increased or decreased by seawater injection. Alteration of temperature and salinity can have profound effects on microbial activity, and this is likely to be most apparent close to the well bore subject to seawater injection. This has led to the concept of a thermal viability shell (TVS) that is incorporated into some reservoir souring models (Eden et al. 1993). The TVS refers to the region surrounding an injector well where the temperature is reduced and provides a gradient of conditions extending from the well bore where both thermophilic and mesophilic organisms may thrive. This can have important consequences for souring and corrosion and their control in high-temperature reservoirs where a reduction in temperature can lead to a microbiologically active zone in the reservoir that was hitherto limited by high temperatures. Reductions in salinity close to the well bore of high-salinity reservoirs could likewise promote a "halo viability shell," promoting the activity of dormant indigenous organisms or permitting organisms introduced with injected fluids to thrive. This effect has been observed in practice where injection of low-salinity water in a Michigan oilfield precipitated reservoir souring, which was attributed to stimulation of sulfate-reducing microorganisms previously held in check by high-salinity conditions (Tinker et al. 1983).

Temperature changes are also inherent in the production infrastructure of an oilfield. Decrease in temperature of produced fluids as they are transported from a high-temperature reservoir to lower temperature above ground infrastructure, such as pipes and oil separators, again can result in stimulation of microbial activity, which may contribute to corrosion or hydrogen sulfide generation but may also contribute to other microbiologically influenced phenomena, such as biofouling (Sanders and Sturman 2005).

It is inevitable that the exploitation of deep geological resources will lead to changes in microbial ecology and activity, which may result in the modulation of beneficial and/or detrimental microbial processes. Our intrusion into the subsurface will therefore always wake the deep, dormant biosphere. Thus, an understanding of the factors that control the deep biosphere is not simply a matter of academic curiosity. Indeed it is, as those working in the business of resource extraction from the Earth know, a matter of significant practical import. The remainder of this treatise summarizes some of the consequences of waking the deep biosphere with respect to souring and corrosion (the central focus of this volume), microbially enhanced oil recovery, and challenges for emerging areas in geoenergy recovery. This sets the stage for more comprehensive treatment of specific issues provided in the following chapters.

3.4 SOURING AND CORROSION AND THEIR CONTROL

A significant, though not the sole, cause of reservoir souring is the stimulation of microbial sulfate reduction (Gieg et al. 2011), which may be exacerbated by injection of sulfate-rich seawater for secondary oil recovery. It is not entirely understood whether the main culprits in reservoir souring are indigenous sulfate-reducing microorganisms (SRM) stimulated by free access to an electron acceptor that was previously limiting or if SRM introduced with the seawater also contribute. Nevertheless, it is clear that many sulfate-reducing taxa have been exclusively isolated from petroleum reservoirs and are likely to be indigenous (Birkeland 2005).

Irrespective of whether the SRM responsible are indigenous or introduced, the hydrogen sulfide produced represents a serious health and safety issue on production platforms as well as reducing the value of the produced oil and contributing to corrosion of the reservoir infrastructure.

Sulfate reduction is thought to be driven by organic compounds in the crude oil, such as organic acids (e.g., acetate, propionate, butyrate), which can be found at levels as high as 20 mM in oilfield waters (Birkeland 2005). It is also possible that crude oil hydrocarbons could act as an electron donor in some circumstances (Hubert et al. 2010; Agrawal et al. 2012; Khelifi et al. 2014). In addition to increasing accessibility to sulfate, the increased fluid movement induced both during primary production and due to water injection for secondary recovery will also increase the mass transport of electron donors as well as providing the sulfate necessary for hydrogen sulfide generation. The fact that secondary oil recovery that does not use injection of high-sulfate seawater (e.g., during produced water reinjection; PWRI) can also result in souring suggests that increasing mass transport alone can be sufficient to stimulate sulfide production (Vance and Thrasher 2005; Bjoernestad et al. 2006; Lysnes et al. 2009) and/or that organisms capable of generating sulfide using substrates other than sulfate (e.g., elemental sulfur-reducing *Thermococcus* and *Thermotoga* spp., or a range of oilfield organisms capable of thiosulfate reduction, e.g., *Thermotoga* spp., *Halanaerobium congolense*, and *Dethiosulfovibrio peptidovorans*) may have a role to play in souring under some circumstances (Ollivier and Cayol 2005).

Corrosion in all its forms has been estimated to cost the oil and gas industry around \$1.4 billion annually in the United States alone (Koch et al. 2002). Microbially influenced corrosion (MIC) is a significant contributor to corrosion in the oil and gas industry (Little et al. 2007). Microbially produced sulfide contributes to MIC, but a range of microbial processes coupled with chemical and electrochemical reactions are involved in the process, which is far from fully understood. Sulfate-reducing bacteria have also been implicated in MIC through what has been termed "cathodic depolarization." This involves the microbial consumption of hydrogen produced electrochemically by reduction of protons with electrons derived from iron oxidation (Hamilton 1985; Larsen et al. 2010). The validity of cathodic depolarization as a key mechanism for MIC has been called into question (Hamilton 1985; Crolet 2005), and in many cases, it seems that prolonged generation of H_2S by SRB and its reaction with metallic iron may be a more significant process in MIC (Enning and Garrelfs 2014). Direct oxidation of metallic iron as an electron donor (known as electrical microbially influenced corrosion or EMIC) for some sulfate-reducing microorganisms has also been shown to be a potentially significant in MIC (Enning and Garrelfs 2014). This process, which has been termed electrical microbially influenced corrosion (EMIC), can result in corrosion rates as high as 0.9 mm/yr, more than sufficient to lead to oilfield equipment failures. Nevertheless, it is likely that multiple processes contribute to corrosion under field conditions and that different mechanisms may dominate in different situations (Angell 1999).

Biocide treatment is often applied to control and reduce souring and MIC. However, these processes are often associated with biofilms, and biocides are generally less effective in biofilms. Treatment with nitrate has therefore been proposed as an alternative control mechanism for souring and corrosion. The precise mechanism of nitrate control of sulfide production may be different under different circumstances. Nitrate treatment may stimulate heterotrophic bacteria that can outcompete sulfate reducers for electron donors. Nitrite generated by nitrate-reducing microorganisms can also be toxic to sulfate reducers and suppress sulfide production, or sulfide-oxidizing nitrate reducers that use sulfide as an electron donor can remove sulfide produced by the sulfate reducers and may also generate toxic nitrite (Hubert 2010). The end product of nitrate reduction may also be different, depending on the organisms present. Most dissimilatory nitrate reducers produce nitrite as an end product (Zumft 1992); however, in many different organisms, the nitrite generated is simply an intermediate, which is further reduced to ammonia, or through nitric oxide with nitrous oxide or dinitrogen gas as end products (Zumft 1992).

Nitrate treatment to control sulfide in petroleum reservoirs is not without its problems, and nitrite and possibly other chemical species generated as a product of nitrate treatment may actually exacerbate corrosion (Beeder et al. 2007; Vik et al. 2007).

Oilfield waters also contain considerable amounts of organic acids (Barth 1991). A number of mechanisms may lead to the generation and presence of compounds such as volatile fatty acids, including acetate, butyrate, and propionate. Thermal maturation of organic matter, either in the source rock from which the oil was derived or thermal alteration of organic matter in the reservoir itself, may explain this (Means and Hubbard 1987). However, fermentation of alkanes or other oil components may also lead to the generation of organic acids, and indeed it is known that during in-reservoir oil biodegradation over geological time scales, many organic acids are generated (Huang and Larter 2005). These include naphthenic acids and other acids structurally more complex than the volatile fatty acids produced by fermentation and generally considered by microbiologists (Huang and Larter 2005). This heterogenous mix of organic acids contributes to the total acid number (TAN) of crude oils, and high-TAN oils are known to lead to problems with corrosion, especially during downstream refining (Kapusta et al. 2004). Despite their significance for corrosion and souring prediction, factors controlling the concentrations of organic acids in petroleum reservoir fluids remain poorly understood, and this is an area worthy of future exploration.

3.5 MICROBIALLY ENHANCED OIL RECOVERY

Waking the petroleum reservoir biosphere can clearly have detrimental consequences by promoting souring and corrosion; however, enhancing mass transport in petroleum reservoirs also has the potential to promote beneficial microbial processes in petroleum reservoirs. Stimulating resident microbial communities (Belyaev and Ivanov 1990; Ivanov et al. 1993) or the addition of exogenous microorganisms (Wang et al. 1993) with a view to enhancing oil production has the potential to use enhanced mass transport to great advantage. While microbially enhanced oil recovery (MEOR) is conceptually attractive, it has a somewhat checkered history and is often viewed with skepticism in the industry. This skepticism stems, in part, from a limited fundamental understanding of the microbiological processes that might lead to enhanced oil recovery, concerns about formation damage, and the difficulty of attributing changes in oil production to a specific intervention on a field scale.

The concept of MEOR can be traced back to 1926 (Lazar et al. 2007), but MEOR was tackled seriously for the first time through the pioneering work of Claude Zobell in 1947 (Zobell 1947), and many of the principle drivers of MEOR were already recognized at that time. The very purpose of MEOR and enhanced oil recovery (EOR) in general is to increase mass transfer, and it is therefore not surprising that implementation of MEOR stimulates the activity of indigenous organisms, which then leads to further improvements in mass transfer. This can occur through modification of porosity and permeability (e.g., by alteration of mineral phases in a reservoir, selective plugging of high permeability "thief zones" with biomass or microbially produced polymers), alteration of wettability and interfacial forces (by production of biosurfactants), or modification of oil viscosity and/or reservoir pressure through gas or solvent production (Lazar et al. 2007). Modification of the oil composition through selective biodegradation of particular components of crude oil may also improve mass transfer and fluid flow (He et al. 2000).

It has also been suggested that stimulating the conversion of residual oil to methane in situ could be a means of enhancing the recovery of otherwise stranded oil (Parkes 1999; Gieg et al. 2008; Jones et al. 2008). This could enhance oil recovery by generating more gas in situ, which would initially dissolve in the oil, reducing its viscosity. When the gas solubility is exceeded, the free gas generated would repressurize the reservoir, improving oil recovery, and finally, if sufficient gas is generated, the gas itself could be recovered as an energy source (Head et al. 2014).

Given the range of processes that may be involved in MEOR, it is perhaps not surprising that it is difficult to reliably evaluate the efficacy of any particular MEOR intervention. Moreover, as noted previously, enhancing mass transfer in a formerly diffusion-driven system has the potential to promote not only beneficial process, but also detrimental processes. A side effect of MEOR treatments can therefore be reservoir souring and increased MIC (Lazar et al. 2007).

3.6 NEW FRONTIERS FOR OIL AND GAS MICROBIAL ECOLOGY

Foremost among new frontiers for geoenergy recovery is the use of hydraulic fracturing (fracking) to recover shale gas. The fact that very low permeability shales retain vast amounts of gas, primarily methane, is testament to the extremely slow mass transport in these systems, and indeed the fundamental basis of fracking is to increase mass transfer of gases trapped in micropores within the shale. The very low levels of mass transport and the extremely small pore sizes (nm scale) in shales (Al Hinaia et al. 2014) make shales inhospitable for microbial life. Nevertheless, microorganisms can inhabit larger pores and the fracture system in intact shales, and so shales are likely not completely devoid of life though microbial abundance is likely to be very low. Since fracking dramatically increases mass transport in shales, it seems likely that there will be equally dramatic alterations in the microbial communities in response. Microbial studies of shales are currently sparse (Mouser et al. 2012; Cluff et al. 2014; Jimenez and Krüger 2014) though the organisms detected fit with the characteristics of the system, being dominated by halophiles presumably from the saline aquifers associated with the shales (Cluff et al. 2014).

Given that increasing mass transport will stimulate the shale-associated deep biosphere, there is considerable potential for microbially mediated problems, such as MIC or souring, to arise in shale gas production facilities. Conversely, managing shale-associated microbial communities could be used beneficially to alter formation porosity and permeability, for example, by transformation of organic matter in the shale and/or alteration of the mineralogy of the formation.

Coal bed methane formation also has a microbiological dimension, and there are many uncertainties about the source of biogenic methane in biologically enhanced coal bed methane formation (Strapoc et al. 2011). The structure of coal also means that microbial conversion to methane may also be mass transfer limited. There is, however, evidence that microorganisms have the capacity to overcome this in novel ways. It has been shown that the phenazine dye, neutral red, at concentrations greater than its aqueous solubility forms conducting organic crystals (Beckmann et al. 2016). These crystals enhance methane production from coal and other materials, and a possible mechanism for this is the direct transfer of negative potential electrons to methanogens by direct electron transport from the coal matrix to the methanogens that can then use the electrons to reduce CO_2 to methane (Beckmann et al. 2016). The realization that direct intercellular electron transport may be a widespread feature of anaerobic ecosystems (Wang et al. 2016) opens the door to many exciting avenues of research and potential applications in low mass transport environments.

REFERENCES

Adkins J. P., Cornell L. A., and Tanner R. S. 1992. Microbial composition of carbonate petroleum reservoir fluids. *Geomicrobiol. J.* 10: 87–97.

Agrawal A., Park H. S., Nathoo S., Gieg L. M., Jack T. R., Miner K., Ertmoed R., Benko A., and Voordouw G. 2012. Toluene depletion in produced oil contributes to souring control in a field subjected to nitrate injection. *Environ. Sci. Technol.* 46: 1285–1292.

Al Hinaia A., Rezaeea R., Estebanb L., and Labania M. 2014. Comparisons of pore size distribution: A case from the Western Australian gas shale formations. *J. Unconventional Oil and Gas Resour.* 8: 1–13.

Angell P. 1999. Understanding microbially influenced corrosion as biofilm-mediated changes in surface chemistry. *Curr. Opin. Biotechnol.* 10: 269–272.

Ball J. S., Whisman M. L., and Wenger W. J. 1951. Nitrogen content of crude, petroleums. *Ind. Eng. Chem.* 43: 2577–2581.

Barth T. 1991. Organic acids and inorganic ions in waters from petroleum reservoirs, Norwegian continental shelf: A multivariate statistical analysis and comparison with American reservoir formation waters. *Appl. Geochem.* 6: 1–15.

Bastin E. 1926. Microorganisms in oilfields. *Science.* 63: 21–24.

Beckmann S., Welte C., Li X., Oo Y. M., Kroeninger L., Heo Y., Zhang M. et al. 2016. Novel phenazine crystals enable direct electron transfer to methanogens in anaerobic digestion by redox potential modulation. *Energy Environ. Sci.* 9: 644–655. doi: 10.1039/C5EE03085D.

Beeder J., Andersen T. R., Liengen T., Drønen K., and Torsvik T. 2007. Corrosion as a side effect during nitrate treatment of produced water and aquifer water injection. 07512 NACE Conference Paper. NACE International. Houston, TX.

Belyaev S. S., and Ivanov V. M. 1990. Contemporary geochemical activity of microorganisms in the oilfields, and the possibility of its regulation. *Geochimiya.* 11: 1618–1625.

Bennett B., Adams J. J., Gray N. D., Sherry A., Oldenburg T. B. P., Huang H., Larter S. R., and Head I. M. 2013. The controls on the composition of biodegraded oils in the deep subsurface—Part 3. The impact of microorganism distribution on petroleum geochemical gradients in biodegraded petroleum reservoirs. *Org. Geochem.* 56: 94–105.

Bernard F. P., Connan J., and Magot M. 1992. Indigenous microorganisms in connate waters of many oil fields: A new tool in exploration and production techniques. SPE 24811. In *Proceedings of the 67th SPE Annual Technical Conference*. Society of Petroleum Engineers, Richardson, TX.

Birkeland N.-K. 2005. Sulfate-reducing bacteria and archaea, pp. 35–54. In Ollivier B., Magot M. eds., *Petroleum Microbiology*. ASM Press, Washington, DC. doi: 10.1128/9781555817589.ch3.

Bjoernestad E. O., Sunde E., and Dinning A. J. 2006. The effect of produced water reinjection on reservoir souring in the Statfjord field. 16th International Oil Field Chemistry Symposium, p. 115. Norway.

Carrales M., and Martin R. W. 1975. Sulfur content of crude oils, Bureau of Mines Information Circular 8676. U.S. Bureau of Mines.

Cluff M. A., Hartsock A., MacRae J. D., Carter K., and Mouser P. J. 2014. Temporal changes in microbial ecology and geochemistry in produced water from hydraulically fractured Marcellus Shale gas wells. *Environ. Sci. Technol.* 48: 6508–6517. doi: 10.1021/es501173p.

Crolet J. 2005. Microbial corrosion in the oil industry: A corrosionist's view, pp. 143–170. In Ollivier B., Magot M. eds., *Petroleum Microbiology*. ASM Press, Washington, DC. doi: 10.1128/9781555817589.ch.

Eden R., Laycock P. J., and Fielder M. 1993. Oilfield Reservoir Souring, Health and Safety Executive, Offshore Technology Report, OTH 92 385, HSE Books, p. 90.

Ehrenberg S. N., and Nadeau P. H. 2005. Sandstone vs. carbonate petroleum reservoirs: A global perspective on porosity-depth and porosity-permeability relationships. *AAPG Bulletin.* 89: 435–445.

Enning D., and Garrelfs J. 2014. Corrosion of iron by sulfate-reducing bacteria: New views of an old problem. *Appl. Environ. Microbiol.* 80: 1226–1236.

Garland G. D. 1971. *Introduction to Geophysics: Mantle, Core, and Crust.* Saunders, Philadelphia.

Gevertz D., Telang A. J., Voordouw G., and Jenneman G. E. 2000. Isolation and characterization of strains CVO and FWKO B, two novel nitrate-reducing, sulfide-oxidizing bacteria isolated from oil field brine. *Appl. Environ. Microbiol.* 66: 2491–2501.

Gieg L. M., Duncan K. E., and Suflita J. M. 2008. Bioenergy production via microbial conversion of residual oil to natural gas. *Appl. Environ. Microbiol.* 74: 3022–3029.

Gieg L. M., Jack T. R., and Foght J. M. 2011. Biological souring and mitigation in oil reservoirs. *Appl. Microbiol. Biotechnol.* 92: 263–282.

Grabowski A., Nercessian O., Fayolle F., Blanchet D., and Jeanthon C. 2005. Microbial diversity in production waters of a low-temperature biodegraded oil reservoir. *FEMS Microbiol. Ecol.* 54: 427–443.

Hamilton W. A. 1985. Sulphate-reducing bacteria and anaerobic corrosion. *Annu. Rev. Microbiol.* 39: 195–217.

He Z., She Y., Xiang T., Xue F., Mei B., Li Y., Ju B., Mei H., and Yen T. F. 2000. MEOR pilot sees encouraging results in Chinese oil field. *J. Oil & Gas.* 98: 46–52.

Head I. M., Gray N. D., and Larter S. R. 2014. Life in the slow lane: Biogeochemistry of biodegraded petroleum containing reservoirs and implications for energy recovery and carbon management. *Front. Microbiol.* 5: 566. doi: 10.3389/fmicb.2014.00566.

Head I. M., Jones D. M., and Larter S. R. 2003. Biological activity in the deep subsurface and the origin of heavy oil. *Nature.* 426: 344–352.

Huang H., and Larter S. 2005. Biodegradation of petroleum in subsurface geological reservoirs, pp. 91–122. In Ollivier B., Magot M., eds., *Petroleum Microbiology.* ASM Press, Washington, DC. doi: 10.1128/9781555817589.ch6.

Hubert C. 2010. Microbial ecology of oil reservoir souring control by nitrate injection pp. 2753–2766. In Timmis, K. N., ed., *Handbook of Hydrocarbon and Lipid Microbiology*. Springer, Berlin, Heidelberg.

Ivanov M. V., Belyaev S. S., Borzenkov I. A., Glumov I. F., and Ibatulin P. B. 1993. Additional oil production during field trials in Russia, pp. 373–381. In Premuzic E., Woodhead A., eds., *Microbial Enhancement of Oil Recovery—Recent Advances*. Elsevier, Amsterdam.

Jeanthon C., Nercessian O., Corre E., and Grabowski-Lux A. 2005. Hyperthermophilic and methanogenic archaea in oil fields, pp. 55–69. In Ollivier B., Magot M., eds., *Petroleum Microbiology*. ASM Press, Washington, DC. doi: 10.1128/9781555817589.ch4.

Jiménez N., and Krüger M. 2014. Effects of Hydraulic Frac Fluids on Subsurface Microbial Communities in Gas Shales. EGU General Assembly 2014, April 27–May 2, 2014, Vienna, Austria, Abstract id.10957.

Jones D. M., Head I. M., Gray N. D., Adams J. J., Rowan A. K., Aitken C. M., Bennett B. et al. 2008. Crude oil biodegradation via methanogenesis in subsurface petroleum reservoirs. *Nature.* 451: 176–180.

Jørgensen B. B., and D'Hondt S. 2006. A starving majority deep beneath the seafloor. *Science.* 314: 932–934.

Jørgensen B. B., and Marshall I. P. G. 2016. Slow microbial life in the seabed. *Annu. Rev. Mar. Sci.* 8: 311–332.

Kallmeyer J., Pockalny R., Adhikari R. R., Smith D. C., and D'Hondt S. 2012. Global distribution of microbial abundance and biomass in subseafloor sediment. *Proc. Natl. Acad. Sci. U.S.A.* 109: 16213–16216.

Kapusta S. D., Ooms A., Smith A., Van den Berg F., and Fort W. 2004. Safe processing of acid crudes. NACE Corrosion 2004. Paper 04637, pp. 1–19.

Khelifi N., Amin Ali O., Roche P., Grossi V., Brochier-Armanet C., Valette O., Ollivier B., Dolla A., and Hirschler-Réa, A. 2014. Anaerobic oxidation of long-chain n-alkanes by the hyperthermophilic sulfate-reducing archaeon, *Archaeoglobus fulgidus. The ISME Journal.* 8: 2153–2166. doi: 10.1038/ismej.2014.58.

Koch G. H., Brongers M. P. H., Thompson N. G., Virmani Y. P., and Payer J. H. 2002. Corrosion Costs and Preventive Strategies in the United States Publication No. FHWA-RD-01-156. NACE International, Houston, TX. https://www.nace.org/uploadedFiles/Publications/ccsupp.pdf.

Larsen J., Kim Rasmussen K., Pedersen H., Sørensen K., Lundgaard T., and Skovhus T. L. 2010. Consortia of MIC bacteria and archaea causing pitting corrosion in top side oil production facilities. NACE-10252, *Corrosion 2010,* March 14–18, San Antonio, TX.

Larter S. R., Wilhelins A., Head I. M., Koopmans M., Aplin A. C., Di Primio R., Zwach C., Erdmann M., and Telnaes N. 2003. The controls on the composition of biodegraded oils in the deep subsurface—Part 1: Biodegradation rates in petroleum reservoirs. *Org. Geochem.* 34: 601–613.

Lazar I., Petrisor I. G., and Yen T. F. 2007. Microbial enhanced oil recovery MEOR. *Petrol. Sci. Technol.* 25: 1353–1366. doi: 10.1080/10916460701287714.

Little B. J., Mansfeld F. B., Arps P. J., and Earthman J. C. 2007. Microbiologically Influenced Corrosion. *Encyclopedia of Electrochemistry*, NACE International, Houston, TX.

Lysnes K., Bødtker G., Torsvik T., Bjørnestad E. Ø., and Sunde E. 2009. Microbial response to reinjection of produced water in an oil reservoir. *Appl. Microbiol. Biotechnol.* 83: 1143–1157.

Magot M. 2005. Indigenous microbial communities in oil fields, pp. 21–34. In Ollivier B., Magot M., eds., *Petroleum Microbiology.* ASM Press, Washington, DC. doi: 10.1128/9781555817589.ch2.

Magot M., Ollivier B., and Patel B. K. 2000. Microbiology of petroleum reservoirs. *Antonie van Leeuwenhoek.* 77: 103–116.

Manning D. A. C., and Hutcheon I. 2004. Distribution and mineralogical controls on ammonium in deep groundwaters. *Appl. Geochem.* 19: 1495–1503.

Means J. L., and Hubbard N. 1987. Short-chain aliphatic acid anions in deep subsurface brines: A review of their origin, occurrence, properties, and importance and new data on their distribution and geochemical implications in the Palo Duro Basin, Texas. *Org. Geochem.* 11: 177–191.

Meckenstock R. U., von Netzer F., Stumpp C., Lueders T., Himmelberg A. M., Hertkorn N., Schmitt-Kopplin P. et al. 2014. Water droplets in oil are microhabitats for microbial life. *Science.* 345: 673–676. doi: 10.1126/science.1252215.

Middelburg J. J. 1989. A simple rate model for organic matter decomposition in marine sediments. *Geochimica et Cosmochimica Acta.* 53: 1577–1581.

Middelburg J. J., Vlug T., Jaco F., and van der Nat W. A. 1993. Organic matter mineralization in marine systems. *Global Planet. Change* 8: 47–58.

Morono Y., Terada T., Nishizawa M., Ito M., Hillion F., Takahata N., Sanoe Y., and Inagaki F. 2011. Carbon and nitrogen assimilation in deep subseafloor microbial cells. *Proc. Natl. Acad. Sci. U.S.A.* 108: 18295–18300. doi: 10.1073/pnas.1107763108.

Mouser P. J., Ansari M., Hartsock A., Lui S., and Lenhart J. 2012. Microbial Community Shifts due to Hydrofracking: Observations from Field-Scale Observations and Laboratory-Scale Incubations. American Geophysical Union, Fall Meeting 2012, abstract #H11C-1196.

Nilsen R. K., Beeder J., Thorstenson T., and Torsvik T. 1996. Distribution of thermophilic marine sulfate reducers in North Sea oil field waters and oil reservoirs. *Appl. Environ. Microbiol.* 62: 1793–1798.

Oldenburg T. B. P., Larter S. R., and Huang, H. 2006. Nutrient supply during subsurface oil biodegradation—Availability of petroleum nitrogen as a nutrient source for subsurface microbial activity. *Energy & Fuels.* 20: 2079–2082.

Ollivier B., and Cayol J. 2005. Fermentative, iron-reducing, and nitrate-reducing microorganisms, pp. 71–88. In Ollivier B., Magot M., eds., *Petroleum Microbiology.* ASM Press, Washington, DC. doi: 10.1128/9781555817589.ch5.

Ollivier B., Fardeau M.-L., Cayol J.-L., Magot M., Patel B. K. C, Prensier G., and Garcia, J.-L. 1998. *Methanocalculus halotolerans* gen. nov., sp. nov., isolated from an oil-producing well. *Int. J. Syst. Evol. Microbiol.* 48: 821–828.

Ollivier B., and Magot M. 2005. *Petroleum Microbiology*. ASM Press, Washington DC, p. 365.

Orphan V. J., Taylor L. T., Hafenbradl D., and DeLong E. F. 2000. Culture-dependent and culture-independent characterization of microbial assemblages associated with high-temperature petroleum reservoirs. *Appl. Environ. Microbiol.* 66: 700–711.

Parkes R. J. 1999. Cracking anaerobic bacteria. *Nature.* 401: 217–218.

Parkes R. J., Cragg B. A., Bale S. J., Getliff J. M., Goodman K., Rochelle P. A., Fry J. C., Weightman A. J., and Harvey S. M. 1994. Deep bacterial biosphere in Pacific-Ocean sediments. *Nature.* 371: 410–413.

Parkes R. J., Cragg B., Roussel E., Webster G., Weightman A., and Sass H. 2014. A review of prokaryotic populations and processes in sub-seafloor sediments, including biosphere: Geosphere interactions. *Mar. Geol.* 352: 409–425.

Parkes R. J., Webster G., Cragg B. A., Weightman A. J., Newberry C. J. et al. 2005. Deep sub-seafloor prokaryotes stimulated at interfaces over geological time. *Nature.* 436: 390–394.

Pedersen K. 1993. The deep subterranean biosphere. *Earth-Sci. Rev.* 34: 243–260.

Pham V. D., Hnatow L. L., Zhang S., Fallon R. D., Jackson S. C., Tomb J.-F., DeLong D. F., and Keeler S. J. 2009. Characterizing microbial diversity in production water from an Alaskan mesothermic petroleum reservoir with two independent molecular methods. *Environ. Microbiol.* 11: 176–187.

Piceno Y. M., Reid F. C., Tom L. M., Conrad M. E., Bill M., Hubbard C. G., Fouke B. W. et al. 2014. Temperature and injection water source influence microbial community structure in four Alaskan North Slope hydrocarbon reservoirs. *Front. Microbiol.* 5: 409. doi: 10.3389/fmicb.2014.00409.

Richter P. P., Ceasser P. D., Meisel S. A., and Offenhauser R. D. 1952. Distribution of nitrogen, according to basicity. *Ind. Eng. Chem.* 44: 2601–2605.

Rogers J. R., Bennett P. C., and Choi W. J. 1998. Feldspars as a source of nutrients for microorganisms. *Am. Mineral.* 83: 1532–1540.

Sanders P., and Sturman P. 2005. Biofouling in the oil industry, pp. 171–198. In Ollivier B., Magot M., eds., *Petroleum Microbiology.* ASM Press, Washington, DC. doi: 10.1128/9781555817589.ch9.

Strąpoć D., Mastalerz M., Dawson K., Macalady J., Callaghan A. V., Wawrik B., Turich C., and Ashby M. 2011. Biogeochemistry of microbial coal-bed methane. *Annu. Rev. Earth Planet. Sci.* 39: 617–656.

Takai K., Nakamura K., Toki T., Tsunogai U., Miyazaki M., Miyazaki J., Hirayama H., Nakagawa S., Nunoura T., and Horikoshi K. 2008. Cell proliferation at 122 degrees C and isotopically heavy CH4 production by a hyperthermophilic methanogen under high-pressure cultivation. *Proc. Natl. Acad. Sci. U.S.A.* 105: 10949–10954.

Tinker G. E., Barnes P. F., Olsen E. E., and Wright M. P. 1983. Geochemical aspects of Michigan waterfloods. SPE 12208. In *Proceedings of the SPE Annual Technical Conference*. Society of Petroleum Engineers, Richardson, TX.

Vance I., and Thrasher D. 2005. Reservoir souring: Mechanisms and prevention, pp. 123–142. In Ollivier B., Magot M., eds., *Petroleum Microbiology*. ASM Press, Washington, DC. doi: 10.1128/9781555817589.ch7.

Vik E. A., Janbu A. O., Garshol F. K., Henninge L. B., Engebretsen S., Kuijvenhoven C., Oliphant D., and Hendriks W. P. 2007. Nitrate based souring mitigation of produced water—Side effects and challenges from the draugen produced water re-injection pilot. 106178-MS SPE Conference. Society of Petroleum Engineers. doi: 10.2118/106178-MS.

Wang L.-Y., Ke W.-J., Sun X.-B., Liu J.-F., Gu J.-D. and, Mu B.-Z. 2014. Comparison of bacterial community in aqueous and oil phases of water-flooded petroleum reservoirs using pyrosequencing and clone library approaches. *Appl. Microbiol. Biotechnol.* 98: 4209–4221. doi: 10.1007/s00253-013-5472-y.

Wang L.-Y., Nevin K. P., Woodard T. L., Mu B.-Z., and Lovley D. R. 2016. Expanding the diet for DIET: Electron donors supporting direct interspecies electron transfer DIET in defined co-culture. *Front. Microbiol.* doi: 10.3389/fmicb.2016.00236.

Wang X., Xue Y., and Xie S. 1993. Characteristics of enriched cultures and their application to MEOR field tests. *Dev. Petrol. Sci.* 39: 335–348.

Wentzel A., Lewin A., Cervantes F. J., Valla S., and Kotlar H. K. 2013. Deep subsurface oil reservoirs as poly-extreme habitats for microbial life. A current review, pp. 439–466. In Seckbach J., Oren A., Stan-Lotter H., eds., *Cellular Origin, Life in Extreme Habitats and Astrobiology*. Springer Netherlands.

Whitman W. B., Coleman D. C., and Wiebe W. J. 1998. Prokaryotes: The unseen majority. *Proc. Natl. Acad. Sci. U.S.A.* 95: 6578–6583.

Wilhelms A., Larter S. R., Head I., Farrimond P., di-Primio R., and Zwach C. 2001. Biodegradation of oil in uplifted basins prevented by deep-burial sterilisation. *Nature.* 411: 1034–1037. doi: 10.1038/35082535.

ZoBell E. C. 1947. Bacterial release of oil from oil-bearing materials Part 1. *World Oil.* 126: 36–47.

Zumft W. G. 1992. The denitrifying prokaryotes, pp. 554–582. In Balows A., Trüper H. G., Dworkin M., Harder W., Schleifer K. H., eds., *The Prokaryotes. A Handbook on the Biology of Bacteria, Ecophysiology, Isolation, Identification, Applications*, 2nd ed. Springer-Verlag, New York.

4 Predominant MIC Mechanisms in the Oil and Gas Industry

Torben Lund Skovhus, Jason S. Lee, and Brenda J. Little

CONTENTS

ABSTRACT

Microbiologically influenced corrosion (MIC) is the terminology applied when biological parameters play a role in the corrosion process. In literature, terms such as microbial corrosion, biocorrosion, microbially influenced/induced corrosion, and biodegradation are often applied. All descriptions express that microorganisms (bacteria, archaea, and fungi) influence the corrosion process of a given material. In this chapter, an overview of the common MIC mechanisms encountered in the oil and gas industry is presented.

4.1 INTRODUCTION

One characteristic of MIC is that it does not produce a unique type of corrosion or morphology of corrosion damage (Little and Lee 2007). However, MIC is often seen as a localized corrosion phenomenon. What defines the MIC mechanism is an interaction of several electrochemical, physical/environmental, operational, and biological factors. Due to the complexity of these interactions, MIC mechanisms are still debated issues in the oil and gas industry.

4.2 CONSTRAINTS OF MICROBIAL LIFE IN ENGINEERED SYSTEMS

Under what conditions can MIC occur? A limiting factor is that microorganisms will need to be alive and active at the exact location and time where the corrosion takes place. The boundaries for microbial life are defined by water activity (a_w), temperature, pH, pressure, and salt content. Fungi are the most desiccant-resistant microorganisms and can remain active down to $a_w = 0.6$, much lower than bacteria, which require at least $a_w = 0.9$ (Little and Lee 2007). Microorganisms have evolved to survive even under the most environmentally hostile conditions, for example, in hot springs, sea ice, or the deep biosphere (Table 4.1). Extreme environmental conditions are typical to the oil, gas, and petroleum-processing industry where MIC attacks have occured. Whether microorganisms significantly contribute to corrosion across the entire "environmental envelope" that supports microbial life is a subject of ongoing research.

A recent example from the Danish offshore industry from 2007 shows how microorganisms caused MIC in one of the types of oilfield facilities that are most commonly found to be subject to microbial degradation (Augustinovic et al. 2012; Skovhus and Eckert 2014). A carbon steel subsea water injection pipeline 32 km in length transports seawater from the offshore Siri oil production platform to the Nini platform (Figure 4.1). A rupture of the pipeline occurred 2 km from the Siri platform. The failure was investigated in detail and the cause of the damage leading to the rupture was determined to be MIC. The investigation revealed that the corrosion deposits at the rupture site were a mixture of iron sulfide and other corrosion products, and contained microorganisms and polysaccharide slime. These deposits allowed sulfate-reducing prokaryotes (SRP) and other troublesome microorganisms to grow protected from biocides. Several actions were taken after the rupture shut down oil production in November 2007. One major step was to establish a robust microbiological monitoring program based on molecular biology methods. The improved microbiological monitoring program was designed to capture selected water samples, pigging solids, and corrosion coupons. Likewise, training of the staff in proper sampling and improving the biocide strategy was implemented.

The example shows that internal corrosion of pipelines at the bottom the sea is a critical issue for oilfield o perators and that corrosion and, in particular, MIC is always a potential risk to the operation.

4.3 TYPICAL MICROBIAL GROUPS INVOLVED IN MIC

Anaerobic organisms active in environments with little or no oxygen are the most cited troublesome microorganisms in the oil and gas industry infrastructure (e.g., oilfield reservoirs and industrial process systems). Aerobic bacteria (and humans) use oxygen as an electron acceptor (respiration) and gain energy from organic matter (electron donor). A typical type of anaerobic microorganism often involved in MIC is sulfate-reducing bacteria (SRB). SRB use sulfate as an electron acceptor and gain energy from organic matter, H_2, and under specific environmental conditions even metallic iron (Fe^0). The majority of MIC studies focus on SRB. MIC from SRB

TABLE 4.1
Examples of Some Extreme Living Conditions of Individual Microorganisms

Parameter	Descriptive Term	Habitat	Genus/Species	Minimum	Optimum	Maximum
Temperature (high)	Hyperthermophile	Hot, deep-sea hydrothermal vents	*Pyrolobus fumarii*	90°C	106°C	113°C
Temperature (low)	Psychrophile	Sea ice	*Polaromonas vacuolata*	0°C	4°C	12°C
pH (high)	Alkaliphile	Soda lakes	*Natronobacterium gregoryi*	8.5	10	12
pH (low)	Acidophile	Acidic hot springs	*Picrophilus oshimae*	–0.06	0.7	4
Pressure	Barophile	Deep sea sediments	MT41 (Mariana Trench 41)	500 atm	700 atm	>1000 atm
Salt content	Halophile	Salt lakes	*Halobacterium salinarum*	15%	25%	32% (saturation)

Source: Inspiration from M. Madigan, K. M. Martinko, and J. Parker, *Brock Biology of Microorganisms* 13th Edition. New Jersey: Pearson Education, 2013.

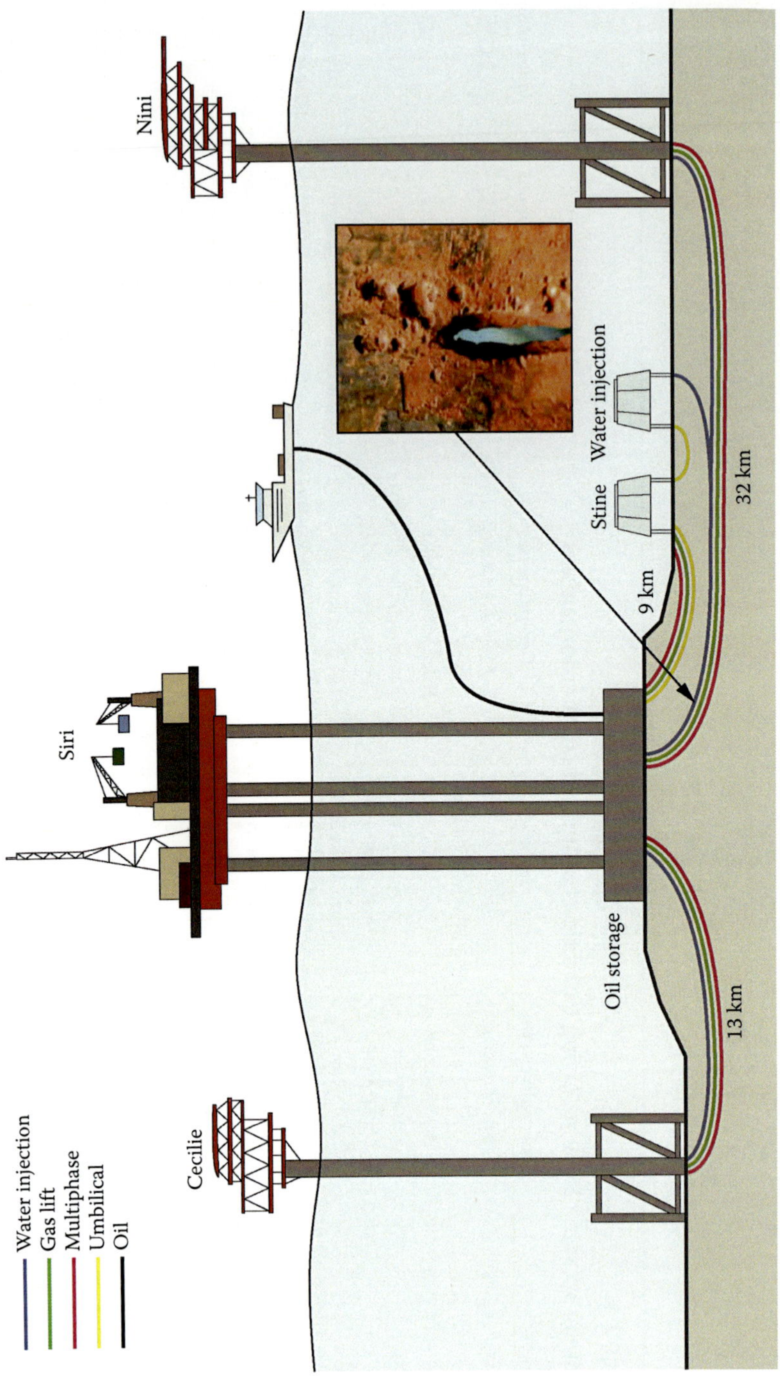

FIGURE 4.1 Overview of the Siri oil production asset from the Danish sector of the North Sea. The inset image shows the rupture of the water injection pipeline.

TABLE 4.2
Microbial Groups Involved in MIC in Oil and Gas Installations

Microbial Group	Example of Genera	Habitats and Mode of Action
Sulfate-reducing bacteria (SRB)	*Desulfobacter* *Desulfococcus* *Desulfovibrieo*	Topside oilfield systems, wastewater treatment plants, almost any anaerobic containments with water and organic matter. Anaerobic respiration of sulfate with the production of hydrogen sulfide.
Sulfate-reducing archaea (SRA)	*Archaeoglobus*	Hot oilfields. Growth optimum of 75°C–90°C. Anaerobic respiration of sulfate with the production of hydrogen sulfide.
Methanogenic archaea (MA)	*Methanococcus* *Methanosarcina*	Oil wells, deep-sea hydrothermal vents, and oxygen-depleted sediments. Anaerobic respiration of carbon dioxide (CO_2) with the production of methane (CH_4).
Iron-reducing bacteria (IRB)	*Geobacter* *Shewanella*	Soils, sediments, and the subsurface. External corrosion of pipelines in soil. Fe(III) is reduced to Fe(II), and an organic molecule delivers the energy (electrons).

activity produces distinct by-products from the corrosion of carbon steel: iron sulfides (Fe_xS_y) and the distinct smell of rotten egg (free hydrogen sulfide gas).

Table 4.2 gives a list of typical microbial groups involved in MIC in industrial installations. The list only contains a small fraction of the known microbiological groups that are found to be involved in MIC.

4.4 MIC AND THE BIOFILM CONCEPT

To understand how MIC mechanisms work, it is essential to understand the concept of a biofilm. A biofilm consists of living cells, extracellular polymeric substances (EPS), signaling molecules (like hormones), DNA, proteins/enzymes, corrosion products, minerals, and water.

Biofilm development on a surface has been intensively investigated over the past decades. Figure 4.2 explains the individual phases of how a biofilm develops on a metal surface. The mature biofilm consists of many different types of microorganisms, and they interact and communicate (with signaling molecules) with each other as is the case in living tissue of animals and plants.

The physical and chemical characteristics of the mature biofilm are very heterogeneous. If the surrounding media (the electrolyte) contains oxygen (as in seawater), oxygen will diffuse into the outer layer of the biofilm, making an aerobic zone. Due to oxygen diffusion resistance and aerobic respiration, anaerobic zones will form deeper in the biofilm and near the metal surface. Figure 4.3 shows the early stage of a biofilm on a metal surface in the scanning electron microscope (SEM). Figure 4.4 shows how a new carbon steel coupon may look after it has been exposed to anaerobic oilfield brine for several months and then cleaned again. This is a typical form of

FIGURE 4.2 The biofilm development and detachment model. First, single cells from the planktonic phase attach to the metal surface (1). This leads to colonization (2) and extensive growth (3) of several kinds of microorganisms. When the climax (4) has been reached, the mature biofilm starts to release single cells to the surroundings (5), and the process starts at a new location. (From L. Hall-Stoodley et al., *Nature Reviews Microbiology,* 2, 95–108, 2004.)

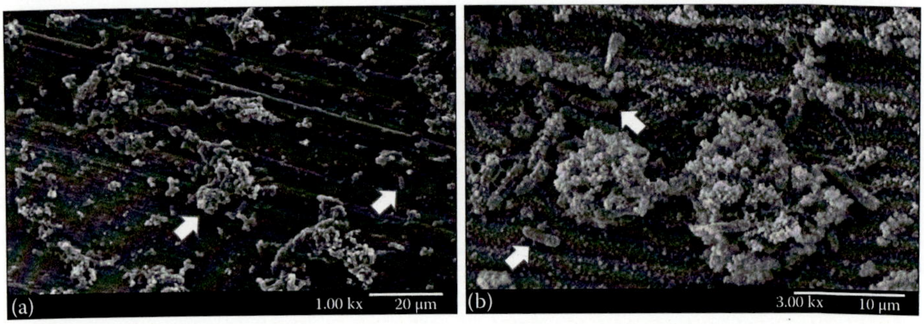

FIGURE 4.3 Biofilm (with water removed) on a steel surface as seen in the scanning electron microscope (SEM). (a) The early stage of a biofilm at 1000× magnification showing both single cells and biofilm clusters. (b) Single cells (rod shaped) in the biofilm at 3000× magnification. (Photos obtained from commercial laboratory.)

corrosion initiation morphology for MIC attack on a carbon steel surface, which has been simulated under controlled laboratory conditions.

4.5 COMMON MIC MECHANISMS

4.5.1 Concentration Cells

A metal surface in the natural environment is not covered evenly with biofilm. Some areas have a dense biofilm while others have a thin or even no biofilm. Therefore, areas covered with biofilm have a lower oxygen concentration and become anodic, and those with less biofilm exhibit higher oxygen concentrations and become

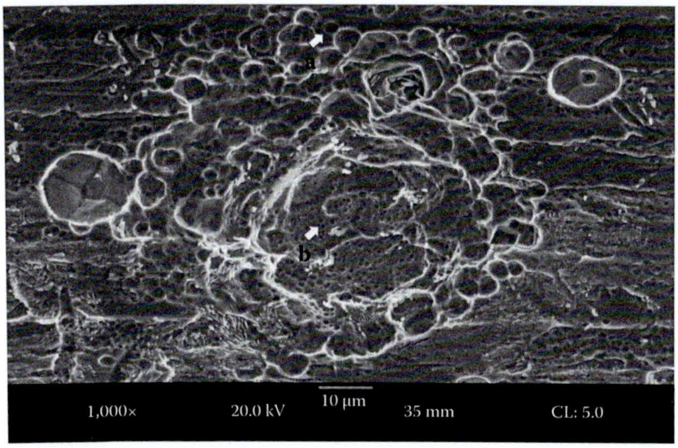

FIGURE 4.4 MIC attack of a carbon steel surface after removing the biofilm. (a) Minor MIC attack at the edge of the biofilm with a diameter of around 2–5 µm. (b) Deeper corrosion attack at the center of the biofilm (where the smaller pits have grown together) with a diameter of 30–40 µm. The image is obtained with a magnification of 1000× in the SEM. (Photo obtained from commercial laboratory.)

cathodic. Figure 4.5 shows how anodic and cathodic sites are fixed at the metal surface. This MIC mechanism is corrosion as a result of differential aeration cells.

In addition to biofilms, microorganisms can create dense surface deposits that initiate localized corrosion. Similar to the situation described above, a deposit on the surface will exclude oxygen from immediately under the deposit (occluded volume), creating an anodic site. The anodic reaction produces metal ions that are hydrolyzed by water to form metal hydroxides and H^+, thereby decreasing the local pH. The cathodic reaction is the reduction of oxygen on the surrounding metal surface. In

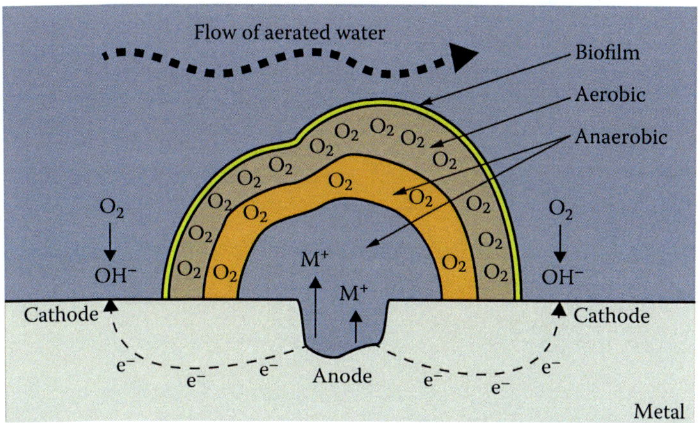

FIGURE 4.5 Biofilm heterogeneity results in differential aeration cells. The figure shows pit initiation (like the one in Figure 4.4) due to oxygen depletion under a biofilm. (Idea obtained from S. Borenstein, *Microbiologically Influenced Corrosion*, Woodhead Publishing, 1994.)

chloride-containing environments (e.g., seawater), chloride is concentrated within occluded volumes to balance the charge produced by the formation of metal hydroxides. The restrictive geometry of under the deposit results in build-up of chloride within the occluded solution volume in excess of the bulk concentration. Once a critical chloride concentration (dependent upon temperature, metallurgy, and pH) is reached, the protective passive film becomes unstable, and active corrosion is initiated. The autocatalytic nature of these corrosion processes further increases the chloride concentration and decreases solution pH resulting in higher corrosion rates. The restrictive geometry of the occluded site prevents dilution of the highly concentrated acidic chloride solution by the bulk solution, and thereby the occluded solution chemistry becomes independent of the bulk. The concentrated solution will spread to adjacent areas localized to the initial attack site and become susceptible to attack.

4.5.2 Microbial Activity in Biofilms Generate Corrosive Substances

A classic corrosion example is MIC related to the activity of SRB. As discussed above, SRB gain energy from organic matter (electron donor) and produce hydrogen sulfide by reducing sulfate ions (electron acceptor). With the reaction of carbon steel, Fe_xS_y is produced as a black corrosion product. In the presence of oxygen, the oxidation of Fe_xS_y will produce highly corrosive elemental sulfur (S^0). Figure 4.6 shows the main reactions and the effect of oxygen on the SRB corrosion of carbon steel.

Enning and Garrelfs (2014) reported that there were species–specific differences among SRB related to corrosion mechanisms and rates. In describing SRB-influenced corrosion of iron, Enning et al. (2012) defined two fundamentally different SRB corrosion mechanisms. "Chemical microbially influenced corrosion" (CMIC) was used to describe the corrosive effects of biogenic H_2S. All SRB in the presence of sulfate

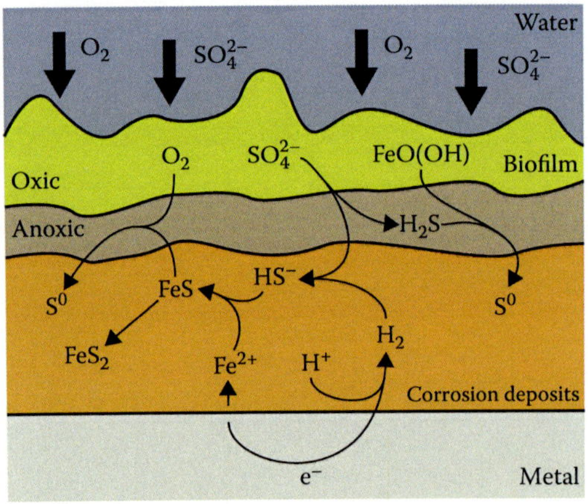

FIGURE 4.6 MIC caused by sulfate-reducing bacteria (SRB) on carbon steel. The influence of oxygen on the reactions is shown. (With kind permission from Springer Science+Business Media: *Marine and Industrial Biofouling*. Flemming, H.-C. et al., Mechanisms of microbially influenced corrosion, Z. Lewandowski and H. Beyenal, 2009.)

and suitable electron donors can cause CMIC of iron. In contrast, "electrical microbially influenced corrosion" (EMIC) is only observed in the presence of specialized lithotrophic SRB, capable of accepting electrons directly from iron.

4.5.3 Complex Consortia of Active Microorganisms in Biofilms Generate Accelerated Corrosion

In many engineered systems, a biofilm has a diverse microbiological community with microorganisms that have neither been characterized nor their roles in MIC defined. A good example is the internal corrosion that takes place in crude oil pipelines. It has now been discovered that alongside sulfate-reducing prokaryotes (SRP) another group of microorganisms has been found that contribute to corrosion: methanogenic archaea (MA). Under the right growth conditions, MA will form methane from carbon dioxide. This process requires electrons that can be obtained from the metal by dissolution of iron (Fe). Recent research has shown that high numbers of MA were present in corrosion products from crude oil systems in the Danish offshore sector (Skovhus et al. 2011, 2012). MA are not easily measured with traditional culture media detection tools. However, after molecular biology tools were introduced to monitoring programs in the oil and gas industry, MA were found in high numbers (Skovhus et al. 2012). Figure 4.7 shows the reactions of anaerobic MIC from both MA and SRP. The net reactions for the described MIC mechanism (Figure 4.7) in the presence of bicarbonate precipitating excess of ferrous iron are the following:

$$\text{Sulfate reduction: } 4Fe^0 + SO_4^{2-} + 3HCO_3^- + H_2O^+ \text{->} FeS + 3\,FeCO_3 + 5\,OH^- \quad (4.1)$$

$$\text{Methane production: } 4Fe^0 + 5\,HCO_3^- + 2H_2O \text{->} 4FeCO_3 + CH_4 + 5OH^- \quad (4.2)$$

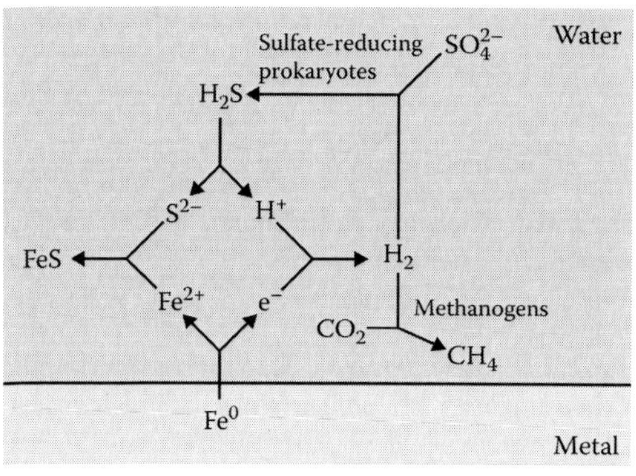

FIGURE 4.7 MIC caused by methanogenic archaea (MA) and sulfate-reducing prokaryotes (SRP) on carbon steel in a crude oil production system. (Idea obtained from T. L. Skovhus, K. B. Sørensen, and J. Larsen, Problems Caused by Microbes and Treatment Strategies—Rapid Diagnostics of Microbiologically Influenced Corrosion (MIC) in Oilfield Systems with a DNA-Based Test Kit. In *Applied Microbiology and Molecular Biology in Oilfield Systems*, eds. C. Whitby and T. L. Skovhus. New York: Springer Publisher, 2011.)

Enning et al. (2012) conducted experiments with metallic iron as the only source of electrons and CO_2 as the sole carbon source. Such SRB grew in intimate contact with the iron surface and severely accelerated corrosion kinetics compared to "traditional" hydrogen-utilizing SRB as well as sterile controls. The postulated corrosion mechanism assumes a direct uptake of electrons from metallic iron coupled to the reduction of sulfate. EMIC has been observed in a limited number of SRB strains. Direct electron uptake involves outer membrane redox proteins that interact with extracellular electron donors and acceptors. The anodic dissolution of iron results from electron consumption by sulfate reduction, a cathodic reaction that is kinetically impossible at room temperature in the absence of biological catalysis. In EMIC, microbial oxidation of 4 moles of Fe^0 to Fe^{2+} is coupled to the reduction of one mole of sulfate, which precipitates as one mole of FeS. The remaining 3 moles Fe^{+2} precipitate as a carbonate, that is, siderite. The result is an electrically conductive deposit through which direct uptake of electrons between methanogens and iron has also been demonstrated as a mechanism for corrosion (Venzlaff et al. 2013).

4.6 SUMMARY

Microbiologically influenced corrosion is corrosion resulting from the presence and activities of microorganisms. Microorganisms influence corrosion by creating concentration cells, for example, oxygen concentration cells, producing corrosive metabolites, for example, sulfides, producing a surface layer that is not protective or removing a protective film. Microorganisms can accelerate rates of partial reactions in corrosion processes and shift the mechanism for corrosion. Most MIC is localized corrosion resulting from the localized presence and activities of microorganisms where deep penetration into the base metal is observed and can take the form of pitting, crevice corrosion, and under-deposit corrosion. Causative microorganisms include bacteria and archaea (e.g., methanogens and sulfate-reducing archaea) and are often defined by the electron acceptor they use, for example, sulfate-reducing prokaryotes.

REFERENCES

Augustinovic, Z., Birketveit, Ø., Clements, K., Freeman, M., Gopi, S., Ishoey, T., and Jackson, G. et al. 2012. Microbes—Oilfield enemies or allies? *Oilfield Review*. 24(2): 4–17.

Borenstein, S. 1994. *Microbiologically Influenced Corrosion*, Woodhead Publishing New York: Industrial Press.

Enning, D. and Garrelfs, J. 2014. Corrosion of iron by sulfate-reducing bacteria: New views of an old problem. *Appl. Environ. Microbiol.* 80(4): 1226–1236.

Enning, D., Venzlaff, H., Garrelfs, J., Dinh, H. T., Meyer, V., Mayrhofer, K., Hassel, A. W., Stratmann, M., and Widdel, F. 2012. Marine sulfate-reducing bacteria cause serious corrosion of iron under electroconductive biogenic mineral crust. *Environ. Microbiol.* 14: 1772–1787.

Hall-Stoodley, L., Costerton, J. W., and Stoodley, P. 2004. Bacterial biofilms: From the natural environment to infectious diseases. *Nat. Rev. Microbiol.* 2: 95–108.

Lewandowski, Z. and Beyenal, H. 2009. Mechanisms of microbially influenced corrosion. In *Marine and Industrial Biofouling*. ed. Flemming, H.-C., Murthy, P. S., Venkatesan, R., and Cooksey, K. E. Berlin: Springer-Verlag, pp. 35–64.

Little, B. and Lee, J. 2007. *Microbiologically Influenced Corrosion*. New Jersey: Wiley.

Madigan, M., Martinko, K. M., and Parker, J. 2013. *Brock Biology of Microorganisms* 13th Edition. New Jersey: Pearson Education.

Skovhus, T. L. and Eckert, R. B. 2014. Practical Aspects of MIC Detection, Monitoring and Management in the Oil & Gas Industry. *Corrosion 2014*. Paper C2014-3920. Houston, TX: NACE International.

Skovhus, T. L., Holmkvist, L., Andersen, K., Pedersen, H., and Larsen, J. 2012. MIC Risk Assessment of the Halfdan Oil Export Spool (SPE-155080). *SPE International Conference and Exhibition on Oilfield Corrosion 2012* held in Aberdeen, UK, May 28–29, 2012.

Skovhus, T. L., Sørensen, K. B., and Larsen, J. 2011. Problems Caused by Microbes and Treatment Strategies—Rapid Diagnostics of Microbiologically Influenced Corrosion (MIC) in Oilfield Systems with a DNA-Based Test Kit. In *Applied Microbiology and Molecular Biology in Oilfield Systems*, eds. C. Whitby and T. L. Skovhus. New York: Springer Publisher, pp. 133–140.

Venzlaff, H., Enning, D., Srinivasan, J., Mayrhofer, K. J. J., Hassel, A. W., Widdel, F., and Stratmann, M. 2013. Accelerated cathodic reaction in microbial corrosion of iron due to direct electron uptake by sulphate-reducing bacteria. *Corros. Sci.* 66: 88–96.

5 Corrosion Risks Associated with (Bio)Chemical Processes in Sour Systems due to Nitrate Injection or Oxygen Ingress

Sven Lahme and Casey Hubert

CONTENTS

ABSTRACT

The oil and gas industry faces high annual costs due to corrosion of metal infrastructure. To some extent, these corrosion issues are caused by growth of microorganisms and referred to as microbiologically influenced corrosion (MIC). Microbial activity influences the overall chemistry of a given system, which, in turn, influences the electrochemical processes at the metal–environment interface. Various microbial processes (e.g., sulfidogenesis, fermentation, acetogenesis, methanogenesis) can contribute to MIC in oil and gas industrial systems and infrastructure. Sulfate-reducing microorganisms have garnered significant attention due to their frequent association with corrosive biofilms in anoxic settings. Additional problems can arise due to SRM-mediated production of toxic and corrosive hydrogen sulfide (H_2S), which contributes to reservoir souring during seawater flooding. Bioengineering strategies, such as nitrate injection, have often proven successful for controlling souring and subsequent corrosion during in vitro and in situ trials by stimulating the activity of nitrate-reducing microorganisms. Despite reported successes, nitrate-based bioengineering has come under scrutiny in light of recent studies showing that nitrate can lead to severe corrosion under some conditions. Nitrate treatment or unintentional ingress of oxygen can lead to redox potential shifts and subsequent formation of highly corrosive inorganic compounds (e.g., elemental sulfur, S^0), which can severely damage metal infrastructure. In recent years, our understanding of MIC in oil systems has gradually increased, and bioengineering strategies, such as injection of nitrate, have been successful in some cases but not in others. This chapter provides an overview of potential corrosion mechanisms related to nitrate injection and ingress of oxygen in sour systems. Microbial and chemical transformations during oil production are discussed to explain and help predict corrosion hazards in sour oil systems.

5.1 IMPLICATIONS OF MICROBES IN CORROSION PROCESSES IN THE OIL AND GAS INDUSTRY

Corrosion of metal infrastructure in the oil and gas industry is critically important due to its potential negative consequences for the environment and associated high repair and replacement costs. Corrosion can be influenced by several factors affecting the overall kinetics of the electrochemical redox processes. Microbiologically influenced corrosion refers to the involvement of microorganisms in these scenarios. Microorganisms have evolved different ways to harness energy from a wide range of redox reactions during which electrons are transferred from electron donor compounds to thermodynamically suitable electron acceptor compounds. The electron donors used by microorganisms to generate energy can be either organic (e.g., sugars, fatty acids, aromatic and aliphatic hydrocarbons) as used by organotrophic microorganisms or inorganic (e.g., H_2, H_2S, S^0, Fe^{2+}, NH_4^+) as used by lithotrophic microorganisms. On the other hand, electron acceptors are usually inorganic compounds (e.g., O_2, NO_3^-, SO_4^{2-}, Fe^{3+}) and only in some cases are organic (e.g., fumarate, DMSO) (Jonkers et al. 1996; Kröger et al. 2002). By transforming chemical compounds, the growth of microorganisms changes the overall chemistry of a system, which, in turn, influences the redox potential and electrochemical processes at the metal–environment interface. Various groups of microorganisms have been shown to potentially cause MIC. These include not only sulfidogens but also fermentative, acetogenic, methanogenic, and metal-reducing microorganisms (e.g., Little et al. 1998; Dinh et al. 2004; Duncan et al. 2009; Mand et al. 2014; Kato et al. 2015; Vigneron et al. 2016). MIC in the oil and gas industry has been traditionally linked to the activity of sulfate-reducing microorganisms (SRMs) based on measuring high abundances of SRM in corrosive biofilms and detecting their characteristic corrosion product FeS (Enning and Garrelfs 2014). SRM also contribute to reservoir souring, causing additional problems as outlined below.

Reservoir souring is the generation of hydrogen sulfide (H_2S) in subsurface oil and gas formations. The corrosive and toxic nature of H_2S poses significant problems for the oil and gas industry. High costs arise from necessary upgrading of infrastructure to sour service grade materials and from mitigation efforts to reduce sulfur or prevent souring and thereby increase the product value (Gieg et al. 2011). H_2S formed during souring also bears serious threats to worker health and safety (Tang et al. 2009; Williamson 2010). Sourings can be the result of (thermo)chemical formation or linked to biological activity (Khatib and Salanitro 1997; Seto and Beliveau 2000; Gieg et al. 2011). The biological formation of H_2S is ascribed in most cases to the activity of SRM, but other sulfidogens, such as sulfur-, thiosulfate-, polythionate-reducing and sulfur-disproportionating microorganisms can potentially contribute to the production of H_2S as well (Barrett and Clark 1987; Thamdrup et al. 1993; Magot et al. 1997). Oil reservoirs are rich in organic carbon sources that offer carbon and energy sources to fuel the metabolism of SRM and other organisms. Seawater, rich in sulfate $\left(SO_4^{2-}\right)$, is frequently injected into oilfields to enhance oil recovery by maintaining in situ pressure, hence introducing an excellent electron acceptor into electron donor–rich environments. The half reaction for sulfide generation via SRM metabolism is shown in Equation 5.1. The electrons derive from suitable electron donors as described above and later on in the text.

$$SO_4^{2-} + 9\,H^+ + 8\,e^- \rightarrow HS^- + 4\,H_2O \tag{5.1}$$

MIC related to the activity of sulfate-reducing microorganisms has been extensively studied and recently reviewed (Enning and Garrelfs 2014). On the other hand, little is known about the impact of nitrate on MIC (referred to here as nitrate-MIC) in oil and gas exploitations. Due to the widespread use of nitrate injection in souring control, it is essential to identify potential corrosion problems associated with nitrate-reducing microorganisms (NRM) before nitrate injection strategies are implemented. H_2S can be rapidly oxidized to highly corrosive partially oxidized sulfur species either by microbial activity or ingress of oxygen (see below). The latter can result from, for example, insufficient deaeration of injection waters, faulty equipment, replacement work, and/or (un)loading during transport and storage of sour oil. This chapter discusses the potential impacts on corrosion of nitrate or oxygen entering sour systems and thereby changing the overall redox potential from reducing to oxidizing conditions.

5.2 MITIGATION STRATEGIES TO CONTROL SOURING AND CORROSION

Microbiological studies have revealed a high microbial diversity associated with oilfields and oil production operations. These microorganisms belong to various physiological groups, such as sulfate reducers, fermenters, methanogens, metal reducers, and nitrate reducers (Youssef et al. 2009; Gieg et al. 2011; Hubert et al. 2012; Hu et al. 2016). Several strategies have been proposed to prevent and counteract souring, including treatment with biocides (e.g., glutaraldehyde) (Reinsel et al. 1996; Greene et al. 2006; Xue and Voordouw 2015), specific SRM inhibitors (e.g., nitrite, molybdate) (Reinsel et al. 1996; Nemati et al. 2001b; Hubert et al. 2003), and introducing alternative electron acceptors (e.g., nitrate, (per)chlorate) to divert the prevailing metabolism toward "noncorrosive" microorganisms that do not catalyze souring (Jenneman et al. 1986; Telang et al. 1997; Myhr et al. 2002; Hubert et al. 2003; Engelbrektson et al. 2014; Liebensteiner et al. 2014).

In recent years, the injection of nitrate as an alternative bioengineering strategy has proven effective in controlling H_2S production in various lab studies (e.g., Jenneman et al. 1986; McInerney et al. 1996; Hubert et al. 2003; Callbeck et al. 2013) and field settings (e.g., Telang et al. 1997; Jenneman et al. 1999; Bødtker et al. 2008; Gittel et al. 2009; Voordouw et al. 2009). In contrast to biocide treatment, which undertakes to inhibit or kill SRM and other microorganisms, nitrate is a more easily handled compound that works through promoting activity by indigenous or introduced (e.g., by seawater injection) microbial populations (Figure 5.1) (Hubert 2010).

Nitrate $\left(NO_3^-\right)$, as an energetically more favorable electron acceptor compared to sulfate $\left(SO_4^{2-}\right)$, counteracts souring by (i) promoting growth of organotrophic nitrate-reducing bacteria (oNRM) that compete with SRM for available electron donors (i.e., so-called biocompetitive exclusion), (ii) removing the souring agent

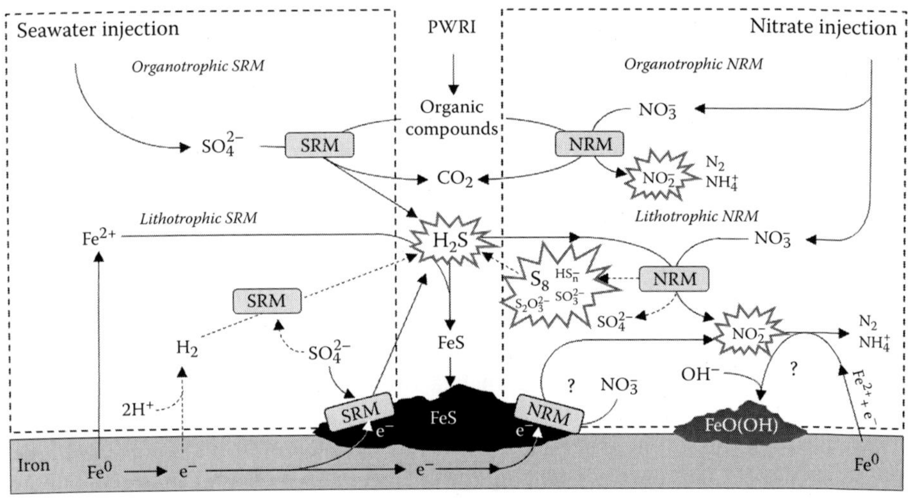

FIGURE 5.1 Microbial processes potentially involved in corrosion during seawater injection (left) or nitrate injection (right) at oilfield production operations. Produced water reinjection (PWRI) into the reservoir for pressure maintenance potentially increases availability of organic compounds (e.g., acetate), which can sustain organotrophic SRM or NRM. Lithotrophic SRM or NRM can be stimulated by, for example, hydrogen, iron, or sulfide, leading to the production of corrosive S or N compounds (outlined). Further details are provided in the text.

H_2S through reoxidation to SO_4^{2-} by lithotrophic sulfur-oxidizing NRM (soNRM), and (iii) NRM-mediated nitrite $\left(NO_2^-\right)$ formation that can inhibit SRM activity (Greene et al. 2003). Concentrations of 0.5 mM NO_2^- have been shown to transiently inhibit SRM whereas 1 mM permanently inhibited SRM in lab experiments (Grigoryan et al. 2008). However, some SRMs can overcome inhibition by detoxifying nitrite via enzymatic reduction to ammonium $\left(NH_4^+\right)$, and in these instances, higher concentrations of nitrite (>5 mM) are required to inhibit sulfide production (Greene et al. 2003).

5.2.1 Potential Electron Donors for NRM in Sour Oil Systems

Seawater has a relatively low organic carbon content, yet injection of sulfate-rich seawater in an electron donor–rich oilfield increases the risk of souring in the reservoir and corrosion problems in downstream facilities (Vance and Thrasher 2005). On the other hand, if produced water (potentially high in additional substrates, such as organic acids) gets reinjected into the reservoir together with seawater, souring and corrosion problems can be further exacerbated (Vance and Thrasher 2005; Beeder et al. 2007; Vik et al. 2007; Lysnes et al. 2009). Therefore, nitrate injection is considered as a souring mitigation strategy during produced water reinjection (PWRI) (Beeder et al. 2007; Vik et al. 2007; Bødtker et al. 2009).

NRM (or SRM) electron donors found in produced waters from oilfields can be volatile fatty acids (VFAs; e.g., acetate, propionate, butyrate), alcohols, and/or hydrogen. These compounds are most likely produced by fermentative microorganisms during their degradation of hydrocarbons and/or other complex organic matter (Barth 1991; Zengler et al. 1999; Head et al. 2003; Grigoryan et al. 2008). Interestingly, SRM isolated from oilfields typically do not show hydrocarbon degradation activity in laboratory tests (Magot 2005) whereas oilfield oNRM with hydrocarbon degradation abilities have been discovered (Ollivier and Cayol 2005; Agrawal et al. 2012). This might indicate a broader substrate range among NRM compared to SRM in oil systems. Additional NRM carbon sources would, in principle, not contribute to souring control via biocompetitive exclusion but could sustain inhibitory NO_2^- levels and promote longer-lasting inhibition of SRMs after depletion of, for example, VFAs via biocompetitive exclusion.

In already sour systems where H_2S has been produced and accumulated, NO_3^- can potentially promote growth of soNRM (Figure 5.1) (Voordouw et al. 1996, 2009; Telang et al. 1997; Eckford and Fedorak 2002; Bødtker et al. 2008, 2009; Lysnes et al. 2009). These organisms are ubiquitous and play important roles in biogeochemical sulfur cycling (Jørgensen 1990; Zopfi et al. 2004; Ghosh and Dam 2009; Hubbard et al. 2014). The ability to oxidize sulfur compounds with NO_3^- as electron acceptor is widespread among members of *Alphaproteobacteria* (e.g., *Magnetospirillum* spp., *Paracoccus* spp.), *Betaproteobacteria* (e.g., *Thiobacillus* spp.), *Gammaproteobacteria* (e.g., *Beggiatoa* spp.), and *Epsilonproteobacteria* (e.g., *Arcobacter* spp., *Sulfurospirillum* spp., *Sulfurimonas* spp.) (Ghosh and Dam 2009). In particular, *Epsilonproteobacteria*—a bacterial lineage comprising several sulfur-oxidizing genera (Campbell et al. 2006)—are frequently found in oilfield samples (Gevertz et al. 2000; Gittel et al. 2012; Hubert et al. 2012).

5.2.2 Potential Microbial Metabolisms Contributing to Nitrate-MIC

Nitrate-MIC has so far only been reported in offshore PWRI tests (Beeder et al. 2007; Vik et al. 2007; Drønen et al. 2014) and during lab tests with sour systems (Nemati et al. 2001a; Hubert et al. 2005; Rempel et al. 2006). Corrosion rates associated with nitrate-MIC in field trials or lab experiments can easily reach levels that are as high (or even higher) than those observed during souring or SRM activity. This can be seen in Figure 5.2, which compares several reports of MIC in field and lab-based studies. It is becoming increasingly evident that care must be taken when considering nitrate injection during PWRI, especially into sour systems, given the risks surrounding nitrate-MIC.

Oilfield soNRM are prominent candidates for catalyzing nitrate-MIC, having been linked to increased corrosion observed in souring control studies (Nemati et al. 2001a; Hubert et al. 2005; Rempel et al. 2006) and identified in field tests showing nitrate-MIC (Beeder et al. 2007; Drønen et al. 2014). These microorganisms oxidize sulfide to SO_4^{2-} or S^0 but can produce other partially oxidized sulfur species, such as thiosulfate $\left(S_2O_3^{2-}\right)$ or sulfite $\left(SO_3^-\right)$ as free intermediates (Figure 5.1) (Kelly et al.

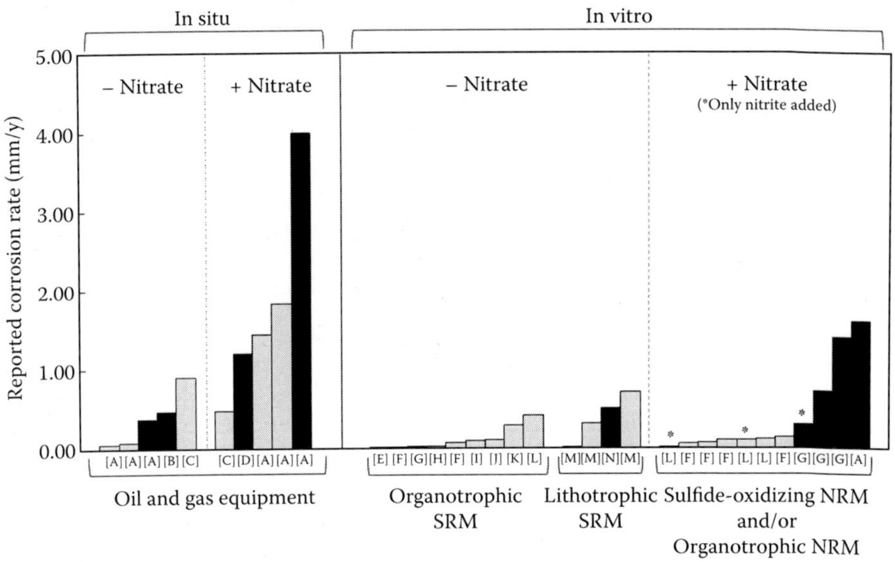

FIGURE 5.2 Comparison of corrosion rates associated with (or without) nitrate addition under in situ conditions (e.g., oil pipeline, water injection systems) or in lab experiments (in vitro) with examples of the latter grouped according to putative MIC candidates (e.g., SRM, NRM). Corrosion rates are calculated based on reported weight loss (gray bars) or electrochemical measurements (black bars) provided in the corresponding published reports (denoted by letters below each example) as follows: (A) Beeder et al. 2007, (B) Li et al. 2000, (C) Bødtker et al. 2008, (D) Vik et al. 2007, (E) Hardy and Brown 1984, (F) Nemati et al. 2001a, (G) Rempel et al. 2006, (H) Gaylarde 1992, (I) Beech et al. 1994, (J) Johnston and Voordouw 2012, (K) Bell and Lim 1981, (L) Hubert et al. 2005, (M) Enning et al. 2012, (N) Pankhania et al. 1986.

1997; Ghosh and Dam 2009). These sulfur species are considered highly aggressive toward metallic iron (Fe^0) (MacDonald et al. 1978; Schaschl 1980; Schmitt 1991; Choudhary et al. 2015).

The initial step in soNRM-mediated H_2S oxidation is the two-electron reduction to polysulfide $\left(HS_n^-\right)$ or elemental sulfur (S^0) as depicted in Figure 5.1 (Griesbeck et al. 2002; Ghosh and Dam 2009; Berg et al. 2014). The subsequent oxidation of these compounds all the way to sulfate can proceed via other partially oxidized intermediates and is considered a bottleneck due to the low solubility, crystallization, and sedimentation of excreted S^0 (Ghosh and Dam 2009). Crystallization and sedimentation of S^0 are governed by physicochemical properties (e.g., temperature, pressure, pH) of a given system (Steudel 1996; Steudel and Eckert 2003). Metabolic end products are determined by an organism's physiology (Gevertz et al. 2000; Ghosh and Dam 2009) combined with external parameters, such as the nitrate-to-sulfide ratio (N/S) (Greene et al. 2003; An et al. 2010), for example, as determined by the nitrate dose chosen to engineer souring control. Under electron acceptor limitation (low N/S ratio) or conditions promoting high growth rates, soNRM tend to oxidize sulfide to S^0 (Stefess et al. 1996; Alcántara et al. 2004; An et al. 2010; Wang et al. 2015). Therefore, accumulation of S^0 in sour systems treated with nitrate is perhaps not unexpected. Whether such accumulation is permanent or transient can depend on electron acceptor availability (i.e., the N/S ratio).

Nemati and colleagues (2001a) showed that corrosion rates doubled (from 0.07 to 0.13 mm/y, Figure 5.2F) and pitting corrosion was enhanced when NO_3^- alone or NO_3^- together with the soNRM *Sulfurimonas* sp. strain CVO were added to sulfidogenic-produced water enrichment cultures. The nitrate dose had a greater influence on corrosion rates than the initial H_2S level, and corrosion was attributed to accumulation of S^0 and NO_2^- (Nemati et al. 2001a). In follow-up experiments with similar sour enrichments in up-flow sand pack columns, Hubert et al. (2005) observed high but general corrosion (0.4 mm/y, Figure 5.2L) near the reactor outlet (i.e., simulating a production well) during souring whereas columns treated with nitrate and hence with no souring showed lower corrosion rates (0.1 mm/y, Figure 5.2L) but with incidence of pitting near the inlet (analogous to an oilfield injection well). Putative soNRM, belonging to the *Epsilonproteobacteria*, were found near the inlet of the nitrate-treated columns and could have produced partially oxidized S species (Hubert et al. 2005). Although (weight loss–based) corrosion rates were four times lower during nitrate treatment, pitting is a major cause for concern related to failures. The high and variable corrosion rates observed near the outlet of parallel untreated (sour) up-flow columns corresponded with discrepancies in total S recovery (as either H_2S or SO_4^{2-}), leading to the hypothesis that oxygen ingress could have oxidized sulfide to S^0 resulting in higher corrosion (Hubert et al. 2005).

Similar observations have been made at the pilot and field scales. Bødtker et al. (2008) monitored long-term nitrate injection into the Gullfaks and Veslefrikk oilfields in the North Sea and observed that average corrosion rates and H_2S levels decreased after the onset of nitrate injection (Figure 5.2C; Bødtker et al. 2008). *Sulfurimonas*-like bacteria (i.e., well-known Epsilonproteobacterial soNRM) formed a major part of the microbial population near the injection wellhead of Veslefrikk field. Vik et al. (2007) reproducibly demonstrated during a pilot study for nitrate injection in the

Draugen field (North Sea) that corrosion rates increased (up to 1.2 mm/y; Figure 5.2D) every time nitrate was injected into sidestream test columns and decreased directly after biocide treatment, thus linking MIC in these test columns to NRM. Inspection of the corrosion test coupons at the end of the pilot revealed formation of a yellow sulfur layer, and soNRM-mediated formation of S^0 or HS_n^- was implicated as the cause of the increased corrosion rates observed during nitrate injection (Vik et al. 2007). However, the presence of soNRM in corrosive biofilms was not confirmed in this study.

Severe corrosion related to nitrate injection has also been reported for other oilfields in the Norwegian sector of the North Sea (Beeder et al. 2007). During an offshore sidestream test for produced water/aquifer water injection, average corrosion rates were 10 times higher (4 mm/y; Figure 5.2A) during nitrate injection compared to unamended tests (0.37 mm/y; Figure 5.2A). Interestingly, sulfate and organic acid concentrations were relatively low in the water mixture, but the aquifer water contained S^0 crystals. In follow-up work, Drønen et al. (2014) speculated that the corrosion proceeded in different phases: (i) initially, SRM were inhibited by oNRM-mediated NO_2^- formation; then (ii) S^0-reducing *Desulfuromonas* spp. consumed electrons directly from iron, initiating severe corrosion. In this system, it was also suggested that nitrite oxidized H_2S to S^0 and that H_2S was reformed via microbial S^0 disproportionation. Putative soNRM belonging to *Arcobacter* and *Sulfurospirillum* spp. were detected and could have contributed to S^0 formation during nitrate treatment (Drønen et al. 2014).

It is noteworthy that in all nitrate-MIC reports soNRM are present and S^0 is suggested to play a role (Nemati et al. 2001a; Hubert et al. 2005; Beeder et al. 2007; Vik et al. 2007; Drønen et al. 2014). Many soNRM can also grow aerobically, coupling H_2S oxidation to oxygen reduction (Gevertz et al. 2000; Ghosh and Dam 2009). Thus, microbial activity together with rapid chemical oxidation of H_2S by oxygen (see Section 5.4.5) could enhance production of corrosive S species during periods of oxygen ingress in sour oil systems.

An interesting aspect of MIC is the ability of some microorganisms to consume electrons from solid-state material (e.g., carbon or metal cathodes) and couple this to the reduction of suitable electron acceptors (e.g., O_2, SO_4^{2-}, or NO_3^-) (Lovley 2011). It has been shown that direct electron consumption from metallic iron (Fe^0) by SRM can result in high corrosion of up to 0.7 mm/y (Figures 5.1 and 5.2M; Dinh et al. 2004; Enning et al. 2012; Venzlaff et al. 2013). Recently, Iino et al. (2015) reported the isolation of non-hydrogenotrophic lithoheterotrophic *Prolixibacter* sp. strain MIC1-1 from crude oil (Akita Prefecture, Japan). This strain was able to grow with nitrate or oxygen as the electron acceptor and metallic iron (Fe^0) as the electron donor. It was suggested that strain MIC1-1 consumed electrons derived from both Fe^0 and Fe^{2+} (Iino et al. 2015). While further research is needed to verify these discoveries, they suggest that nitrate-dependent iron oxidizing (Fe^0- or Fe^{2+}-dependent) microorganisms could play a role in corrosion during oilfield nitrate injection for souring control (Figure 5.1).

Alternative electron acceptors for microbial souring control such as (per)chlorate (ClO_4^- and ClO_3^-) have also garnered discussion recently as substitutes to nitrate injection based on demonstrations of their effectiveness as souring control agents

(Engelbrektson et al. 2014; Gregoire et al. 2014; Carlson et al. 2015). Nonetheless, the effect of these compounds on MIC is largely unknown. Microbial iron oxidation (Fe^0 or Fe^{2+}) coupled to perchlorate reduction is possible and thus could potentially lead to increased corrosion (Son et al. 2006; Yu et al. 2006). Chlorite $\left(ClO_2^-\right)$, as a toxic intermediate produced during perchlorate reduction, reacts with H_2S to corrosive partially oxidized sulfur species (e.g., S^0, $S_2O_3^{2-}$) (Mao et al. 2009; Liebensteiner et al. 2013). The relevance of these processes with respect to MIC should be further evaluated under in vitro and in situ test systems, prior to field applications.

5.3 CORROSION MECHANISMS RELEVANT TO NITRATE INJECTION OR OXYGEN INGRESS IN SOUR SYSTEMS

This section provides an overview of the current view of corrosion by sulfur and nitrogen species. A brief introduction to the corrosion under aerobic conditions is followed by corrosion mechanisms of S and N compounds. In addition, putative chemical transformations during nitrate treatment or oxygen ingress in sour systems are presented.

5.3.1 General Considerations in Aerobic and Anaerobic Systems

Corrosion is an electrochemical process involving anodic (oxidation) and cathodic (reduction) reactions. Iron is a base metal and in aqueous settings its dissolution releases Fe^{2+} (into the surrounding environment) and electrons (into the metal lattice) according to Equation 5.2 (Revie and Uhlig 2008a):

$$Fe^0 \rightarrow Fe^{2+} + 2e^- \ (E^0 = -0.47\ V) \tag{5.2}$$

In the absence of oxygen (anoxic conditions), the corresponding cathodic reaction is the reduction of protons to hydrogen according to Equation 5.3:

$$2H^+ + 2e^- \rightarrow H_2 \ (E^0 = 0.00\ V;\ E^{0'} = -0.41\ V) \tag{5.3}$$

At neutral pH, the reaction is rather slow due to limited cathodic reactants (H^+), and corrosion is thus under cathodic control (Revie and Uhlig 2008a; Enning and Garrelfs 2014). Under oxic conditions, the electrons are consumed by oxygen according to Equation 5.4:

$$0.5\ O_2 + 2H^+ + 2e^- \rightarrow H_2O\ (E^{0'} = 0.81\ V) \tag{5.4}$$

Under anoxic conditions and circumneutral pH, severe corrosion phenomena are usually associated with activity of microorganisms (MIC). Various aspects of SRM-mediated corrosion have been recently reviewed (Enning and Garrelfs 2014). This section therefore focuses on corrosion caused by other compounds potentially formed in H_2S-containing systems due to nitrate injection or the unintentional ingress of oxygen.

5.3.2 Elemental Sulfur

Elemental sulfur (S^0) can exist in a variety of molecular forms. At ambient pressure and temperature (i.e., <95°C) the majority of S^0 would consist mainly of thermodynamically stable cyclooctasulfur $\left(S_8^0\right)$ (Steudel and Eckert 2003). Elemental sulfur has been long known to cause severe corrosion, including pitting and stress corrosion cracking (SCC) (Schmitt 1991; Gee and Chen 1995). The corrosion mechanism as described below is based on observations at moderate temperatures (<95°C, i.e., below the melting point of sulfur) and ambient pressure (other effects might influence corrosion caused by S^0 at increased temperature and pressure; Schmitt 1991).

Iron corrosion by S^0 in solution affects anodic as well as cathodic reactions and proceeds autocatalytically when a conductive FeS layer forms (MacDonald et al. 1978; Schaschl 1980; Schmitt 1991). Direct contact between metal (or a FeS layer) and S^0 is required to promote severe corrosion events. This can be realized by sedimentation of S^0 or its solubilization as HS_n^- (MacDonald et al. 1978; Schaschl 1980; Schmitt 1991). In that regard S_8^{2-}, S_6^{2-}, and HS_2^- would be the dominant polysulfide species in a $SO_4^{2-}/S_8^0/HS^-$ system between pH 5 and 9 (Rickard and Luther 2007). The corrosion process can be simplified as followed (see also Figure 5.3):

$$S_8^0 + (2n)e^- + nH_2O \overset{FeS}{\rightarrow} S_{8-n}^{2-} + nHS^- + nOH^- (\text{with } n = 1-8) \tag{5.5}$$

$$Fe^{2+} + HS^- \rightarrow FeS + H^+ \tag{5.6}$$

$$\text{Overall: } 8\ Fe^0 + S_8^0 \rightarrow 8\ FeS \tag{5.7}$$

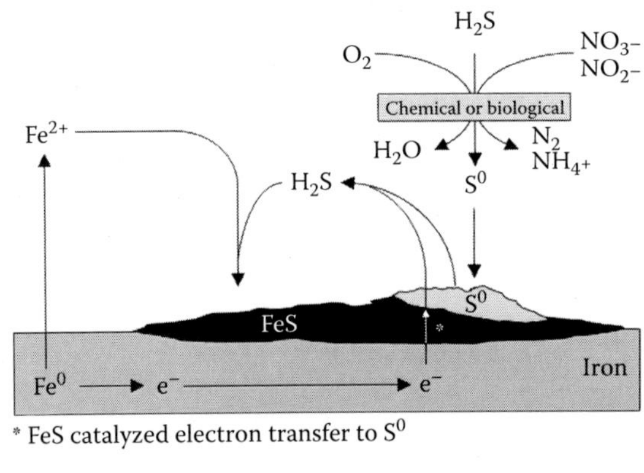

FIGURE 5.3 Schematic of the corrosion mechanism by elemental sulfur (S^0). Formation of S^0 can result from microbial or chemical sulfide oxidation by oxygen or nitrite. Contact between S^0 and Fe^0 will initiate corrosion by iron sulfide (FeS) catalyzed S^0 reduction. For details, see Section 5.3.2.

Electrons from the dissolution of iron (Equation 5.2) fuel the contact-mediated cathodic reduction of S^0 or polysulfides (Equation 5.5). Fe^{2+} produced during iron dissolution (Equation 5.2) readily reacts with the resulting sulfide to form iron sulfides (FeS; Equation 5.6). This results in a very simplified overall reaction according to Equation 5.7 (for further details, see Schmitt 1991). The conductive FeS that forms seems to play a key role as electron mediator between Fe^0 and S^0 (Figure 5.3), and upon contact with sulfur can accelerate corrosion rates to as high as >20 mm/y (MacDonald et al. 1978; Küster et al. 1984; Schmitt 1991). Due to its conductivity, the FeS layer provides an additional cathodically reactive surface area, which may further accelerate Fe^0 dissolution (Figure 5.3). In sour systems, the required FeS for severe autocatalytic S^0 corrosion would form readily and likely reduces the induction time required for the onset of severe corrosion. In sulfide-free systems, sulfur disproportionation has been suggested to initiate sulfide-dependent FeS formation, according to Equation 5.8 (Maldonado-Zagal and Boden 1982; Küster et al. 1984; Schmitt 1991):

$$S^0 + 4\,H_2O \rightarrow HS^- + SO_4^{2-} + 5\,H^+ \tag{5.8}$$

This reaction is slow at ambient temperatures but could be catalyzed by S^0 disproportionating microorganisms, which could therefore induce corrosion by S^0 (Bak and Cypionka 1987; Thamdrup et al. 1993; Finster et al. 1998; Drønen et al. 2014).

Corrosion by S^0 is pH-dependent, and maximal corrosion rates (>20 mm/y) have been observed in systems that are initially pH 5–6 (MacDonald et al. 1978; Küster et al. 1984; Schmitt 1991). On the other hand, alkaline pH lessens S^0 corrosion, which is completely prevented at pH >12 (Küster et al. 1984). Analogous to differential aeration cells, local pH differences are expected in S^0 corrosion due to the separation of anodic and cathodic reactions, possibly explaining the observed localized attacks often associated with S^0 corrosion (MacDonald et al. 1978; Küster et al. 1984; Schmitt 1991; Fang et al. 2008, 2011; Boivin et al. 2011). Neutral salts, such as sodium chloride (NaCl), sodium sulfate (Na_2SO_4), or sodium perchlorate ($NaClO_4$), have been shown to accelerate corrosion (Küster et al. 1984; Schmitt 1991). The latter is particularly important, given perchlorate is gaining attention as an alternative electron acceptor for promoting souring control (see Section 5.3.2) and microbial enhanced oil recovery (Liebensteiner et al. 2014; Carlson et al. 2015).

It should be noted that the particle sizes of S globules have an influence on the corrosion reaction. Smaller S aggregates markedly reduced the induction period of severe corrosion under anaerobic conditions (MacDonald et al. 1978). S globules produced by microorganisms have an initial diameter of <1 μm (Kleinjan et al. 2003). It is therefore expected that microbial sulfur particles greatly reduce the induction time for corrosion. Additionally, microbial-produced sulfur particles have generally a lower density and are more hydrophilic as opposed to chemically produced sulfur particles (Kleinjan et al. 2003). The hydrophilic character and small particle size would also greatly benefit the mass transfer of sulfur towards the metal/FeS surface.

5.3.3 Thiosulfate and Other Partially Oxidized S Species

In addition, other more oxidized S compounds, for example, thiosulfate $\left(S_2O_3^{2-}\right)$ have long been known as corrosive compounds (Isaacs et al. 1982; Choudhary et al. 2015). In sour systems, formation of $S_2O_3^{2-}$ can result from chemical oxidation of H_2S following oxygen exposure (see Section 5.4.5) or during microbial H_2S oxidation (see Section 5.3.2). Once formed, $S_2O_3^{2-}$ causes localized corrosion (e.g., pitting and crevice corrosion) and SCC of steel (e.g., Isaacs et al. 1982; Newman et al. 1982; Park et al. 1997; Choudhary et al. 2015). Thiosulfate is metastable and can be reduced (to S^0 and/or H_2S), oxidized (to SO_3^{2-}, SO_4^{2-}), or react via disproportionation (to HS^- and SO_4^{2-}), depending on the pH and redox state of the system and the microbial population present (Choudhary et al. 2015). The corrosion mechanism involves initial reduction of $S_2O_3^{2-}$ to S^0 (Equation 5.9) by Fe^0-derived electrons (Newman et al. 1982; Kappes et al. 2012).

$$4\,S_2O_3^{2-} + 24\,H^+ + 16\,e^- \rightarrow S_8^0 + 12\,H_2O \tag{5.9}$$

Once S^0 is formed, additional corrosion is promoted as described above (Equations 5.5 through 5.7). It should be noted that polythionates (e.g., tetrathionate, $S_4O_6^{2-}$) have been shown to corrode steel and cause SCC as well, following similar reduction and/or disproportionation reactions (Newman et al. 1982; Horowitz 1983; Abd Elhamid et al. 2001; Yonezu et al. 2012). Also noteworthy is that greater reactivity of thiosulfate with iron is expected in low pH systems whereas tetrathionate is relatively stable and more reactive at high pH (Choudhary et al. 2015).

5.3.4 Nitrate and Nitrite

Metallic iron (Fe^0) has the ability to reduce nitrate chemically to ammonium $\left(NH_4^+\right)$, nitrite $\left(NO_2^-\right)$, nitric oxide (NO), or dinitrogen (N_2) (Huang et al. 1998; Kielemoes et al. 2000; Huang and Zhang 2005; Ahn et al. 2008). The end product is suggested to depend on the redox potential of the system (Huang and Zhang 2006). Rapid chemical nitrate reduction by Fe^0 occurs mainly at pH <5 (Huang et al. 1998; Huang and Zhang 2004) but can be enhanced, for example, by microorganisms (Till et al. 1998; Kielemoes et al. 2000), in the presence of high Fe^{2+} concentrations (Huang and Zhang 2005) or temperatures >50°C (Ahn et al. 2008).

In sour systems, the formation of nitrite is generally considered beneficial given its effectiveness as a SRM inhibitor (Greene et al. 2003; Hubert 2010). However, care should be taken as nitrite plays a dual role in corrosion. It is often applied as a corrosion inhibitor in water-cooling systems, and its performance can depend on concentrations of NO_2^- and Cl^- as well as temperature and pH (Wachter 1945; Uhlig and King 1959; Sherwood 1993; Mustafa et al. 1997; Kielemoes et al. 2000; Hu et al. 2001; Karim et al. 2010). Nitrite has been shown to promote corrosion at concentrations <3.5 mM but protects at levels >10 mM (Sherwood 1993; Mustafa et al. 1997; Kielemoes et al. 2000). Sufficiently high concentrations of nitrite will produce a thin passivating iron (hydr)oxide layer (e.g., Fe_2O_3) on the surface of mild steel

(Equation 5.10), preventing further dissolution of iron (as Fe^{2+}) and therefore acting as an anodic corrosion inhibitor (Revie and Uhlig 2008b).

$$2\ Fe^{2+} + 2\ OH^- + 2\ NO_2^- \rightarrow 2\ NO + Fe_2O_3 + H_2O \tag{5.10}$$

However, this layer becomes unstable at pH <6 such that nitrite corrosion rates accelerate at low pH (Hu et al. 2001; Revie and Uhlig 2008b; Karim et al. 2010). The passivation via nitrite can additionally be influenced by the presence of chloride ions, which accelerate iron dissolution (Wachter 1945; Berke and Hicks 2004).

In sour systems, nitrite (NO_2^-) has been reported to chemically react with H_2S to form S^0 according to Equation 5.11 (Kaster et al. 2007; Johnston et al. 2009).

$$3\ HS^- + NO_2^- \rightarrow 3\ S^0 + NH_4^+ + 2\ H_2O \tag{5.11}$$

To which extent this contributes to corrosion in sour systems treated with nitrate needs to be further investigated.

5.3.5 Oxygen

Ingress of oxygen can greatly affect corrosion processes in H_2S-containing systems by promoting rapid chemical transformation of H_2S or FeS into highly corrosive S compounds (Hardy and Brown 1984; Lee et al. 1993a,b; Nielsen et al. 1993; Lewandowski 2009; Boivin and Oliphant 2011). During the O_2-mediated oxidation of sulfide to sulfate, several reaction intermediates have been described, including, for example, S^0, SO_3^{2-}, and $S_2O_3^{2-}$ (Avrahami and Golding 1968; Cline and Richards 1969; Chen and Morris 1972; Zhang and Millero 1993; Zopfi et al. 2004). The amount of oxidation product strongly depends on temperature, pH, and the sulfide-to-oxygen ratio (Zhang and Millero 1993). Sulfide is initially oxidized to sulfite (Equation 5.12), which can be further oxidized to sulfate (Equation 5.13). In the presence of additional sulfide, thiosulfate can form (Equation 5.14). This reaction could be envisioned in systems with a high sulfide load and neutral-to-alkaline pH, potentially leading to accumulation of thiosulfate (Avrahami and Golding 1968; Zhang and Millero 1993).

$$HS^- + 1.5\ O_2 \rightarrow SO_3^{2-} + H^+ \tag{5.12}$$

$$SO_3^{2-} + 0.5\ O_2 \rightarrow SO_4^{2-} \tag{5.13}$$

$$HS^- + SO_3^{2-} + 0.5\ O_2 \rightarrow S_2O_3^{2-} + OH^- \tag{5.14}$$

Trace metals (e.g., Fe, Mn, Cu) act as catalysts and not only increase the oxidation rate of sulfide and sulfite, but also affect the amount of individual intermediates formed (Zhang and Millero 1993; Steudel 1996; Yao and Millero 1996). Fe^{2+} in that regard has been shown to be the most effective catalyst for oxygen-dependent sulfide oxidation (Zhang and Millero 1993).

Formation of polysulfides and elemental sulfur (Equation 5.15) can be expected in systems with excess sulfide compared to oxygen (Chen and Morris 1972; Zhang and Millero 1993). The elemental sulfur formed can react with sulfite (Equation 5.16) or sulfide (Equation 5.17) (Zhang and Millero 1993; Zopfi et al. 2004).

$$2\,HS^- + O_2 \rightarrow 2\,S^0 + 2\,OH^- \tag{5.15}$$

$$S^0 + SO_3^{2-} \rightarrow S_2O_3^{2-} \tag{5.16}$$

$$(n)S^0 + HS^- \rightarrow HS^-_{(n+1)} \tag{5.17}$$

The formed sulfur species eventually react with the iron according to the corrosion mechanisms described in the preceding sections. Corrosion increased and severe pitting was measured (up to 50 mm/y pitting rate) when SRM-corroded iron specimens were exposed to air (Hardy and Brown 1984; Nielsen et al. 1993). The severe corrosion attributed to ingress of oxygen in sour systems has been mainly ascribed to the formation of S^0 by oxidation of H_2S and FeS (Lee et al. 1993a,b; Nielsen et al. 1993; Johnston and Voordouw 2012). While most SRM are sensitive to oxygen, sulfate-reducing layers can form beneath aerobic surficial layers even at high oxygen tension outside the biofilm (Lee et al. 1993a,b).

FeS layers can oxidize to iron (hydr)oxides and zero-valent sulfur when exposed to air (Equation 5.18). Additional S^0 can be formed in the presence of high H_2S according to Equation 5.19 (Schmitt 1991; Lewandowski 2009).

$$4\,FeS + 3\,O_2 + 2\,H_2O \rightarrow S^0 + FeO(OH) \tag{5.18}$$

$$4\,FeO(OH) + 3\,H_2S \rightarrow 2\,FeS + S^0 + 4\,H_2O \tag{5.19}$$

The cyclic transformation from iron sulfides and iron (hydr)oxides can weaken protective scale structures. Sherar et al. (2013) reported formation of iron (hydr)oxides underneath FeS, indicating that the structure (e.g., porosity) of FeS can allow oxygen penetration and subsequently iron corrosion. This effect might greatly promote pitting as the metal surface becomes locally exposed to the surrounding environment.

Thus, both ingress of oxygen as well as formation of nitrite by NRM in sour systems can contribute to the formation of highly corrosive partially oxidized S species and therefore increase corrosion risk. Most soNRM associated with oil production systems (e.g., *Epsilonproteobacteria*) are facultative anaerobes and, as discussed above, can use oxygen as an electron acceptor (Ghosh and Dam 2009). This could, in principle, increase and accelerate formation of partially oxidized sulfur species by additive action of chemical and biological sulfide oxidation.

5.4 CONCLUSIONS

Understanding of the microbiology and biogeochemistry in oil systems is steadily improving. Widespread adoption of nitrate injection to control souring and corrosion during oil production demonstrates this. Numerous case studies, including those reviewed here, have addressed various questions around nitrate treatment and helped to improve this technology. However, under certain conditions, nitrate treatment has failed or resulted in increased corrosion in lab tests and real-world systems. Careful inspection of instances of nitrate-MIC suggests involvement of sulfur cycling and production of corrosive S intermediates (e.g., S^0) in the elevated corrosion. Putative S^0-producing soNRM are consistently found in these examples and therefore likely play a key role in nitrate-MIC. Still, our understanding of nitrate-MIC remains limited, and microbial processes, such as S^0 disproportionation (Drønen et al. 2014) or NRM-mediated Fe^0 or Fe^{2+} oxidation (Iino et al. 2015), may also contribute in ways that are not yet fully elucidated. The impact of other potentially corrosive chemicals such as nitrite (Nemati et al. 2001a; Hubert et al. 2005) and CO_2 (Beeder et al. 2007) and parameters such as pH, temperature, and salinity must also be carefully evaluated for their potential influence on microbial activity and associated corrosion. In this regard, the potential of high-temperature oilfields to exhibit nitrate-MIC has not been investigated. *Epsilonproteobacteria*, including soNRM like *Arcobacter* and *Sulfurospirillum* spp. have been detected in a nitrate-treated high-temperature oilfield (Halfdan in the North Sea), but little is known about their activity in situ (Gittel et al. 2009, 2012). Likewise, corrosion during periodic ingress of oxygen has been studied mainly under in vitro conditions (Hardy and Brown 1984; Nielsen et al. 1993; Johnston and Voordouw 2012). However, the effect of oxygen ingress on microbial communities and their response needs to be further addressed in studies of oil production systems.

Future studies need to move beyond microbial community structure and diversity and begin to address key players and understand their functional potential, activity, and responses to changes that can be realistically anticipated under in situ conditions. A better understanding of microbial processes during oil production will help to better predict and mitigate MIC events but also yield valuable information to consider in light of implementation of new technologies (e.g., perchlorate injection). In this way, souring control strategies that also consider and prevent harmful MIC can be developed and prioritized.

ACKNOWLEDGMENTS

This work was supported by funding from the UK Engineering & Physical Sciences Research Council (EP/J002259/1 and EP/L001942/1), ExxonMobil Upstream Research Company, and the EU's Marie Skłodowska Curie Actions (660401–MOLMIC). The authors wish to thank Dennis Enning and Kelly Hawboldt for providing constructive reviews that improved this contribution.

REFERENCES

Abd Elhamid, M. H., B. G. Ateya, K. G. Weil, and H. W. Pickering. 2001. Effect of Thiosulfate and Sulfite on the Permeation Rate of Hydrogen through Iron. *Corrosion* 57: 428–32.

Agrawal, A., H. Park, S. Nathoo, L. M. Gieg, T. R. Jack, K. Miner, R. Ertmoed, A. Benko, and G. Voordouw. 2012. Toluene Depletion in Produced Oil Contributes to Souring Control in a Field Subjected to Nitrate Injection. *Environ. Sci. Technol.* 46: 1285–92.

Ahn, S. C., S. Oh, and D. K. Cha. 2008. Enhanced Reduction of Nitrate by Zero-Valent Iron at Elevated Temperatures. *J. Hazard. Mater.* 156: 17–22.

Alcántara, S., A. Velasco, A. Muñoz, J. Cid, S. Revah, and E. Razo-Flores. 2004. Hydrogen Sulfide Oxidation by a Microbial Consortium in a Recirculation Reactor System: Sulfur Formation under Oxygen Limitation and Removal of Phenols. *Environ. Sci. Technol.* 38: 918–23.

An, S., K. Tang, and M. Nemati. 2010. Simultaneous Biodesulphurization and Denitrification Using an Oil Reservoir Microbial Culture: Effects of Sulphide Loading Rate and Sulphide to Nitrate Loading Ratio. *Water Res.* 44: 1531–41.

Avrahami, M., and R. M. Golding. 1968. The Oxidation of the Sulphide Ion at Very Low Concentrations in Aqueous Solutions. *J. Chem. Soc. A.* 0: 647–51.

Bak, F., and H. Cypionka. 1987. A Novel Type of Energy Metabolism Involving Fermentation of Inorganic Sulphur Compounds. *Nature* 326: 891–2.

Barrett, E. L., and M. A. Clark. 1987. Tetrathionate Reduction and Production of Hydrogen Sulfide from Thiosulfate. *Microbiol. Rev.* 51: 192–205.

Barth, T. 1991. Organic Acids and Inorganic Ions in Waters from Petroleum Reservoirs, Norwegian Continental Shelf: A Multivariate Statistical Analysis and Comparison with American Reservoir Formation Waters. *Appl. Geochem.* 6: 1–15.

Beech, I. B., C. W. S. Cheung, C. S. P. Chan, M. A. Hill, R. Franco, and A. Lino. 1994. Study of Parameters Implicated in the Biodeterioration of Mild Steel in the Presence of Different Species of Sulphate-Reducing Bacteria. *Int. Biodeterior. Biodegrad.* 34: 289–303.

Beeder, J., T. R. Andersen, T. Liengen, K. Drønen, and T. Torsvik. 2007. Corrosion as a Side Effect during Nitrate Treatment of Produced Water and Aquifer Water Injection. Paper no. 07512, NACE International Corrosion Conference, Nashville, TN.

Bell, R. G., and C. K. Lim. 1981. Corrosion of Mild and Stainless Steel by Four Tropical *Desulfovibrio desulfuricans* Strains. *Can. J. Microbiol.* 27: 242–5.

Berg, J. S., A. Schwedt, A. Kreutzmann, M. M. M. Kuypers, and J. Milucka. 2014. Polysulfides as Intermediates in the Oxidation of Sulfide to Sulfate by *Beggiatoa* spp. *Appl. Environ. Microbiol.* 80: 629–36.

Berke, N. S., and M. C. Hicks. 2004. Predicting Long-Term Durability of Steel Reinforced Concrete with Calcium Nitrite Corrosion Inhibitor. *Cem. Concr. Compos.* 26: 191–8.

Bødtker, G., K. Lysnes, T. Torsvik, E. Ø. Bjørnestad, and E. Sunde. 2009. Microbial Analysis of Backflowed Injection Water from a Nitrate-Treated North Sea Oil Reservoir. *J. Ind. Microbiol. Biotechnol.* 36: 439–50.

Bødtker, G., T. Thorstenson, B. P. Lillebø, B. E. Thorbjørnsen, R. H. Ulvøen, E. Sunde, and T. Torsvik. 2008. The Effect of Long-Term Nitrate Treatment on SRB Activity, Corrosion Rate and Bacterial Community Composition in Offshore Water Injection Systems. *J. Ind. Microbiol. Biotechnol.* 35: 1625–36.

Boivin, J., and S. Oliphant. 2011. Sulfur Corrosion due to Oxygen Ingress. Paper no. 11120, NACE International Corrosion Conference, Houston, TX.

Boivin, J., Z. Wen, X. Li, J. Ding, and M. Lu. 2011. Effects of Depositing Characteristic and Temperature on Elemental Sulfur Corrosion. Paper no. 11121, NACE International Corrosion Conference, Houston, TX.

Callbeck, C. M., A. Agrawal, and G. Voordouw. 2013. Acetate Production from Oil under Sulfate-Reducing Conditions in Bioreactors Injected with Sulfate and Nitrate. *Appl. Environ. Microbiol.* 79: 5059–68.

Campbell, B. J., A. S. Engel, M. L. Porter, and K. Takai. 2006. The Versatile *Epsilonproteobacteria*: Key Players in Sulphidic Habitats. *Nat. Rev. Microbiol.* 4: 458–68.

Carlson, H. K., J. V. Kuehl, A. B. Hazra, N. B. Justice, M. K. Stoeva, A. Sczesnak, M. R. Mullan et al. 2015. Mechanisms of Direct Inhibition of the Respiratory Sulfate-Reduction Pathway by (Per)chlorate and Nitrate. *ISME J.* 9: 1295–305.

Chen, K. Y., and J. C. Morris. 1972. Kinetics of Oxidation of Aqueous Sulfide by Oxygen. *Environ. Sci. Technol.* 6: 529–37.

Choudhary, L., D. D. Macdonald, and A. Alfantazi. 2015. Role of Thiosulfate in the Corrosion of Steels: A Review. *Corrosion* 71: 1147–68.

Cline, J. D., and F. A. Richards. 1969. Oxygenation of Hydrogen Sulfide in Seawater at Constant Salinity, Temperature and pH. *Environ. Sci. Technol.* 3: 838–43.

Dinh, H. T., J. Kuever, M. Mußmann, A. W. Hassel, M. Stratmann, and F. Widdel. 2004. Iron Corrosion by Novel Anaerobic Microorganisms. *Nature* 427: 829–32.

Drønen, K., I. Roalkvam, J. Beeder, T. Torsvik, I. H. Steen, A. Skauge, and T. Liengen. 2014. Modeling of Heavy Nitrate Corrosion in Anaerobe Aquifer Injection Water Biofilm: A Case Study in a Flow Rig. *Environ. Sci. Technol.* 48: 8627–35.

Duncan, K. E., L. M. Gieg, V. A. Parisi, R. S. Tanner, S. Green Tringe, J. Bristow, and J. M. Suflita. 2009. Biocorrosive Thermophilic Microbial Communities in Alaskan North Slope Oil Facilities. *Environ. Sci. Technol.* 43: 7977–84.

Eckford, R. E., and P. M. Fedorak. 2002. Planktonic Nitrate-Reducing Bacteria and Sulfate-Reducing Bacteria in Some Western Canadian Oil Field Waters. *J. Ind. Microbiol. Biotechnol.* 29: 83–92.

Engelbrektson, A., C. G. Hubbard, L. M. Tom, A. Boussina, Y. T. Jin, H. Wong, Y. M. Piceno et al. 2014. Inhibition of Microbial Sulfate Reduction in a Flow-through Column System by (Per)chlorate Treatment. *Front. Microbiol.* 5 (June): 315. doi: 10.3389/fmicb.2014.00315.

Enning, D., and J. Garrelfs. 2014. Corrosion of Iron by Sulfate-Reducing Bacteria: New Views of an Old Problem. *Appl. Environ. Microbiol.* 80: 1226–36.

Enning, D., H. Venzlaff, J. Garrelfs, H. T. Dinh, V. Meyer, K. Mayrhofer, A. W. Hassel, M. Stratmann, and F. Widdel. 2012. Marine Sulfate-Reducing Bacteria Cause Serious Corrosion of Iron under Electroconductive Biogenic Mineral Crust. *Environ. Microbiol.* 14: 1772–87.

Fang, H., B. Brown, D. Young, and S. Nešić. 2011. Investigation of Elemental Sulfur Corrosion. Paper no. 11398, NACE International Corrosion Conference, Houston, TX.

Fang, H., D. Young, and S. Nešić. 2008. Corrosion of Mild Steel in the Presence of Elemental Sulfur. Paper no. 08637, NACE International Corrosion Conference, New Orleans, Louisiana.

Finster, K., W. Liesack, and B. Thamdrup. 1998. Elemental Sulfur and Thiosulfate Disproportionation by *Desulfocapsa sulfoexigens* sp. nov., a New Anaerobic Bacterium Isolated from Marine Surface Sediment. *Appl. Environ. Microbiol.* 64: 119–25.

Gaylarde, C. C. 1992. Sulphate-Reducing Bacteria Which Do Not Induce Accelerated Corrosion. *Int. Biodeterior. Biodegrad.* 30: 331–8.

Gee, R., and Z. Y. Chen. 1995. Hydrogen Embrittlement during the Corrosion of Steel by Wet Elemental Sulphur. *Corr. Sci.* 37: 2003–11.

Gevertz, D., A. J. Telang, G. Voordouw, and G. E. Jenneman. 2000. Isolation and Characterization of Strains CVO and FWKO B, Two Novel Nitrate-Reducing, Sulfide-Oxidizing Bacteria Isolated from Oil Field Brine. *Appl. Environ. Microbiol.* 66: 2491–501.

Ghosh, W., and B. Dam. 2009. Biochemistry and Molecular Biology of Lithotrophic Sulfur Oxidation by Taxonomically and Ecologically Diverse *Bacteria* and *Archaea*. *FEMS Microbiol. Rev.* 33: 999–1043.

Gieg, L. M., T. R. Jack, and J. M. Foght. 2011. Biological Souring and Mitigation in Oil Reservoirs. *Appl. Microbiol. Biotechnol.* 92: 263–82.

Gittel, A., M. V. W. Kofoed, K. B. Sørensen, K. Ingvorsen, and A. Schramm. 2012. Succession of *Deferribacteres* and *Epsilonproteobacteria* through a Nitrate-Treated High-Temperature Oil Production Facility. *Syst. Appl. Microbiol.* 35: 165–74.

Gittel, A., K. B. Sørensen, T. L. Skovhus, K. Ingvorsen, and A. Schramm. 2009. Prokaryotic Community Structure and Sulfate Reducer Activity in Water from High-Temperature Oil Reservoirs with and without Nitrate Treatment. *Appl. Environ. Microbiol.* 75: 7086–96.

Greene, E. A., C. Hubert, M. Nemati, G. E. Jenneman, and G. Voordouw. 2003. Nitrite Reductase Activity of Sulphate-Reducing Bacteria Prevents Their Inhibition by Nitrate-Reducing, Sulphide-Oxidizing Bacteria. *Environ. Microbiol.* 5: 607–17.

Greene, E. A., V. Brunelle, G. E. Jenneman, and G. Voordouw. 2006. Synergistic Inhibition of Microbial Sulfide Production by Combinations of the Metabolic Inhibitor Nitrite and Biocides. *Appl. Environ. Microbiol.* 72: 7897–901.

Gregoire, P., A. Engelbrektson, C. G. Hubbard, Z. Metlagel, R. Csencsits, M. Auer, M. E. Conrad, J. Thieme, P. Northrup, and J. D. Coates. 2014. Control of Sulfidogenesis through Bio-Oxidation of H_2S Coupled to (Per)chlorate Reduction. *Environ. Microbiol. Rep.* 6: 558–64.

Griesbeck, C., M. Schütz, T. Schödl, S. Bathe, L. Nausch, N. Mederer, M. Vielreicher, and G. Hauska. 2002. Mechanism of Sulfide-Quinone Reductase Investigated Using Site-Directed Mutagenesis and Sulfur Analysis. *Biochemistry* 41: 11552–65.

Grigoryan, A. A., S. L. Cornish, B. Buziak, S. Lin, A. Cavallaro, J. J. Arensdorf, and G. Voordouw. 2008. Competitive Oxidation of Volatile Fatty Acids by Sulfate- and Nitrate-Reducing Bacteria from an Oil Field in Argentina. *Appl. Environ. Microbiol.* 74: 4324–35.

Hardy, J. A., and J. L. Brown. 1984. The Corrosion of Mild Steel by Biogenic Sulfide Films Exposed to Air. *Corrosion* 40: 650–4.

Head I. M., D. M. Jones, S. R. Larter. 2003. Biological Activity in the Deep Subsurface and the Origin of Heavy Oil. *Nature* 426: 344–52.

Horowitz, H. H. 1983. Chemical Studies of Polythionic Acid Stress-Corrosion Cracking. *Corr. Sci.* 23: 353–62.

Hu, H.-Y., N. Goto, and K. Fujie. 2001. Effect of pH on the Reduction of Nitrite in Water by Metallic Iron. *Water Res.* 35: 2789–93.

Hu, P., L. Tom, A. Singh, B. C. Thomas, B. J. Baker, Y. M. Piceno, G. L. Andersen, and J. F. Banfield. 2016. Genome-Resolved Metagenomic Analysis Reveals Roles for Candidate Phyla and Other Microbial Community Members in Biogeochemical Transformations in Oil Reservoirs. *mBio* 7 (January): e01669–15. http://mbio.asm.org/content/7/1/e01669-15.long.

Huang, C., H. Wang, and P. Chiu. 1998. Nitrate Reduction by Metallic Iron. *Water Res.* 32: 2257–64.

Huang, Y. H., and T. C. Zhang. 2004. Effects of Low pH on Nitrate Reduction by Iron Powder. *Water Res.* 38: 2631–42.

Huang, Y. H., and T. C. Zhang. 2005. Effects of Dissolved Oxygen on Formation of Corrosion Products and Concomitant Oxygen and Nitrate Reduction in Zero-Valent Iron Systems with or without Aqueous Fe^{2+}. *Water Res.* 39: 1751–60.

Huang, Y. H., and T. C. Zhang. 2006. Nitrite Reduction and Formation of Corrosion Coatings in Zerovalent Iron Systems. *Chemosphere* 64: 937–43.

Hubbard, C. G., Y. Cheng, A. Engelbrekston, J. L. Druhan, L. Li, J. B. Ajo-Franklin, J. D. Coates, and M. E. Conrad. 2014. Isotopic Insights into Microbial Sulfur Cycling in Oil Reservoirs. *Front. Microbiol.* 5 (September): 480. doi: 10.3389/fmicb.2014.00480.

Hubert, C. 2010. Microbial Ecology of Oil Reservoir Souring and Its Control by Nitrate Injection. In *Handbook of Hydrocarbon and Lipid Microbiology*, ed. K. N. Timmis, 2753–66. Springer, Berlin, Heidelberg.

Hubert, C., M. Nemati, G. Jenneman, and G. Voordouw. 2003. Containment of Biogenic Sulfide Production in Continuous up-Flow Packed-Bed Bioreactors with Nitrate or Nitrite. *Biotechnol. Prog.* 19: 338–45.

Hubert, C., M. Nemati. G. Jenneman, and G. Voordouw. 2005. Corrosion Risk Associated with Microbial Souring Control Using Nitrate or Nitrite. *Appl. Microbiol. Biotechnol.* 68: 272–82.

Hubert, C. R. J., T. B. P. Oldenburg, M. Fustic, N. D. Gray, S. R. Larter, K. Penn, A. K. Rowan et al. 2012. Massive Dominance of *Epsilonproteobacteria* in Formation Waters from a Canadian Oil Sands Reservoir Containing Severely Biodegraded Oil. *Environ. Microbiol.* 14: 387–404.

Iino, T., K. Ito, S. Wakai, H. Tsurumaru, M. Ohkuma, and S. Harayama. 2015. Iron Corrosion Induced by Nonhydrogenotrophic Nitrate-Reducing *Prolixibacter* sp. Strain MIC1-1. *Appl. Environ. Microbiol.* 81: 1839–46.

Isaacs, H. S., B. Vyas, and M. W. Kendig. 1982. The Stress Corrosion Cracking of Sensitized Stainless Steel in Thiosulfate Solutions. *Corrosion* 38: 130–6.

Jenneman, G. E., M. J. McInerney, and R. M. Knapp. 1986. Effect of Nitrate on Biogenic Sulfide Production. *Appl. Environ. Microbiol.* 51: 1205–11.

Jenneman, G. E., P. D. Moffitt, G. A. Bala, and R. H. Webb. 1999. Sulfide Removal in Reservoir Brine by Indigenous Bacteria. *SPE Prod. Facil.* 14: 219–25.

Johnston, S., S. Lin, P. Lee, S. M. Caffrey, J. Wildschut, J. K. Voordouw, S. M. da Silva, I. A. C. Pereira, and G. Voordouw. 2009. A Genomic Island of the Sulfate-Reducing Bacterium *Desulfovibrio vulgaris* Hildenborough Promotes Survival under Stress Conditions While Decreasing the Efficiency of Anaerobic Growth. *Environ. Microbiol.* 11: 981–91.

Johnston, S. L., and G. Voordouw. 2012. Sulfate-Reducing Bacteria Lower Sulfur-Mediated Pitting Corrosion under Conditions of Oxygen Ingress. *Environ. Sci. Technol.* 46: 9183–90.

Jonkers, H. M., M. J. E. C. van Der Maarel, H. van Gemerden, and T. A. Hansen. 1996. Dimethylsulfoxide Reduction by Marine Sulfate-Reducing Bacteria. *FEMS Microbiol. Lett.* 136: 283–7.

Jørgensen, B. B. 1990. A Thiosulfate Shunt in the Sulfur Cycle of Marine Sediments. *Science* 249: 152–4.

Kappes, M., G. S. Frankel, N. Sridhar, and R. M. Carranza. 2012. Reaction Paths of Thiosulfate during Corrosion of Carbon Steel in Acidified Brines. *J. Electrochem. Soc.* 159: 195–204.

Karim, S., C. M. Mustafa, M. Assaduzzaman, and M. Islam. 2010. Effect of Nitrite Ion on Corrosion Inhibition of Mild Steel in Simulated Cooling Water. *Chem. Eng. Res. Bull.* 14: 87–91.

Kaster, K. M., A. Grigoriyan, G. Jenneman, and G. Voordouw. 2007. Effect of Nitrate and Nitrite on Sulfide Production by Two Thermophilic, Sulfate-Reducing Enrichments from an OilField in the North Sea. *Appl. Microbiol. Biotechnol.* 75: 195–203.

Kato, S., I. Yumoto, and Y. Kamagata. 2015. Isolation of Acetogenic Bacteria That Induce Biocorrosion by Utilizing Metallic Iron as the Sole Electron Donor. *Appl. Environ. Microbiol.* 81: 67–73.

Kelly, D. P., J. K. Shergill, W. Lu, and A. P. Wood. 1997. Oxidative Metabolism of Inorganic Sulfur Compounds by Bacteria. *Antonie Van Leeuwenhoek* 71: 95–107.

Khatib, Z. I., and J. R. Salanitro. 1997. Reservoir Souring: Analysis of Surveys and Experience in Sour Waterfloods. SPE paper no. 38795, SPE Annual Technical Conference and Exhibition, San Antonio, Texas.

Kielemoes, J., P. De Boever, and W. Verstraete. 2000. Influence of Denitrification on the Corrosion of Iron and Stainless Steel Powder. *Environ. Sci. Technol.* 34: 663–71.

Kleinjan, W. E., A. de Keizer, and A. J. H. Janssen. 2003. Biologically Produced Sulfur. In *Elemental Sulfur and Sulfur-Rich Compounds* I, ed. R. Steudel, *Top. Curr. Chem.* 230: 167–88. Springer, Berlin, Heidelberg.

Kröger, A., S. Biel, J. Simon, R. Gross, G. Unden, and C. R. D. Lancaster. 2002. Fumarate Respiration of *Wolinella succinogenes*: Enzymology, Energetics and Coupling Mechanism. *Biochim. Biophys. Acta* 1553: 23–38.

Küster, W., H. Schlerkmann, G. Schmitt, W. Schwenk, and D. Steinmetz. 1984. Korrosionsreaktionen von Elementarem Schwefel Mit Unlegiertem Stahl in Wäßrigen Medien. *Mater. Corros.* 35: 556–65.

Lee, W., Z. Lewandowski, M. Morrison, W. G. Characklis, R. Avci, and P. H. Nielsen. 1993a. Corrosion of Mild Steel underneath Aerobic Biofilms Containing Sulfate-reducing Bacteria Part II: At High Dissolved Oxygen Concentration. *Biofouling* 7: 217–39.

Lee, W., Z. Lewandowski, S. Okabe, W. G. Characklis, and R. Avci. 1993b. Corrosion of Mild Steel underneath Aerobic Biofilms Containing Sulfate-reducing Bacteria Part I: At Low Dissolved Oxygen Concentration. *Biofouling* 7: 197–216.

Lewandowski, Z., and H. Beyenal. 2009. Mechanisms of Microbially Influenced Corrosion. In *Marine and Industrial Biofouling*. eds. H. Flemming, P. S. Murthy, R. Venkatesan and K. E. Cooksey. 35–65. Springer, Berlin, Heidelberg.

Li, S. Y., Y. G. Kim, K. S. Jeon, and Y. T. Kho. 2000. Microbiologically Influenced Corrosion of Underground Pipelines under the Disbonded Coatings. *Met. Mater.* 6: 281–6.

Liebensteiner, M. G., M. W. H. Pinkse, P. J. Schaap, A. J. M. Stams, and B. P. Lomans. 2013. Archaeal (Per)chlorate Reduction at High Temperature: An Interplay of Biotic and Abiotic Reactions. *Science* 340: 85–7.

Liebensteiner, M. G., N. Tsesmetzis, A. J. M. Stams, and B. P. Lomans. 2014. Microbial Redox Processes in Deep Subsurface Environments and the Potential Application of (Per)chlorate in Oil Reservoirs. *Front. Microbiol.* 5: 428.

Little, B., P. Wagner, K. Hart, R. Ray, D. Lavoie, K. Nealson, and C. Aguilar. 1998. The Role of Biomineralization in Microbiologically Influenced Corrosion. *Biodegradation* 9: 1–10.

Lovley, D. R. 2011. Powering Microbes with Electricity: Direct Electron Transfer from Electrodes to Microbes. *Environ. Microbiol. Rep.* 3: 27–35.

Lysnes, K., G. Bødtker, T. Torsvik, E. O. Bjørnestad, and E. Sunde. 2009. Microbial Response to Reinjection of Produced Water in an Oil Reservoir. *Appl. Microbiol. Biotechnol.* 83: 1143–57.

MacDonald, D. D., B. Roberts, and J. B. Hyne. 1978. The Corrosion of Carbon Steel by Wet Elemental Sulphur. *Corros. Sci.* 18: 411–25.

Magot, M. 2005. Indigenous Microbial Communities in OilFields. In *Petroleum Microbiology*, eds. B. Ollivier and M. Magot, 21–34. ASM Press, Washington.

Magot, M., G. Ravot, X. Campaignolle, B. Ollivier, B. K. C. Patel, M.-L. Fardeau, P. Thomas, J.-L. Crolet, and J.-L. Garcia. 1997. *Dethiosulfovibrio peptidovorans* gen. nov., sp. nov., a New Anaerobic, Slightly Halophilic, Thiosulfate-Reducing Bacterium from Corroding Offshore Oil Wells. *Int. J. Syst. Evol. Micr.* 47: 818–24.

Maldonado-Zagal, S. B., and P. J. Boden. 1982. Hydrolysis of Elemental Sulphur in Water and Its Effect on the Corrosion of Mild Steel. *Brit. Corros. J.* 17: 116–20.

Mand, J., H. S. Park, T. R. Jack, and G. Voordouw. 2014. The Role of Acetogens in Microbially Influenced Corrosion of Steel. *Front. Microbiol.* 5 (June). doi: 10.3389/fmicb.2014.00268.

Mao, S., Q. Gao, H. Wang, J. Zheng, and I. R. Epstein. 2009. Oscillations and Mechanistic Analysis of the Chlorite-Sulfide Reaction in a Continuous-Flow Stirred Tank Reactor. *J. Physic. Chem.* 113: 1231–4.

McInerney, M. J., N. Q. Wofford, and K. L. Sublette. 1996. Microbial Control of Hydrogen Sulfide Production in a Porous Medium. *Appl. Biochem. Biotechnol.* 57–58: 933–44.

Mustafa, C. M., S. M. Shahinoor, and I. Dulal. 1997. Corrosion Behaviour of Mild Steel in Moderately Alkaline to Acidic Simulated Cooling Water Containing Molybdate and Nitrite. *Brit. Corros. J.* 32: 133–7.

Myhr, S., B.-L. Lillebø, E. Sunde, J. Beeder, and T. Torsvik. 2002. Inhibition of Microbial H2S Production in an Oil Reservoir Model Column by Nitrate Injection. *Appl. Microbiol. Biotechnol.* 58: 400–8.

Nemati, M., G. E. Jenneman, and G. Voordouw. 2001a. Impact of Nitrate-Mediated Microbial Control of Souring in Oil Reservoirs on the Extent of Corrosion. *Biotechnol. Progr.* 17: 852–9.

Nemati, M., T. J. Mazutinec, G. E. Jenneman, and G. Voordouw. 2001b. Control of Biogenic H(2)S Production with Nitrite and Molybdate. *J. Ind. Microbiol. Biotechnol.* 26: 350–5.

Newman, R. C., K. Sieradzki, and H. S. Isaacs. 1982. Stress-Corrosion Cracking of Sensitized Type 304 Stainless Steel in Thiosulfate Solutions. *Metall. Trans. A.* 13: 2015–26.

Nielsen, P. H., W. Lee, Z. Lewandowski, M. Morison, and W. G. Characklis. 1993. Corrosion of Mild Steel in an Alternating Oxic and Anoxic Biofilm System. *Biofouling* 7: 267–84.

Ollivier, B., and J. Cayol. 2005. Fermentative, Iron-Reducing, and Nitrate-Reducing Microorganisms. In *Petroleum Microbiology*, eds. B. Ollivier and M. Magot, 71–88. ASM Press, Washington, DC.

Pankhania, I. P., A. N. Moosavi, and W. A. Hamilton. 1986. Utilization of Cathodic Hydrogen by *Desulfovibrio vulgaris* (Hildenborough). *Microbiology* 132: 3357–65.

Park, J. O., M. Verhoff, and R. Alkire. 1997. Microscopic Behavior of Single Corrosion Pits: The Effect of Thiosulfate on Corrosion of Stainless Steel in NaCl. *Electrochim. Acta.* 42: 3281–91.

Reinsel, M. A., J. T. Sears, P. S. Stewart, and M. J. McInerney. 1996. Control of Microbial Souring by Nitrate, Nitrite or Glutaraldehyde Injection in a Sandstone Column. *J. Ind. Microbiol.* 17: 128–36.

Rempel, C. L., R. W. Evitts, and M. Nemati. 2006. Dynamics of Corrosion Rates Associated with Nitrite or Nitrate Mediated Control of Souring under Biological Conditions Simulating an Oil Reservoir. *J. Ind. Microbiol. Biotechnol.* 33: 878–86.

Revie, R. W., and H. H. Uhlig. 2008a. Inhibitors and Passivators. In *Corrosion and Corrosion Control*, 303–16. Wiley & Sons, Inc., New York.

Revie, R. W., and H. H. Uhlig. 2008b. Iron and Steel. In *Corrosion and Corrosion Control*, 115–48. John Wiley & Sons, Inc., New York.

Rickard, D., and G. W. Luther. 2007. Chemistry of Iron Sulfides. *Chem. Rev.* 107: 514–62.

Schaschl, E. 1980. Elemental Sulfur as a Corrodent in Deaerated Neutral Aqueous Solutions. *Mater. Performance.* 19: 9–12.

Schmitt, G. 1991. Effect of Elemental Sulfur on Corrosion in Sour Gas Systems. *Corrosion* 47: 285–308.

Seto, C. J., and D. A. Beliveau. 2000. Reservoir Souring in the Caroline Field. SPE paper no. 59778, SPE/CERI Gas Technology Symposium, Calgary, Alberta.

Sherar, B. W. A., P. G. Keech, and D. W. Shoesmith. 2013. The Effect of Aerobic Corrosion on Anaerobically-Formed Sulfide Layers on Carbon Steel in Dilute near-Neutral pH Saline Solutions. *Corros. Sci.* 77: 257–64.

Sherwood, P. M. A. 1993. Corrosion Inhibitor Surface Chemistry Studied by Core and Valence Band Photoemission. *J. Vac. Sci. Technol. A.* 11: 2280–85.

Son, A., J. Lee, P. C. Chiu, B. J. Kim, and D. K. Cha. 2006. Microbial Reduction of Perchlorate with Zero-Valent Iron. *Water Res.* 40: 2027–32.

Stefess, G. C., R. A. M. Torremans, R. de Schrijver, L. A. Robertson, and J. G. Kuenen. 1996. Quantitative Measurement of Sulphur Formation by Steady-State and Transient-State Continuous Cultures of Autotrophic *Thiobacillus* Species. *Appl. Microbiol. Biotechnol.* 45: 169–75.

Steudel, R. 1996. Mechanism for the Formation of Elemental Sulfur from Aqueous Sulfide in Chemical and Microbiological Desulfurization Processes. *Ind. Eng. Chem. Res.* 35: 1417–23.

Steudel, R., and B. Eckert. 2003. Solid Sulfur Allotropes Sulfur Allotropes. In *Elemental Sulfur and Sulfur-Rich Compounds I*, ed. R. Steudel, 1–80. Topics in Current Chemistry 230. Springer, Berlin, Heidelberg.

Tang, K., V. Baskaran, and M. Nemati. 2009. Bacteria of the Sulphur Cycle: An Overview of Microbiology, Biokinetics and Their Role in Petroleum and Mining Industries. *Biochem. Eng. J.* 44: 73–94.

Telang, A. J., S. Ebert, J. M. Foght, D. Westlake, G. E. Jenneman, D. Gevertz, and G. Voordouw. 1997. Effect of Nitrate Injection on the Microbial Community in an OilField as Monitored by Reverse Sample Genome Probing. *Appl. Environ. Microbiol.* 63: 1785–93.

Thamdrup, B., K. Finster, J. W. Hansen, and F. Bak. 1993. Bacterial Disproportionation of Elemental Sulfur Coupled to Chemical Reduction of Iron or Manganese. *Appl. Environ. Microbiol.* 59: 101–8.

Till, B. A., L. J. Weathers, and P. J. J. Alvarez. 1998. Fe(0)-Supported Autotrophic Denitrification. *Environ. Sci. Technol.* 32: 634–9.

Uhlig, H. H., and P. F. King. 1959. The Flade Potential of Iron Passivated by Various Inorganic Corrosion Inhibitors. *J. Electrochem. Soc.* 106: 1–7.

Vance I., and D. R. Thrasher. 2005. Reservoir Souring: Mechanisms and Prevention. In *Petroleum Microbiology*, eds. B. Ollivier and M. Magot, 123–42. American Society for Microbiology, Washington.

Venzlaff, H., D. Enning, J. Srinivasan, K. J. J. Mayrhofer, A. W. Hassel, F. Widdel, and M. Stratmann. 2013. Accelerated Cathodic Reaction in Microbial Corrosion of Iron due to Direct Electron Uptake by Sulfate-Reducing Bacteria. *Corros. Sci.* 66: 88–96.

Vigneron, A., E. B. Alsop, B. Chambers, B. P. Lomans, I. M. Head, and N. Tsesmetzis. 2016. Complementary Microorganisms in Highly Corrosive Biofilms from an Offshore Oil Production Facility. *Appl. Environ. Microbiol.* 82: 2545–54.

Vik, E. A., A. O. Janbu, F. K. Garshol, L. B. Henninge, S. Engebretsen, C. Kuijvenhoven, D. Oliphant, and W. P. Hendriks. 2007. Nitrate Based Souring Mitigation of Produced Water—Side Effects and Challenges from the Draugen Produced Water Re-Injection Pilot. Paper SPE 106178. SPE International Symposium on Oilfield Chemistry, Houston, Texas.

Voordouw, G., S. M. Armstrong, M. F. Reimer, B. Fouts, A. J. Telang, Y. Shen, and D. Gevertz. 1996. Characterization of 16S rRNA Genes from OilField Microbial Communities Indicates the Presence of a Variety of Sulfate-Reducing, Fermentative, and Sulfide-Oxidizing Bacteria. *Appl. Environ. Microbiol.* 62: 1623–9.

Voordouw, G., A. A. Grigoryan, A. Lambo, S. Lin, H. S. Park, T. R. Jack, D. Coombe et al. 2009. Sulfide Remediation by Pulsed Injection of Nitrate into a Low Temperature Canadian Heavy Oil Reservoir. *Environ. Sci. Technol.* 43: 9512–8.

Wachter, A. 1945. Sodium Nitrite as Corrosion Inhibitor for Water. *Ind. Eng. Chem.* 37: 749–51.

Wang, X., Y. Zhang, J. Zhou, T. Zhang, and M. Chen. 2015. Regeneration of Elemental Sulfur in a Simultaneous Sulfide and Nitrate Removal Reactor under Different Dissolved Oxygen Conditions. *Bioresour. Technol.* 182: 75–81.

Williamson, N. 2010. Problems Caused by Microbes and Treatment Strategies Health and Safety Issues from the Production of Hydrogen Sulphide. In *Applied Microbiology and Molecular Biology in Oilfield Systems*, eds. C. Whitby and T. L. Skovhus, 151–7. Springer, the Netherlands.

Xue, Y., and G. Voordouw. 2015. Control of Microbial Sulfide Production with Biocides and Nitrate in Oil Reservoir Simulating Bioreactors. *Front. Microbiol.* 6 (December): 1387. doi: 10.3389/fmicb.2015.01387.

Yao, W., and F. J. Millero. 1996. Oxidation of Hydrogen Sulfide by Hydrous Fe(III) Oxides in Seawater. *Mar. Chem.* 52: 1–16.

Yonezu, A., R. Kusano, and X. Chen. 2012. On the Mechanism of Intergranular Stress Corrosion Cracking of Sensitized Stainless Steel in Tetrathionate Solution. *J. Mater. Sci.* 48: 2447–53.

Youssef, N., M. S. Elshahed, and M. J. McInerney. 2009. Microbial Processes in OilFields: Culprits, Problems, and Opportunities. *Adv. Appl. Microbiol.* 66:141–251.

Yu, X., C. Amrhein, M. A. Deshusses, and M. R. Matsumoto. 2006. Perchlorate Reduction by Autotrophic Bacteria in the Presence of Zero-Valent Iron. *Environ. Sci. Technol.* 40: 1328–34.

Zengler, K., H. H. Richnow, R. Rosselló-Mora, W. Michaelis, and F. Widdel. 1999. Methane Formation from Long-Chain Alkanes by Anaerobic Microorganisms. *Nature* 401: 266–9.

Zhang, J., and F. J. Millero. 1993. The Products from the Oxidation of H_2S in Seawater. *Geochim. Cosmochim. Ac.* 57: 1705–18.

Zopfi, J., T. G. Ferdelman, and H. Fossing. 2004. Distribution and Fate of Sulfur Intermediates—Sulfite, Tetrathionate, Thiosulfate, and Elemental Sulfur—In Marine Sediments. *Geo. S. Am. S.* 379: 97–116.

6 Effects of Reservoir Souring on Materials Performance

David Fischer, Monica Canalizo-Hernandez, and Amit Kumar

CONTENTS

ABSTRACT

Reservoir souring is a phenomenon associated with the growth and activity of specific microorganisms within a reservoir, which leads to increasing levels of hydrogen sulfide (H_2S) production from a reservoir that did not initially produce H_2S. As a result of the introduction of H_2S into production fluids, reservoir souring poses both significant safety and integrity risks to assets—particularly for those that were not originally designed for sour service. With this concern in mind, this chapter provides a background on reservoir souring and addresses corrosion and environmental cracking issues associated with materials when exposed to souring conditions. Strategies for managing the risks associated with reservoir souring are discussed, including topics on reservoir souring mitigation, corrosion control, and sour service materials selection. This chapter also describes considerations for performing fit-for-service evaluations of materials exposed to mildly sour environments. Finally, the chapter closes with a summary on general best practices to asset integrity management for souring fields.

6.1 INTRODUCTION

Carbon and low-alloy steels are the most commonly used materials for the production, processing, and transportation of produced hydrocarbons. Hydrogen sulfide generated due to reservoir souring induces corrosion of materials (sour corrosion) and also introduces the risk of environmental cracking. The H_2S corrosion that occurs as a result of reservoir souring has the potential to be very localized if conditions lead to incomplete or porous iron sulfide film formation (Greco and Wright 1962; Ikeda et al. 1985; Videm and Kvarekval 1995). Hydrogen generated from sour corrosion can lead to various modes of environmental cracking via hydrogen embrittlement. Both localized corrosion attack and hydrogen embrittlement due to reservoir souring make materials prone to catastrophic failure. Proper knowledge of reservoir souring mitigation methods and sour service materials performance is required to ensure optimized materials selection and asset integrity in souring fields. Overdesign of facilities that are expected to experience souring can generate an unnecessary increase in capital expenditure (CAPEX) or result in the cancellation of an otherwise economically viable project. Underdesign of souring facilities can ultimately lead to failures that pose high safety risks and long production downtime. Jordan et al. (2005) summarized the following reasons for how reservoir souring can result an increase in production costs:

- Requirement for the use of expensive sour service materials based on corrosion and cracking considerations.
- Chemical treatments may need to be implemented to reduce H_2S to acceptable levels.
- Additional well work-over activities to address any scaling and downhole corrosion issues.
- Shutting-in of affected wells if facility is not designed or fit for sour service.
- Health and safety considerations (e.g., toxicity, risk of explosion, etc.).

This chapter provides a background on reservoir souring and addresses the risks associated with materials in souring systems that were not designed for such environments. Available reservoir souring mitigation strategies are discussed, coupled with corrosion control and materials design guidelines for sour systems. The chapter also covers considerations related to fit-for-service evaluations of materials exposed to reservoir souring conditions.

6.2 WHAT IS RESERVOIR SOURING?

Reservoir souring can be defined as the phenomenon in which levels of H_2S increase over time in a reservoir that initially did not produce any H_2S (sweet reservoir). The increasing levels of H_2S are a result of the growth of microorganisms within the reservoir, specifically due to the growth of sulfate-reducing bacteria (SRB) or, in more general terms, of sulfate-reducing microorganisms (SRMs). These organisms reduce sulfate $\left(SO_4^{2-}\right)$ to sulfide (S^{2-}) via a redox reaction where sulfate is the electron acceptor and a number of substrates form the electron donor. The anaerobic (anoxic) conditions in the reservoir allow sulfate-reducing microorganisms to thrive while environmental parameters, such as temperature and pressure, affect the level of sulfide generation. The electron donors used by the SRMs range from light oil components (alkanes, aromatics) to small organic acids, usually referred to as volatile fatty acids (VFAs) contained in the reservoir formation water. Some SRMs have even been found to use iron as the electron donor, suggesting a diverse metabolism for this group of microorganisms (Enning and Garrelfs 2014). A detailed review of sulfate-reducing microorganisms is outside of the scope of this chapter and has been covered in detail elsewhere (Muyzer and Stams 2008).

Reservoir souring is associated with seawater injection or waterflooding operations performed as a secondary oil recovery method for both new and producing oilfields. Although waterflooding enhances hydrocarbon production, it can have the deleterious effect of stimulating SRM growth within the reservoir by optimizing a number of critical conditions, namely temperature and nutrient availability (German 2015). As the temperature of seawater is generally lower than the temperature of reservoir fluids, the injection of seawater can reduce the temperature of the reservoir. Depending on the initial reservoir temperature and the reservoir connectivity and permeability, the "cooling effect" can be pronounced and spread beyond the wellbore or "cooling zone" (Figure 6.1), optimizing the downhole temperature to below the proposed temperature limit of 80°C for SRM growth (Vance and Thrasher 2005; Gieg et al. 2011).

It has been hypothesized that the SRMs that cause reservoir souring might not only be introduced with the seawater into the reservoir during waterflooding, but that they also might already be present within the reservoir in a dormant state (Nilsen et al. 1996). Regardless of the exact source of SRMs, a field that undergoes waterflooding should be considered at risk of souring, and a technical assessment should be performed to determine the probability of souring and the best mitigation option—with mitigation being important even if it is determined that the probability of souring is low (Maxwell and Spark 2005).

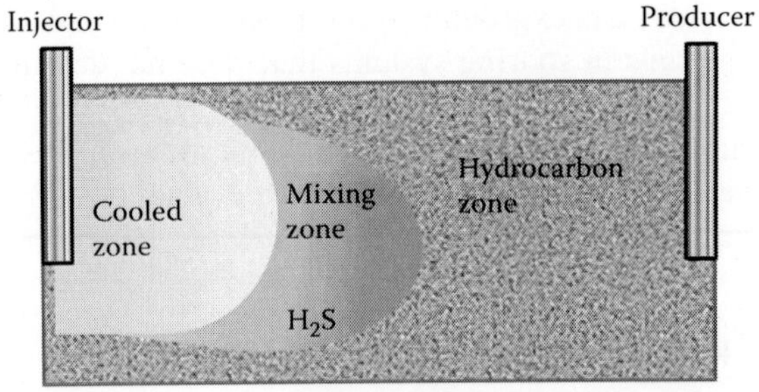

FIGURE 6.1 Simplified diagram of reservoir souring. The seawater injected decreases the temperature in what is called the cooling zone in the near wellbore. The mixing zone is proposed to be the area where the seawater mixes with the formation water. It is hypothesized that the organism activity and thus the H_2S production occurs in this zone. The H_2S will then travel to the producer with the sweeping fluids.

Sulfate, the key metabolite needed for SRM respiration and H_2S production, is introduced into the reservoir in large quantities during seawater injection for which sulfate concentration is approximately 2800 mg/L for the North Sea and Gulf of Mexico and up to 3600 mg/L for the Arabian Gulf (Bader 2007). Sulfate can also be present in the formation water but often at much lower concentrations than for seawater. Injected seawater could also provide SRMs with other micronutrients (phosphate, trace metals, nitrogen compounds) although it is also possible that these nutrients are available within the reservoir mineralogy (Gadd 2010).

The extent and timeline of souring for a specific field also depends greatly on the particular reservoir characteristics, including temperature, pressure, formation and seawater chemistry, reservoir connectivity, water sweeping, field development (e.g., injector/producer placement), and the capacity of reservoir minerals to scavenge H_2S. Each of these factors determines when and how much H_2S will be detected at the producer and therefore the potential mitigation strategies that can be implemented (Burger and Jenneman 2009; Evans et al. 2015).

6.3 RESERVOIR SOURING MITIGATION STRATEGIES

The number of commercially available mitigation strategies for reservoir souring are limited. Moreover, the efficacy of these technologies in preventing or mitigating souring is not consistent in the field as shown in the limited cases reported in the literature and that are referenced in this section. The current understanding of each mitigation technology and examples of field use are summarized below.

6.3.1 Sulfide Scavengers

Sulfide scavengers are compounds that, as their name indicates, react with sulfide ions and actively remove them from the gas stream. Alkanolamines react with H_2S in a reversible manner and are commonly used to treat sour gas in a regenerative and cyclic process (Kucka et al. 2003). Nonregenerative H_2S scavengers include metallic compounds, aldehydes, carboxilates, and amine-based products with triazine being widely used (Aften and Roberts 2011; Madsen et al. 2014). These can be injected into the production fluids at the wellhead and/or downstream of the wellhead. Although these products have been developed to react in a range of conditions, including low pH and high temperature, their efficacy when used in downhole injections remains under investigation (Mendez et al. 2009; Vera et al. 2016). As a result, even if the levels of H_2S being produced are manageable with scavengers on the topside and in processing, the risk of materials failure downhole still remains. In addition, the amount and associated cost of H_2S scavengers can rapidly increase as the levels of H_2S increase over time as a result of reservoir souring. The high cost that can be incurred if scavengers are selected as the mitigation technology can be prohibitive (Kelland 2014).

6.3.2 Biocide Injection

Biocides, or compounds that kill bacteria, are widely used in oil and gas operations to prevent and mitigate microbially influenced corrosion (MIC). They are used in the treatment of injection seawater to control bacterial growth that could occur in topside facilities. The efficacy of biocides, particularly glutaraldehyde and bis[tetrakis(hydroxymethyl)phosphonium] sulfate (THPS), to prevent or mitigate reservoir souring remains controversial with a limited number of reports claiming successful control of souring by adding biocides to the injected seawater (Littman and McLean 1987; Talbot et al. 2000; Jones et al. 2011). Specifically, the injection of THPS reportedly reduced the levels of H_2S in a North Sea field. It has been hypothesized that this particular case was unique in that the high reservoir temperature may have prevented widespread souring. As a result, most of the bacterial activity could have been isolated to regions near the injectors where it could be reached and affected by the injected biocide (Macleod et al. 1994).

A number of biocides are reported to be stable under conditions relevant to the reservoir (high temperature, high salinity) and are shown to be efficacious against planktonic and sessile SRMs (Larsen et al. 2014; Williams and Schultz 2015). However, further laboratory and field data are needed to assess whether the efficacy persists for longer periods of time with respect to long water transport times from the injector to producer (in the range of years for some cases) and higher temperatures. Interestingly, biocides that appear to have more prolonged effects have been shown to be effective in hydraulic fracturing; however, it remains to be investigated if they could be applicable to waterflooding operations (Raymond et al. 2014).

6.3.3 Sulfate Removal

Membrane filtration of sulfate ions from seawater is a process widely used in waterflooding operations to prevent scale formation in reservoirs that contain barium and/or strontium in their formation water. Removal of sulfate using sulfate removal units

(SRUs) is an effective reservoir souring prevention approach as it will limit the key nutrient needed for SRMs to produce H_2S unless sulfate is present in greater quantities in the formation water (Robinson et al. 2010). As such, SRUs are a more feasible mitigation option for greenfields than for brownfields that have been performing waterflooding for some time. There is at least one case reported in the literature where an SRU was selected specifically for reservoir souring mitigation and not for scaling prevention (Keasler et al. 2012).

In order to ensure that sulfate is not injected in the reservoir, filtration units have to operate continuously. Additionally, they need to be adequately maintained to prevent potential biofouling to which they are prone. Two of the main drawbacks of SRUs are the high capital investment required for installation and their large footprint; the latter of which is expected to be lessened as the development of subsea units continues (Hegdal and Pinchin 2015). It has been noted that the mixing of treated seawater with produced water can lead to additional sulfate being introduced back into the treated seawater. Thus, the effect of mixing waters on the final sulfate concentration in the water for injection should be carefully assessed (Evans et al. 2015).

6.3.4 Nitrate Injection

The injection of nitrate to mitigate reservoir souring is perhaps the most widely used technology for this purpose. In spite of the inconsistency of nitrate efficacy in field reports, the large quantities of nitrate theoretically needed for injection for effective mitigation, and an evolving understanding of the underlying mitigation mechanism, there are several fields around the world that perform this treatment (Larsen 2002; Hubert 2010). The nitrate-based souring mitigation mechanism has been extensively studied in academic papers, and its understanding continues to evolve (Hubert et al. 2009). The mechanisms of action proposed to date include the following:

Growth stimulation of competing organisms: Nitrate stimulates the growth of nitrate reducing bacteria (NRB), which act as competing organisms by consuming the limited nutrients within the reservoir and inhibiting the growth and metabolism of SRMs. The main premise of this mode of action is that a key nutrient will be limited in the reservoir to a level low enough to be fully consumed by the NRBs.

SRB metabolism switching to nitrate reduction: Some SRBs are capable of reducing both nitrate and sulfate. By introducing nitrate to the reservoir, the SRBs are driven to switch to consumption of nitrate instead of sulfate due to the thermodynamic advantages involved with this reaction.

Nitrate and nitrite inhibitory effect on SRBs: Nitrite is an intermediary formed during nitrate reduction. Nitrate interferes with the final step in sulfate reduction (by competitively binding to the enzyme) and thus inhibits sulfide generation (Haveman et al. 2004).

Growth stimulation of organisms that oxidize sulfide: Bacteria that can reduce nitrate and oxidize sulfide to sulfate in the same redox reaction are expected to be stimulated with nitrate injection. These types of organisms are referred to as nitrate-reducing sulfide-oxidizing bacteria (NRSOB).

Cases where nitrate injection has successfully prevented reservoir souring in fields that were thought to be at risk are reported in the literature. The Halfdan field in the North Sea has performed continuous nitrate injection since it initiated waterflooding operations, and the levels of H_2S have reportedly been kept at 0.1 ppm (Gittel et al. 2009). Other fields in the North Sea have reported mixed results over time as is the case for Gullfaks, which may be attributed to either complications in data collection or lack of sufficient data needed to determine nitrate efficacy. Whether the extent of souring would have been greater without nitrate injection for this case is not known, casting additional doubts on the efficacy of nitrate (Mitchell et al. 2010). Application of nitrate has also been reported to increase corrosion in topside facilities in a number of cases. This increase in corrosion is proposed to be due to the formation of elemental sulfur produced by incomplete sulfide oxidation carried out by NRSOB, and it is thought to be associated with inadequate nitrate dosing (Beeder et al. 2007). As more data is collected in the fields that are conducting nitrate injection, its efficacy (or lack of) and performance dependency on field conditions should become clearer.

6.3.5 Materials Upgrade

For the development of greenfields, the levels and timeline of souring can be estimated through simulation and modeling tools (Evans et al. 2015). Based on the expected levels of souring, alloys that will withstand the H_2S levels under field conditions can be selected. The upgrade of materials from carbon steel will most likely increase the capital cost but can reduce overall lifecycle costs by reducing operating and maintenance expenses (e.g., asset integrity chemicals) and avoiding potential workover or replacement of components due to failure. Despite recent advancements in the capabilities of reservoir souring prediction software, the simulations carry a degree of uncertainty as they depend on historical field data, accurate reservoir models, and accurate water chemistry parameters. Considerations for materials selection and evaluation in the context of reservoir souring are covered in the subsequent sections.

6.4 CORROSION ISSUES IN SOURING FIELDS

Selection of materials to combat corrosion relies mainly on the type of corrosion anticipated (e.g., whether general or localized) and the confidence in predicting the rate and type of corrosion, risk of failure, and lifecycle cost. While the primary concern in the selection of materials for H_2S-containing systems is environmental fracture due to hydrogen embrittlement, the issue of corrosion should not be underestimated. Typical operational parameters affecting corrosion include the following:

- Pressure
- Temperature
- Mole fraction of acidic gases such as CO_2, H_2S, and O_2
- CO_2/H_2S ratio
- Fluid chemistry (water chemistry, pH, organic acids, water cut, oil wettability, phase ratios, etc.)
- Hydrocarbon phase
- Flow characteristics and fluid velocity
- Steel chemistry and surface condition

Both CO_2 and H_2S are acid gases, and when produced with the hydrocarbon phase, can render the associated water (condensed or formation) corrosive and lead to severe degradation. Corrosion resulting from each of these two acidic gases has its unique characteristics and, as a result, has received considerable industry attention—both to understand the corrosion mechanisms associated with the particular acid gas and the options available to mitigate the resulting corrosion (Kermani et al. 2006). The respective corrosion mechanisms that occur from exposure to each of the two acid gases can be vastly different. Each of these gases occurs naturally in some of the producing reservoirs or may result from external contamination of the reservoir, such as the case of reservoir souring. It is important to note that the synergistic effects of these gases are extremely influential in materials design and should be a primary point of consideration. A simple addition of respective damage rates does not necessarily lead to the overall damage as the complementary corrosion processes are very complex (Kermani et al. 2006). Furthermore, such a simple correlation does not consider localized types of attack wherein the damage rate can be significantly higher than the overall rate. The factors that govern corrosion behavior for steel in CO_2/H_2S environments are discussed in the subsequent sections.

6.4.1 Factors Affecting Corrosion Rates in Souring Fields

With respect to corrosion in reservoir souring conditions, hydrogen sulfide can affect CO_2 corrosion in different ways. The combination of H_2S and CO_2 modifies the corrosion characteristics significantly as compared to damage caused in the sole presence of either of the two acid gases. While very small concentrations of H_2S tend to suppress corrosion through the formation of a protective scale (Choi et al. 2010), the presence of H_2S can also increase corrosion rates and lead to localized corrosion, depending on the type and properties of iron sulfide (FeS) scale. Therefore, it is prudent to understand the relationship between iron sulfide scale formation and corrosion behavior in mixed CO_2/H_2S systems.

In contrast to pure CO_2 corrosion where only one type of iron carbonate ($FeCO_3$) corrosion scale may form, there are several different types of iron sulfide corrosion scales that may form during CO_2/H_2S corrosion, such as amorphous iron sulfide, mackinawite, pyrrhotite, cubic iron sulfide, troilite, marcasite, pyrite, and so on (Smith and Miller 1975; Shoesmith et al. 1980; Taylor 1980; Shoesmith 1981; Smith and Wright 1994; Smith and Joosten 2006). In CO_2/H_2S environments, the type and characteristic of iron sulfide scale will depend on both thermodynamics and kinetics of the scale formation. Thermodynamics determines whether a specific FeS species may be stable in a given environment (Anderko and Shuler 1998; Sun et al. 2008b). Kinetics determines the FeS growth rate and hence which of the thermodynamically stable FeS species actually grows into the scale. Scale composition varies as a function of H_2S concentration, temperature, and exposure time (Sun et al. 2006, Sun and Nešic 2007; Sun et al. 2008a). Additionally, the corrosion scales are not always simple structures made up of one form of FeS. Thin films and scales formed in short exposure times are often composed of one form of FeS. However, heavier scales formed during extended exposures are frequently more complicated, layered structures composed of one or more types of FeS (Meyer et al. 1958).

Pots et al. (2002) introduced a notion of a CO_2-to-H_2S ratio based on partial pressures of the gases (denoted as CO_2/H_2S) and considered three different corrosion domains based on the primary governing corrosion mechanism, which is driven by the dominating acid gas as shown in Figure 6.2. Dunlop et al. (1983) proposed that as a general rule for a ratio of CO_2/H_2S >500, corrosion is dominated by CO_2 and $FeCO_3$ will form; when CO_2/H_2S <20, FeS scale will form and H_2S corrosion dominates. Smith and Pacheco (2002) identified that the corrosion rate-limiting step is determined by the type of corrosion product that forms as a result of the chemical reactions of CO_2 and H_2S. They also developed a relationship that determines the boundaries of the iron carbonate and iron sulfide scales by extending the Dunlop correlation and proposing the equilibrium boundary between different scales (Figure 6.3).

The key parameters in determining the scale morphology are the $FeCO_3$ and FeS supersaturation values. It has been claimed that the mechanisms of corrosion product film growth can produce multilayer films with some leading to increased corrosion rate and probability of localized corrosion (Brown and Nešic 2005). Smith and Wright (1994) also postulated an equation to calculate the minimum H_2S required

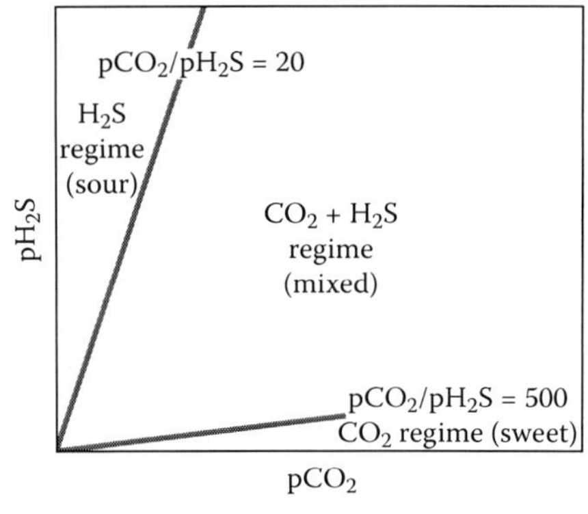

FIGURE 6.2 CO_2–H_2S corrosion domains based on ratio of acid gas partial pressures. (Adapted from Pots, B. F., Kapusta, S. D., John, R. C., Thomas, M. J. J., Rippon, I. J., Whitham, T. S. and Girgis, M., Improvements on de Waard-Milliams corrosion prediction and applications to corrosion management, *CORROSION 2002*, Denver, CO: NACE International, 2002.)

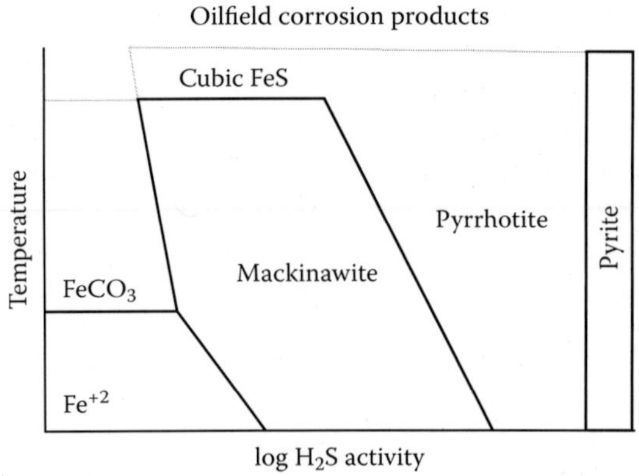

FIGURE 6.3 CO_2/H_2S corrosion product phase map. (Reproduced from Smith, S. N. and Pacheco, J. L., Prediction of Corrosion in Slightly Sour Environments, *CORROSION 2002*, Denver, CO: NACE International, 2002.)

for mackinawite formation as the corrosion product rather than as ferrous ions. This equation can be used as a guiding tool to determine whether H_2S generated by reservoir souring is sufficient to form an iron sulfide scale.

Solution composition is also an important factor that affects the corrosion rate in acid gas environments. The solution composition parameters that have been deemed relevant to sour corrosion include pH, organic acid concentration, Fe^{2+} concentration, chloride concentration, total ionic concentration, and brine composition. There is extensive literature available on sour corrosion (Rogers and Rowe Jr. 1955; Meyer et al. 1958; Berner 1967; Smith and Miller 1975; Shoesmith et al. 1980; Taylor 1980; Shoesmith 1981; Smith 1993; Smith and Wright 1994; Kvarekval et al. 2005; Kermani et al. 2006; Smith and Joosten 2006). However, most of the CO_2/H_2S corrosion studies reported in literature focus on highly sour environments, which differ from the mildly sour conditions experienced after reservoir souring.

6.4.2 Studies on Steel Corrosion in Mildly Sour Environments

To understand the effects of mild H_2S concentrations on corrosion in mixed CO_2/H_2S environments, Smith and Pacheco (2002) investigated CO_2/H_2S corrosion in the presence of small amounts of H_2S and reported that the corrosion rate decreased with increase in H_2S concentration. Smith and DeWaard (2005) proposed an H_2S reduction factor in their corrosion rate prediction model for CO_2 corrosion. The factor is a function of the partial pressure ratio of H_2S and CO_2. Nevertheless, it must be emphasized that H_2S may also form a nonprotective layer and that steels may experience some form of rapid, localized corrosion in the presence of H_2S.

Some early research in mildly sour environments has suggested that a low concentration of H_2S (<30 ppm) in a CO_2-saturated water solution can accelerate the corrosion rate significantly in comparison to corrosion in a similar CO_2 environment without H_2S (Ikeda et al. 1985). In general, the reasons behind the "H_2S effect" on CO_2 corrosion are not entirely understood. It has been speculated that adsorbed sulfide species and/or sulfide films affect the corrosion rate of mild steel through a catalytic or a galvanic effect (Kvarekval 1997). Brown et al. (2004) studied the effect of low levels of H_2S on CO_2 corrosion and showed that in the absence of protective iron carbonate and iron sulfide scales, very small amounts of H_2S (<10 ppm) in the gas phase can lead to rapid and significant reduction in the CO_2 corrosion rate (Brown et al. 2004). The trend is arrested and somewhat reversed at higher H_2S concentration; an observation that has also been reported elsewhere (Greco and Wright 1962; Videm and Kvarekval 1995). Omar et al. (2005) reported on the impact of high fluid flow velocities on localized corrosion, which is attributed to the local breakdown of the protective films due to the high shear stresses or flow eddies close to the film. This study again highlights the issue of pitting corrosion under conditions where film breakdown may occur (in this case, due to high flow rates).

One major difference between existing literature on CO_2/H_2S corrosion (Rogers and Rowe Jr. 1955; Meyer et al. 1958; Berner 1967; Smith and Miller 1975; Shoesmith et al. 1980; Taylor 1980; Shoesmith 1981; Smith 1993; Smith and Wright 1994; Kermani et al. 2006; Smith and Joosten 2006) and reservoir souring is the timing of which H_2S enters the production system. Typical sour fields contain both H_2S and CO_2 together from the start of production, which is not the case with reservoir souring. In reservoir souring, the producing field is initially under sweet conditions and then transitions into a sour environment as the field matures. To assess the effect of reservoir souring on corrosion rates, the present authors conducted laboratory testing that simulated the reservoir souring process. In these experiments, initial corrosion testing was conducted in a sweet environment, followed by an addition of 50 ppm H_2S for two different CO_2/H_2S gas conditions. The results shown in Table 6.1 demonstrate that the general corrosion rate decreased after reservoir souring. However, significant localized corrosion was observed on coupons for $CO_2/H_2S = 100$ after simulation of the souring effect whereas it was not present for $CO_2/H_2S = 16$. This was attributed to differences in the corrosion scale products formed during these tests. These results indicate that reservoir souring may lead to localized corrosion depending on brine chemistry and gas composition.

Russ and Rainsford (2004) noticed a similar effect of the CO_2/H_2S ratio on localized corrosion in pipelines. They reported that a change in the CO_2/H_2S ratio from 3 mole% CO_2/700 ppm H_2S to 3.75 mole% CO_2/350 ppm H_2S resulted in significant

TABLE 6.1
Lab Corrosion Rates for Two Different CO_2/H_2S Ratios Before and After Souring

	Before Reservoir Souring (Sweet Corrosion Test)	After Reservoir Souring (Mildly Sour Corrosion Test)
Test #A:	Corrosion rate: 55 ± 2 mpy	Corrosion rate: 37 ± 7 mpy
$CO_2/H_2S = 16$	No localized/pitting corrosion	No localized/pitting corrosion
Test #B:	Corrosion rate: 144 ± 11 mpy	Corrosion rate: 44 ± 9 mpy
$CO_2/H_2S = 100$	No localized/pitting corrosion	Significant localized/pitting corrosion

increase in corrosion of oil pipelines. In-line inspection revealed that the section of the line with a CO_2/H_2S partial pressure ratio of 43 had low or nondetectable corrosion while the one that had a ratio of 88 showed localized corrosion areas representing 2.5% of the length of the line with pits 20%–35% of the wall thickness.

In summary, the impact of souring on CO_2/H_2S corrosion is dependent on the dominant acid gas and corrosion scale type, which is a function of the CO_2/H_2S partial pressure ratio. Compared to fully sweet systems, there may be an increased risk for localized corrosion due to the preferential formation of a nonprotective iron sulfide corrosion product, which disrupts the formation of a protective iron carbonate corrosion scale. At the same time, the presence of H_2S can decrease corrosion rate by promoting the formation of protective iron sulfide film in place of a less protective iron carbonate film or through the formation of a combined protective layer of iron sulfide and iron carbonate. To appropriately manage corrosion in souring fields, it is suggested that the impact of reservoir souring on corrosion be assessed through laboratory investigations under simulated operating conditions or through corrosion monitoring in the field.

6.5 ENVIRONMENTALLY ASSISTED CRACKING DUE TO SOURING

While reservoir souring can have a significant impact on corrosion of materials, the evolution of H_2S can have even more severe implications on materials integrity due to environmentally assisted cracking (EAC). Corrosion of steel when exposed to water and H_2S leads to the formation of atomic hydrogen through the following reactions:

$$H_2S + Fe^{2+} \rightarrow FeS + 2H^+ \tag{6.1}$$

$$2H^+ + 2e^- \rightarrow 2H^0 \tag{6.2}$$

The atomic hydrogen that is produced at the steel surface (through Equation 6.2) can either recombine to form molecular hydrogen (H_2) or it can diffuse into the steel. Due the presence of sulfide, the hydrogen recombination reaction is poisoned, thereby promoting more of the nascent hydrogen to diffuse into the metal. Once the hydrogen permeates into the metal, it can cause severe mechanical degradation of the material through a process generally classified as hydrogen embrittlement (Elboujdaini 2006). Various forms of hydrogen embrittlement include hydrogen-induced cracking or stepwise cracking (HIC/SWC), stress-oriented hydrogen-induced cracking (SOHIC), sulfide stress cracking (SSC), and soft zone cracking (SZC). In addition, corrosion fatigue can become accelerated in sour environments.

Hydrogen-induced cracking is related to an increase of internal pressure inside of the steel caused by accumulation of the atomic hydrogen at defects or inclusions, which then recombine to form molecular hydrogen. These types of cracks tend to propagate parallel to the surface of the steel along elongated grains, banded microstructures, or nonmetallic inclusions, which are aligned with the rolling direction of the steel plate. If multiple HIC cracks form in the material, they may link up in a stepwise manner, leading to through-wall cracking (known as SWC). The HIC mechanism does not require any form of stress to occur and depends entirely on the level of hydrogen intake. The steel microstructures that are susceptible to HIC are primarily found in carbon and low-alloy steels whereas they are not commonly present in corrosion-resistant alloys (CRAs). Figure 6.4a shows an example of HIC/SWC in a piece of carbon steel tubing where the hydrogen accumulated around the mid-wall as a result of centerline banding of the steel's microstructure during processing. Similar to HIC, SOHIC is due to the build-up of molecular hydrogen inside of the steel. Unlike HIC, however, SOHIC does require a source of stress to occur. In this case, microfissures caused by the trapped molecular hydrogen are linked up in a direction perpendicular to the applied or residual stress, causing through-wall failure of the material (Elboujdaini 2006).

Sulfide stress cracking is due to embrittlement (i.e., a reduction in fracture toughness) of the steel as a direct result of the absorbed atomic hydrogen. In sour, corrosive environments, this type of hydrogen embrittlement causes fracture of the steel under tensile loading at stress levels much less than would be required to fracture the material in an inert environment. Susceptibility to SSC is heavily linked to the material's mechanical properties, such as strength and hardness. Figure 6.4b illustrates SSC through a seam weld region of the same carbon steel tubing as in Figure 6.4a, where the welding fusion zone contained high-hardness regions susceptible to hydrogen embrittlement. The SSC mechanism can occur in carbon and low-alloy steels as well as in CRAs. Soft zone cracking is a form of SSC that can occur when a steel contains a local area of lower yield strength that is plastically deformed, leading to increased sensitivity to SSC. This form of cracking is usually associated with welded microstructures in carbon steels (Renpu 2011).

Corrosion fatigue refers to the initiation and propagation of cracking as a result of cyclic loading in a corrosive media. In sour service, hydrogen embrittlement enhances fatigue crack growth, leading to a reduction in fatigue strength (i.e., number of cycles that the material can endure for a given cyclic stress range) and potential elimination of an endurance limit (i.e., cyclic stress range below which fatigue failure will not occur). This type of failure is of particular concern for carbon steel flowlines and risers used in offshore applications that are exposed to wave- and current-induced motions (Buitrago et al. 2008).

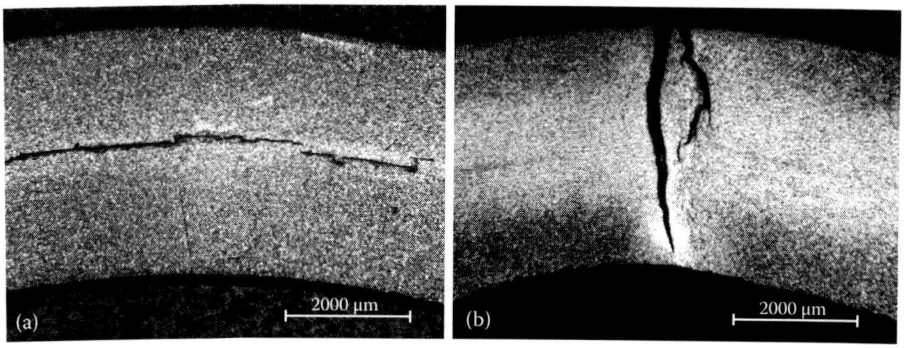

FIGURE 6.4 Examples of (a) centerline HIC and (b) SSC in welded carbon steel tubing.

6.6 DEFINING THE SEVERITY OF RESERVOIR SOURING FOR MATERIALS SELECTION

Since each of the aforementioned forms of EAC in the previous section are dependent on the level of hydrogen entering the material, the severity of the sour environment is most commonly defined by the pH and H_2S partial pressure (P_{H_2S}), which have been demonstrated to control the level of absorbed hydrogen in steels (Asahi et al. 1994; Asahi and Nose 1999). In fact, the standard used by the oil and gas industry for materials selection in sour service, ANSI/NACE MR0175/ISO 15156 (2015), defines various regions of SSC environmental severity based on these two parameters as illustrated in Figure 6.5. For CRAs, it should be noted that the standard also defines maximum chloride concentration limits since salinity plays a significant role in governing cracking of these materials. The H_2S partial pressure below which an operating environment is typically defined as "non-sour" is less than 0.3 kPa (0.05 psi). However, as is discussed in subsequent sections, precautions still need to be made for materials with very high levels of susceptibility to either HIC- or SSC-type failure mechanisms.

In order to directly apply the standard (to be subsequently referred to here as NACE MR0175) to define the environmental severity levels that can result from the effects of reservoir souring, it is necessary to relate microbial H_2S generation in the water phase to the equilibrated concentration, or mole fraction, of H_2S in the gas phase. If the mole fraction of H_2S (X_{H_2S}) is known, then P_{H_2S} can be calculated using the following equations provided in NACE MR0175 Annex C:

$$P_{H_2S} = P_{total} \times X_{H_2S} \tag{6.3a}$$

$$P_{H_2S} = P_{bubble\ point} \times X_{H_2S} \tag{6.3b}$$

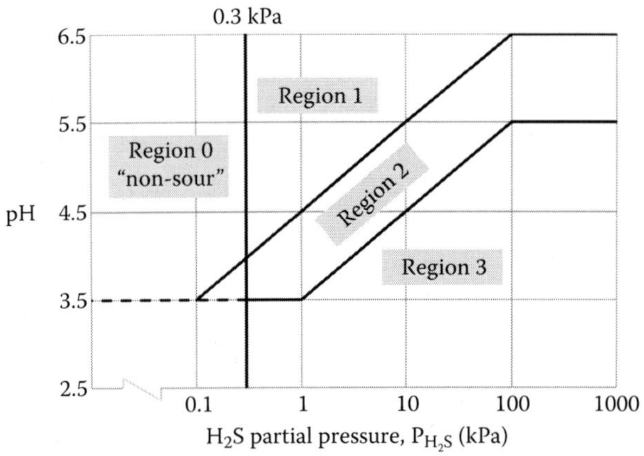

FIGURE 6.5 Environmental severity regions for SSC as defined by ANSI/NACE MR0175/ISO 15156. (Adapted from ANSI/NACE MR0175/ISO 15156. 2015.)

where Equation 6.3a is used for systems with an equilibrium gas phase (P_{total} is the total absolute pressure of the system) and Equation 6.3b is used for gas-free liquid systems ($P_{bubble\ point}$ is the bubble point pressure). However, translation of the H_2S production from reservoir souring to the expected H_2S gas partial pressures throughout a production facility is not trivial. Unless H_2S concentration is able to be physically measured in the field, reservoir souring models are required to first determine the amount of H_2S generated at reservoir conditions. Then, the H_2S gas partial pressure at a specific point in production (downhole, pipelines, and topside) is calculated using equilibrium thermodynamic models at the given operating conditions (Vera et al. 2015). Prediction of the H_2S concentrations in the produced fluids is very complex as it depends on several characteristics of the reservoir, including the total system pressure, temperature, oil-to-gas ratios, water cut, and pH (Schofield and Stott 2012). Moreover, H_2S partitioning between the oil and water phases must be considered. A good example of the complexities in determining the severity of souring is the effect of system pressure. While an increase in total system pressure would tend to increase the effective H_2S partial pressure (per Equation 6.3), it would also tend to decrease SRB activity and H_2S generation in the reservoir (Dunsmore et al. 2006). Thus, higher-pressure systems do not necessarily equate to more severe sour environments. For relatively low-pressure systems (e.g., topsides), reservoir souring generally transitions a nonsour environment from Region 0 to Region 1 or Region 2. For higher pressure environments (e.g., downhole), it is possible that even small amounts of H_2S can lead to severely sour conditions (Region 3), particularly in the presence of a low pH, non-buffered water phase (e.g., condensed water).

6.7 ENVIRONMENTAL CRACKING OF MATERIALS IN MILDLY SOUR ENVIRONMENTS

This section provides examples of lab studies and field cases that demonstrate the impact of mildly sour conditions on environmental cracking susceptibility of various oilfield materials. These examples highlight the concern of reservoir souring conditions for material applications that were initially intended for sweet service.

6.7.1 Downhole Materials

High-strength carbon and low-alloy steels used for downhole casing and tubing are very susceptible to hydrogen embrittlement even under mild H_2S conditions. Therefore, careful consideration of potential reservoir souring levels must be considered for downhole materials selection. Skogsberg (2004) performed fit-for-purpose testing on various non-sour service grade high-strength, low-alloy steels (8630M, P-110, 4130M, and Q-125) selected for well design in deep water Gulf of Mexico, which was initially considered to be in sweet service. Atmospheric laboratory test conditions were P_{H_2S} = 0.55 kPa (0.08 psi) at pH = 4–6 to simulate 20 ppm H_2S gas at actual well pressure. Uniaxial tensile testing showed that all alloys with the exception of P-110 were susceptible to cracking under these test conditions and thus not suitable for service under the assumed souring conditions. Sulfide stress fracture mechanics testing on a Cr-Mo type P-110 in mildly sour environments showed similar SSC fracture toughness values (K_{ISSC}) at conditions of P_{H_2S} = 0.34 kPa (0.05 psi)/pH = 4 and P_{H_2S} = 3.4 kPa (0.5 psi)/pH = 5.0, suggesting that a unit of pH change has the same effect on SSC as a magnitude of H_2S partial pressure change at mildly sour conditions (Huang et al. 2012).

In addition to carbon and low-alloy steels, some CRAs used for downhole applications may also be susceptible to hydrogen embrittlement in mildly sour conditions. Stainless steels containing ferritic and martensitic microstructures tend to have higher susceptibility to SSC compared to stainless steels that are fully austenitic. A study on SSC of 13% Cr martensitic stainless steels (commonly used as production tubing in highly corrosive, sweet environments) showed cracking of a particular American Petroleum Institute (API) grade 13% Cr steel chemistry (having high carbon and low Ni) at P_{H_2S} = 1.0 kPa (0.15 psi) and pH ≤ 4 (Hara and Asahi 2000). In this case, hydrogen embrittlement was related to depassivation of the stainless steel and subsequent hydrogen ingress under the given pH conditions. Precipitation hardening (performed to increase material strength) of martensitic stainless steels greatly increases SSC susceptibility, especially if the alloy chemistry and heat treatment are not carefully controlled. A review of SSC testing on 17-4 PH, a martensitic stainless steel with Cu precipitation, under P_{H_2S} = 4.1 kPa (0.6 psi) showed failure of the material even at applied stress levels much lower than the material's yield strength (Cassagne et al. 2003). While NACE MR0175 currently permits the use of 17-4 PH steel within H_2S and pH limits of P_{H_2S} ≤ 3.4 kPa (0.5 psi) and pH ≥ 4.5, precautions still need to be taken to avoid the overstressing of these materials in mildly sour conditions.

6.7.2 Line Pipe Steels

Due to their relatively low strength and hardness, line pipe steels are regarded as more resistant to sour cracking than the higher strength grades used for downhole applications. In industry practice, environmental cracking is typically of secondary concern for very mild H_2S conditions with corrosion being a primary focus. However, both SSC and HIC may pose a threat to these materials even at low H_2S concentrations. According to NACE MR0175, the hardness of carbon and low-alloy steels and weldments should be kept below 22 HRC (250 H_V) to avoid cracking in sour environments. However, local hard zones can develop during steel processing and/or welding, which lead to enhanced crack susceptibility. Treseder and Badrak (1997) described a study on API 5L X60 pipe in mildly sour environments that showed that local hard zones above 327 H_V and 340 H_V were susceptible to SSC at pH = 4 and H_2S partial pressures above 2 kPa (0.3 psi) and 0.7 kPa (0.1 psi), respectively. Smith et al. (2013) demonstrated the potential for increased HIC susceptibility in heritage line pipe steels (compared to modern seam-welded line pipe) in mildly sour conditions due to the higher impurity levels attained from older steel processing routes. The authors defined a preliminary safe operation limit of P_{H_2S} = 0.5 kPa (0.07 psi) at pH = 5.2 with a lower threshold H_2S partial pressure anticipated at lower pH. Cold working, which may be performed intentionally to increase steel strength (for sweet service applications) or may be an unintentional result of mechanical damage, can also lead to increased SSC susceptibility (Treseder and Swanson 1968; Mack et al. 2000; Sourmail 2007). Strain-based design pipelines, which are purposely designed to withstand plastic deformation, are not intended for sour service. In cases of unexpected reservoir souring, the effects of cold work or plastic deformation on sweet service line pipe steels should be reevaluated through fit-for-service assessments.

6.7.3 Risers and Flowlines

Steel catenary risers (SCR) and flowlines are used to tie back production from subsea pipelines to a central facility, such as a floating production storage and offloading system (FPSO) or fixed oil production platform. Corrosion fatigue is of particular concern for riser and flowline components in sour, offshore applications. Gui et al. (2012) studied the effect of sour environments on the corrosion fatigue of API 5L X65 (C-Mn) pipe steel. Tests performed in solutions of P_{H_2S} = 0.83 kPa (0.12 psi) and pH = 5 showed an increase in fatigue crack growth rate (FCGR) from 40 to 130 times that as experienced in air with the FCGR increasing with decreasing frequency of the cyclic loading. Enhanced crack growth rates were attributed mainly to bulk hydrogen charging from the test solution with some contribution also coming from trapped hydrogen in the steel. Buitrago et al. (2008) assessed the sour corrosion fatigue performance of girth welds commonly used for connections on SCRs and flowlines. Even at relatively mild conditions of P_{H_2S} = 0.63 kPa (0.09 psi) and pH = 4, knockdown factors on fatigue life were found to be 10–100 (i.e., 10–100 less than the fatigue life in air) with failures occurring at the weld root toe. It is evident from studies such as these that the combination of cyclic loading and hydrogen embrittlement can significantly exacerbate fracture of risers and flowlines under mildly sour conditions.

6.8 MATERIALS SELECTION FOR SOUR SERVICE

Selection of materials for downhole, wellhead, and subsea applications is typically controlled by the need for resistance to both corrosion and cracking. For low-to-moderate corrosive conditions, low-alloy carbon steels with continuous inhibition and H_2S scavengers have proven to be successful in certain conditions. However, such systems can often prove impractical or too costly over project life—in these cases, it may be more economical (in terms of total lifecycle costs) to upgrade materials to CRAs. For proper materials selection to avoid environmental cracking, the NACE MR0175 standard provides a list of acceptable sour service grade materials for various levels of H_2S severity—Part 2 for carbon and low-alloy steels and Part 3 for CRAs. Generally, if a material is selected to operate within its sour service envelope, no failure should be expected. Therefore, reservoir souring predictions are important to identify the potential presence and maximum amount of H_2S generation to ensure proper materials selection. In the absence of accurate reservoir souring data or models, conservative estimates on H_2S generation are warranted. If unexpected levels of reservoir souring cause an environment to approach a material's sour service limit, fit-for-service testing is recommended as is described in the following section. In addition to its effects on metallic alloys, H_2S can cause severe degradation to nonmetallic elastomers and increase the propensity for elastomer embrittlement and explosive decompression (Ray and Ivey 1979; Williford et al. 1999). For potential reservoir souring conditions, it is generally recommended to select seals and elastomers based on sour environments using standards such as NORSOK M-710 (2014) or NACE TM0187 (2011).

6.9 FIT-FOR-SERVICE TESTING OF MATERIALS

In the context of reservoir souring, fit-for-service (FFS) testing typically refers to the need to demonstrate the integrity of a component operating in a sour environment for which it was not originally designed. This type of evaluation is especially critical for non-sour service grade materials that have limited SSC resistance. While there are well-defined test methods for the sour service qualification testing of materials, adaptation of these methods to simulated reservoir souring conditions requires certain precautions.

6.9.1 Sulfide Stress Crack Testing

High-strength carbon and low-alloy steels used for downhole applications are typically qualified for sour service using one of the SSC test methods described in the ANSI/NACE TM0177 (2016) standard. The methods contained within the standard (to be subsequently referred to here as NACE TM0177) are classified as either constant load or constant displacement tests (e.g., tensile, bent beam, and C-ring tests) or fracture mechanics tests (e.g., double cantilever beam [DCB]). The former type provides qualitative pass/fail results based on whether the material shows signs of SSC initiation, whereas the latter provides a quantitative measure of SSC fracture toughness (commonly denoted as K_{ISSC}). The standard test environment prescribed in NACE TM0177 is acidified brine saturated by 100% H_2S at 75°F and ambient pressure. For SSC tests, NACE TM0177 recommends use of either Solution "A" with an initial pH of 2.6–2.8, or Solution "B" with an initial pH of 3.4–3.6. While these standard test environments are useful for quality assurance testing, they do not represent service conditions generally associated with reservoir souring. Therefore, materials performance evaluation in the presence of reservoir souring requires an accurate simulation of the expected environment using anticipated worst-case SSC conditions (highest P_{H_2S}, lowest pH, and lowest temperature).

As discussed earlier, the SSC mechanism is governed by the absorption of hydrogen, which is related to both P_{H_2S} and pH of the environment. It has been demonstrated that the effect of both P_{H_2S} and pH on SSC resistance of high-strength materials can be significant under mild H_2S conditions, whereas P_{H_2S} tends to dominate SSC response under more severe sour conditions (Huang et al. 2013). Thus, for FFS testing in mildly sour environments, both the P_{H_2S} and pH need to be carefully controlled for the given field conditions. In addition, the type (e.g., chemistry and crystal structure) and morphology (e.g., porous vs. compact) of corrosion scale that is formed on the surface of the test specimen will vary depending on the specific test environment. This, in turn, can affect hydrogen ingress into the steel, resulting in different apparent SSC toughness measurements. To attain the minimum K_{ISSC} for a given FFS test condition, the specimen exposure time in solution should be chosen to ensure bulk hydrogen saturation in the steel and allow for hydrogen interaction with the growing crack front. This exposure time may be longer than those prescribed in qualification test standards. The required exposure times to achieve hydrogen saturation may be estimated through hydrogen permeation testing defined by ASTM G148-97 (2011) by measuring the time to reach steady-state hydrogen flux through a particular steel grade when exposed to the test environment.

For CRAs, the slow strain rate (SSR) test method defined by NACE TM0198 (2011) is sometimes used to evaluate SSC resistance. In this type of test, specimens are dynamically strained at a very slow, constant extension rate (~10^{-9} to 10^{-7} m/s) until failure. The SSC resistance is then measured by comparing results of the test in the corrosive environment to that in air (e.g., defined by ratios in time to failure, plastic strain to failure, reduction in area ratio, etc.). This test is useful for CRAs since plastic straining accelerates the disruption of the passive film, which is required for hydrogen diffusion into the material. However, because application of such strains is typically unrealistic in actual operations, the testing is best suited as a screening tool to compare the SSC behavior of different CRA materials rather than as an absolute measurement. For SSR testing, the extension rate must be chosen to be slow enough to ensure the hydrogen embrittlement process takes place, allowing for the worst-case materials performance to be captured. As the H_2S concentration of the test environment decreases, slower extension rates may need to be used to allow for the incubation time required for hydrogen embrittlement.

6.9.2 HIC Testing

While the oil and gas industry has put considerable focus into the effects of souring on SSC performance of high-strength steels used for high pressure (e.g., downhole) applications, the effect of low levels of H_2S on the HIC resistance of lower strength steels (e.g., for pipelines and pressure vessels) is not well understood. NACE MR0175 requires that the user consider HIC resistance in materials selection even for cases where trace amounts of H_2S are present. Testing for HIC is defined by ANSI/NACE TM0284 (2011) and involves exposing a section of unstressed material to one of the two NACE standard test Solutions "A" or "B" as defined in the previous section. HIC susceptibility is then measured by examining specified cross-sectional areas of the test specimen and comparing the length, width, and area of the HIC regions to the respective dimensions of the examined cross-section. The standard test exposure time of 96 hours defined by the test standard (to be subsequently referred to here as NACE TM0284), however, may be inadequate for assessing HIC susceptibility in souring environments with mild H_2S concentrations. In fact, NACE TM0284 explicitly states that the standard method is not intended to duplicate service conditions as is done in FFS testing. Kittel et al. (2010) performed hydrogen permeation testing to demonstrate that the required time to reach the critical hydrogen concentration for HIC depends on the severity of the environment as determined by the P_{H_2S} and pH. The data from their HIC testing suggests that up to 4 weeks may be needed to initiate cracking for lower H_2S partial pressures and high pH combinations.

6.9.3 Fatigue Testing

Unlike for SSC and HIC testing, there are no well-defined test standards for performing fatigue testing under sour service conditions. In general, conventional fatigue testing approaches developed for inert environments are adapted for sour service testing with some potential alterations made to specimen design and testing equipment to account for the sour conditions (McMaster et al. 2008). Sour service fatigue data taken on small-scale base or weld metal specimens (e.g., compact tension, "dog-bone" strip specimens) is then used to define a "knockdown" factor for the reduction in fatigue life of the material in the sour service environment as compared to in an inert environment. Like other environmental fracture tests, steps need to be taken to ensure that the test specimen is saturated with hydrogen and that enough time is allowed for hydrogen embrittlement mechanisms to take place during the course of the test. For fatigue testing, hydrogen saturation is typically achieved by exposing the specimen to the test solution for several hours or days prior to running the fatigue test. After charging the specimen with hydrogen, fatigue testing is immediately performed either in the same test solution in which it was charged or the specimen is removed from the test solution and tested in air. In the latter case, extreme care needs to be taken that considerable amounts of diffusible hydrogen are not removed from the specimen once exposed to air. If this occurs, the FCGR will be measured to be much lower than expected in actual sour service conditions (Gui et al. 2012). In addition, corrosion fatigue rates are also highly sensitive to the loading frequency. Fatigue crack growth rates in sour environments tend to increase with decreasing test frequency since lower frequencies allow more time for hydrogen embrittlement effects to take place between loading cycles. Unless a realistic loading frequency for a given application is known, the test frequency should be chosen to provide the maximum FCGR and thus provide a conservative measure of the knockdown factor on fatigue life for a specific H_2S environment.

6.9.4 Simulating a Fit-for-Service H_2S Environment

Materials performance evaluation requires an accurate simulation of field conditions in the laboratory, which typically includes partial pressure of acid gases, brine chemistry, pH, and temperature. Generally, FFS tests are conducted in anticipated worst-case conditions expected for the anticipated damage mechanism(s). For example, in SSC testing, worst-case test conditions are typically the highest H_2S partial pressure, lowest temperature, and lowest solution pH. Conventional FFS tests simulate only the H_2S and CO_2 partial pressures that exist at field operating conditions. In this case, the H_2S (or CO_2) partial pressures used for testing are calculated by multiplying the mole fraction of H_2S (or CO_2) in the gas by the total operating pressure. However, it is well accepted that the H_2S in the gas phase does not directly lead to hydrogen embrittlement mechanisms, such as SSC. Rather, it is the H_2S dissolved in the aqueous phase that enables the hydrogen embrittlement processes to initiate through the chemical reactions that occur during corrosion. As such, FFS laboratory tests should be conducted so that the aqueous test solution contains the same concentration of dissolved H_2S as is expected to exist at actual operating conditions—this is particularly important for high-pressure applications (e.g., downhole) as will be discussed.

Under ideal conditions, solubility (or equilibrium concentration) of a gas in an aqueous solution is directly proportional to the partial pressure of that gas as defined by Henry's law. Application of Henry's law is limited to dilute systems at relatively low pressure. As pressure and solute concentration increase, non-ideal behavior of both the gas and water phases begin to affect solubility. Three corrections need to be made to account for these non-ideal behaviors of gas and water phases. These are based on the fugacity coefficient (ϕ), the activity coefficient (γ), and Poynting's correction $\left(v_i^\infty\right)$ for changes in partial molar volume of the solutes at infinite dilution. For high pressure–high temperature (HPHT) testing, Kumar et al. (2014) proposed a methodology to select representative test conditions for fit-for-service material evaluations. They developed a thermodynamic model based on the ensemble Henry's law and showed that laboratory testing conducted using partial pressure of H_2S at ambient pressure may be more severe (i.e., result in more dissolved H_2S) than actual oil and gas fields with high total pressure:

$$\gamma_i x_i H_{ij} \exp\left[\int_{p_j^o}^{p}\left(\frac{v_i^\infty}{RT}\right)dp\right] = \phi_i y_i P_{tot} \tag{6.4}$$

where γ_i is the activity coefficient; H_{ij} is the Henry constant; x_i and y_i are the mole fractions of the H_2S in water and vapor phases, respectively; P_{tot} is the total pressure; ν^∞ is molar volume; R is the universal gas constant; and T is temperature. Subscripts *i* and *j* represent solute (i.e., H_2S) and solvent (i.e., water), respectively.

Figure 6.6 compares the different forms of Henry's law in a high-pressure system. It can be seen that, at the same total pressure, differences in the calculated

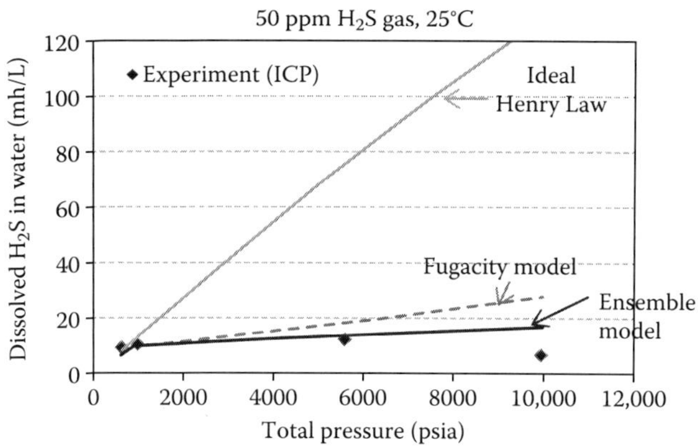

FIGURE 6.6 Comparison of predicted dissolved H_2S using three different thermodynamic models (conditions: 50 ppm H_2S gas at a temperature of 25°C). Experimentally measured H_2S was performed using inductively coupled plasma (ICP) analysis. (Reproduced from Kumar, A., Pacheco, J. L., Desai, S. K., Huang, W., Reddy, R. V., Sun, W. and Haarseth, C. A., Selecting representative laboratory test conditions for mildly sour sulfide stress corrosion (SSC) testing *CORROSION 2014*, San Diego, TX: NACE International, 2014.)

concentration of dissolved H_2S in solution can be significant depending on the form of Henry's law that is used. For example, at a total pressure of 8000 psi, the dissolved H_2S content using ideal Henry's law is nearly four times and eight times those obtained using the fugacity-corrected Henry's law and ensemble Henry's law, respectively. Using the ideal Henry's law to predict dissolved H_2S in high-pressure systems can lead to significant errors and be overly conservative. For FFS testing in mildly sour conditions, caution should be taken to maintain the H_2S concentrations at the desired levels. In some cases, H_2S may need to be replenished based on consumption rate due to corrosion. Furthermore, special precautions are needed to avoid significant pH drift during the test (e.g., use of buffers, solution recirculation, etc.).

6.10 ASSET INTEGRITY MANAGEMENT FOR SOURING FIELDS

As discussed throughout this chapter, effective asset integrity management of a production unit that is expected to, or is experiencing, reservoir souring starts with strategies for reservoir souring prediction and/or monitoring for H_2S production in the field. Once the level of reservoir souring is determined, various approaches to souring mitigation (Section 6.3) can be evaluated to prevent or minimize the extent of H_2S production, and/or the sour production fluids can be handled through a robust corrosion management program (Section 6.4), appropriate materials selection (Sections 6.5 through 6.8), and FFS performance evaluation (Section 6.9). For fields that are expected to sour, it is important that facilities include multiple sampling points for fluid/gas compositional analysis, which provide the information on field conditions required for accurate lab-based FFS evaluations of material corrosion or cracking susceptibility. Another good practice in asset design is to include a provision for access ports throughout the facility to allow future retrofit of additional or alternative corrosion monitoring probes to track the effect of souring and iron-sulfide scaling on production system integrity (Jordan et al. 2005). Additionally, asset integrity management should include an adequate plan to baseline nondestructive testing (NDT) measurements for subsequent reference should corrosion risks for particular components change following the transition from sweet to sour production. With respect to this point, it should be confirmed that the NDT tools are effective in the presence of any corrosion scale products (e.g., iron sulfides) that may have been formed on the surfaces of components to be inspected. Through proactive asset integrity management of greenfield or brownfield developments affected by reservoir souring, capital and operating costs can be optimally managed while avoiding the risk of hazardous and costly failures during the intended lifecycle of the asset.

REFERENCES

Aften, C. W. and Roberts, G. 2011. New Compounds for Hydrogen Sulfide Scavenging and Iron Sulfide Control. In *SPE International Symposium on Oilfield Chemistry*. Society of Petroleum Engineers.

Anderko, A. and Shuler, P. J. 1998. Modeling the Formation of Iron Sulfide Scales Using Thermodynamic Simulation Software. In *CORROSION 1998*. San Diego, CA: NACE International.

ANSI/NACE MR0175/ISO 15156. 2015. Petroleum, Petrochemical, and Natural Gas Industries—Materials for Use in H_2S-Containing Environments in Oil and Gas Production. NACE International.

ANSI/NACE TM0177. 2016. Laboratory Testing of Metals for Resistance to Sulfide Stress Cracking and Stress Corrosion Cracking in H_2S Environments. NACE International.

ANSI/NACE TM0284. 2011. Evaluation of Pipeline and Pressure Vessel Steels for Resistance to Hydrogen-Induced Cracking. NACE International.

Asahi, H. and Nose, K. 1999. Effects of Environmental Conditions on SSC resistance of C125 OCTG. In *CORROSION 1999*. San Antonio, TX: NACE International.

Asahi, H., Ueno, M. and Yonezawa, T. 1994. Prediction of sulfide stress cracking in high-strength tubulars. *Corrosion Science*, *50*(7), 537–545.

ASTM G148-97. 2011. Standard Practice for Evaluation of Hydrogen Uptake, Permeation, and Transport in Metals by an Electrochemical Technique. ASTM International.

Bader, M. S. H. 2007. Sulfate removal technologies for oilfields seawater injection operations. *Journal of Petroleum Science and Engineering*, *55*(1), 93–110.

Beeder, J., Andersen, T. R., Liengen, T., Drønen, K. and Torsvik, T. 2007. Corrosion as a side effect during nitrate treatment of produced water and aquifer water injection. In *CORROSION 2007*. Nashville, TN: NACE International.

Berner, R. A. 1967. Thermodynamic stability of sedimentary iron sulfides. *American Journal of Science*, *265*(9), 773–785.

Brown, B. and Nešic, S. 2005. CO_2/H_2S Corrosion Under Scale Forming Conditions. In *CORROSION 2005*. Houston, TX: NACE International.

Brown, B., Parakala, S. R. and Nešic, S. 2004. CO_2 Corrosion in the Presence of Trace Amounts of H_2S. In *CORROSION 2004*. New Orleans, LA: NACE International.

Buitrago, J., Hudak, S. and Baxter, D. 2008. High-Cycle and Low-Cycle Fatigue Resistance of Girth Welds in Sour Service. In *ASME 2008 27th International Conference on Offshore Mechanics and Arctic Engineering*. American Society of Mechanical Engineers.

Burger, E. D. and Jenneman, G. E. 2009. Forecasting the Effects of Reservoir Souring From Waterflooding a Formation Containing Siderite. In *SPE International Symposium on Oilfield Chemistry*. Society of Petroleum Engineers.

Cassagne, T., Bonis, M., Duret, C. and Crolet, J. L. 2003. Limitations of 17-4 PH Metallurgical, Mechanical and Corrosion Aspects. In *CORROSION 2003*. San Diego, CA: NACE International.

Choi, Y. S., Nešic, S. and Ling, S. 2011. Effect of H_2S on the CO_2 corrosion of carbon steel in acidic solutions. *Electrochimica Acta*, *56*(4), 1752–1760.

Dunlop, A. K., Hassell, H. L. and Rhodes, P. R. 1983. Fundamental Considerations in Sweet Gas Well Corrosion. In *CORROSION 1983*. Anaheim, CA: NACE International.

Dunsmore, B., Evans, P. J., Jones, M., Burton, S. and Lappin-Scott, H. M. 2006. When Is Reservoir Souring a Problem for Deepwater Projects? In *Offshore Technology Conference*. Offshore Technology Conference.

Elboujdaini, M. 2006. Hydrogen-Induced Cracking and Sulfide Stress Cracking. In *Uhlig's Corrosion Handbook*, ed. R. W. Revie, 205–220. Hoboken, NJ: John Wiley & Sons, Inc.

Enning, D. and Garrelfs, J. 2014. Corrosion of iron by sulfate-reducing bacteria: New views of an old problem. *Applied and Environmental Microbiology*, *80*(4), 1226–1236.

Evans, P., Nederlof, E. and Richmond, W. 2015. Souring Development Associated with PWRI in a North Sea Field. In *SPE Produced Water Handling & Management Symposium*. Society of Petroleum Engineers.

Gadd, G. M. 2010. Metals, minerals and microbes: Geomicrobiology and bioremediation. *Microbiology*, *156*(3), 609–643.

German, M. 2015. Reviewing Waterflood Projects Over 12 Years: What Have We Learned? In *SPE Kuwait Oil and Gas Show and Conference*. Society of Petroleum Engineers.

Gieg, L. M., Jack, T. R. and Foght, J. M. 2011. Biological souring and mitigation in oil reservoirs. *Applied Microbiology and Biotechnology*, *92*(2), 263–282.

Gittel, A., Sørensen, K. B., Skovhus, T. L., Ingvorsen, K. and Schramm, A. 2009. Prokaryotic community structure and sulfate reducer activity in water from high-temperature oil reservoirs with and without nitrate treatment. *Applied and Environmental Microbiology*, *75*(22), 7086–7096.

Greco, E. C. and Wright, W. B. 1962. Corrosion of Iron in an H_2S-CO_2-H_2O System. *Corrosion*, *18*(3), 119t–124t.

Gui, F., Ramgopal, T. and Muller, M. G. 2012. Role of sour environments on the corrosion fatigue growth rate of X65 pipe steel. *Corrosion*, *68*(8), 730–738.

Hara, T. and Asahi, H. 2000. Conditions under which cracks occur in modified 13% chromium steel in wet hydrogen sulfide environments. *Corrosion*, *56*(5), 533–542.

Haveman, S. A., Greene, E. A., Stilwell, C. P., Voordouw, J. K. and Voordouw, G. 2004. Physiological and gene expression analysis of inhibition of *Desulfovibrio vulgaris* Hildenborough by nitrite. *Journal of Bacteriology*, *186*(23), 7944–7950.

Hegdal, T. and Pinchin, D. 2015. Subsea Membrane Water Treatment Progresses Towards Application. In *Offshore Technology Conference*.

Huang, W., Asmann, M. and Asher, S. 2012. Evaluation of a Cr-Mo Type P110 Steel for Application in Mild Sour Environments. In *CORROSION 2012*. Salt Lake City, UT: NACE International.

Huang, W., Asmann, M., Cao, F., Srinivasan, R., Ma, N., Mueller, R. and Ayer, R. 2013. SSC Resistance and Cracking Behavior of 125ksi Grade Low Alloy OCTG in Sour Environments. In *CORROSION 2013*. Orlando, FL: NACE International.

Hubert, C. 2010. Microbial Ecology of Oil Reservoir Souring and Its Control by Nitrate Injection. In *Handbook of Hydrocarbon and Lipid Microbiology*, eds. K. Timmis, T. McGenity, J. R. van der Meer, and V. de Lorenzo, 2753–2766. Heidelberg: Springer-Verlag.

Hubert, C., Voordouw, G. and Mayer, B. 2009. Elucidating microbial processes in nitrate-and sulfate-reducing systems using sulfur and oxygen isotope ratios: The example of oil reservoir souring control. *Geochimica et Cosmochimica Acta*, *73*(13), 3864–3879.

Ikeda, A., Ueda, M. and Mukai, S. 1985. Influence of Environmental Factors on Corrosion in CO_2 Source Well. In *Advances in CO_2 Corrosion*, Vol. 2, 1–22. Houston, TX: NACE International.

Jones, C., Downward, B., Edmunds, S., Curtis, T. and Smith, F. 2011. A Novel Approach To Using THPS for Controlling Reservoir Souring. In *CORROSION 2011*. Houston, TX: NACE International.

Jordan, L. C., Landry, J. P., Duhon, H. and Gibson, G. T. 2005. The Impact of Reservoir Souring upon Decision Processes Made During the Design of New Deepwater Developments. In *CORROSION 2005*. Houston, TX: NACE International.

Keasler, V., Bennett, B., Keller, C., Kuijvenhoven, C., Mahon, T. and James, S. 2012. Multi-Faceted Approach for Optimizing a Microbial Control Program. In *SPE International Conference & Workshop on Oilfield Corrosion*. Society of Petroleum Engineers.

Kelland, M. A. 2014. *Production chemicals for the oil and gas industry*, 2nd Edition. Boca Raton, FL: CRC Press/Taylor & Francis.

Kermani, B., Martin, J. W. and Esaklul, K. A. 2006. Materials Design Strategy: Effects of H_2S/CO_2 Corrosion on Materials Selection. In *CORROSION 2006*. San Diego, CA: NACE International.

Kittel, J., Smanio, V., Fregonese, M., Garnier, L. and Lefebvre, X. 2010. Hydrogen induced cracking (HIC) testing of low alloy steel in sour environment: Impact of time of exposure on the extent of damage. *Corrosion Science*, *52*(4), 1386–1392.

Kucka, L., Müller, I., Kenig, E. Y. and Górak, A. 2003. On the modelling and simulation of sour gas absorption by aqueous amine solutions. *Chemical Engineering Science*, *58*(16), 3571–3578.

Kumar, A., Pacheco, J. L., Desai, S. K., Huang, W., Reddy, R. V., Sun, W. and Haarseth, C. A. 2014. Selecting representative laboratory test conditions for mildly sour sulfide stress corrosion (SSC) testing. In *CORROSION 2014*. San Diego, TX: NACE International.

Kvarekval, J. 1997. The Influence of Small Amounts of H_2S on CO_2 Corrosion of Iron and Carbon Steel. In *EUROCORR '97*. European Corrosion Congress.

Kvarekval, J., Dugstad, A., Omar, I. H. and Gunaltun, Y. 2005. H_2S Corrosion of Carbon Steel under Simulated Kashagan Field Conditions. In *CORROSION 2005*. Houston, TX: NACE International.

Larsen, J. 2002. Downhole nitrate applications to control sulfate reducing bacteria activity and reservoir souring. In *CORROSION 2002*. Denver, CO: NACE International.

Larsen, J., Tang, L., Thomsen, U. S. and Sørensen, K. B. 2014. Testing Continuous Biocide Injection in a Seawater Injection Scenario. In *CORROSION 2014*. San Antonio, TX: NACE International.

Littman, E. S. and McLean, T. L. 1987. Chemical control of biogenic H_2S in producing formations. In *SPE Production Operations Symposium*. Society of Petroleum Engineers.

Mack, R. D., Kendziora, L., Filippov, A. G. and Ring, L. 2000. In-Situ Expansion of Casing and Tubing: Effect on Mechanical Properties and Resistance to Sulfide Stress Cracking. In *CORROSION 2000*. Orlando, FL: NACE International.

Macleod, N., Bryan, T., Buckley, A. J., Talbot, R. E. and Veale, M. A. 1994. A Novel Biocide for Oilfield Applications. Society of Petroleum Engineers.

Madsen, H. T., Jensen, C. V. and Søgaard, E. G. 2014. Triazine-based H_2S scavenging: Development of a conceptual model for the understanding of fouling formation. *Petroleum Science and Technology*, *32*(23), 2803–2806.

Maxwell, S. and Spark, I. 2005. Souring of Reservoirs by Bacterial Activity during Seawater Waterflooding. In *SPE International Symposium on Oilfield Chemistry*. Society of Petroleum Engineers.

McMaster, F., Bowman, J., Thompson, H., Zhang, M. and Kinyon, S. 2008. Sour Service Corrosion Fatigue Testing of Flowline and Riser Welds. In *ASME 2008 27th International Conference on Offshore Mechanics and Arctic Engineering*. American Society of Mechanical Engineers.

Mendez, C., Magalhaes, A. A. O., Altoe, P. and Acosta, E. 2009. H_2S Scavengers Injection: A Novel Evaluation Protocol to Enhance the Integrity of Production Lines. In *Offshore Technology Conference*.

Meyer, F. H., Riggs, O. L., McGlasson, R. L. and Sudbury, J. D. 1958. Corrosion products of mild steel in hydrogen sulfide environments. *Corrosion*, *14*(2), 69–75.

Mitchell, A. F., Anfindsen, H., Harvik, A. M. B. and Hustad, B. M. 2010. A Review of Reservoir Souring for Three North Sea Fields. In *CORROSION 2010*. San Antonio, TX: NACE International.

Muyzer, G. and Stams, A. J. 2008. The ecology and biotechnology of sulphate-reducing bacteria. *Nature Reviews Microbiology*, *6*(6), 441–454.

NACE TM0187. 2011. Evaluating Elastomeric Materials in Sour Gas Environments. NACE International.

NACE TM0198. 2011. Slow Strain Rate Test Method for Screening Corrosion-Resistant Alloys for Stress Corrosion Cracking in Sour Oilfield Service. NACE International.

Nilsen, R. K., Beeder, J., Thorstenson, T. and Torsvik, T. 1996. Distribution of thermophilic marine sulfate reducers in North Sea oilfield waters and oil reservoirs. *Applied and Environmental Microbiology*, *62*(5), 1793–1798.

NORSOK M-710. 2014. *Qualification of non-metallic materials and manufacturers—Polymers*, 3rd Edition. *Norsk Sokkels Konkuranseposisjon*.

Omar, I. H., Gunaltun, Y., Kvarekval, J. and Dugstad, A. 2005. H_2S Corrosion of Carbon Steel under Simulated Kashagan Field Conditions. In *CORROSION 2005*. Houston, TX: NACE International.

Pots, B. F., Kapusta, S. D., John, R. C., Thomas, M. J. J., Rippon, I. J., Whitham, T. S. and Girgis, M. 2002. Improvements on de Waard-Milliams Corrosion Prediction and Applications to Corrosion Management. In *CORROSION 2002*. Denver, CO: NACE International.

Ray, T. W. and Ivey, C. E. 1979. Evaluation of seal materials for high-temperature H_2S service. *Journal of Canadian Petroleum Technology*, *18*(01), 91–95.

Raymond, J., Parnell, E. and Fichter, J. 2014. Determining Effective Antimicrobial Treatments for Long-Term Protection of Hydrocarbon Reservoirs. In *CORROSION 2014*. San Antonio, TX: NACE International.

Renpu, W. 2011. *Advanced Well Completion Engineering*, 3rd Edition. Waltham: Gulf Professional Publishing.

Robinson, K., Ginty, W. R., Samuelsen, E. H., Lundgaard, T. and Skovhus, T. L. 2010. Reservoir Souring in a Field with Sulphate Removal: A Case Study. In *SPE Annual Technical Conference and Exhibition*. Society of Petroleum Engineers.

Rogers, W. F. and Rowe Jr., A. 1955. Corrosion Effects of Hydrogen Sulphide and Carbon Dioxide in Oil Production. In *4th World Petroleum Congress*.

Russ, P. R. and Rainsford, C. 2004. Modifying a Pipeline Corrosion Management Strategy to Changing Conditions. In *SPE Asia Pacific Oil and Gas Conference and Exhibition*. Society of Petroleum Engineers.

Schofield, M. and Stott, J. 2012. Assessing the magnitude and consequences of reservoir souring. *Journal of Petroleum Technology*, *64*(05), 76–79.

Shoesmith, D. W. 1981. Formation, Transformation, and Dissolution of Phases Formed on Surfaces, Lash Miller Award Address. In *Electrochemical Society Meeting*. The Electrochemical Society.

Shoesmith, D. W., Taylor, P., Bailey, M. G. and Owen, D. G. 1980. The formation of ferrous monosulfide polymorphs during the corrosion of iron by aqueous hydrogen sulfide at 21 C. *Journal of the Electrochemical Society*, *127*(5), 1007–1015.

Skogsberg, L. 2004. Fitness for Purpose Sulfide Stress Corrosion Testing for Alloys in Existing Wells in Deep Water Gulf of Mexico. In *CORROSION 2004*. New Orleans, LA: NACE International.

Smith, S. N. 1993. A Proposed Mechanism for Corrosion in Slightly Sour Oil and Gas Production. In *12th International Corrosion Congress*. NACE International.

Smith, L. and DeWaard, K. 2005. Corrosion Prediction and Materials Selection for Oil and Gas Producing Environments. In *CORROSION 2005*. Houston, TX: NACE International.

Smith, S. N. and Joosten, M. W. 2006. Corrosion of Carbon Steel by H_2S in CO_2 Containing Oilfield Environments. In *CORROSION 2006*. San Diego, CA: NACE International.

Smith, J. S. and Miller, J. D. A. 1975. Nature of sulphides and their corrosive effect on ferrous metals: A review. *British Corrosion Journal*, *10*(3), 136–143.

Smith, S. N. and Pacheco, J. L. 2002. Prediction of Corrosion in Slightly Sour Environments. In *CORROSION 2002*. Denver, CO: NACE International.

Smith, S. N. and Wright, E. J. 1994. Prediction of Minimum H_2S Levels Required for Slightly Sour Corrosion. In *CORROSION 1994*. NACE International.

Smith, V., Martin, J. W., Hinds, G., Street, N., Bosch, C., Haase, T. and Jäger, S. 2013. HIC Resistance of Heritage Pipelines Exposed to Mildly Sour Environments. In *CORROSION 2013*. Orlando, FL: NACE International.

Sourmail, T. 2007. A Review of the Effect of Cold-Work on Resistance to Sulphide Stress Cracking. In *CORROSION 2007*. Nashville, TN: NACE International.

Sun, W. and Nešic, S. 2007. A Mechanistic Model of H_2S Corrosion of Mild Steel. In *CORROSION 2007*. Nashville, TN: NACE International.

Sun, W., Nešic, S. and Papavinasam, S. 2006. Kinetics of Iron Sulfide and Mixed Iron Sulfide/Carbonate Scale Precipitation in CO_2/H_2S Corrosion. In *CORROSION 2006*. San Diego, CA: NACE International.

Sun, W., Nešic, S. and Papavinasam, S. 2008a. Kinetics of corrosion layer formation. Part 2-Iron sulfide and mixed iron sulfide/carbonate layers in carbon dioxide/hydrogen sulfide corrosion. *Corrosion*, *64*(7), 586–599.

Sun, W., Nešic, S., Young, D. and Woollam, R. C. 2008b. Equilibrium expressions related to the solubility of the sour corrosion product mackinawite. *Industrial and Engineering Chemistry Research*, *47*(5), 1738–1742.

Talbot, R. E., Larsen, J. and Sanders, P. F. 2000. Experience with the Use of Tetrakishydroxymethylphosphonium Sulfate (THPS) for the Control of Downhole Hydrogen Sulfide. In *CORROSION 2000*. Orlando, FL: NACE International.

Taylor, P. 1980. The stereochemistry of iron sulfides; a structural rationale for the crystallization of some metastable phases from aqueous solution. *American Mineralogist*, *65*(9–10), 1026–1030.

Treseder, R. S. and Badrak, R. P. 1997. Effect of Cold Working on SSC Resistance of Carbon and Low Alloy Steels—A Review. In *CORROSION 1997*. New Orleans, LA: NACE International.

Treseder, R. S. and Swanson, T. M. 1968. Factors in sulfide corrosion cracking of high strength steels. *Corrosion*, *24*(2), 31–37.

Vance, I. and Thrasher, D. R. 2005. Reservoir Souring: Mechanisms and Prevention. In *Petroleum Microbiology*, eds. B. Ollivier, and M. Magot, 123–142. Washington, DC: ASM Press.

Vera, J. R., Franco, R. and Suarez, H. 2015. Sulfide Stress Cracking Susceptibility Assessment of Topside and Subsea Facilities. In *CORROSION 2015*. Dallas, TX: NACE International.

Vera, J. R., Mendez, C. and Suarez, H. 2016. Field Evaluation of H_2S Scavenger Efficiency Injected in a Multiphase Subsea System. In *CORROSION 2016*. Vancouver, BC, Canada: NACE International.

Videm, K. and Kvarekval, J. 1995. Corrosion of carbon steel in carbon dioxide-saturated solutions containing small amounts of hydrogen sulfide. *Corrosion*, *51*(4), 260–269.

Williams, T. M. and Schultz, C. M. 2015. Effect of Environmental Parameters on the Stability and Performance of Oil and Gas Biocides. In *SPE International Symposium on Oilfield Chemistry*. Society of Petroleum Engineers.

Williford, J., Rice, P. and Ray, T. 1999. Selection of Metallurgy and Elastomers Used in Completion Products to Achieve Predicted Product Integrity for the HP/HT Oil and Gas Fields of Indonesia. In *SPE Asia Pacific Oil and Gas Conference and Exhibition*. Society of Petroleum Engineers.

Section III

Microbiologically Influenced Corrosion Management

7 Management of MIC in the Oil and Gas Industry

Torben Lund Skovhus and Richard B. Eckert

CONTENTS

ABSTRACT

Effective management of microbiologically influenced corrosion (MIC) in upstream oil and gas assets begins in the design stage and continues through construction, commissioning, and operations until the point of abandonment and decommissioning.

Corrosion management (CM) is a structured process that is applied concurrently to abiotic and biotic internal corrosion threats throughout all stages of asset life. The essence of the corrosion management process consists of the assessment of corrosion threats, the identification of mitigation and preventive measures to reduce the severity and likelihood of those threats, and monitoring the performance and the effectiveness of the control measures.

In addition to MIC, oil and gas assets are often exposed to abiotic internal corrosion threats, resulting from CO_2, H_2S, oxygen, and other corrosive species as well as flow-assisted corrosion, erosion, under-deposit corrosion, and cracking. These corrosion threats may occur concurrently or in synergy with one another, and they may change over time in the different stages of asset life. To respond to changes in the types or severity of corrosion threats, the corrosion management process must be a continuing cycle of assessment, mitigation, and monitoring, and it must also incorporate learning from failures.

While microorganisms exist in essentially all upstream production assets, determining their role in internal corrosion and developing effective corrosion management strategies requires a consistent and methodical approach. In the oil and gas industry, MIC assessment efforts have historically focused on enumeration of planktonic rather than sessile microorganisms in biofilms, thereby discounting surface phenomena that directly influence both localized abiotic corrosion and MIC. Preventive and mitigative measures that are based on an understanding of biofilms, surface microbiological consortia and activity, and their relationship to corrosion are significantly more effective in managing MIC than measures based only on planktonic culture data (Larsen et al. 2008; Eckert 2015).

A number of recent case histories illustrate the benefits of improved MIC diagnostics based on integrating data from molecular microbiological methods (MMM) with corrosion monitoring and operational information. Although the industry is witnessing growing use of MMM for managing MIC, a procedure for clearly linking microbiological conditions with electrochemical corrosion initiation and propagation mechanisms is still emerging.

This chapter discusses a holistic approach to MIC management that incorporates the latest diagnostic methods and data integration within the corrosion management process, improving the operator's ability to manage MIC threats.

7.1 CORROSION MANAGEMENT SYSTEMS AND MIC MANAGEMENT

Corrosion management has been described by Morshed as "The process of reviewing the existing integrity management measures, regular monitoring of their performance and the assessment of their effectiveness post-commissioning" (Morshed 2013).

Additionally, corrosion management has been defined (Capcis 2001) as being "...part of the overall management system, which is concerned with the development, implementation, review and maintenance of the corrosion policy."

The oil and gas industry generally agrees that a fundamental objective of the CM process is to establish processes to manage the threats associated with corrosion. Therefore, corrosion management is focused on reducing the likelihood of corrosion threats resulting in an unintended event (e.g., a leak or failure) rather than addressing consequence or overall risk (i.e., likelihood × consequence). At the core of corrosion management is the process of assessing corrosion threats, identifying preventive and mitigative barriers, and monitoring the effectiveness of those barriers. The core elements of the corrosion management process are shown in Figure 7.1.

Molecular microbiological methods (Skovhus et al. 2007, 2014; Whitby and Skovhus 2011; Juhler et al. 2012; Passman et al. 2013; TM0212-2012) can be incorporated within each of these fundamental corrosion management processes, improving the

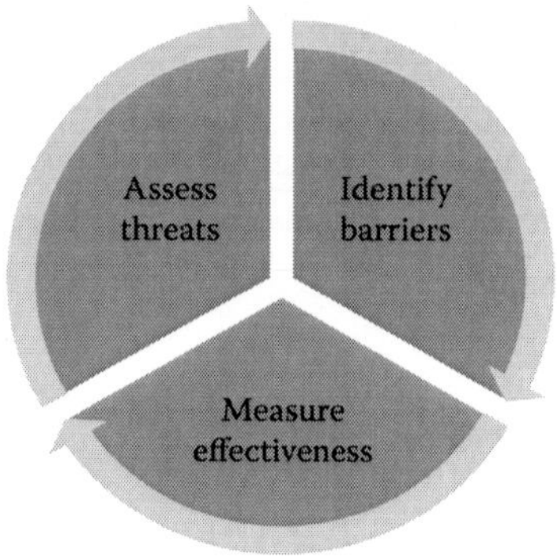

FIGURE 7.1 Fundamental process of managing corrosion threats including MIC.

operator's ability to manage MIC threats by providing accurate, meaningful microbiological data.

7.1.1 Assess Threats

In the first step, the likelihood of each potential corrosion threat mechanism is evaluated on each system and circuit for the asset. During this step, corrosion engineers review design, process, operation, chemical treatment, corrosion monitoring, fluid sampling, previous failure/leak data, and inspection data to help identify the likely corrosion threats affecting different locations in the system. Pipelines and circuits may be further broken down by segments in which different operating conditions or fluid composition change the severity or type of the applicable corrosion threats. In assessing the potential for MIC, it is important to establish a clear relationship between the biofilm and localized corrosion. MMM can be used in this step to characterize baseline microbiological conditions; to look for associations between the biofilm, the biofilm composition, and corrosion; and to relate biofilm characteristics with operating conditions. Interaction between various corrosion threats must also be considered, for example, biofilm establishment may be enhanced by solid deposits that promote under-deposit corrosion. When adequate data is available, programmatic integration and analysis can help verify the predicted corrosion threats. Investigation of corrosion damage or failures using formalized field and laboratory analysis procedures may also be used to verify the corrosion mechanisms in a pipeline.

7.1.2 Identify Barriers

In this step, the types of preventive and mitigative barriers needed to manage the corrosion threats are selected. Typically, the choices for internal corrosion mitigation include the use of treatment chemicals (corrosion inhibitor, biocide, oxygen scavenger, etc.), mechanical cleaning (maintenance pigging), velocity control, and control of fluid quality parameters, such as dissolved oxygen or oil in water. When assessing mitigation options prior to full implementation, MMM are beneficial for evaluating the effect of candidate treatments on corrosion and the biofilm and to monitor for shifts in the types or numbers of microorganisms after treatment is applied. Evaluation of mitigation targeted at controlling the biofilm is best performed using a sidestream test or in a limited portion of the process system where actual (not simulated) fluids are present. Data provided by MMM are also useful in determining the optimum approach to MIC mitigation and for selecting the appropriate biocides or antimicrobial chemicals.

7.1.3 Measuring Effectiveness

The final step in the process is measuring the effectiveness of the barriers employed to reduce the likelihood and/or severity of the corrosion threats. Corrosion monitoring typically provides information about the rate of metal loss due to corrosion, not the mechanism of corrosion. However, review and integration of operating condition data; monitoring data from coupons, probes, and inspection; chemical/microbiological fluid and deposit analysis results; and output from flow regime and corrosion models can lead to a better understanding of the corrosion mechanisms driving the corrosion damage. MMM can provide meaningful data to determine ongoing efficacy of biocide treatment. The extended laboratory analysis of corrosion coupons (Eckert et al. 2003; Zintel et al. 2003) can provide direct evidence of the active corrosion mechanisms in the system.

7.2 METHODS AND STANDARDS FOR MIC MANAGEMENT IN THE OIL AND GAS INDUSTRY

Several standards and guidance documents have been published over the past 15 years on the topics of corrosion management, risk-based inspection, and integrity management in the oil, gas, and petroleum industry (Table 7.1).

However, only the most recent recommended practice DNV-RP-F116 from February 2015 takes a more up-to-date view on potential groups of microorganisms

TABLE 7.1
Published Standards on Oilfield Corrosion Management and Inspection

Standard/Guidance Document	Source
DNV-RP-F116, Integrity management of submarine pipeline systems (2015)	DNV GL
EFC 64, Recommended Practice for Corrosion Management of Pipelines in Oil & Gas Production and Transportation (2012)	European Federation of Corrosion
DNV-RP-G101, Risk Based Inspection of Offshore Topsides Static Mechanical Equipment (2010)	DNV GL
Guidance for Corrosion Management in Oil and Gas Processing (2008)	Energy Institute
Review of Corrosion Management for Offshore Oil and Gas Processing (2001)	Health and Safety Executive

that could cause MIC, such as sulfate-reducing archaea, nitrate-reducing bacteria, and methanogens. To be effective, these documents will need to link to more detailed standards and industry protocols that focus more on sampling, measuring, and monitoring corrosion processes, including MIC. Methods used to monitor MIC must demonstrate that microorganisms and their activities provide the predominant influence over the corrosion mechanism as opposed to abiotic mechanisms that may be present. In general, this means that monitoring must integrate measurements of surface/biofilm characteristics with localized corrosion. In this regard, MMM are valuable for monitoring long-term and short-term effects of mitigation and changes in chemical treatment effectiveness relative to biofilm characteristics. Separating the contributions of abiotic and biotic environmental and surface characteristics in regard to corrosion is perhaps the most challenging aspect of managing MIC because there is no single method that can provide adequate information to fully discern the predominant corrosion mechanism. Typically, the integration of microbiological and corrosion monitoring data is required.

A significant development has taken place in this field over the past 10 years, initiated by the introduction of MMM as a new diagnostic tool to the oil and gas industry (Skovhus et al. 2007, 2010). In particular, two key documents from NACE International (TM0194-2014 and TM0212-2012) have now both been updated with the latest MMM technology (Table 7.2).

However, other standards and guidance documents at the American Petroleum Institute (API), NACE International, DNV GL, and the Energy Institute are still lacking the latest technological developments and will soon require an update.

A detailed comparison of analytical methods was made in a research report for the Energy Institute: "A practical evaluation of 21st century microbiological techniques for the upstream oil and gas industry" (Juhler et al. 2012). Six different often-applied analytical methods were tested on four different types of relevant samples from the oil recovery process: (1) injection seawater, (2) produced fluid from a separator,

TABLE 7.2
Published Standards with the Topic of Oilfield Microbiology and MIC

Standard/Guidance Document	Source
TM0106-2016, Detection, Testing, and Evaluation of Microbiologically Influenced Corrosion (MIC) on External Surfaces of Buried Pipelines (2016)	NACE International
TM0194-2014, Standard Test Method Field Monitoring of Bacterial Growth in Oil and Gas Systems (2014)	NACE International
TM0212-2012, Standard Test Method Detection, Testing, and Evaluation of Microbiologically Influenced Corrosion on Internal Surfaces of Pipelines (2012)	NACE International
A Practical Evaluation of 21st Century Microbiological Techniques for the Upstream Oil and Gas Industry (2012)	Energy Institute
SP0499-2012, Standard Practice Corrosion Control and Monitoring in Seawater Injection Systems (2012)	NACE International
Technical Report Publication 31205, Selection, Application, and Evaluation of Biocides in the Oil and Gas Industry (2006)	NACE International

(3) pigging debris from an oil export pipeline, and (4) corrosion products/biofilm from a failure investigation case. Pros and cons for each of the methods were discussed. Additionally, the chapter *Molecular Methods in Microbiologically Influenced Corrosion Research, Monitoring and Control* by Muyzer and Marty (2014) gives an excellent up-to date overview of various MMM available to the industry today.

A limited number of references have been published on how to obtain biofilm samples of high value for microbiological analyses. However, three publications stand out with some useful guidance. Keasler and Chatterjee (2014) pointed out the importance of combining measures (weight loss and pitting formation) from corrosion coupons from the field with quantitative polymerase chain reaction (qPCR) and denaturing gradient gel electrophoresis (DGGE) data. Galvão and Lutterbach (2014) showed the importance of obtaining biofilm samples from the inside of a fuel storage tank suffering from MIC. In a recent paper by Eroini et al. (2015), sampling and shipment of biofilm samples was explained in detail. The biofilm/corrosion products were transferred to the laboratory in a glass bottle closed by screw cap with two hose connections. The atmosphere was purged with inert gas that limited ingress of oxygen to avoid additional oxidation and reduced further growth of any aerobic microorganisms. The obtained samples were analyzed with qPCR and X-ray diffraction and fluorescence (XRD/XRF) when they arrived in the laboratory.

7.3 INCREASED AMOUNT OF DATA—BUT WHAT DOES IT TELL US?

The price of molecular biology data has seen a dramatic decrease over the past decades after the human genome was fully sequenced in 2003. In September 2001, the price for sequencing the complete (human) genome was estimated to be around $95,000,000 USD. To generate the same amount of data 14 years later (October 2015), the price was reduced to $1,245 USD (www.genome.gov/sequencingcosts/). With the technological breakthrough of DNA sequencing and automation of laboratory processes, molecular biology data has now become a commodity. The challenge is no longer to obtain data at a good price—it is to get relevant meaning out of the obtained data with the available bioinformatics tools.

Over the same period, we have seen an increase in technical and scientific literature related to oilfield microbiology and biocorrosion. The online database onepetro.org (http://onepetro.org) contains input from 27 industry and technical journals all related to the oil, gas, and offshore industry. When onepetro.org was searched for the key words "corrosion" AND "mic" AND "oil" covering the past 20 years, 1,087 papers were found. Of these papers, 70% (764 papers) were published in the period from 2005 to 2015 (Figure 7.2). Half of the papers (547) were published by NACE International.

Figure 7.2 shows an increasing trend of published technical and scientific literature in the field of oilfield microbiology and biocorrosion. An interesting trend is that years 2010 (87), 2012 (113), and 2014 (100) have the highest number of published conference papers in the 10-year period. The reason for this could be that every second year two popular technical symposia are hosted by NACE International: Microbiological Influenced Corrosion (TEG-187X) and Control of Problematic Microorganisms in the Oil & Gas Industry (TEG-286X). In addition to this trend is the increased attendance to both International Symposium on Applied Microbiology

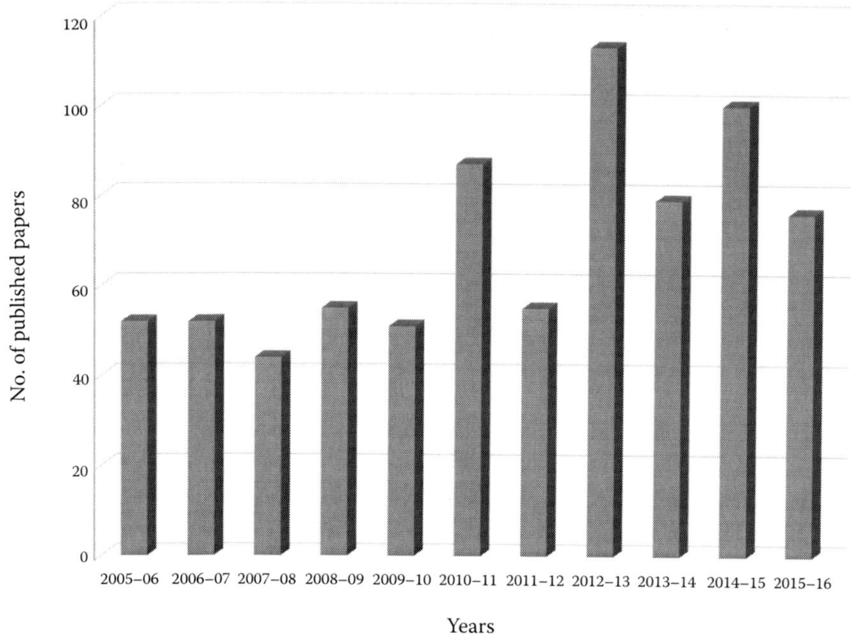

FIGURE 7.2 Review of published papers during the past 10 years in specialized industry literature. A total of 764 papers were found in 27 journals when the search contained the following key words: "corrosion" AND "mic" AND "oil." Onepetro.org was applied as the search tool.

and Molecular Biology in Oil Systems (ISMOS) and Reservoir Microbiology Forum (RMF) conferences with an estimated 2,500 participants combined over the past 10 years.

Hence, we have a trend of cheaper and more available molecular biology data combined with an increased industrial awareness, publication frequency, and conference activity in the field of oilfield microbiology and biocorrosion over the past 10 years.

Currently, the oil and gas industry has been collecting molecular biology data/information as never before seen in history. However, we are still lacking tools to transform the data and gathered information into knowledge from which we can make properly informed corrosion management decisions (Figure 7.3).

Until the industry develops corrosion management systems that can digest, transform, and interpret the obtained MMM data, it will continue to be of relatively low value to the pipeline operators (Skovhus and Eckert 2014; Eckert et al. 2015).

FIGURE 7.3 Route from sporadic observations to knowledge from where operators can take decisions.

7.4 INDUSTRY CASES ILLUSTRATING THE IMPORTANCE OF CORROSION MANAGEMENT SYSTEMS

The following published case studies illustrate the approach to identifying MIC, selecting mitigation, and monitoring effectiveness. The case studies below are good examples of how molecular biology data are put into a corrosion management context to facilitate management decisions on the overall corrosion status in the system.

7.4.1 MIC Assessment (Corrosion Management Phase 1)

In a recent field case from the North Sea, Rodrigues and Akid (2014) investigated the conflicting results of measured corrosion in the topside oil production system versus low value data obtained from traditional culturing methods (serial dilution). The authors reported that throughout the service life of the Otter oil production spool, internal corrosion was identified during the inspection programs, indicating severe metal loss of up to 1.7 mm/y for the oil topside production spool. Internal corrosion and degradation were also detected in the subsea oil production pipeline during 2010 and 2012. In-line inspection (ILI) programs detected metal loss values of 12.5 mm depth. Corrosion coupons and probes confirmed that internal corrosion was active. Furthermore, the microbial surveys performed using the most probable number (MPN) method reported *low* presence of microorganisms within the system. Failure investigation of the spools when removed from topside, confirmed severe internal corrosion damage that aligned with inspection data. Samples of black deposits directly in contact with the carbon steel were taken for molecular microbiological analysis (qPCR). The qPCR results showed *high* numbers of microorganisms were present, which could not be identified during service life using the MPN method. The data indicated that MIC was, in fact, the primary mechanism responsible for the corrosion damage found.

Based on the results obtained in the paper, the authors discussed how the degree of internal corrosion was previously *underestimated* and how the microorganisms responsible for MIC were not properly identified and quantified. The authors concluded, "The Oil & Gas industry should work towards less, but high quality data (like the MMM) to be able to predict and mitigate the MIC threat. We are spending too much time and money on data that gives false security and makes operators take the wrong decisions."

Also in the North Sea, Mitchell et al. (2012) investigated the conflicting microbiology results obtained from traditional culturing methods (serial dilution) and qPCR. The system encompassed dual 10-inch nominal diameter flowlines transporting unprocessed hydrocarbons and water 10.7 km from a subsea field to the host platform. A mixture of organic and inorganic solids had been causing fouling and plugging problems in the process system. The corrosion rate was estimated to be 1–3 mm/y. Comparative microbial analyses were performed on the recovered solids from the system. The results obtained by culture and molecular techniques both showed low levels of sulfate-reducing bacteria (SRB) ($<10^6$ cells/g), but the molecular technique provided a more definite indication of methanogen and sulfate-reducing archaea (SRA) abundance (both groups in the range of 10^7–10^9 cells/g). It is possible that if the culture technique enumeration was used in isolation, there could be a risk of false negative results. The authors concluded that with serial dilution results alone it would have been challenging to diagnose and mitigate MIC. Also it was concluded by the authors that MIC was one of the causes of the high observed corrosion rates leading to the need to replace the flowlines. For future work, they suggested that the trends in microbial numbers obtained with MMM should be a key part of the routine monitoring program once the pipeline is recommissioned to ensure that an effective mitigation strategy is in operation.

7.4.2 MIC Mitigation (Corrosion Management Phase 2)

Keasler et al. (2010) showed the benefit of selecting the correct biocide in the lab using a dynamic flow loop in combination with MMM. A severe corrosion problem was identified as being caused by microorganisms in an 18-year-old offshore pipeline system in Nigeria. The main cause was attributed to MIC. After identifying the problem, it was demonstrated that the combination of molecular tools (i.e., DGGE and DNA sequencing) and a dynamic flow loop setup could be used to provide insight to the bacterial genera/species present in the system and allow for a scientific basis for biocide selection, dosage, and contact time. Specifically, the work identified a synergistic two-component product that enhanced biofilm penetration, increased bacterial kill, and delayed biofilm regrowth in laboratory studies. Field trials were initiated to evaluate the effectiveness of this new treatment.

In the book chapter *Biocide Testing Using Single and Multispecies Biofilms* (Skovhus 2014), a newly developed methodology was presented to screen the effectiveness of biocides both in static and dynamic laboratory tests and in the field (Figure 7.4).

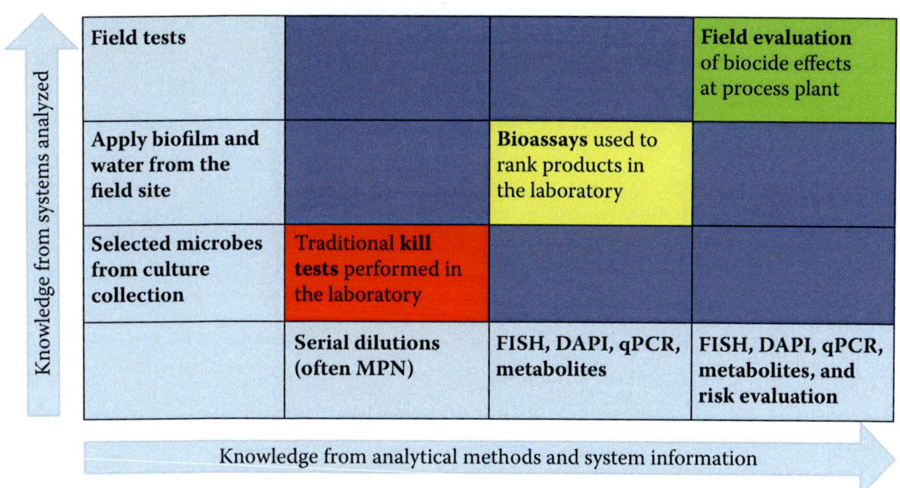

FIGURE 7.4 Overview of ways to test the effect of biocides toward microorganisms and the gained knowledge obtained at each level of the investigation. Traditional kill tests are often performed by the biocide manufacturer in the development stage of the product while bioassays and field evaluations are often carried out as a third-party test for the operator.

The simplest tests used were traditional **kill tests** in which selected laboratory strains are dosed with different concentrations of a biocide. From these tests, planktonic kill curves (survival as a function of concentration or contact time) were generated using the MPN technique with serial dilutions. Traditional kill tests should be seen as preliminary tests in the development of a new biocide formulation, and such tests do not reflect the actual field environment in which the biocide must perform (Holmkvist et al. 2011). The concept of **bioassays** (Hansen et al. 2009) and **field evaluations** (Skovhus et al. 2011) of biocides were used to rank the performance of different biocides against each other and to define dosing procedures for field use, such as contact time × dosage concentration × dosage frequency. Pros and cons of the three different testing approaches (kill tests, bioassays, and field evaluations) were discussed, and a detailed protocol for laboratory-based bioassays is provided in the original text (Skovhus 2014).

7.4.3 MIC Monitoring (Corrosion Management Phase 3)

Jensen et al. (2013) described in a recent field case how a robust monitoring system based on MMM could highlight the increasing amount of biomass in a deaerator, which could promote a MIC threat if not treated correctly. On the South Arne platform, the injection water was prepared from seawater that was run through a system of coarse and fine filters, followed by a membrane filtration package for reduction of the sulfate content. The oxygen was then removed in a deaerator tower. The challenge in this case was reducing biofouling of the injection water system, managing the threat of MIC, and preventing the potential loss of injectivity to the oil reservoir.

Historically, biocides had been applied upstream of the deaerator tower in order to control biofouling, but the expected beneficial effect was not previously verified by adequate analyses. The microbial status of the injection water upstream and downstream of the deaerator tower at South Arne began being evaluated once a year using MMM in April 2009. An increase in planktonic numbers of both total microorganisms and SRB across the deaerator tower was observed. This increase in the abundance of microorganisms is often seen since the packing material in the deaerator constitutes a very large surface area where a biofilm can establish and create dense layers of biomaterial that, in effect, works as a protective barrier toward most biocides. Generally, when a biofilm has been established, it can be difficult to regain control by means of chemicals, and therefore, mechanical cleaning will typically be needed.

As a follow-up on the observations of increased planktonic numbers, the deaerator was visually inspected during a planned shutdown in May 2011. The visual inspection revealed a massive biofilm build-up from which solid samples were collected (Figure 7.5).

The biofilm samples from the packer material were analyzed using MMM and wavelength dispersive X-ray fluorescence spectroscopy. The results revealed high bacteria and SRB numbers and showed that the build-up consisted mainly of organic material, which is usually associated with biofilm growth. The inorganic fraction consisted mostly of iron, calcium, and silicon.

FIGURE 7.5 Packer material covered in biofilm from the deaerator (May, 2011). (Obtained with permission from M. Jensen, HESS Denmark.)

Based on monitoring and inspection data, it was concluded that the applied biocide strategy was *insufficient* to prevent biofilm build-up. Therefore, it was decided to soak the deaerator tower with a batch treatment of biocide during the upcoming water injection shutdown. It was also decided to add new packing material to the deaerator on the next possible occasion. The deaerator was opened in May 2012, and the packing material appeared to be much cleaner than was observed in the previous year. A sample of deposit was recovered and analyzed for microbial abundance. The data showed that the abundance of bacteria and SRB in the solids was similar to the previous measurements. Knowing that the solid sample contained similar concentrations of bacteria and SRB as one year earlier simply emphasized the need for a biocide to keep microorganisms at a low planktonic level. Biofilm samples are generally of higher value in MIC threat assessment, and the high concentration of cells in the biomaterial indicates a high growth potential. Therefore, continued focus on the biocide strategy was important. The deaerator was cleaned and the packer material was replaced in May 2012. It was expected that a clean system would make it possible to keep planktonic and sessile microbial growth under control. Moreover, water quality was to be monitored over time to catch possible increases in microbial loading to the reservoir. In summary, the following initiatives were taken by the operator based on the MMM findings from this field case:

- Replacement of heavily fouled packing material in the deaerator tower
- Annual biocide batch soaking of the deaerator tower
- Routine monitoring of planktonic levels of microorganisms upstream and downstream of the deaerator with MMM in order to verify biocide efficacy
- Increased focus on consistency in biocide treatment
- Application of monitoring and inspection data as feedback to biocide optimization strategy

7.5 IMPLEMENTING CORROSION MANAGEMENT OF MIC: WHAT TO CONSIDER?

As seen in the case studies above, a number of very diverse and complex problems caused by microbial activity have been investigated and solved with the combined approach of good corrosion management practice and the application of MMM. The approach of using MMM for improving the corrosion management program can be applied both in the lab and in the field.

Management of MIC as a corrosion threat is based on the process of threat assessment, mitigation selection and implementation, and monitoring of mitigation effectiveness as shown in Figure 7.1. In each of these steps, the use of MMM has been demonstrated to be a particularly useful tool to operators.

As the oil and gas industry gains more experience with MMM, knowledge of the strengths and weaknesses of MMM in relation to managing MIC will continue to increase. Data produced through molecular methods must be interpreted and applied properly to truly take advantage of these techniques, and molecular microbiological data is not directly comparable to culture test results with which the industry is most familiar.

Because both corrosion and microbiological activity may occur directly on the internal surfaces of a pipeline, the analysis of samples (e.g., corrosion products, biofilm, etc.) associated with surfaces typically provides the most meaningful information. Establishing meaningful correlations between localized corrosion and surface microbiology has led to practical improvements in operators' corrosion management programs.

MIC diagnostics can be incorporated into the processes of MIC threat assessment, identification of mitigation or barriers, and monitoring of barrier effectiveness, improving the operator's ability to manage MIC threats. Further, MIC diagnostics have been demonstrated in multiple case studies to support operators' corrosion management systems. These case studies have demonstrated a number of key concepts that operators should consider in their MIC and corrosion management strategies, including the following:

- Establishing a corrosion management system that includes a strategy for MIC assessment, mitigation, and monitoring over the life of the asset (Figure 7.1)
- Evaluating the microbiological baseline of the system and defining the optimal sampling frequency to monitor the changes over time
- Reducing the impact of microorganisms in the design phase through the inclusion of pigging facilities, chemical injection nozzles, facilities for coupon pulling, access points for monitoring, fluid sampling, velocity control, etc.
- Including the right experts in the development of the MIC management strategy, including oilfield microbiologists, corrosion and materials engineers, chemical treatment specialists, and staff familiar with system operation
- Providing a means for efficient communication between technical disciplines when working to assess and manage the MIC threat as part of the corrosion management system

- Implementing the proper mitigative treatment only when the problem has been properly diagnosed (both from a microbiology and corrosion perspective)
- Knowing the strengths and weaknesses of the selected sample types (water, solids, and removable surfaces) and locations as well as analytical approaches used (MMM vs. culture-based methods) before building a monitoring program
- Establishing proof-of-concept through laboratory or limited field studies where possible before fully implementing new mitigation measures for MIC
- Monitoring using multiple key parameters (microbiology, chemistry, operation, corrosion, etc.) to ensure the applied MIC mitigation strategy is effective
- Investing in better understanding of high-quality data, not in increased amounts of data of low quality
- Always providing a detailed technology qualification of new analytical detection methods
- Continuing to become educated and being aware of advancements in technology and their optimal application and correct interpretation
- Using published standards (NACE, Energy Institute, American Petroleum Institute, DNV GL, etc.) to help develop efficient MIC management strategies

REFERENCES

Capcis Limited. 2001. *Review of Corrosion Management for Offshore Oil and Gas Processing*. Manchester, UK: HSE Books.

DNV-RP-F116. *Integrity Management of Submarine Pipeline Systems*. 2015. Høvik: DNV GL.

DNV-RP-G101. *Risk Based Inspection of Offshore Topsides Static Mechanical Equipment*. 2010. Høvik: DNV GL.

Eckert, R B. 2015. Emphasis on biofilms can improve mitigation of microbiologically influenced corrosion in the oil and gas industry. *Corrosion Engineering Science and Technology*. 50(3): 163–168.

Eckert, R B, Aldrich, H C, Edwards, C A, and Cookingham, B A. 2003. Microscopic differentiation of internal corrosion initiation mechanisms in natural gas pipeline systems. *Corrosion 2003*. Paper no. 03544, Houston, TX: NACE International.

Eckert, R B, Graver, B, and Skovhus, T L. 2015. Corrosion management of MIC contributes to pipeline integrity. *Corrosion 2015*. Paper no. 05729, Houston, TX: NACE International.

Energy Institute Guidance for Corrosion Management in Oil and Gas Processing. 2008. London: Energy Institute.

Eroini, V, Anfindsen, H, and Mitchell, A F. 2015. Investigation, classification and remediation of amorphous deposits in oilfield systems (SPE-173719). *SPE International Symposium on Oilfield Chemistry 2015*, The Woodlands, Texas, April 13–15, 2015.

Galvão, M, and Lutterbach, M. 2014. Application of the qPCR technique for SRB quantification in samples from the oil and gas industries. In *Applications of Molecular Microbiological Methods*, eds. T L Skovhus, S Caffrey and C Hubert, 69–76. Norfolk, UK: Caister Academic Press.

Hansen, L H, Larsen, J, Jensen, M, Thomsen, U S, Sørensen, K, Lundgaard, T, and Skovhus, T L. 2009. The application of bioassays for evaluating in-situ biocide efficiency in offshore oil Production in the North Sea (SPE 121656). *SPE International Symposium on Oilfield Chemistry 2009*, The Woodlands, Texas, April 20–22, 2009.

Holmkvist, L, Thomsen, U S, Larsen, J, Jensen, M, and Skovhus, T L. 2011. Problems Caused by Microbes and Treatment Strategies: Monitoring Microbial Responses to Biocides; Bioassays—A Concept to Test the Effect of Biocides on Both Archaea and Bacteria in Oilfield Systems. In *Applied Microbiology and Molecular Biology in Oilfield Systems*, eds. C Whitby and T L Skovhus, 117–124. New York: Springer Publisher.

Jensen, M L, Jensen, J, Lundgaard, T, and Skovhus, T L. 2013. Improving risk based inspection with molecular microbiological methods. *Corrosion 2013*. Paper no. 02247, Houston, TX: NACE International.

Juhler, S, Skovhus, T L, and Vance, I. 2012. *A Practical Evaluation of 21st Century Microbiological Techniques for the Oil and Gas Industry*. London: Energy Institute.

Keasler, V and Chatterjee, I. 2014. Using the power of molecular microbiological methods in oilfield corrosion management to diagnose MIC. In *Applications of Molecular Microbiological Methods*, eds. T L Skovhus, S Caffrey and C Hubert, 23–32. Norfolk, UK: Caister Academic Press.

Keasler, V, Bennett, B, Bromage, B, Franco, R J, Lefevre, D, Shafer, J, and Moninuola, B. 2010. Bacterial characterization and biocide quantification for full wellstream crude oil pipelines. *Corrosion 2010*. Paper no. 10250, Houston, TX: NACE International.

Kermani, B and Chevrot, T. 2012. Recommended practice for corrosion management of pipelines in oil & gas production and transportation (EFC 64). European Federation of Corrosion. Belgium.

Larsen, J, Skovhus, T L, Saunders, A M, Højris, B, and Agerbæk, M. 2008. Molecular identification of MIC bacteria from scale and produced water: Similarities and differences. *Corrosion 2008*. Paper no. 08652, Houston, TX: NACE International.

Mitchell, A F, Liengen, T, Anfindsen, H, and Molid, S. 2012. Experience of molecular monitoring techniques in upstream oil and gas operations. *Corrosion 2012*. Paper no. 01756, Houston, TX: NACE International.

Morshed, A. 2013. The evolution of the corrosion management concept. *Materials Performance*. 52(8): 66.

Muyzer, G and Marty, F. 2014. Molecular methods in microbiologically influenced corrosion research, monitoring and control. In *Applications of Molecular Microbiological Methods*, eds. T L Skovhus, S Caffrey and C Hubert, 7–22. Norfolk, UK: Caister Academic Press.

Passman, F J, Skovhus, T L, and Whitby, C. 2013. *Special Issue—3rd International Symposium on Applied Microbiology and Molecular Biology in Oil Systems*. Elsevier. 81: 1–146. Elsevier, BV, The Netherlands.

Rodrigues, E and Akid, R. 2014. Internal corrosion assessment of the otter oil production spool (SPE-169603). *SPE International Conference and Exhibition on Oilfield Corrosion 2014*, Aberdeen, UK, May 12–13, 2014.

Skovhus, T L. 2014. Biocide testing using single and multispecies biofilms. In *Biofouling Methods*, eds. S Dobretsov, D N Williams and J C Thomason, 76–86. New Jersey: Wiley-Blackwell.

Skovhus, T L and Eckert, R B. 2014. Practical aspects of mic detection, monitoring and management in the oil & gas industry. *Corrosion 2014*. Paper no. 03920, Houston, TX: NACE International.

Skovhus, T L, Højris, B, Saunders, A M, Thomsen, T R, Agerbæk, M, and Larsen, J. 2007. Practical use of new microbiology tools in oil production (SPE 109104). *SPE Offshore Europe Conference 2007*, Aberdeen, UK, September 4–7, 2007.

Skovhus, T L, Sørensen, K B, Larsen, J, Rasmussen, K, and Jensen, M. 2010. Rapid determination of MIC in oil production facilities with a DNA-based diagnostic kit (SPE 130744). *SPE International Conference on Oilfield Corrosion 2010*, Aberdeen, UK, May 24–25, 2010.

Skovhus, T L, Thomsen, U, Gydesen, B, and Hansen, L H. 2011. Concept for evaluating the effect of biocides in offshore field tests: A case study. TEKNA Oilfield Chemistry Symposium, Geilo, Norway. Petroleum Abstracts. University of Tulsa, OK.

Skovhus, T L, Caffrey, S, and Hubert, C. 2014. *Applications of molecular microbiological methods*. Norfolk, UK: Caister Academic Press.

SP0499-2012. *Standard Practice Corrosion Control and Monitoring in Seawater Injection Systems*. 2012. Houston, TX: NACE International.

Technical Report Publication 31205. *Selection, Application, and Evaluation of Biocides in the Oil and Gas Industry*. 2006. Houston, TX: NACE International.

TM0106-2016. *Detection, Testing, and Evaluation of Microbiologically Influenced Corrosion (MIC) on External Surfaces of Buried Pipelines*. 2016. Houston, TX: NACE International.

TM0194-2014. *Standard Test Method Field Monitoring of Bacterial Growth in Oil and Gas Systems*. 2014. Houston, TX: NACE International.

TM0212-2012. *Standard Test Method Detection, Testing, and Evaluation of Microbiologically Influenced Corrosion on Internal Surfaces of Pipelines*. 2012. Houston, TX: NACE International.

Whitby, C and Skovhus, T L. 2011. *Applied Microbiology and Molecular Biology in OilField Systems*. New York: Springer Publisher.

Zintel, T P, Kostuck, D A, and Cookingham, B A. 2003. Evaluation of chemical treatments in natural gas systems vs. MIC and other forms of internal corrosion using carbon steel coupons. *Corrosion 2003*. Paper no. 03574, Houston, TX: NACE International.

8 Diagnosing Microbiologically Influenced Corrosion

Jason S. Lee and Brenda J. Little

CONTENTS

ABSTRACT

No single analysis can be used to diagnose microbiologically influenced corrosion (MIC). Instead, diagnosing MIC requires an interpretation of a combination of microbiological, metallurgical, and chemical data. The following sections provide a discussion of available techniques, their advantages and disadvantages, and, most importantly, their limitations.

8.1 INTRODUCTION

An accurate diagnosis of MIC requires the following: (1) a sample of the corrosion product or affected surface that has not been altered by collection or storage, (2) identification of a corrosion mechanism that is consistent with the vulnerabilities of the material being examined, (3) identification of microorganisms capable of growth and maintenance of the corrosion mechanism in the particular environment, and (4) demonstration of an association of the microorganisms with the observed corrosion. Three types of evidence are used to diagnosis MIC: metallurgical, chemical, and microbiological. The objective is to have three independent types of measurements that are consistent with a mechanism for MIC. It is essential to collect a sample from an area that is not corroded and perform the same diagnostic testing to demonstrate that conclusions are based on the reaction of the corroded material with microorganisms and not just an environmental sample. Much has been written about sample collection, the necessity for sterile sample containers, and precautions to prevent contamination (Little et al. 1997; Eckert 2003). That information will not be repeated here. New information about sample storage and abiotic transformations is included in the appropriate sections.

8.2 IDENTIFICATION OF CAUSATIVE MICROORGANISMS

One of the first steps in diagnosing corrosion as MIC is identification of the specific microorganisms associated with corrosion products (sessile cells). Four approaches can be used to identify microorganisms: (1) culture microorganisms on solid or in liquid media, (2) extract and quantify a particular cell constituent, (3) demonstrate/measure some cellular activity, and (4) demonstrate a spatial relationship between microbial cells and corrosion products using microscopy.

8.2.1 Culture Techniques

A method commonly used for detecting and enumerating groups of bacteria is serial dilution to extinction using selective culture media. To culture microorganisms, a small amount of liquid or a suspension of a solid (the inoculum) is added to a solution or solid that contains nutrients (culture medium). There are three considerations when growing microorganisms: type of culture medium, incubation temperature, and length of incubation. The present trend in culture techniques is to attempt to culture several physiological groups, including heterotrophic aerobic bacteria; facultative anaerobic bacteria; sulfate-reducing bacteria (SRB); and acid-producing bacteria (APB). Growth is detected as turbidity or a chemical reaction (e.g., color change) within the culture medium. Traditional SRB media contain sodium lactate as the carbon source (API 1975; Postgate 1979). When SRB are present in the sample, sulfate is reduced to sulfide, which reacts with iron (either in solution or solid) to produce black ferrous sulfide. Culture media are typically observed for approximately 30 days. There have been several attempts to improve culture media and to grow higher numbers of bacteria or to shorten the time required for some indication of growth. A complex SRB medium was developed containing multiple carbon sources that can be degraded to both acetate and lactate. In comparison tests, the complex medium produced higher counts of SRB from waters and surface deposits among five commercially available media (Scott and Davies 1992). Jhobalia et al. (2005) developed an agar-based culture medium for accelerating the growth of SRB. They noted that over the range of 1.93 to 6.50 g/L sulfate, SRB grew best at the lowest concentration. Cowan (2005) developed a rapid culture technique for SRB based on rehydration of dried nutrients with water from the system under investigation. The author claimed that using system water reduced the acclimation period for microorganisms by ensuring that the culture medium had the same salinity as the system water used to prepare the inoculum. She reported quantification of SRB within 1 to 7 days.

The distinct advantage of culturing techniques to detect specific microorganisms is that low numbers of viable cells can grow to easily detectable higher numbers in the proper culture medium. However, there are numerous limitations for the detection and enumeration of cells by culturing techniques.

It has been recognized for over 80 years that the majority of microbial life cannot be easily cultured (Giovannoni et al. 1990; Ward et al. 1998; Knight et al. 2012). Microorganisms in natural environments can enter into a nonculturable state (Roszak and Colwell 1987). Culture media can be made specific for many types of microorganisms, including iron-oxidizing bacteria (IOB), SRB, and acid-producing bacteria, but culture media cannot approximate the complexity of a natural environment. Growth media tend to be strain-specific. Enning and Garrelfs (2014) concluded that there were species–specific differences among SRB related to corrosion mechanism and rate of corrosion. Incubating at one temperature is further selective. In addition, some microorganisms may be more readily culturable than others, a phenomenon known as culture bias, providing a distorted or false indication of specific microorganisms in a sample. In summary, the medium and the incubation conditions determine the numbers and types of microorganisms that grow in culture.

Using liquid culture techniques, Kilbane (2014) demonstrated significant changes in the quantification of microorganisms resulting from storage conditions when oilfield samples were not analyzed immediately. Kilbane (2014) demonstrated that storage of a biofilm sample at any temperature affected the accuracy of the culture results.

Culture techniques cannot account for the metabolic flexibility and syntropic interactions among microorganisms. Bacteria in biofilms do not behave as individuals with independent unicellular lifestyles. Instead, many bacteria regulate cooperative activities by releasing, sensing, and responding to diffusible signal molecules (Stoodley et al. 2002). Syntropic interactions generally involve two microbial types whose combined metabolism performs reactions not possible by either alone; that is, degradation of a substrate by one species is made thermodynamically possible through the removal of end products by another species. For example, in the absence of sulfate, some SRB ferment organic acids and alcohols, producing hydrogen, acetate, and carbon dioxide, and may rely on hydrogen- and acetate-scavenging methanogens to convert organic compounds to methane (Plugge et al. 2011). In laboratory experiments, Zhang et al. (2003) demonstrated that some methanogens influenced corrosion indirectly by consuming acetate and H_2 produced by the SRB oxidation of lactate. In this demonstration, the SRB were unable to further degrade the acetate.

8.2.2 Biochemical Assays

Biochemical assays have been developed for the detection of specific microorganisms associated with MIC. Unlike culturing techniques, biochemical assays for detecting and quantifying bacteria do not require growth of the bacteria. Instead, biochemical assays measure constitutive properties, including adenosine triphosphate (ATP) (Littmann 1977), phospholipid fatty acids (PLFA) (Franklin and White 1991), and cell-bound antibodies (Pope 1986). Adenosine-5′-phosphosulfate (APS) reductase (Tatnall et al. 1988) and hydrogenase (Boivin et al. 1990) have been used to estimate SRB populations.

Since ATP is a compound found in all living matter, ATP assays estimate the total number of viable organisms in a sample. The procedure requires that a water sample be filtered to remove solids and salts, which may interfere with the test. The filtered sample is added to a reagent to release cell ATP. An enzyme then reacts with the ATP to produce a photochemical reaction. Emitted light is measured with a photometer, and the number of bacterial cells is estimated from the total light emitted. Fichter et al. (2012) used ATP as one method for determining microbial activity in produced waters and fracturing fluids. ATP provides information about biological activity but does not provide specifics about the specific organisms in the sample.

Biofilm community structure can be analyzed using cluster analysis of PLFA profiles (Franklin and White 1991). Phospholipids are found in the membranes of all cells. Under the conditions in natural communities, bacteria contain a relatively constant proportion of their biomass as phospholipids. Phospholipids are not found in storage lipids and have a relatively rapid turnover so that their assay gives a measure of the viable cellular biomass. The phosphate of the phospholipids or the glycerol-phosphate and acid-labile glycerol from phosphatidyl glycerol-like lipids can be assayed to increase the specificity and sensitivity of the phospholipids assay. The ester-linked fatty acids in the phospholipids are both the most sensitive and the most useful chemical measures of microbial biomass and community structure. PLFA profiles for natural biofilms have been shown to be more complex than profiles for laboratory biofilms. None of the laboratory profiles clustered closely with profiles from natural biofilms. In addition, the PLFA profiles for attached bacteria clustered separately from profiles of the same bacteria in the bulk phase, suggesting that either the community or the physiology of attached bacteria differs from that of bulk phase bacteria. Despite the fact the PLFA analysis cannot provide an exact description of each species in a given environment, the analysis does provide a quantitative description of the micro biota in a particular environment. Analysis of other components of the phospholipids fraction gives insight into community structure. For example SRB contain lipids that can be used to identify at least a portion of the class. Dowling et al. (1988) identified unusual fatty acids as biomarkers for two SRB: iso 17:1w7c and branched monoenoics for a hydrogen-oxidizing *Desulfovibrio* sp. and 10 methyl 16:0 for an acetate oxidizing *Desulfobacter* sp.

Both APS reductase, an intercellular enzyme found in all SRB, and hydrogenase, an enzyme present in some SRB (hydrogenase^{+}), can be extracted from liquids or solids, including corrosion products and sludge. In a procedure to quantify APS reductase, cells are lysed to release the enzyme, added to an antibody reagent, and exposed to a color-developing solution. In the presence of APS reductase, a blue color appears whose intensity and development rate is proportional to the amount of enzyme and roughly to the number of cells from which the enzyme was extracted. Similarly, hydrogenase activity may be measured in a procedure where the enzyme is extracted from cells and exposed to hydrogen anaerobically (Boivin et al. 1990). The rationale for relating hydrogenase to MIC is that during corrosion in anaerobic environments, hydrogen gas is produced at the cathode. Some, but not all SRB are hydrogenase^{+}, meaning that they possess the enzyme required to oxidize hydrogen gas. In the assay, hydrogenase reacts with hydrogen and reduces an indicator dye in solution. The activity of hydrogenase is established by development and intensification of a blue color proportional to the rate of hydrogen uptake by the enzyme. The technique does not attempt to estimate specific numbers of SRB. Bryant et al. (1991) suggested that hydrogenase levels were better indicators of MIC than numbers of SRB. Mara and Williams (1972) reported that hydrogenase was more important when the environment contained low concentrations of ferrous ions but was less important in the presence of sufficient ferrous ions to precipitate the sulfide produced by SRB. Other investigators have established that there is no relationship between levels of hydrogenase enzyme and the rate or extent of corrosion (Jones-Meehan et al. 1995).

8.2.3 Cell Activity

Roszak and Colwell (1987) reviewed techniques commonly used to detect microbial activities in natural environments, including transformations of radiolabeled metabolic precursors. Phelps et al. (1991) and Mittleman et al. (1990) used uptake or transformation of ^{14}C-labeled metabolic precursors to examine activities of sessile bacteria in natural environments and in laboratory models. Phelps et al. (1991) used a variety of ^{14}C-labeled compounds to quantify catabolic and anabolic bacterial activities associated with corrosion tubercles in steel natural gas transmission pipelines. They demonstrated that organic acid was produced from hydrogen and carbon dioxide in natural gas by acetogenic bacteria and that acidification could lead to enhanced corrosion of the steel. Mittleman et al. (1990) used measurement of lipid biosynthesis from ^{14}C-acetate, in conjunction with measurements of microbial biomass and extracellular polymer, to study effects of differential fluid shear on physiology and metabolism of *Alteromonas* (formerly *Pseudomonas*) *atlantica*. Increasing shear force increased the rate of total lipid biosynthesis but decreased per cell biosynthesis. Increasing fluid shear also increased cellular biomass and greatly increased the ratio of extracellular polymer to cellular protein. Maxwell (1986) developed a radiorespirometric technique for measuring SRB activity on metal surfaces that involved two distinct steps: incubation of the sample with ^{35}S sulfate and trapping the released sulfide.

Techniques for analyzing microbial metabolic activity at localized sites have been developed. Franklin et al. (1991) incubated microbial biofilms with ^{14}C-metabolic precursors and autoradiographed the biofilms to localize biosynthetic activity on corroding metal surfaces. The localized uptake of labeled compounds was related to localized electrochemical activities associated with corrosion reactions.

Reporter genes can signal when the activity of a specific metabolic pathway is induced. King et al. (1990) engineered the incorporation of a promotorless cassette of lux genes into specific operons of *Pseudomonas* to induce bioluminescence during degradation of naphthalene. Using reporter genes, Marshall et al. (1994) demonstrated that bacteria immobilized at surfaces exhibit physiological properties not found in the same organisms in the aqueous phase. Some genes are turned on at solid surfaces despite not being expressed in liquid or on solid media. It is also likely that other genes are turned off at surfaces. They further demonstrated gene transfers within biofilms even in the absence of imposed selection pressure.

8.2.4 Molecular Microbiological Methods (MMM)

Recent standards for assessing MIC (see NACE Standard TM0212-2012) acknowledge the problems associated with attempting to diagnose or predict MIC based on numbers and types of particular microorganisms and include MMM that do not require culture techniques to characterize microbial populations (Caffrey 2011). MMM analyze molecules that microorganisms possess or generate, including deoxyribonucleic acid (DNA), ribonucleic acid (RNA), proteins, lipids, and cellular metabolites. Muyzer and Marty (2014) reviewed MMM applications in MIC research, monitoring, and control. The following sections provide brief descriptions of MMM used in diagnosing MIC.

8.2.4.1 Community Fingerprinting

The polymerase chain reaction (PCR) is used to generate copies of a particular DNA (double-stranded) or RNA (single-stranded) sequence. A nucleic acid sequence is a series of nucleotides within a DNA or RNA molecule. The DNA PCR method requires repeated cycles of heating and cooling for DNA melting and enzymatic replication. Primers (oligonucleotides or short DNA fragments) containing sequences complementary to the target region along with DNA polymerase enable selective and repeated amplification. The selectivity of PCR results from the use of primers complementary to the DNA region targeted for amplification.

PCR-based techniques are used in combination with genetic fingerprinting techniques, such as denaturing gradient gel electrophoresis (DGGE) and terminal restriction fragment length polymorphism (T-RFLP). For fingerprinting using DGGE, DNA (or RNA) is applied to an electrophoresis gel that contains a denaturing agent that causes DNA to melt (become single stranded) in stages. As a result of the melting, the DNA spreads through the gel and can be analyzed for single components. Each DGGE band is representative of a specific bacterial population, and the number of distinctive bands is indicative of microbial diversity. Comparison of DNA sequences with sequence databases allows the identity of the species within an environmental sample.

T-RFLP is also based on PCR amplification of a target gene. Genomic DNA extracted from microbial communities is amplified with primers labeled with a fluorochrome. The resulting DNA fragment is cut with a restriction enzyme and the fragments analyzed with a capillary sequencer. Only the DNA fragments that are fluorescently labeled are detected, resulting in a profile of the microbial community.

Traditional PCR is used to exponentially amplify target DNA sequences. Reverse transcription polymerase chain reaction (RT-PCR), a variant of PCR, is used to detect RNA expression. RT-PCR is used to qualitatively detect gene expression through creation of complementary DNA (cDNA) transcripts from RNA. RT-PCR is used to clone expressed genes by reverse transcribing the RNA of interest into its DNA complement through the use of reverse transcriptase. Subsequently, the newly synthesized cDNA is amplified using traditional PCR. Quantitative PCR (qPCR) is used to quantitatively measure the amplification of DNA using fluorescent probes.

Direct molecular characterization of natural microbial populations can be accomplished with sequence analysis of 16S ribosomal RNA (rRNA) (Stahl et al. 1984, 1988). Fluorescent dye-labeled oligonucleotide probes have been used for microscopic identification of single cells and characterization of mixed populations. Genetic techniques have been used to identify and quantify microorganisms in natural environments. These techniques involve retrieval of 16S rRNA gene sequences by polymerase chain reaction amplification of extracted and purified nucleic acids. The PRC products can be evaluated using community fingerprinting techniques, such as denaturing gradient gel electrophoresis. In addition to fingerprinting techniques, PCR amplified 16S rRNA gene fragments can also be directly cloned and sequenced. Comparative analysis of these sequences to those stored in public databases can provide the identity of community members.

8.2.4.2 Hybridization Techniques

8.2.4.2.1 Dot-Blot Hybridization and Reverse Sample Genome Probing

rRNA is essential for protein synthesis in all living organisms. In hybridization techniques, DNA or RNA is spotted onto a membrane and hybridized with labeled oligonucleotide probes (dot-blot hybridization). Whole cell hybridization or fluorescence in situ hybridization (FISH) can be used to detect and quantify particular microorganisms microscopically and to demonstrate their spatial relationships. Cells are fixed, incubated with an oligonucleotide probe labeled with a fluorochrome specific for 16S rRNA, and viewed with a fluorescence microscope.

8.2.4.2.2 DNA Microarrays

DNA microarrays are glass slides containing immobilized probes specific for 16S rRNA genes of different microorganisms or for functional genes. Environmental DNA or PCR products labeled with a fluorochrome are incubated on the slide. Homologous sequences will hybridize to the probes, and the results can be measured.

8.2.4.2.3 Functional Gene Markers

Some genes for metabolic pathways implicated in MIC are sufficiently conserved to be used as a metabolic marker for that pathway. In particular, the genes encoding the dissimilatory sulfite reductase (dsrAB), an enzyme in the sulfate-reducing pathway, has been used in MIC studies (Price et al. 2010; Lutterbach et al. 2011; Paisse et al. 2013).

8.2.4.3 Applications

Major contributions of MMM to our understanding of MIC have been (1) the verification that planktonic population in an electrolyte cannot be used to predict the microorganisms in the corrosion product or corrosion mechanism; (2) culture techniques cannot be used on corroded samples that have been stored; and (3) because of culture bias, culture techniques do not provide an accurate representation of microbial groups within a natural population. Wrangham and Summer (2013) used metagenomics from production and injection systems in geographically distant locations to demonstrate that the planktonic microbial population was not indicative of the sessile population. Planktonic and sessile populations from the same locations were as different from each other as samples taken from different locations. Hoffmann et al. (2008) used DGGE and FISH to demonstrate that the sessile population community diversity was greater than that of the planktonic population and that there was no obvious relationship between the two.

Kilbane (2014) used similar techniques to demonstrate that culture techniques could not be used on samples that had been stored but that qPCR could be used to detect and quantify microorganisms in corroded pipe samples removed form service for up to a month. DNA degradation is influenced by temperature and the presence of water. Hoffmann et al. (2008) used DGGE and FISH to make the following observations: the numbers of planktonic SRB in North Sea water column was greater than viable counts.

Zhu et al. (2005) demonstrated dramatic changes in the microbial population from a gas pipeline after samples were introduced into liquid culture media. For example, using MMM techniques they found that methanogens were more abundant in most pipeline samples than denitrifying bacteria and that SRB were the least abundant. However, using culture techniques, SRB dominated the microflora. Similarly, Romero et al. (2005) used MMM monitoring to identify bacterial populations in a seawater injection system. They found that some bacteria present in small amounts in the original waters were enriched in the culture process. Zhu et al. (2004) used genetic techniques to characterize the types and abundance of bacterial species in gas pipeline samples and made similar observations.

Zhu et al. (2006) used real-time qPCR to detect and quantify bacteria that might be involved in MIC, using general and specific taxonomic and functional markers, that is, 16S rRNA for bacteria and archaea, dsrAB, encoding the dissimilatory sulfite reductase in SRB; mrcA, encoding the methyl-coenzyme M reductase in methanogens; and nirS, encoding the nitrite reductase in denitrifying bacteria. By using this approach, they could detect and quantify bacteria in general as well as sulfate-reducing bacteria, denitrifiers, and methanogens in gas pipeline samples. Comparative analysis between qPCR and MPN-counts showed that the concentrations of bacteria determined by qPCR were several orders of magnitude higher than those found by MPN (i.e., 2 to 5 log higher for bacteria in general; 2 to 3 log higher for SRB; 1 to 2 log higher for denitrifiers).

The MPN technique was shown to be inadequate in quantifying SRB levels in oilfield samples with an approximate 500-fold underestimation of their numbers when compared to the numbers determined by the quantitative FISH (qFISH) technique (Price et al. 2010).

MMM have been used to detect causative microorganisms in MIC. For example, Lutterbach et al. (2011) used qPCR on dsrAB to detect SRB in fuel storage tanks. Larsen et al. (2013) used the average number of RNA molecules per cell determined by reverse transcription quantitative PCR as a measure of activity of methanogens and sulfate-reducing prokaryotes within biofilms. Pipelines with more activity were judged to be at higher risk for MIC. Lopez et al. (2006) used clone libraries of PCR-amplified 16S rRNA gene fragments to identify bacteria in a corrosive biofilm in a steel pipeline that was used to inject water into an oilfield in the Gulf of Mexico. They found members belonging to different genera within the Gammaproteobacteria (96.66%), Bacilli (2.67%), and Alphaproteobacteria (0.67%) but, significantly, no SRB. Almahamedh et al. (2011) identified both metal and sulfate-reducing bacteria in oilfield water using 16S rRNA gene sequence analysis for bacteria, archaea, and eucaryotes using database comparisons.

The related suffix "-ome" is used to specify fields of study related the genome, proteome, or metabolome, that is, genomics, proteomics, or metabolomics, respectively. Omics aims at the collective characterization and quantification of pools of biological molecules that translate into the structure, function, and dynamics of an organism or organisms. DNA- and RNA-based molecular methods are collectively referred to as metagenomics. Functional genomics aims at identifying the functions of as many genes as possible of a given organism. It combines different -omics techniques, such as transcriptomics and proteomics, with saturated mutant collections. Metagenomics coupled with metabolomics data provide information on the presence, activation, and expression of metabolic pathways in bacteria and archaea. Metagenomics allows one to interrogate the frequency of functional genes characteristic of specific pathways potentially involved in corrosion reactions, for example, sulfate reduction, sulfur oxidation. One can link the frequency of genes coding for a particular pathway with the prevalence of microbial taxa in a corroding system versus a noncorroding one. Overlapping the metabolome with the metagenome provides insight into the microbial processes likely to be coupled to corrosion.

8.2.4.4 Advantages and Disadvantages

The distinct disadvantage of using DNA to identify and quantify microorganisms related to MIC is that DNA can be related to viable or dead cells and extracellular sources (X-DNA). Several researchers have reported that PCR coupled with DGGE produced higher bacterial numbers than serial dilutions (MPN) of the same samples (Adesina et al. 2015). However, there are also limitations related to the specificity of the primers, detection limits, and reproducibility (Wagner et al. 2007). Summer et al. (2014) identified thiosulfate-reducing bacteria as opposed to SRB in samples from a receiving station of an onshore pipeline. They noted that the thiosulfate-reducing bacteria, lacking genes for sulfate reduction, would not be identified using a current rapid identification screen, such as qPCR or microarrays. Application of FISH to samples from an oil production system is limited because of the inhibiting effect of hydrocarbons on hybridization.

8.2.5 Microscopy

8.2.5.1 Light Microscopy

Investigators (Chamberlain et al. 1995) have demonstrated a relationship between an unusual variety of copper pitting corrosion and gelatinous, polysaccharide-containing biofilms. The films are removed, stained for polysaccharides, and evaluated with light microscopy (Chamberlain et al. 1988).

8.2.5.2 Epifluorescence Microscopy

Immunofluorescence techniques have been developed for the identification of specific bacteria in biofilms (Zambon et al. 1984; Howgrave-Graham and Steyn 1988). Epifluorescence cell surface antibody (ECSA) methods for detecting SRB are based on the use and subsequent detection of specific antibodies, produced in rabbits, that react with SRB cells (Pope 1986). A secondary antibody, produced in goats, is then reacted with the primary rabbit antibodies bound to SRB cells. In some cases, goat antibodies are linked to a fluorochrome, which enables bacterial cells marked with the secondary antibody to be viewed with an epifluorescence microscope. In other cases, goat antibodies are conjugated with an enzyme (alkaline phosphatase) that can then be reacted with a colorless substrate to produce a visible color proportional to the quantity of SRB present. Detection limits for the field test are 10,000 SRB mm^{-2} filter area. The color reagent used for field tests is unstable at room temperature and tends to bind nonspecifically with antibodies adsorbed directly at active sites on the filter, creating a false positive that may interfere with the detection of SRB at levels below 10,000 cells mm^{-2}. Antigenic structures of marine and terrestrial strains are distinctly different, and therefore, antibodies to either strain did not react with the other. Hogan (1990) described a nonisotopic semiquantitative procedure sensitive to 104 organisms ml^{-1} for the detection of *Desulfobacterium* and *Desulfotomaculum* using DNA probes labeled with an acridinium ester.

8.2.5.3 Confocal Laser Scanning Microscopy

Confocal laser scanning microscopy (CLSM) permits one to create three-dimensional images, determine surface contour in minute detail, and accurately measure critical dimensions by mechanically scanning the object with laser light. A sharply focused image of a single horizontal plane within a specimen is formed while light from out-of-focus areas is repressed from view. The process is repeated again and again at precise intervals on horizontal planes and the visual data from all images are compiled to create a single, multidimensional view of the subject. Geesey et al. (1994) used CLSM to produce three-dimensional images of bacteria within scratches, milling lines, and grain boundaries.

8.2.5.4 Atomic Force Microscopy

Atomic force microscopy (AFM) uses a microprobe mounted on a flexible cantilever to detect surface topography by scanning at a subnanometer scale. Repulsion by electrons overlapping at the tip of the microprobe cause deflections of the cantilever that can be detected by a laser beam. The signal is read by a feedback loop to maintain a constant tip displacement by varying voltage to a piezoelectric control.

The variations in the voltage mimic the topography of the sample and together with the movement of the microprobe in the horizontal plane are converted to an image. Telegdi et al. (1998) imaged microorganisms associated with corrosion on several substrata.

8.2.5.5 Electron Microscopy

Many of the conclusions about biofilm development, composition, distribution, and the relationship to substratum/corrosion products have been derived from traditional scanning electron microscopy (SEM) and transmission electron microscopy (TEM). SEM has been used to detent SRB from corrosion products on alloy 904L (Scott and Davies 1989), microorganisms in corroding gas pipelines (Zhu et al. 2003), and iron-oxidizing *Gallionella* in water distribution systems (Ridgeway and Olson 1981). TEM has been used to demonstrate that bacteria are intimately associated with sulfide minerals and that on copper-containing surfaces the bacteria were found between alternate layers of corrosion products and attached to base metal (Blunn 1986).

In traditional SEM, nonconducting samples, including biofilms associated with corrosion products, must be dehydrated and coated with a conductive film of metal before the specimen can be viewed. Traditional TEM methods for imaging biofilms require fixation of biological material, embedding in a resin, and thin sectioning to achieve a section that can transmit an electron beam. Environmental electron microscopy includes both scanning (ESEM) and transmission (ETEM) techniques for the examination of biological materials with a minimum of manipulation, that is, fixation and dehydration. Little et al. (1991) used ESEM to study marine biofilms on stainless steel surfaces. They observed a gelatinous layer in which bacteria and microalgae were embedded. Traditional SEM images of the same areas demonstrated a loss of cellular and extracellular material. Ray and Little (2003) used ESEM to demonstrate sulfide encrusted SRB in corrosion layers on copper alloys and iron-depositing bacteria in tubercles on stainless steels. Little et al. (1991) used environmental TEM to image *P. putida* on corroding iron filings and to demonstrate that the organisms were not directly in contact with the metal. Instead, the cells were attached to the substratum with extracellular material. Design and operation of the ESEM and ETEM have been described elsewhere (Ray and Little 2003).

There are fundamental problems in attempting to diagnose MIC by establishing a spatial relationship between numbers and types of microorganisms in the bulk medium or those associated with corrosion products using any of the techniques previously described. Zintel et al. (2003) established that there were no relationships between the presence, type, or levels of planktonic or sessile bacteria and the occurrence of pits. Because microorganisms are ubiquitous, the presence of bacteria or other microorganisms does not necessarily indicate a causal relationship with corrosion. In fact, microorganisms can nearly always be cultured from natural environments. Little et al. (1996) reported that electrochemical polarization could influence the number and types of bacteria associated with the surface. Artificial crevices created in 304 stainless steels in abiotic seawater were associated with large numbers of bacteria after 5-day exposures to natural seawater. Bacteria did not cause the crevice; instead, bacteria were attracted to the anodic site. Several other investigators have made similar observations. For example, de Sanchez and Schiffrin (1996) demonstrated that Cu(II) and titanium ions were strong attractants for *Pseudomonas*. Detection or demonstration of bacteria associated with corrosion is not diagnostic for MIC.

8.3 PIT MORPHOLOGY

Pope (1990) completed a study of gas pipelines to determine the relationship between extent of MIC and the levels/activities of SRB. They concluded that there was no relationship. Pope (1990) instead found large numbers of APB and organic acids, particularly, lactic acid, and identified the following metallurgical features in carbon steel: (1) large craters from 5 to 8 cm or greater in diameter surrounded by uncorroded metal, (2) cup-type hemispherical pits on the pipe surface or in the craters, (3) striations or contour lines in the pits or craters running parallel to longitudinal pipe axis (rolling direction), and (4) tunnels at the ends of the craters also running parallel to the longitudinal axis of the pipe. They reported that these metallurgical features were "fairly definitive for MIC." However, Pope (1990) did not advocate diagnosis of MIC based solely on pit morphology. Subsequent research has demonstrated that these features can be produced by abiotic reactions (Eckert et al. 2003) and cannot be used to independently diagnose MIC.

Other investigators described ink bottle–shaped pits in 300 series stainless steel that were supposed to be diagnostic of MIC. Borenstein and Lindsay (1988, 1994) reported that dendritic corrosion attack at welds was "characteristic of MIC." Chung and Thomas (1999) compared MIC pit morphology with non-MIC chloride-induced pitting in 304/304L and E308 stainless steel base metals and welds. A faceted appearance was common to both types of pits in 304 and 304L base metal. Facets were located in the dendritic skeletons in MIC and non-MIC cavities of E308 weld metal. They concluded that there were no unique morphological characteristics for MIC pits in these materials. The problem that has resulted from the assumption that pits can be independently interpreted as MIC is that MIC is often misdiagnosed. For example, Welz and Tverberg (1998) reported leaks at welds in a 316L stainless steel hot water system in a brewery after 6 weeks in operation were due to MIC. The original diagnosis was based on the circumstantial evidence of attack at welds and the pitting morphology–scalloped pits within pits. However, after a thorough investigation, MIC was dismissed. There were no bacteria associated with the corrosion sites, and deposits were too uniform to have been produced by bacteria. The hemispherical pits had been produced when CO_2 was liberated and low pH bubbles nucleated at surface discontinuities.

Several investigators have demonstrated that during the initial stages of pit formation due to certain types of bacteria, pits do have unique characteristics. Geiser et al. (2001) found that pits in 316L stainless steel due to the manganese-oxidizing bacterium *Leptothrix discophora* had different morphologies than pits initiated by anodic polarization. Pits initiated by these organisms in a solution of sodium chloride were approximately 10 times longer than wide. Pits produced by microorganisms were much smaller and not nearly as deep as pits produced in the same solution by electrochemical means. Pits had almost identical sizes and aspect ratios as the sizes and aspect ratios of the manganese-oxidizing bacteria. The similarity between the dimensions of the bacterial cells attached to the surface and the dimensions of corrosion pits indicate a possibility that the pits were initiated at the sites where the microbes were attached. Eckert (2003) used API 5L steel to demonstrate micro-morphological characteristics that could be used to identify MIC initiation. Coupons were installed at various points in a pipeline system and were examined by SEM at 1000 and 2000×. They demonstrated that pit initiation and bacterial colonization were correlated and that pit locations physically matched the locations of cells. Telegdi et al. (1998) used atomic force microscopy to image biofilm formation, extracellular polymer production, and subsequent corrosion. Pits produced by *Thiobacillus intermedius* had the same shape as the bacteria. None of these investigators claims that these unique features can be detected with the unaided eye or that the features will be preserved as pits grow, propagate, and merge.

8.4 CHEMICAL TESTING

Analyses for corrosion product chemistry can range from simple field tests to mineralogy and isotope fractionation. Field tests for solids and corrosion products typically include pH and a qualitative analysis for the presence of sulfides and carbonates. A drop of dilute hydrochloric acid placed on a small portion of the corrosion product will indicate the presence of carbonates if noticeable bubbling occurs. When the acid-treated corrosion product is exposed to lead acetate test paper, a white-to-brown color change will occur on the test paper, or it will produce the smell of rotten eggs if sulfide is present.

8.4.1 Elemental Composition

Elements in corrosion deposits can provide information as to the cause of corrosion. Because all living organisms contain ATP, a phosphorus peak in an EDS spectrum can be related to cells associated with the corrosion products. Other sources of phosphorus, for example, phosphate water treatments, must be eliminated. The activities of SRB and manganese-oxidizing bacteria produce surface-bound sulfur and manganese, respectively. Chloride is typically found in crevices and pits and cannot be directly related to MIC.

Energy dispersive X-ray spectroscopy (EDS) coupled with SEM can be used to determine the elemental composition of corrosion deposits. There are several limitations for EDS surface chemical analyses. Samples for EDS cannot be evaluated after heavy metal coating, so EDS spectra must be collected prior to heavy metal coating. It is difficult or impossible to match spectra with exact locations on images. This is not a problem with the ESEM because nonconducting samples can be imaged directly, meaning that EDS spectra can be collected of an area that is being photographed. Little et al. (1991) documented the changes in surface chemistry as a result of solvent extraction of water, a requirement for SEM. Other shortcomings of SEM/EDS include peak overlap. Peaks for sulfur overlap peaks for molybdenum, and the characteristic peak for manganese coincides with the secondary peak for chromium. Wavelength dispersive spectroscopy can be used to resolve overlapping EDS peaks.

Peak heights cannot be used to determine the concentration of elements. It is also impossible to determine the form of an element with EDS. For example, a high sulfur peak may indicate sulfide, sulfate, or elemental sulfur.

8.4.2 Mineralogical Fingerprints

McNeil et al. (1991) used mineralogical data determined by X-ray crystallography, thermodynamic stability diagrams (Pourbaix), and the simplexity principle for precipitation reactions to evaluate corrosion product mineralogy. They concluded that many sulfides under near surface natural environmental conditions could only be produced by microbiological action on specific precursor metals. They reported that djurleite, spinonkopite, and the high-temperature polymorph of chalcocite were mineralogical fingerprints for the SRB-induced corrosion of copper–nickel alloys. They also reported that the stability or tenacity of sulfide corrosion products determined their influence on corrosion.

Jack et al. (1995) maintained that the mineralogy of corrosion products on pipelines could provide insight into the conditions under which the corrosion took place. For example, under anaerobic conditions in the absence of SRB, an iron (II) carbonate was identified in water trapped under defective coatings. Introduction of air caused a rapid discoloration of the white corrosion product to orange iron (III) oxides. In the presence of SRB, indicator minerals are siderite ($FeCO_3$) and iron (II) sulfide in a ratio of 3:1 or more. Mackinawite ($Fe_{1+x}S$) is a sulfur-deficient sulfide that converts to gregite (Fe_3S_4) in a time- and pH-dependent manner. Conversion is only rapid at pH below 6. At pH 6, the conversion is complete in about 3 months. At pH 7, it takes 6 months, and at pH 8, mackinawite persists along with gregite, and pyrrhotite ($Fe_{1-x}S$) may form after 9 months. At aerobic corrosion sites, the minerals are iron (III) oxides: magnetite (Fe_3O_4), hematite (Fe_2O_3), lepidocrocite (γ-FeOOH), and goethite (α-FeOOH). Oxidation rates of seconds to hours have been reported for Fe-deficient sulfides (Howie 1992). Steger and Desjardins (1978) reported that the predominant oxidation products of pyrrhotite were goethite and elemental sulfur. In the GOM samples, goethite dominated the mineralogy, and large concentrations (0.4 to 6.3 wt%) of sulfur were detected by inductively coupled plasma atomic emission spectroscopy (ICP-AES), suggesting that unstable iron sulfides could have formed in the FeRA.

8.4.3 Isotope Fractionation

The stable isotopes of sulfur (^{32}S and ^{34}S) naturally present in any sulfate source are selectively metabolized during sulfate reduction by SRB, and the resulting sulfide is enriched in ^{32}S (Chambers and Trudinger 1979). The ^{34}S accumulates in the starting sulfate as the ^{32}S is removed and becomes concentrated in the sulfide. Little et al. (1993) demonstrated sulfur isotope fractionation in sulfide corrosion deposits resulting from activities of SRB within biofilms on copper surfaces. ^{34}S accumulated in sulfide-rich corrosion products; ^{34}S was concentrated in the residual sulfate in the culture medium. Accumulation of the lighter isotope was related to surface derivatization or corrosion as measured by weight loss. Use of this technique to identify SRB-related corrosion requires sophisticated laboratory procedures.

8.5 SUMMARY

It is essential in diagnosing MIC to demonstrate a spatial relationship between the causative microorganisms and the corrosion phenomena. However, that relationship cannot be independently interpreted as MIC. Pitting due to MIC can initiate as small pits that have the same size and characteristics of the causative organisms. These features are not obvious to the unaided eye and are most often observed with an electron or atomic force microscope. MIC does not produce a macroscopic unique metallographic feature. Metallurgical features previously thought to be unique to MIC, for example, hemispherical pits in 300 series stainless steel localized at weld or tunneling in carbon steel, are consistent with some mechanisms for MIC but cannot be interpreted independently. Bacteria do produce corrosion products that could not be produced abiotically in near surface environments, resulting in isotope fractionation and mineralogical fingerprints. The result is corrosion where none could be anticipated based on the composition of the bulk medium, for example, low-chloride waters and rates that are exceptionally fast. Development of sophisticated genetic and imaging techniques has made it possible to more accurately characterize microorganisms and their spatial relationships to corrosion products and localized corrosion. However, the requirements for diagnosing MIC have not changed.

ACKNOWLEDGMENT

NRL Publication no. NRL/BC/7330—16-2965. Funding support from NRL Base Program.

REFERENCES

Adesina A. Y., I. K. Aliyu, F. M. Al-Abbas. 2015. Microbiologically Influenced Corrosion (MIC) Challenges in Unconventional Gas Fields CORROSION/2015, Dallas, TX. Paper no. 5725. Houston, TX: NACE International.

Almahamedh H. H., C. Williamson, J. R. Spear, B. Mishra, D. L. Olson. 2011. Identification of Microorganisms and Their Effects on Corrosion of Carbon Steels Pipelines CORROSION/2011, Houston, TX. Paper no. 11231. Houston, TX: NACE International.

API RP 38. 1975. API Recommended Practice for Biological Analysis of Subsurface Injection Waters. Dallas, TX: American Petroleum Institute.

Blunn G. 1986. Biological fouling of copper and copper alloys. In: Barry S., Houghton D. R., Llewellyn G. C. and O'Rear C. E., editors. *Biodeterioration VI.* Slough, UK: CAB International. pp. 567–575.

Boivin J., E. J. Laishley, R. D. Bryant, J. W. Costerton. 1990. The influence of enzyme systems on MIC. CORROSION/90, Las Vegas, NV. Paper no. 128. Houston, TX: NACE International.

Borenstein S. W., P. B. Lindsay. 1988. Microbiologically influenced corrosion failure analysis. *Mater Performance*. 27: 51–54.

Borenstein S. W., P. B. Lindsay. 1994. MIC failure analysis. *Mater Performance*. 33: 43–45.

Bryant R. D., W. Jansen, J. Boivin, E. J. Laishley, J. W. Costerton. 1991. Effect of hydrogenase and mixed sulfate-reducing bacterial-populations on the corrosion of steel. *Appl Environ Microbiol*. 57: 2804–2809.

Caffrey S. M. 2011. Application of molecular microbiological methods to the oil industry to analyse DNA, RNA and proteins. In: Whitby C. and Skovhus T. L., editors. *Applied Microbiology and Molecular Biology in Oilfield Systems.* New York: Springer Science. pp. 19–26.

Chamberlain A. H. L., P. Angell, H. S. Campbell. 1988. Staining procedures for characterizing biofilms in corrosion investigations. *Brit Corros J.* 23: 197–198.

Chamberlain A. H. L., W. R. Fisher, U. Hinze, H. H. Paradies, C. A. C. Sequeira, H. Siedlarek, M. Thies, D. Wagner, J. N. Wardell. 1995. An interdisciplinary approach for microbiologically influenced corrosion of copper. Proceedings of Microbial Corrosion; March 13–16; Estoril, Portugal. London: The Institute of Materials.

Chambers A., P. A. Trudinger. 1979. Microbiological fractionation of stable sulfur isotopes: A review and critique. *Geomicrobiol J.* 1: 249–293.

Chung Y., L. K. Thomas. 1999. Comparison of MIC pit morphology with non-MIC chloride induced pits in types 304/304L/E308 stainless steel base metal/welds. CORROSION/99, San Antonio, TX. Paper no. 159. Houston, TX: NACE International.

Cowan J. K. 2005. Rapid enumeration of sulfate-reducing bacteria. CORROSION/2005, Houston, TX. Paper no. 05485. Houston, TX: NACE International.

de Sanchez S. R., D. J. Schiffrin. 1996. Bacterial chemo-attractant properties of metal ions from dissolving electrode surfaces. *J Electroanal Chem.* 403: 39–45.

Dowling N. J. E., P. D. Nichols, D. C. White. 1988. Phospholipid Fatty-Acid and Infrared Spectroscopic Analysis of a Sulfate-Reducing Consortium. *FEMS Microbiol Ecol.* 53: 325–334.

Eckert R. 2003. *Field Guide for Investigating Internal Corrosion of Pipelines.* Houston, TX: NACE International.

Eckert R. B., H. C. Aldrich, C. A. Edwards, B. A. Cookingham. 2003. Microscopic differentiation of internal corrosion Initiation mechanisms in natural gas pipeline systems. CORROSION/2003, San Diego, CA. Paper no. 03544. Houston, TX: NACE International.

Enning D., J. Garrelfs. 2014. Corrosion of iron by sulfate-reducing bacteria: New views of an old problem. *Appl Environ Microbiol.* 80: 1226–1236.

Fichter J., K. Wunch, R. Moore, E. Summer, S. Braman, P. Holmes. 2012. How Hot is Too Hot for Bacteria? A Technical Study Assessing Bacterial Establishment in Downhole Drilling, Fracturing and Stimulation Operations. CORROSION/2012, Salt Lake City, UT. Paper no. C2012-0001310. Houston, TX: NACE International.

Franklin M. J., D. C. White. 1991. Biocorrosion. *Curr Opin Biotech.* 2: 450–456.

Franklin M. J., J. B. Guckert, D. C. White, H. S. Isaacs. 1991. Spatial and temporal relationships between localized microbial metabolic activity and electrochemical activity of steel. CORROSION/91, Cincinnati, OH. Paper no. 115. Houston, TX: NACE International.

Geesey G. G., Z. Lewandowski, H. C. Flemming, editors. 1994. *Biofouling and Biocorrosion in Industrial Water Systems.* Boca Raton, FL: CRC Press.

Geiser M., R. Avci, Z. Lewandowski. 2001. Pit initiation on 316L stainless steel in the presence of bacteria *Leptothrix discophora*. CORROSION/2001, Houston, TX. Paper no. 01257. Houston, TX: NACE International.

Giovannoni S. J., T. B. Britschgi, C. L. Moyer, K. G. Field. 1990. Genetic diversity in Sargasso Sea bacteriopiankton. *Nature*. 344: 60–63.

Hoffmann H., C. Devine, S. Maxwell. 2008. Application of DGGE and FISH Analyses in Seawater Injection Systems. CORROSION/2008, New Orleans, LA. Paper no. 08513. Houston, TX: NACE International.

Hogan J. J. 1990. A rapid, non-radioactive DNA probe for detection of SRBs. *Institute of Gas Technology Symposium on Gas, Oil, Coal, and Environmental Biotechnology.* Chicago: IGT.

Howgrave-Graham A. R., P. L. Steyn. 1988. Application of the fluorescent-antibody technique for the detection of *Sphaerotilus natans* in activated sludge. *Appl Environ Microbiol.* 54: 799–802.

Howie F. M., editor. 1992. *The Care and Conservation of Geological Material: Minerals, Rocks, Meterorites and Lunar Fields.* Burlington, MA: Butterworth-Heinemann.

Jack T. R., M. J. Wilmott, R. L. Sutherby. 1995. Indicator minerals formed during external corrosion of line pipe. *Mater Performance.* 34: 19–22.

Jhobalia C. M., A. Hu, T. Gu, S. Nesic. 2005. Biochemical engineering approaches to MIC. CORROSION/2005, Houston, TX. Paper no. 05500. Houston, TX: NACE International.

Jones-Meehan J., J. W. Cofield, B. Little, R. Ray, P. Wagner, M. McNeil, J. McCay. 1995. Microbiologically Influenced Corrosion of Steels: Weight Loss Measurements, ESEM/EDS and XRD Analyses. International Conference on Microbially Influenced Corrosion, New Orleans, LA. Paper no. 29. Houston, TX: NACE International.

Kilbane J. 2014. Effect of Sample Storage Conditions on Oilfield Microbiological Samples. CORROSION/2014, San Antonio, TX. Paper no. 3788. Houston, TX: NACE International.

King J. M. H., P. M. Digrazia, B. Applegate, R. Burlage, J. Sanseverino, P. Dunbar, F. Larimer, G. S. Sayler. 1990. Rapid, sensitive bioluminescent reporter technology for naphthalene exposure and biodegradation. *Science.* 249: 778–781.

Knight R., J. Jansson, D. Field, N. Fierer, N. Desai, J. A. Fuhrman, P. Hugenholtz et al. 2012. Unlocking the potential of metagenomics through replicated experimental design. *Nat Biotechnol.* 30: 513–520.

Larsen J., J. V. Sorensen, S. Juhler, D. S. Pedersen. 2013. The Application of Molecular Microbiological Methods for Early Warning of MIC in Pipelines. CORROSION/2013, Orlando, FL. Paper no. 2029. Houston, TX: NACE International.

Little B. J., P. A. Wagner, R. I. Ray, R. K. Pope, R. Scheetz. 1991. Biofilms: An ESEM evaluation of artifacts introduced during SEM preparation. *J Ind Microbiol.* 8: 213–222.

Little B. J., P. A. Wagner, R. I. Ray, M. B. McNeil, J. Jones-Meehan. 1993. Indicators of microbiologically influenced corrosion in copper alloys. *OEBALIA.* 19: 287–294.

Little B. J., P. A. Wagner, K. R. Hart, R. I. Ray. 1996. Spatial Relationships between Bacteria and Localized Corrosion. CORROSION/96, Denver, CO. Paper no. 278. Houston, TX: NACE International.

Little B. J., P. A. Wagner, F. Mansfeld. 1997. *Corrosion Testing Made Easy: Microbiologically Influenced Corrosion.* Houston, TX: NACE International.

Littmann E. S. 1977. Use of ATP extraction in oil field waters. In: Wright C. C., Cross, Ostroff A. G. and Stanford J. R., editors. *ASTM STP 641 Oil Field Subsurface Injection of Water.* Philadelphia: ASTM. p. 79.

Lopez M. A., F. J. Z. D. de la Serna, J. Jan-Roblero, J. M. Romero, C. Hernandez-Rodriguez. 2006. Phylogenetic analysis of a biofilm bacterial population in a water pipeline in the Gulf of Mexico. *FEMS Microbiol Ecol.* 58: 145–154.

Lutterbach M. T. S., L. S. Contador, M. M. Galvao, V. de Oliveira, M. Araujo, G. de Souza Pimenta. 2011. Real-time PCR Applied to Quantification of SRB in Bottom Water from Fuel Tanks. CORROSION/2011, Houston, TX. Paper no. 11233. Houston, TX: NACE International.

Mara D. D., D. J. A. Williams. 1972. Influence of the microstructure of ferrous metals on the rate of microbial corrosion. *Brit Corros J.* 7: 139–142.

Marshall K. C., K. N. Power, M. L. Angles, R. P. Schneider, A. E. Goodman. 1994. Analysis of bacterial behavior during biofouling of surfaces. In: Geesey G. G., Lewandowski Z. and Flemming H.-C., editors. *Biofouling and Biocorrosion in Industrial Water Systems.* Boca Raton, FL: CRC Press, Inc. pp. 15–26.

Maxwell S. 1986. Assessment of sulfide corrosion risks in offshore systems by biological monitoring. *SPE Prod Eng.* 1: 363–368.

McNeil M. B., J. M. Jones, B. J. Little. 1991. Mineralogical fingerprints for corrosion process induced by sulfate-reducing bacteria. CORROSION/91, Cincinnati, OH. Paper no. 580. Houston, TX: NACE International.

Mittleman M. W., D. E. Nivens, C. Low, D. C. White. 1990. Differential adhesion, activity, and carbohydrate: Protein ratios of *Pseudomonas Atlantica* monocultures attaching to stainless steel in a linear shear gradient. *Microb Ecol*. 19: 269–278.

Muyzer G., F. Marty. 2014. Molecular methods in microbiologically influenced corrosion: Research, monitoring and control. In: Skovhus T. L., Caffrey S. M. and Hubert C. R. J., editors. *Applications of Molecular Microbiological Methods*. Norfolk: Caister Academic Press. pp. 7–22.

NACE Standard TM0212-2012. 2012. Detection, Testing, and Evaluation of Microbiologically Influenced Corrosion on Internal Surfaces of Pipelines. Houston, TX: NACE International, 36.

Paisse S., J. F. Ghiglione, F. Marty, B. Abbas, H. Gueune, J. M. S. Amaya, G. Muyzer, L. Quillet. 2013. Sulfate-reducing bacteria inhabiting natural corrosion deposits from marine steel structures. *Appl Microbiol Biot*. 97: 7493–7504.

Phelps T. J., R. M. Schram, D. B. Ringelberg, N. J. E. Dowling, D. C. White. 1991. Anaerobic microbial activities including hydrogen-mediated acetogenesis within natural gas transmission lines. *Biofouling*. 3: 265–276.

Plugge C. M., W. Zhang, C. M. Johannes, A. J. M. Stams. 2011. Metabolic flexibility of sulfate-reducing bacteria. *Front Microbiol*. 2. doi: 10.3389/fmicb.2011.00081.

Pope D. H. 1986. Discussion of methods for the detection of microorganisms involved in microbiologically influenced corrosion. In: Dexter S. C., editor. *Biologically Induced Corrosion*. Houston, TX: NACE International. pp. 275–281.

Pope D. H. 1990. *GRI-90/0299 Field Guide: Microbiologically Influenced Corrosion (MIC): Methods of Detection in the Field*. Chicago: Gas Research Institute.

Postgate J. R. 1979. *The Sulfate Reducing Bacteria*. New York: Cambridge University Press.

Price A., L. A. Alvarez, C. Whitby, J. Larsen. 2010. Detection of SRP Activity by Quantification of mRNA for the Dissimilatory (bi) Sulfite Reductase Gene (dsrA) by Reverse Transcription Quantitative PCR. CORROSION/2010, San Antonio, TX. Paper no. 10253. Houston, TX: NACE International.

Ray R., B. Little. 2003. Environmental electron microscopy applied to biofilms. In: Lens P., Moran A. P., Mahony T., Stoodley P. and O'Flaherty V., editors. *Biofilms in Medicine, Industry and Environmental Biotechnology*. London: IWA Publishing. pp. 331–351.

Ridgeway H. F., B. H. Olson. 1981. Scanning electron microscope evidence for bacterial colonization of a drinking-water distribution system. *Appl Environ Microbiol*. 41: 274–287.

Romero J. M., E. Velazquez, G.-V. J. L., M. Amaya, S. Le Borgne. 2005. Genetic Monitoring of Bacterial Populations in a Seawater Injection System. Identification of Biocide Resistant Bacteria and Study of Their Corrosive Effect. CORROSION/2005, Houston, TX. Paper no. 05483. Houston, TX: NACE International.

Roszak D. B., R. R. Colwell. 1987. Survival strategies of bacteria in the natural environment. *Microbiol Rev*. 51: 365–379.

Scott P. J. B., M. Davies. 1989. Microbiologically influenced corrosion of alloy 904L. *Mater Performance*. 28: 57–63.

Scott P. J. B., M. Davies. 1992. Survey of field kits for sulfate-reducing bacteria. *Mater Performance*. 31: 64–68.

Stahl D. A., B. Flesher, H. R. Mansfield, L. Montgomery. 1988. The use of phylogenetically based hybridization probes for studies of ruminal microbial ecology. *Appl Environ Microbiol*. 54: 1079–1084.

Stahl D. A., D. J. Lane, G. L. Olsen, N. R. Pace. 1984. Analysis of hydrothermal vent-associated symbiots by ribosomal RNA sequence. *Science*. 224: 409–411.

Steger H. F., L. E. Desjardins. 1978. Oxidation of sulfide minerals, 4. pyrite, chalcopyrite and pyrrhotite. *Chem Geol*. 23: 225–237.

Stoodley P., K. Sauer, D. G. Davies, J. W. Costerton. 2002. Biofilms as complex differentiated communities. *Annu Rev Microbiol*. 56:187–209.

Summer E. J., S. Duggleby, C. Janes, M. Liu. 2014. Microbial populations in the O&G: Application of this knowledge. CORROSION/2014, San Antonio, TX. Paper no. 4376. Houston, TX: NACE International.

Tatnall R. E., K. M. Stanton, R. C. Ebersole. 1988. Methods of Testing for the Presence of Sulfate-Reducing Bacteria. CORROSION/88, St. Louis, MO. Paper no. 88. Houston, TX: NACE International.

Telegdi J., Z. Keresztes, G. Paalinkas, E. Kalaman, W. Sand. 1998. Microbially influenced corrosion visualized by atomic force microscopy. *Applied Physics A*. 66: S639–S649.

Wagner M., H. Smidt, A. Loy, J. Z. Zhou. 2007. Unravelling microbial communities with DNA-microarrays: Challenges and future directions. *Microb Ecol*. 53: 498–506.

Ward D. M., M. J. Ferris, S. C. Nold, M. M. Bateson. 1998. A natural view of microbial diversity within hot spring cyanobacterial mat communities. *Microbiol Mol Biol R*. 62: 1353–1370.

Welz J. C., J. T. Tverberg. 1998. Case history: Corrosion of a stainless steel hot water system in a brewery. *Mater Performance*. 37: 66–72.

Wrangham J. B., E. J. Summer. 2013. Planktonic Microbial Population Profiles Do Not Accurately Represent Same Location Sessile Population Profiles. CORROSION/2013, Orlando, FL. Paper no. 2248. Houston, TX: NACE International.

Zambon J. J., P. S. Huber, A. E. Meyer, J. Slots, M. S. Fornalik, R. E. Baier. 1984. In-situ identification of bacterial species in marine microfouling films by using an immunofluorescence technique. *Appl Environ Microbiol*. 48: 1214–1220.

Zhang T., H. H. P. Fang, B. C. B. Ko. 2003. Methanogen population in a marine biofilm corrosive to mild steel. *Appl Microbiol Biot*. 63: 101–106.

Zhu X., J. Lubeck, J. J. Kilbane. 2003. Characterization of microbial communities in gas industry pipelines. *Appl Environ Microbiol*. 69: 5354–5363.

Zhu X., J. Lubeck, K. Lowe, A. Daram, J. J. Kilbane. 2004. Improved Method for Monitoring Microbial Communities in Gas Pipelines. CORROSION/2004, New Orleans, LA. Paper no. 04592. Houston, TX: NACE International.

Zhu X., A. Ayala, H. Modi, J. J. Kilbane. 2005. Application of Quantitative, Real-Time PCR in Monitoring Microbiologically Influenced Corrosion (MIC) in Gas Pipelines. CORROSION/2005, Houston, TX. Paper no. 05493. Houston, TX: NACE International.

Zhu X. Y., H. Modi, A. Ayala, J. J. K. II. 2006. Technical note: Rapid detection and quantification of microbes related to microbiologically influenced corrosion using quantitative polymerase chain reaction. *Corrosion*. 62: 950–955.

Zintel T. P., D. A. Kostuck, B. A. Cookingham. 2003. Evaluation of Chemical Treatments in Natural Gas Systems versus MIC and Other Forms of Internal Corrosion Using Carbon Steel Coupons. CORROSION/2003, San Diego, CA. Paper no. 03574. Houston, TX: NACE International.

9 MIC Detection and Assessment

A Holistic Approach

Mohita Sharma and Gerrit Voordouw

CONTENTS

ABSTRACT

MIC is caused by a variety of microorganisms, such as bacteria, archaea, fungi, algae, and others, acting together to create conditions resulting in metal loss and is typically observed as a type of localized corrosion. However, MIC is not known to produce any type of corrosion that is easily diagnosed by specific tests or observations. Therefore, a holistic approach must be taken when suspecting MIC by looking at microbiological, chemical, metallurgical, and operational indicators that support a MIC root cause and eliminate those abiotic factors that are not major contributors. This chapter describes various practical techniques and methods that can be used to diagnose MIC from samples collected from the field. Many types of samples, including, for example, water, solids, pipe, coupons, and pig sludges, can be collected and analyzed in the field or laboratory. The proper collection and preservation of these samples, however, underscores their usefulness to the investigator in making a good MIC assessment. These tests include the analysis of microorganisms by culture-dependent or culture-independent methods, biochemicals, inorganic and organic ions, corrosion products, and metallurgical observations that try to link the activity of the microorganisms to the corrosion. Tests and techniques to analyze these samples can encompass a wide range of relatively simple chemical/biological tests to more sophisticated tests requiring specialized technical expertise and expensive equipment. Various inspection technologies and MIC assessment models can also be used to screen the likelihood that MIC is a risk. In addition, both field monitoring and laboratory flow cells can be set up that mimic aspects of the subject site and try to provide a replica of the corrosion event. Genomic techniques that can identify microbial communities and activities in field samples provide particularly powerful tools to confirm if processes conducted in the lab relate to those in the field. This chapter discusses some of the pros and cons of these methods and approaches and provide few case histories to illustrate their application in the field.

9.1 INTRODUCTION TO MIC AND MIC DETECTION

Microorganisms and their associated effect on equipment, wells, reservoirs, and transportation pipelines are well known in the oil and gas industry in the form of biofouling, fuel degradation, modification of fluid viscosities, and microbiologically influenced corrosion (MIC). A good understanding of MIC, detection methods, and the factors that contribute to corrosion plays a crucial role in preventing any corrosion-related failure and the associated economic losses. Monitoring and assessment of corrosion damage have been facilitated with advanced and sophisticated technology-based sensors and other equipment currently available in the market. Concurrently, it becomes important to collect samples from the failure site with precaution and suitably apply detection methods onsite and in the laboratory so that a good judgment is made during adoption of subsequent mitigation measures.

9.2 TYPES OF BACTERIA THAT CAUSE MIC—BIOFILM AND PLANKTONIC POPULATIONS: WHAT ARE WE TRYING TO DETECT?

Microorganisms thrive on the surface of metal and equipment used in the oil industry and storage facilities, especially where water is present even in trace quantities (Williamson et al. 2015). The microbial community composition is influenced by the geochemical and physical properties of the reservoir as well as the production practices of the facility (Stevenson et al. 2011). Voordouw et al. (1993, 1996, 1999, 2011) did a comprehensive study on characterization of 16S rRNA genes from oilfield microbial communities and found a variety of sulfate-reducing, fermentative, nitrate-reducing, and sulfide-oxidizing bacteria to be inherently present in the oilfield environment (Gevertz et al. 2000). From a MIC perspective, it should be noted that the thermodynamics and kinetics of the microbial attack on the surface of steel is mainly dependent on the physicochemical environment at the microbial–metal interface (Al Abbas et al. 2012). It starts with the formation of a biofilm. This alters the chemistry by changing the pH, oxygen levels, nutrients, and formation of extrapolymeric substances (EPS) promoting/inhibiting the corrosion of the underlying surface based on the type of biofilm formed on the surface (Wadood et al. 2015). In cases where differential concentration cells are produced on the metal's surface, corrosive substances are generated that may inactivate corrosion inhibitors. Anion ratios are altered leading to change in pH and, in some cases, the acceleration of MIC (Zarasvand and Rai 2014). Several mechanisms have been proposed in the literature to explain MIC, and these are not discussed here (Beech and Sunner 2004; Videla and Herrera 2005; Lewandowski and Beyenal 2008; Enning et al. 2012). Kip and Veen (2015) and Williamson et al. (2015) provide a comprehensive account of various acid-producing bacteria (APB), sulfate-reducing bacteria (SRB) and methanogens related to MIC failures. Conventionally, monitoring programs were mainly focused on the assessment of planktonic microbial populations, but they do not represent the number and type of microorganisms living in the biofilms on the metal surface, which directly contribute to MIC (Videla et al. 1994). Biofilm-associated microorganisms are more likely to cause a localized attack of the metal surfaces and are considered more corrosive than planktonic populations. The exopolysaccharide (EPS) layer they form is not easily penetrated by biocide. Such EPS layers are resistant to antimicrobial agents and have strong adherence to metal surfaces (Labena et al. 2015). Organic inhibitors can also be biodegraded by biofilms, and they can act as carbon and energy sources for planktonic microorganisms in the long run (Starosvetsky et al. 2001; Muthukumar et al. 2003).

The collection of samples from oilfield operations should be methodically performed in cooperation with the oilfield operators. Once the collected samples are brought to the laboratory for subsequent testing, all relevant physical, chemical, electrochemical, analytical, and microbial tests should be performed to get a holistic picture of conditions that contributed to MIC as shown in Figure 9.1.

Xue et al. (2015) recently documented a step-by-step guide from sample collection to laboratory detection of field samples, including water chemistry, microbiological assays, and incubation methods, to study microbial souring and control in the

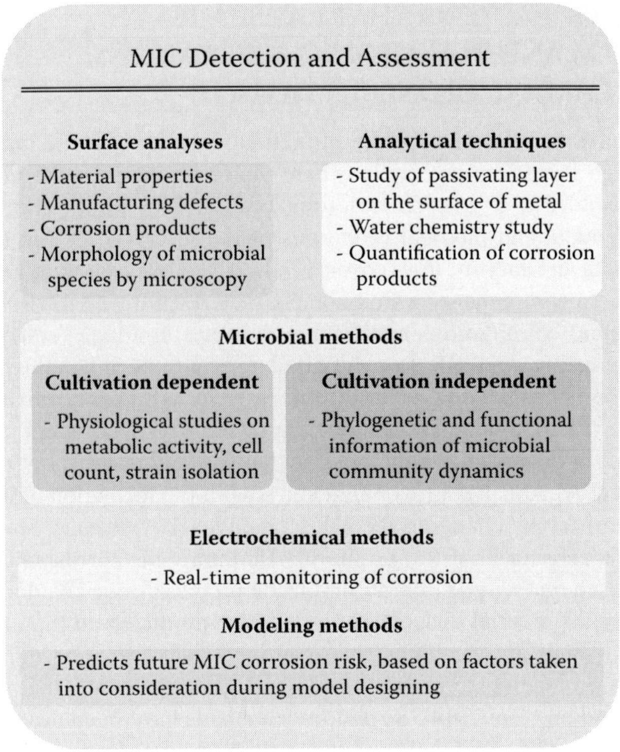

FIGURE 9.1 Different methods that should be used together for a holistic understanding of MIC.

laboratory environment using bioreactors. ASTM D4057 provides standard practices and procedures for the collection of petroleum-related samples. Muthukumar et al. (2003) also provided an extensive review on MIC in petroleum product pipelines along with factors that influence oil degradation and its control.

In the following sections, we briefly describe the main microbial groups considered as key players in MIC.

9.2.1 Sulfate-Reducing Bacteria (SRB)

SRB are responsible for sulfide generation by dissimilatory sulfate reduction. Sulfide can react with metal surfaces, causing iron depolarization and subsequent corrosion. Sulfide can be reoxidized to sulfur or sulfate by sulfide-oxidizing bacteria like *Thiomicrospira*, forming sulfur and polysulfide, which are also corrosive (Lin et al. 2008; Enning and Garrelfs 2014). The energy metabolism of SRB has been studied by Voordouw et al. (Voordouw 1995, 2000; Grigoryan and Voordouw 2008) in part through characterization of *Desulfovibrio* species, which are among the most corrosive SRB. Sulfide can be further oxidized to polysulfide, thiosulfate, and elemental sulfur due to oxygen ingress in pipelines, and these can increase the risk of biocorrosion to metal infrastructure. Fardeau et al. (1993) reported a *Thermoanaerobacter* strain, isolated from an oil-producing well, which is capable of oxidizing H_2 by reducing thiosulfate to sulfide. *Fusibacter paucivorans* is another thiosulfate-reducing, halotolerant strain isolated from an African saline oil-producing well. Magot et al. (1997) reported the study of a highly corrosive thiosulfate-reducing bacterium *Dethiosulfovibrio peptidovorans* SEBR 4207T, isolated from corroding offshore oil wells in the Congo where high concentrations of thiosulfate were detected. This bacterium showed active pitting corrosion with penetration rates of up to 4 mm/yr. Some SRB, like *Desulfovibrio vulgaris* Hildenborough, can also reduce elemental sulfur and polysulfide back to sulfide (Johnston and Voordouw 2012). A corrosion rate of 0.5 mm/yr assessed through electrochemical assays was reported for this freshwater strain (Pankhania et al. 1986). Some others, such as *Desulfopila corrodens* ("*Desulfobacterium corrodens*" IS4) and *Desulfovibrio ferrophilus*, can reduce sulfate with metallic iron as the sole electron donor (Dinh et al. 2004; Enning et al. 2012). Kato (2016) has recently provided a comprehensive review of the relevance of extracellular electron transfer mechanisms to MIC. Lee et al. (1995) provided an extensive review of the role of SRB in corrosion of mild steel and also emphasized the importance of process analysis (defining the system, identification of factors contributing to corrosion, and evaluation of a rate-controlling step) and careful examination of factors that lead to corrosion failure in the field. Understanding these factors will help in better design of laboratory experiments to bridge the gap between the two. Electrochemical and weight-loss studies along with scanning electron microscopy (SEM) and energy dispersive spectroscopy (EDS) measurements were performed by Fonseca et al. (1997) on *Desulfovibrio desulfuricans*. They reported through instantaneous current density and initial weight-loss measurements an enhancement in the corrosion rate in the presence of SRB in lactate–sulfate but not in lactate–nitrate medium. The current density in the presence of SRB was 30 $\mu A/cm^2$ in lactate–sulfate medium as compared to 20 $\mu A/cm^2$ in lactate–nitrate medium. The biofilm formed by SRB on the surface of the metal altered the electrochemical process and pH, produced H_2S, and introduced multiple cathodic reactions that also promoted the anodic dissolution process. Iron sulfide also forms galvanic cells on the surface of metal, resulting in localized galvanic attack of the iron surface adjacent to deposits of iron sulfide (Al Abbas et al. 2012).

9.2.2 Sulfur-Oxidizing Bacteria (SOB)

SOB, like *Thiobacillus* spp., oxidize reduced forms of sulfur to sulfate-producing acid and cause severe corrosion damage to steel surfaces (Zaveri et al. 2007). Accelerated low-water corrosion of a steel piling in a harbor in Southern England where sulfate-reducing and sulfur-oxidizing bacteria were found to propagate together has also been reported by Beech and Campbell (2008). Dang et al. (2011) reported a study focused on understanding the microbial diversity associated with putative early carbon steel surface colonizing MIC-related microorganisms in coastal seawater. *Sulfurimonas*-related bacteria and *Arcobacter* were reported to be early surface colonizers, which participated in surface microbiota development on submerged carbon steel surfaces contributing to MIC. Phototrophic SOB of the orders *Chromatiales* and *Rhodobacterales* were also abundantly present on the carbon steel surfaces, indicating that SOB play an important role in sulfur cycling in marine environments.

9.2.3 Acid-Producing Bacteria (APB)

Though SRB are considered to be the main microorganisms to cause corrosion, fermenting bacteria like APB and acetoclastic methanogens can also attack the metal surface and accelerate corrosion. APB convert carbohydrates to organic acids and CO_2, which hydrates in water to form carbonic acid. Carbonic acid further disassociates into carbonate and bicarbonate ions, which react with ferrous iron giving $FeCO_3$ as a corrosion product (Papavinasam 2013).

$$Fe + H_2CO_3 \rightarrow FeCO_3 + H_2 \quad (9.1)$$

Besides this, the formation of an $FeCO_3$ layer is dependent on other in situ conditions like the partial pressure of CO_2, pH, temperature, and the presence of organic acids, like acetic acid, which can temporarily decrease corrosion by forming a passivating layer on the surface of the metal. Pitting corrosion can still easily occur once this layer is disturbed. Williamson et al. (2015) reported the presence of strictly anaerobic APB like *Acetobacter* and *Gluconobacter* to be abundantly present in fuel-grade ethanol capable of oxidizing ethanol to acetic acid and further to CO_2. *Acetobacter* species have also been reported to cause rapid propagation of fatigue cracks of carbon steel grade used for pipelines and tanks (Sowards et al. 2014). APB can cause corrosion of pumping equipment at fueling stations and can accelerate corrosion of cathodically protected stainless steel (Little et al. 1988; Pollock 2012). APB like *Pseudomonas* are known for producing extracellular polymeric substances (EPS), which bind with iron oxides, iron phosphates, and phosphate minerals on the metal surface to form an iron–EPS complement. Beech et al. (1991) reported a comprehensive study on the composition of EPS formed by *Desulfovibrio desulfuricans* and *Pseudomonas fluorescens* on steel coupons and highlighted the potential of increased corrosion by these biofilms due to accumulation of organic acids on the metal surfaces, resulting in alteration of electrochemical processes on the biofilm–metal interface. EPS of *Pseudomonas alcaligenes* was also reported to be acidic in a separate study, resulting in accelerated corrosion (Chongdar et al. 2005).

9.2.4 Manganese and Iron Oxidizers

Manganese-oxidizing bacteria and iron-oxidizing bacteria form mineral deposits like oxides and hydroxides that cover the metal surface and provide oxygen-depleted zones for other anaerobes like SRB to proliferate (Chukhrov and Gorshkov 1981). Some of the commonly reported ones include *Leptothrix*, *Siderocapsa*, *Gallionella*, and *Sphaerotilus*. Novel iron-oxidizing bacteria isolated from deep-sea hydrothermal vents and coastal ocean waters (mainly *Marinoprofundus*) have been classified as *Zetaproteobacteria* and are associated with accelerated low water biocorrosion in coastal areas (McBeth et al. 2010; Dang et al. 2011). Some other genera like *Bacillus* and *Pseudomonas* and some fungi are also known for mineral deposition. These actively participate in localized corrosion attacks on the surface of metals by causing electrochemical alterations (Starosvetsky et al. 2001; Shi et al. 2002). These microorganisms can promote accumulation of chloride ions leading to formation of ferric and manganese chlorides, which further contribute to pitting corrosion (Videla and Characklis 1992; Chamritski et al. 2004). The Mn oxidizer *Leptothrix discophora* was reported to contribute to pitting corrosion by formation of Mn oxides, shifting the open circuit potential of stainless steel from –100 to +300 mV (SCE) (Dickinson and Lewandowski 1996; Dickinson et al. 1997). Iron-oxidizing bacteria can also cause under-deposit corrosion by creation of oxygen concentration zones and partitioning of the metal surface into smaller anodic sites and larger surrounding cathodic areas (Starosvetsky et al. 2001).

9.2.5 Iron-Reducing Bacteria

The presence of iron-reducing bacteria in oilfields can be either positive or detrimental depending on the niche environment. These microorganisms can protect metal surfaces and can contribute to corrosion by reducing the oxide layers deposited on the surface of the metal, exposing it to corrosive medium (Beech and Gaylarde 1999; Herrera and Videla 2009). IRB like *Pseudomonas* and *Shewanella* are also capable of outcompeting SRB and methanogenic *Archaea* for H_2 as an energy source (Fredrickson and Gorby 1996). Wang et al. (2012) have demonstrated that electroactive *Shewanella* can dissolve protective ferric oxides for anaerobic respiration. However, once a protection layer was formed by synergistic interactions between the iron-oxidizing bacterium (IOB) *Sediminibacterium* and the IRB *Shewanella*, it prevented corrosion by inhibiting the further dissolution of corrosion products. Corrosion products containing Fe (III) like magnetite can be used by hydrogenotrophic IRB, which promote mineral displacement reactions in the presence of hydrogen, altering the protective magnetite layer for the reactivation of the corrosion process (Little et al. 1998; Esnault et al. 2011).

9.2.6 Nitrate-Reducing Bacteria

Nitrate injection into reservoirs is a widely used method to mitigate SRB-based souring. The growth of nitrate-reducing bacteria can outcompete that of SRB, inhibiting H_2S production (Voordouw 2011). Reduction of nitrate injected into a pipeline leads to production of nitrite with potential for enhanced pitting corrosion. Non-hydrogenotrophic nitrate-reducing bacteria, which belong to the phylum *Bacteroidetes*, genus *Prolixibacter* sp., have been reported to cause MIC by coupling Fe^0 oxidation to nitrate reduction (Iino et al. 2015) with sixfold higher corrosion as compared to an aseptic control. Xu et al. (2013) also reported that *Bacillus licheniformis* caused a 14.5 μm maximum pit depth and 0.89 mg/cm^2 normalized weight loss in 1 week, demonstrating a very aggressive pitting attack of C1018 carbon steel under anaerobic conditions. Nitrate attack of Fe^0 is thermodynamically favorable as nitrate is a stronger oxidant than sulfate (Gu et al. 2015).

9.2.7 Methanogens

Acetoclastic and hydrogenotrophic methanogens are two major types that can be directly or indirectly associated with MIC. While hydrogenotrophic methanogens like *Methanogenium*, *Methanoplanus*, *Methanocalculus*, and *Methanobacterium* can cause mild steel corrosion by direct elemental iron oxidation, acetoclastic methanogens like *Methanosaeta* can syntrophically thrive with SRB (Zhang et al. 2003). Accelerated corrosion rate by additional corrosion product deposition was directly related to the proliferation of methanogenic biomass in pure methogenic culture incubations by Daniels et al. (1987). *Methanothermobacter thermoautotrophicus* and *Methermicoccus* have been found in the pigging solid samples of Alaska North slope oil production facilities (Stevenson et al. 2011). A recent study by Okoro and Amund (2015) also reported methanogenic activities related to MIC in onshore oil-producing facilities in Nigeria. An isolate obtained from the sludge of a crude oil storage tank, *Methanococcus maripaludis*, has been reported to oxidize iron and cause corrosion (Mori et al. 2010; Uchiyama et al. 2010). Williamson et al. (2015) also reported the presence of acetoclastic *Methanobacterium* in fuel-grade ethanol environments. Dinh et al. (2004) described marine methanogenic *Methanobacterium* and *Methanobrevibacter* like archaea that were capable of producing methane faster with iron than other hydrogenotrophic methanogens. Beese-Vasbender et al. (2015) found that the corrosive methanogen *Methanobacterium* sp. strain IM1 can microbially electrosynthesize methane at an electrode potential of –0.4 V versus SHE demonstrating extracellular electron transfer capabilities of *Methanobacterium* sp. strain IM1.

9.2.8 General Heterotrophic Bacteria

General heterotrophic bacteria play a significant role in MIC as they promote the attachment of other corrosive species like SRB to the metal surface. Cheng et al. (2009) investigated the corrosion behavior of *Vibrio natriegens* using microscopic atomic force microscopy (AFM), SEM, and EDX and electrochemical techniques like open-circuit potential (OCP), electrochemical impedance spectroscopy (EIS), and potentiodynamic polarization (PDP) for a period of 14 days. Biofilm attachment to the surface of 303 SS coupons was studied using AFM and EIS whereas formation of pits on the surface of the metal due to microbial attack was studied using SEM. EDX analysis showed the presence of C, S, and P, further supporting the formation of porous biofilms on the surface of the metal. In the presence of *Vibrio natriegens*, an increase in corrosion current densities was observed using potentiodynamic polarization curves and as a decrease in charge transfer resistance using EIS. Hence, this study effectively demonstrated how the information collected with multiple techniques helped in concluding that the corrosion of stainless steel was mainly due to microbial attack. Bermont-Bouis et al. (2007) investigated a corrosive marine biofilm using both cultivation-dependent and -independent methods. *Gammaproteobacteria* of the genus *Vibrio* were detected by molecular methods whereas *Enterobacteriaceae* of the genera *Citrobacter*, *Klebsiella*, and *Proteus* were detected by cultivation methods. This study highlighted the pros and cons of using culture-dependent and -independent methods and showed that more than one technique should be used to identify all microbial community members involved in corrosion.

9.2.9 Syntrophic Interactions in MIC

Syntrophy of methanogens, acetogens, sulfate reducers, and other bacteria involves interspecies electron transfer or successive metabolism of compounds, neither of which can be metabolized in isolation (Usher et al. 2015). This allows microorganisms to grow under thermodynamically limiting conditions. Recycling of carbon, sulfur, and iron can occur between members of microbial communities. Ramos-Padron et al. (2011) found carbon and sulfur cycling by microbial communities as a function of depth in oil sand tailing ponds routinely treated with gypsum. Multispecies biofilms develop as a result of syntrophic interactions, leading to a cascade of biochemical reactions between species, which can accelerate corrosion (Kip and Veen 2015 and references therein). Usher et al. (2015) demonstrated such syntrophy in corrosion with metallic iron as the sole electron donor under nutrient-limiting conditions. Syntrophic microorganisms from groundwater and soil increased the corrosion rate of mild steel by 45.5% and 41.9%, respectively, as compared to sterile controls. Siderite ($FeCO_3$) and magnetite (Fe_3O_4) iron (II) compounds were identified using XRD and Raman spectroscopy as components of the microbially produced corrosion products. These contributed to iron cycling by direct interspecies electron transfer. Siderite and magnetite are formed by SRB by direct enzymatic reduction of Fe^{3+}, creating galvanic cells on the surface of metals and accelerating corrosion. Syntrophy of manganese and iron oxidizers with SRB allows the former to create oxygen-depleted environments in which the latter can proliferate. Syntrophic community members are often difficult to culture, and hence, 16S rRNA gene–based metagenomic studies can help in understanding their role for better design of corrosion-mitigation strategies (Park et al. 2011). This involves use of ribosomal RNA gene sequence analysis pipeline tools (Soh et al. 2013), metagenome rapid annotation using subsystem technology (MG-RAST, Glass et al. 2010), mothur (Schloss et al. 2009), or quantitative insights into microbial ecology (QIIME, Kuczynski et al. 2012), which all allow handling of the large data sets obtained.

9.3 METHODS FOR MIC-RELATED MICROORGANISM DETECTION

The microbial community composition of an oilfield environment should be determined to control undesired microbial activities like corrosion, souring, and biofouling in a timely manner (McInerney et al. 2007). Establishment of microbial trends and knowledge of microbial dynamics through regular monitoring is of particular importance to decide whether activity has reached critical levels and treatment is necessary. Traditional cultivation methods have been used for this purpose. However, because many microorganisms in nature are not cultivable by standard techniques, new methods of microbial evaluation like next-generation sequencing (NGS), quantitative PCR (qPCR), and fluorescent in situ hybridization (FISH) are now used to complement cultivation techniques (Bermont-Bouis et al. 2007 and references therein). Use of cultivation methods and NGS of samples collected from the field provides information on all possible microbial processes. Methods used in previous studies are summarized in Table 9.1.

TABLE 9.1
Different Methods Used for Studying MIC

Name of the Technique	Focus	Reference
	Microscopic and Surface Analysis Methods	
SEM: Scanning electron microscopy	Morphology of microbial cells, EPS, nature and surface topography and structure of corrosion products	Wadood et al. 2015; Usher et al. 2015; Lee and Characklis 1993; Bermont-Bouis et al. 2007; Bryant et al. 1991; Starosvetsky et al. 2001; Nguyen et al. 2008; Sherar et al. 2011
ESEM: Environmental scanning electron microscopy	Helps in structural analysis of hydrated specimens of biofilms	Prack et al. 1993
Phase contrast microscopy	Provides high-contrast images of microbial specimens and helps in studying biological process in real time	Usher et al. 2015; Tardy-Jacquenod 1996; Uchiyama et al. 2010
Transmission electron microscopy	High-resolution images of thin specimen, provide crystalline, microstructure, topographical, compositional, and morphological information and for study of material properties and defects	Usher et al. 2015
AFM: Atomic force microscopy	Topographical studies, properties of surfaces and single molecules and forces within the molecules, imaging of polymers, nanostructures, and study of hydrated biofilms in real time	Cheng et al. 2009; Telegdi et al. 1998; Wadood et al. 2015; Yuan et al. 2007; Nguyen et al. 2008
Scanning auger microprobe	Good for elemental analysis of the top surface and provides topographical 3-D information with high spatial resolution	Chen et al. 1988; Avci et al. 2013
DIC: Reflected differential interference microscopy	Useful only when the metal surface provides adequate reflective properties to resolve attached microorganisms	Beech 2014
Microautoradiography	Functional bacterial group quantification using radioactively labeled substrate as a measure of active bacterial count	Larsen et al. 2005; Ito et al. 2002
CLSM: Confocal laser scanning microscopy	Study biofilm structures under in situ conditions	Kuehn et al. 1998; Walker et al. 1998
Epifluorescence microscopy and FISH	Cell morphology and identification of different bacteria based on specific primers used for hybridization	Fichter et al. 2015; Larsen et al. 2005; Poulsen et al. 1993; Ito et al. 2002; Gittel et al. 2009; Kumaraswamy et al. 2011

(Continued)

TABLE 9.1 (CONTINUED)
Different Methods Used for Studying MIC

Name of the Technique	Focus	Reference
XRD: X-ray diffraction	Crystal structures, chemical composition of the exopolysaccharides, corrosion products; limitation: lacks high spatial resolution, mineral signatures formed due to MIC attack	Usher et al. 2015
EDS/EDX: Energy dispersive spectroscopy/X-ray analysis	Elemental composition of the corrosion products and offers high spatial resolution	Wadood et al. 2015; Usher et al. 2015; Lee and Characklis 1993; Starosvetsky et al. 2001
PIXE: Particle-induced X-ray emission spectroscopy	Measures X-ray of the sample, generates less secondary radiation, more sensitive to detect trace element distribution	Govil 2001; Beech 2004
AES: Auger electron spectroscopy	Provides elemental abundance of top few nm of surface at very high spatial resolution, good for mapping of corrosion products on a metal surface that experienced localized attack	Beech and Gaylarde 1999
Laser profilometer	For scanning the surface topography, measurement of pit depth and the distribution of pits	Turris et al. 2014
Scanning vibrating electrode technique	Determines the potential gradients developed on a surface using a reference electrode	Iken et al. 2008
Mössbauer spectroscopy	For detection of iron-containing compounds	Beech 2004
AEM: Auger electron microscopy	Can map the corrosion across the metal surface with localized attack	
ATR/FT-IR Attenuated total reflectance Fourier transform infrared spectroscopy	Detects the changes in the thickness of the metal films exposed to flow or stagnant water (with or without microorganisms) conditions	Wadood et al. 2015
	Analytical Methods	
AAS: Atomic absorption spectroscopy	Require large amount of samples	
ICP: Induced coupled plasma spectroscopy	Require large amount of samples	
TOF-SIMS: Time of flight secondary ionization mass spectroscopy	Characterization of microbial cells and bio adhesives (only few ng sample required)	Beech et al. 1999; Beech and Sunner 2004; Wang et al. 2009
Raman	For analysis of corrosion products	Usher et al. 2015; Sherar et al. 2011

(Continued)

TABLE 9.1 (CONTINUED)
Different Methods Used for Studying MIC

Name of the Technique	Focus	Reference
XPS: X-ray photoelectromicroscopy	Helps in studying the passivating layers over the metal surface, distribution of alloying elements within the layer (only few ng sample required)	Yuan et al. 2007; Beech et al. 1999; Johansson et al. 1999; Nguyen et al. 2008
Cell enumeration using adenosine triphosphate (ATP) quantification assay/Adenosine-5′-phosphosulfate reductase (APS) assay	Luminescence-based assays that are easy to use, but there can be interference in measurements due to contaminants present in the samples	Fichter et al. 2015
Isotopic measurements	Radiolabeled sulfate and sulfur isotopic signatures can differentiate biological versus nonbiological souring in reservoirs	Ulrich et al. 1997; Khatib and Salanitro 1997; Gieg et al. 2010; Rozanova et al. 2001; Cavallaro et al. 2005
	Molecular Methods	
Genome probing (RGP)	Determine changes in composition of culturable component of microbial communities	Voordouw et al. 1991, 1992, 1993, 2002; Telang et al. 1997; Shen 1998; Greene and Voordouw 2003
Quantitative PCR	Quantification of the number of microbial 16S rRNA gene copies/ml can be done throughout the amplification cycles and the fluorescence intensity emitted reflects the amplicon concentration in real time	Stevenson et al. 2011; Dang et al. 2011; Zhu et al. 2005; Larsen et al. 2005, 2009, 2010; Guan et al. 2013; Agarwal and Lal 2009; Gittel et al. 2009; Von Der Weid et al. 2008; Wei et al. 2010
Denaturing gradient gel electrophoresis profiles	Visually shows any community shifts	Stevenson et al. 2011; Duncan et al. 2009; Jurelevicius et al. 2008
Culture-independent molecular techniques like next-generation sequencing (NGS)	Provides complete phylogenetic information about the microbial community dynamics	Voordouw et al. 1995; Williamson et al. 2015; Fichter et al. 2015; Bermont-Bouis et al. 2007; Dang et al. 2011; Mand et al. 2014; Stevenson et al. 2011; Usher et al. 2015
Culture-dependent cultivation method	Includes microbial cultivation in selective growth medium and enumeration of bacterial numbers per unit volume (MPN), helps in understanding physiological characteristics	Oblinger and Koburger 1975; Shen and Voordouw 2015; Larsen et al. 2005; Bhagobaty 2014

(*Continued*)

TABLE 9.1 (CONTINUED)
Different Methods Used for Studying MIC

Name of the Technique	Focus	Reference
	Electrochemical Methods	
Open circuit potential	Nondestructive method, potential measurement of the material under study when no net current is passed through the system and can be broadly used for distinguishing typical metal behavior: passive (constant potential) and localized pitting corrosion (oscillating potential)	Fonseca et al. 1998
Tafel extrapolation	Destructive method, measured with keeping either current or potential constant, and the response is used for determining the E_{corr} (corrosion potential) and I_{corr} (corrosion current). $R = 0.13\ i_{corr}\ k/\rho$, where R = corrosion rate (mm/yr), i_{corr} = corrosion density (μA/cm^2), k = equivalent weight of metal, ρ = density of the metal (g/cm^3). Besides Tafel slopes, primary passivation potential, critical current density, passive current, breakdown potential, protection potential also obtained using this method	Cheng et al. 2009; Fonseca et al. 1998; Starosvetsky et al. 2001; Wadood et al. 2015; ATM G5; Papavinasam 2013
Linear polarization resistance (LPR)	One of the most widely used electrochemical techniques in the field. Related to general corrosion of metals at or near their corrosion potential where a small potential scan is applied to metal with respect to corrosion potential (±10 to 20 mV), and Rpi and corrosion current is measured as slope of a potential versus current density plot	ASTM G96; ASTM G3; ASTM G5; ASTM G59; NACE 3T 199; ASTM G102
Electrochemical noise analysis	Nondestructive technique for measuring corrosion processes, specially localized corrosion processes, and pitting factor by statistically studying the fluctuations in current and potential measurements. Interference from other noise sources during measurement should be strictly prevented. $R_n = \sigma_E A/\sigma_I$ where σ_E and σ_I are the standard deviation of potential noise and current noise, respectively, A is the surface area of the specimen, and R_n is considered equal to Rp	Zaveri et al. 2007, 2008; NACE 3T 199; ASTM G199; Tan 2009

(*Continued*)

TABLE 9.1 (CONTINUED)
Different Methods Used for Studying MIC

Name of the Technique	Focus	Reference
Electrochemical impedance spectroscopy	Nondestructive method for calculation of polarization resistance (used in corrosion current calculation) and also for studying the metal–electrolyte interface where frequency response of an AC potential to the metal is studied	Cheng et al. 2009; Nguyen et al. 2008; Wadood et al. 2015; Lee and Characklis 1993; Finkenstadt et al. 2011; ASTM G106
Electrical resistance probe	Nondestructive method, used for monitoring general corrosion $R = \sigma_{metal} L/A$ Where R is the resistance, σ_{metal} is resistivity of metal, and L is the length and A is the cross-sectional surface area	ASTM G96; Li et al. 2007
Four probe sensor	This contains three regular electrodes for corrosion rate determination (WE, CE, RE) and a fourth enzymatic probe for determination of SRB activity	Papavinasam 2013
	Weight Loss Method	
Weight loss method	Destructive method for general corrosion rate determination and can be used for collecting and evaluation of sessile microorganisms present in the biofilms developed over the surface of the coupon and surface analysis of corrosion products	NACE Standard RP0775; ASTM G4-95; ASTM G1; ASTM G31; ASTM G-46

Fichter et al. (2015) compared different culture-dependent and culture-independent techniques like NGS, ATP-based enumeration, most probable number (MPN), and microscopy for samples collected from natural gas and oil production operations. This highlighted that the importance of these tests is based on ease of use, accuracy, cost, and time required to obtain the results. Based on the results from several case studies, they concluded that more than one method is always required to understand the impact of microorganisms in oilfield operations (Skovhus et al. 2009). The MPN method is not effective for samples collected from extreme environments unless the environmental conditions like pH, temperature, salinity, and nutrient composition are matched when the test is conducted. Results from microbial culture–based enumeration, ATP detection assays, and DNA yields can differ by orders of magnitude as the former two only detect live bacterial cells while the DNA yield detects both live and dead bacteria present in the sample. Many of the techniques also require cell lysis, which may not be equally effective for all microorganisms, especially those in biofilms. It should also be noted that many of the NGS-based detection methods suffer from PCR bias. Larsen et al. (2005) applied both cultivation-dependent (MPN) and -independent methods (FISH and microautoradiography [MAR]) to pigging samples collected from the Halfdan field and observed that the MPN technique gave hundredfold lower counts as compared to MAR and FISH techniques and concluded that cultivation-independent methods are more rapid and more accurate and identify many more different bacteria in the samples. This is because MAR and FISH when applied together help in simultaneous in situ determination of identities, ecophysiology, metabolic activities, and specific substrate profiles of individual bacterial cells within complex mixed microbial communities (Lee et al. 1999).

Surface analysis (e.g., XRD) and microscopic methods (e.g., scanning electron microscopy) are now often used to study the role of biofilms in corrosion of metal surfaces. Surface, type of microorganisms, and medium characteristics play a significant role in the adhesion process and the formation of biofilms (Al Abbas et al. 2012). Bacterial gene expression can alter with the surface on which the microorganisms grow. The surface can either promote growth by facilitating the production and attachment of EPS to the metal or can inhibit growth by causing toxicity and impeding attachment to the surface (Beech 2004). In the former case, bacterial attachment can lead to formation of electrochemical cells on the metal surface, resulting in accelerated corrosion. Zinkevich et al. (1996) reported that distinct protein expression was observed when SRB were grown on steel surfaces. Lee et al. (2010) studied alteration of gene expression in the sulfate-reducing bacterium *Desulfovibrio vulgaris* Hildenborough by application of different biocides. NACE TM0212-2012 and NACE TM0194-2014 provide standard methods for the detection, testing, and evaluation of MIC in the oil and gas industry. A comprehensive list of different MIC detection techniques is provided in Table 9.1, and commonly used techniques are briefly described in the following subsections.

9.3.1 MPN Method

Culture-based assays are widely used in the oil and gas industry to determine the number of microorganisms present per unit volume or weight. This method is routinely used for APB and SRB detection as they are considered main groups responsible for MIC-related failures (Shen and Voordouw 2015). This method involves statistical evaluation of the number of cultivable cells of the microorganism of interest per unit volume (usually conducted in triplicate) by use of a selective medium required for the detection of microorganisms of interest (like APB and SRB) and by tenfold serial dilution of a given sample in order to determine the MPN (Oblinger and Koburger 1975). Shen and Voordouw (2015) recently described a scaled-down, cost-effective, and microtiter plate–based MPN method, which is as accurate as the traditional MPN bottle tests. This scaled-down method is very cost-effective for determining MPNs of field samples in the lab. It is not yet easily used in the field.

9.3.2 Enzymatic Assays

Adenosine triphosphate (ATP, present in all viable cells)–and adenosine-5′-phosphosulfate APS (a key metabolite involved in sulfate reduction)–based enumeration is available as portable test kits, which can be used at field sites. These luminescence-based assays are based on the principle that every picogram of ATP or APS measured is directly proportional to a defined number of bacterial cells. However, it should be noted that contaminating particulates like oil, sand, salts, or other organic residuals might also be present in these samples while these assays are conducted. This can compromise the accuracy and sensitivity of these assays. Landoulsi et al. (2008) provided a comprehensive review of enzymes involved in MIC on stainless steels and explained how determination of oxygen concentrations in biofilms and other biochemical indicators like catalases, peroxidases, and glucose oxidases from aerobic biofilms and hydrogenases, APS reductase, and dissimilatory sulfite reductase (Dsr) from SRB can help in the understanding of MIC mechanisms.

9.3.3 PCR-Based Methods

Specific DNA primers can help in identification of key organisms involved in MIC and also help in tracing the presence of particular microorganisms like SRB from an oilfield. In these methods, DNA is extracted from an oilfield environment and amplified using "universal" or specific bacterial and/or archaeal 16S rRNA gene primers (Gieg et al. 2011). Ribosomal RNA genes have highly conserved regions in microorganisms, allowing primer design for PCR amplification, but also have more variable regions that allow identification of different taxa in characterization of microbial communities (Soh et al. 2013 and references therein).

The reverse genome probing technique helped in early studies of the diversity of SRB in Canadian oilfields (Voordouw et al. 1990, 1991, 1992, 1993, 1996, 2000). Study of hydrogenase genes present in *Desulfovibrio* also allowed identification of species from the oilfield environment (Voordouw et al. 1989). Various rapid PCR-based detection methods have been reported for identification of SRB in oil ecosystems by specifically targeting the *dsrAB* genes coding for Dsr, required for the final step of the sulfate reduction pathway (Agrawal and Lal 2009; Gittel et al. 2009; Guan et al. 2013) and the *aps* gene encoding adenosine-5-phosphosulfate reductase (Jurelevicius et al. 2008; Von Der Weid et al. 2008; Wei et al. 2010). Shen and Voordouw (2015) have also recently published a protocol to detect the presence of SRB in oil reservoirs using primers (DSRp2060F and DSR4R) for *dsrAB* genes. Use of qPCR for the study of microbial communities present in the samples from oilfields has been extensively done by Larsen et al. (2005, 2009, 2010). Zhu et al. (2005) used qPCR for quantification of the number of copies of a specific gene to study the distribution, abundance, and functional diversity present in microbial communities related to MIC in oil and gas pipelines.

9.3.4 Microscopy Techniques for Evaluation of Corrosive Biofilms

Surman et al. (1996) compared microscopy techniques for the study of biofilms. Atomic force microscopy has the advantage of in situ high-resolution imaging of the surface for the study of corrosion and biofilm formation in aqueous environments (Telegdi et al. 1998; Cheng et al. 2009). Confocal laser scanning microscopy (CLSM) is a noninvasive technique that allows in situ imaging of hydrated biofilms with minimum artifacts and loss of out-of-focus information. This technique has been used for the evaluation of biofilms in biocorrosion studies. Walker et al. (1998) used CLSM for the study of MIC on copper plumbing tubes and established a correlation of the type of biofilm with stimulation of pitting corrosion. Table 9.1 lists commonly used microscopy techniques used for MIC and surface analyses of the material under study.

9.3.5 Surface and Corrosion Product Analyses as Proxies for MIC (Chemical Indicators)

New techniques allow the study of metal binding to microorganisms and related biomolecules using a small sample size (Beech et al. 1999). These have indicated a correlation between the corrosion products and the associated microbial action contributing to microbially influenced pitting corrosion (Table 9.1). These include the microbial deposition of Fe oxides and Mn oxides and hydroxides by manganese and iron-oxidizing bacteria that leads to tubercle formation (Dickinson and Lewandowski 1996). Predominance of iron-reducing bacteria (IRB) will increase the fraction of Fe^{2+} in the oxide layer. IOB, on the other hand, oxidize Fe^{2+} to Fe^{3+}, forming an insoluble Fe_2O_3 layer that may protect the metal surface from corrosion depending on how intact this layer is (Papavinasam 2013). The presence of magnetite, mackinawite, and greigite among the corrosion products indicates the participation of SRB in corrosion. Corrosion tubercles with siderite and iron sulfide were found to be associated with a MIC failure site of a gas transmission pipeline, which had disbonded tape coating (Jack et al. 1995). Enning et al. (2012) found the deposition of siderite and amorphous iron sulfide and coprecipitated minerals like calcite in the corrosion crust of experimental and field samples incubated with marine SRB where metallic iron was provided as the sole electron donor for sulfate reduction. Sherar et al. (2011) reported that corrosion product morphology (in the form of nonstoichiometric tetragonal mackinawite, stoichiometric hexagonal troilite, or cubic ferrous sulfide) on the surface of the metal is dependent on the pH, nutrients, and the concentration of sulfide formed (Sherar et al. 2011). They also found mackinawite to be the dominant iron phase formed in both biotic (SRB consortium) and abiotic conditions. Chalcocite, covellite, djurleite, and copper sulfides could also be formed by SRB during corrosion of copper and its alloys (Beech and Gaylarde 1999 and references therein).

9.4 MIC STUDY AND ASSESSMENT METHODS IN THE LABORATORY AND IN THE FIELD

Laboratory incubations often use synthetic medium for propagating microbial biomass, which may not be a true representation of oilfield environmental conditions (Gieg et al. 2011). However, information collected through both in situ and laboratory-based incubations can help in designing predictive models for oilfield environments. Isotopic and radiolabeling methods can be performed both in situ and in the laboratory to distinguish biological versus abiotic souring in oilfields as the reduction of $^{35}S\text{-}SO_4^{2-}$ to sulfide, or sulfur can be traced in reservoir fluids

(Gieg et al. 2010, 2011 and references therein). Nondestructive techniques do not alter the surface of the sample under study and can be used repeatedly to follow MIC corrosion during the experiment. This includes EIS and open circuit potential or electrochemical noise measurement. Destructive techniques include potentiodynamic polarization, which alters the surface of the metal coupon and can thus be used only once.

9.4.1 Weight-Loss Method

The weight loss method is commonly used in the industry to evaluate the corrosion rate of the material used for fabrication of pipelines and other infrastructure in the field. This method is easy to perform both at the field site and in the laboratory. Inspection coupons are generally installed in the infrastructure itself and removed from time to time to measure corrosion. However, it should be kept in mind that the placement of coupons has a huge impact on the weight-loss corrosion rates. Hence, it is crucial to place them in a position that truly represents the corrosion under study, that is, at the top, middle, or bottom of the pipeline to determine major factors influencing corrosion. These different positions could represent gaseous, oil, and aqueous phases, respectively. For weight loss determination of the corrosion rate, the coupons are pretreated by polishing and cleaning. They are then weighed and incubated for a certain time based on experimental conditions. At the end of the experiment, the coupons are again cleaned and reweighed to measure the weight loss and calculate corrosion rate (Rcorr) using the following equation:

$$\mathrm{Rcorr} = (\mathrm{K} \times \Delta \mathrm{W})/\mathrm{D} \times \mathrm{A} \times \mathrm{T} \tag{9.2}$$

K is a constant, ΔW is the weight loss (g) at the end of incubation, D is the density of the coupon material, A is the surface area of the coupon, and T is the incubation time (h).

The coupons can yield three kinds of related information, including the general corrosion rate, the pitting corrosion rate by measuring pit depth (using a pit gauge or laser profilometer), and the pitting factor by calculating the ratio of the corrosion rate of the deepest pit to the general corrosion rate (Papavinasam 2013). The coupons should be made from the same structural metal as used in the field. Another advantage of using the weight loss method is that the biofilms on the surface of the coupons can be collected and analyzed further using microscopic and molecular methods in order to characterize the corrosion products and microorganisms involved in localized corrosion. This can be compared to the planktonic microorganisms in water samples in contact with the studied structure. Recently, carbon steel ball bearings have also been used in a number of MIC-related studies for weight-loss measurements, and they are uniform and available in various sizes (Sharma et al. 2016; Voordouw et al. 2016). A number of theoretical models have been developed, which include various environmental factors and data collected from the field operations, to predict the future risk of MIC. Some of these MIC models and the factors these take into account are listed in Table 9.2.

TABLE 9.2
MIC-Associated Models

	MIC Models	
Name of the Model	**Basis of the Model**	**Mainly Developed for**
Checkworks model	For comparison of MIC at different locations and takes into account material–aqueous interface, temperature, operation mode, biocide dose, distance between the biocide dosage site and site of interest, discontinuity to incorporate the factor of weld and crevice points	Cooling water and industrial water systems
Union Electric Callaway index	Provides likelihood of MIC occurrence based on field data and includes data related to the growth of *SRB*, *Clostridia*, and *Gallionella*, include information related to sludge and silt deposits, visual inspection of deposit or tubercle formation	
Lutey and Stein model	Emphasis on which type of bacteria are present: SRB, slime formers, IOB, MOB, APB, includes data related to fouling, scaling, sediments, material, flow rate	
Pots model	Includes environmental factors that can promote MIC like pH, TDS, temperature, sulfate, total carbon, nitrogen, C:N ratio, flow rate, pigging frequency, deposits, oxygen ingress, biocide, age of pipeline	Oil and gas industry
Mawell and Campbell model	Modification of Pot's model and also includes additional factors like cell numbers and growth rate of SRB, sulfide production, water content, pigging efficiency, biofilm formation, lag period before biofilm development, non-MIC corrosion rate	
Sooknah model	Based on flow rate, pH, pressure, Langemuir Saturation Index, to predict formation of calcium carbonate scale, temperature, presence of water, TSS, redox potential, TDS, sulfur content	

Source: S. Papavinasam, *Corrosion control in the oil and gas industry*, ISBN: 9780123970220, Gulf Professional Publishing Elsevier, 2013, and references therein.

9.4.2 Electrochemical Methods

Electrochemical methods in the laboratory and in the field use the material under study as the working electrode, together with the reference electrode and counter electrode during measurements in a three-electrode setup. NACE standard 3T199 on techniques for monitoring corrosion and related parameters in field applications provides an exhaustive list of internal corrosion-monitoring methods. Many of the electrochemical methods used in the lab and in the field, like Tafel extrapolation, linear polarization resistance (LPR), EIS, and other techniques are briefly discussed in Table 9.1. Two commonly observed electrochemical related MIC processes are described here.

9.4.2.1 Ennoblement due to Microbial Biofilms

Ennoblement refers to a positive shift in the open circuit or free corrosion potential due to microbial action (Little et al. 2013). It is considered as one of the first indicators of the influence of biofilms on the electrochemical behavior of metals (Landoulsi et al. 2008) and can be detected by monitoring the shifts in anodic or cathodic processes. The passivity of some alloys like stainless steel can get altered due to changes in potential and anion concentrations of the electrolyte due to microbial actions especially that of SRB (Landoulsi et al. 2008).

9.4.2.2 Electrochemical Interference: Cathodic Protection

Esquivel et al. (2011) evaluated XL 52 steel in the presence of SRB to determine if the applied potential of –0.950V versus Cu/$CuSO_4$ was sufficient for cathodic protection. They performed physicochemical (H_2S production, iron concentration, electrolyte alkalinity), microbial (MPN), and electrochemical evaluation (PR, EIS) and concluded that this potential was insufficient for cathodic protection of steel in the presence of SRB; in fact, localized corrosion was seen on the surface of the steel along with generation of cathodic hydrogen. Yu et al. (2013) demonstrated that the presence of SRB like *Desulfovibrio caledoniensis* can accelerate anaerobic corrosion and that these microorganisms obtained electrons from a Q235 carbon steel surface polarized at –0.74V versus SCE. Cetin and Aksu (2009) studied the role of the gram-positive SRB *Desulfotomaculum* in the corrosion process and observed a cathodic shift and deeper pits on the coupon surface, which correlated with an increase in corrosion rates. Systems inoculated with *Pseudomonas aeruginosa* and *Bacillus subtilis* showed higher solution and charge transfer resistance in impedance measurements as compared to sterile controls mainly due to accumulation of microbial metabolic products and the formation of a biofilm–metal oxide layer on the surface of the metal, respectively. This led to formation of a passivation layer and in reduction of biocorrrosion by formation of compact and homogenous biofilms (Wadood et al. 2015).

9.4.3 Batch versus Flow-Through Systems

Batch incubations are normally performed in serum bottles, and the samples collected from the field are used for enrichment of microorganisms that are constituents of the microbial community of the field samples. On the other hand, for continuous mode experiments, bioreactors are used, which have solid packing through which liquid (medium, produced waters collected from the field, etc.) is passed to simulate a well environment. This approach has been used to study a souring environment and perform mitigation tests with chemical injections (Kaster et al. 2007; Voordouw 2007; Callbeck et al. 2011; Pinnock and Voordouw 2016). Lee and Characklis (1993) studied mixed population biofilms containing SRB in continuous flow reactors to simulate continuous industrial processes. They focused on understanding the effect of substrate loading rate on corrosion of mild steel, correlation of SRB activity with a cathodic protection reaction, and the effect of iron sulfide film and of suspended iron sulfide on the corrosion of mild steel in the presence of biofilm. They used biological (MPN method), water chemistry (total iron, dissolved oxygen, lactate, acetate, sulfate, total dissolved solids, and redox potential), and electrochemical tests (corrosion potential, cathodic polarization, pitting potential, alternating current impedance), weight loss, and surface analysis techniques (SEM and EDX). The major findings were that accumulation of SRB biofilm was strongly dependent on the substrate-loading rate. SRB biofilm accumulation increased generalized and localized corrosion at defect sites of the protective iron sulfide film. Intergranular attack was associated with loose iron sulfide deposition. Flow cell–based studies more closely mimic the field environment but are more challenging to design and operate as they require continuous monitoring (Xue et al. 2015).

9.5 STUDY OF MIC ON SAMPLES COLLECTED FROM OIL PRODUCTION FACILITIES AND SAMPLING DEVICES

Because many measurements cannot be done by online monitoring, samples are collected and sent for offline monitoring. The results of offline methods will thus depend on how samples were collected, shipped, and processed in the test facility. The time between sample collection and lab testing should be minimized (Gieg et al. 2011). Van der Kraan (2010) have previously described correct sampling strategy for the study of oilfield ecosystems and elaborately described methods for sampling and treatment of cores. Table 9.3 provides a list of samples collected from different oil production facilities and the corresponding microbial communities detected.

These studies help in identification of key players, which can be further quantified using more rapid PCR assays to determine MIC risk (Larsen et al. 2005; Dang et al. 2011; Stevenson et al. 2011). Water samples may not provide information about the microbial community present on the pipeline walls (Kip and Veen 2015). Under-deposit corrosion can only be studied with samples such as pipeline solids. As Videla (1994) previously described, the main characteristics of a sampling device should include a similar flow regime at the surface of the metal coupon as in the rest of the surface of the pipe; it should be affordable and easy to fabricate, retrofit, install, and disassemble in the conventional devices used in pipeline and laboratory flow loops. It would also be advantageous to have multiple coupon holding assemblies to allow duplication and simultaneous evaluation of different parameters along with compatibility with high-pressure fittings when used for evaluation in high-pressure lines. Many new portable sensors for online monitoring have been recently introduced in the oil and gas industry (Yang et al. 2002; Davis et al. 2000a). A thin wall electrical resistance–type corrosion monitoring tool was developed by Waterman and colleagues (1997) for remote monitoring of the wellbore. Li et al. (2007) developed a thin film electrical resistance (TFER) sensor that can measure corrosion rates in low corrosive environments in the presence of a corrosion inhibitor or cathodic protection. This probe can be connected with a temperature sensor and positioned within the wellbore to allow the fluids to flow over the sensor at the same rate as the fluid flows in the tubing. The probe can be removed anytime for data retrieval and inserted back again. Davis et al. (2000b, 2002) developed an in situ corrosion sensor that is based on the EIS technique, which can monitor material degradation in real time and allow condition-based maintenance, reducing the risk of corrosion-related failures. Wire beam electrodes are multipiece electrodes enabling the measurement

TABLE 9.3
Some MIC Case Studies and Associated Microbial Communities Detected

No.	Sample Type	Location	Water Type Used	Associated Microbial Communities Detected	Corrosion Rate	Treatment Used	Reference
1	IW, PW, PS	Onshore oil production facility	Low sulfate groundwater (0–7 ppm)	SRB, hNRB, methanogens, *Methanolobus*	0.001–0.002 mm/yr	NM	Okoro et al. 2014
2	IW, PW, PS	Offshore oil production facility	High-sulfate seawater (2,200 ppm)	SRB, sulfide oxidizing SRB, hNRB, Marine bacteria like *Parvibaculum*, *Thalassobaculum*, *Clostridiales*	0.045–0.049 mm/yr	THPS (450 ppm for 2 h), twice weekly	Okoro et al. 2014
3	PS, PW, PWRI	North Sea oil producing platforms	High-sulfate seawater (2,600 ppm in PWRI)	*Desulfovibrio* and *Desulfuromonadales*, *Arcobacter*, *Caminicella*, *Methanothermococcus*, *Defluvibacter*, *Fermicutes*, *Marinobacterium*, *Desulfuromonas*	0.46 mm/yr for PW and 0.3 mm/yr for PWRI (LPR)	Gluteraldehyde twice weekly (600 ppm)	Mand et al. 2014
4	Lab test	Lab test to study sulfur corrosion under O_2 ingress	Widdel Pfenning medium	*Desulfovibrio vulgaris* strain Hildenborough	0.1	NA	Johnston and Voordouw 2012
5	Facility water	(SAGD) steam assisted gravity drainage operation	Bicarbonate-rich water	Methanogens, planktonic microbial community: *Acidovorax, Geobacter Desulfovibrio, Rhodocyclaceae and Desulfocapsa* after SBS injection; biofilm microbial community: *Pseudomonas*, *Methanosaeta*, and *Methanobacterium*	0.073 mm/yr (WL) and 0.59 mm/yr (LPR)	Sodium bisulfite addition as oxygen scavenger	Park et al. 2013

(Continued)

TABLE 9.3 (CONTINUED)
Some MIC Case Studies and Associated Microbial Communities Detected

No.	Sample Type	Location	Water Type Used	Associated Microbial Communities Detected	Corrosion Rate	Treatment Used	Reference
6	Lab test	Study of hydrogenases involved in SRB mediated iron corrosion and hydrogen oxidation	Widdel Pfenning medium	*Desulfovibrio vulgaris* strain Hildenborough	NM	NA	Caffrey et al. 2007, 2008
7	PS, PW, IW	Alaska's North Slope	NM	*Thermoanaerobacter pseudethanolicus*, *Deltaproteobacteria*, *Desulfomicrobium thermophilum*, *Synergistetes*, *Thermotoga*, *Thermoacetogenium*, *Proteiniphilum*, *Thermovirga lienii*, *Methanothermobacter thermautotrophicus*, *Methermicoccus*, *Thermococcus alcaliphilus*, *Firmicutes*, *Marinobacterium*, *Pseudomonas*	NM	NM	Stevenson et al. 2011
8	Lab tests using groundwater bore and soil samples	To understand role of syntrophy in corrosion with metallic iron as the sole electron source	LML medium	*Desulfovibrio mexicanus*, *Sulfurospirillum*, *Acetobacterium*	45.5% higher that abiotic control	NA	Usher et al. 2015

(*Continued*)

TABLE 9.3 (CONTINUED)
Some MIC Case Studies and Associated Microbial Communities Detected

No.	Sample Type	Location	Water Type Used	Associated Microbial Communities Detected	Corrosion Rate	Treatment Used	Reference
9	Gypsum-treated oil sand tailing ponds	Athabasca oil sands region in northeastern Alberta, Canada	Pond water	*Syntrophs* (*Pelotomaculum*, *Syntrophus*, and *Smithella* spp.), sulfate- and sulfur-reducing bacteria (SRB, *Desulfocapsa* and *Desulfurivibrio* spp.), acetate- and H_2-using methanogens, *Methanosaeta*, *Methanoregula*, *Brachymonas*, *Thiobacillus*, and *Cellulomonas*	NM	NM	Ramos-Padron et al. 2010
10	Fuel-grade ethanol and accumulated water, solids	Ethanol containment tanks that collect fuel-grade ethanol and water	NM	*Alphaproteobacteria* like *Acetobacter*, *Acidobacterium*, *Methanobacterales*, *Clostridiales, Methylocystis* *Sediminibacterium*, *Denitratisoma*, *Desulfosporosinus*	NM	NM	
11	Continuous flow system	Role of substrate loading rate and iron sulfide on corrosion	NM	*Desulfovibrio desulfuricans*	NM	NA	Lee and Characklis 1993

(Continued)

TABLE 9.3 (CONTINUED)
Some MIC Case Studies and Associated Microbial Communities Detected

No.	Sample Type	Location	Water Type Used	Associated Microbial Communities Detected	Corrosion Rate	Treatment Used	Reference
12	Lab test	Qingdao coastal seawater	Coastal seawater	*Acidobacteria*, *Actinobacteria*, *Bacteriodetes*, *Roseobacter*, *Chloroflexi*, *Cyanobacteria*, *Firmicutes*, *Lentisphaerae*, *Planctomyces* and *Verrucomicrobia*, *Proteobacteria*	30% of the coupon's surface after 1 week	NA	Dang et al. 2011
13	Lab test and field study	To study marine SRB that grow lithotrophically with metallic iron as the only electron donor	CO_2/bicarbonate buffered artificial seawater medium	*Desulfovibrio ferrophilus*, *Desulfopila corrodens*	0.7 mm/yr	NA	Enning et al. 2012

Note: IW: injection water; NA: not applicable; NM: not mentioned; PS: pigging solids; PW: produced water; PWRI: produced water reinjection.

of heterogeneous electrochemical corrosion processes from local areas over a working electrode surface by means of wires connected to those areas, for example, in nonuniform organic films or cathodic protection conditions (Tan 1998; Tan et al. 2000). Ansuini and Howe (1988) developed a photofabrication-based corrosion sensor, which has the capacity to measure both instantaneous and averaged corrosion rates. Papavinasam (2013) developed a patented four-probe biosensor, which can be utilized for simultaneous monitoring of corrosion rate and enzymatic activity of SRB. Besides the three electrodes used for corrosion rate measurements, namely the working electrode (sample coupon), the counter electrode, and the reference electrode, the fourth probe is an enzymatic probe, which is used in combination with counter and reference electrodes to detect sulfide production. The enzymatic probe is fabricated using a paste of sulfide oxidase (SO) enzyme and ferrocene immobilized on the tip of a graphite electrode, which is poised with an applied overpotential of –300 mV. SRB produce sulfide ions, and the SO enzyme (present on the enzymatic sensor probe) oxidizes this sulfide to sulfur with the release of additional electrons, which reduces ferrocene. Reduced ferrocene reoxidizes at the surface of the electrode, and the current generated in this process is proportional to the SRB activity. Hence, it allows simultaneous monitoring of corrosion rate and of enzymatic activity of SRB. Pang et al. (2015) recently reported an amperometric sensor fabricated with single-walled carbon nanotubes (SWCNTs) functionalized with a conducting polymer poly (3,4-ethylenedioxythiopene)-poly(styrenesulfonate) (PEDOT-PSS) to facilitate signal transduction. This probe is claimed to have a large linear detection range, short response time, and high sensitivity to the detection of sulfide through direct oxidation. Since the oxidation of sulfide in this biosensor does not require the assistance of any enzyme or mediator, this makes it ideal for online biosensors for fast and reliable online monitoring of MIC-related SRB activities.

9.6 CONCLUSIONS

A multipronged approach should be adopted when designing a monitoring program where microbiological (planktonic and biofilm microbial community, MPNs, characteristics of biofilm), chemical (water chemistry), electrochemical (EIS, LPR, etc.), and surface analysis (composition of corrosion products) studies are conducted along with the use of in situ monitoring methods to design effective corrosion prevention and mitigation strategies.

REFERENCES

Agrawal, A., and B. Lal. 2009. Rapid detection and quantification of bisulfite reductase genes in oilfield samples using real-time PCR. *FEMS Microbiol. Ecol.* 69(2): 301–312.

Al Abbas, F. M., J. R. Spear, A. Kakpovbia, N. M. Balhareth, D. L. Olson, and B. Mishra. 2012. Bacterial attachment to metal substrate and its effects on microbiologically-influenced corrosion in transporting hydrocarbon pipelines. *J. Pipeline Eng.* 11: 63.

Ansuini, F. J., and R. E. Howe. 1988. *Corrosion sensor for measuring the corrosion loss and the instantaneous corrosion rate*. U.S. Patent 4,780,664.

ASTM D4057-12. 2012. Standard Practice for Manual Sampling of Petroleum and Petroleum Products, ASTM International, West Conshohocken, PA.

ASTM G1-03. 2003. Standard Practice for Preparing, Cleaning, and Evaluating Corrosion Test Specimens.

ASTM G102-89. 1999. Standard Practice for Calculation of Corrosion Rates and Related Information from Electrochemical Measurements. Reapproved 1999.

ASTM G106-89. 2010. Standard Practice for Verification of Algorithm and Equipment for Electrochemical Impedance Measurements.

ASTM G3-89. 2010. Standard Practice for Conventions Applicable to Electrochemical Measurements in Corrosion Testing.

ASTM G31-72. 2004. Standard Practice for Laboratory Immersion Corrosion Testing of Metals. West Conshohocken, PA: ASTM International.

ASTM G4-95. 2002. Standard Guide for Conduction Corrosion Coupon Tests in Field Applications. West Conshohocken, PA: ASTM.

ASTM G46-94. 2005. Standard Guide for Examination and Evaluation of Pitting Corrosion.

ASTM G5-94. 2004. Standard Reference Test Method for Making Potentiostatic and Potentiodynamic Anodic Polarization Measurements.

ASTM G59-97. 2003. Standard Test Method for Conducting Potentiodynamic Polarization Resistance Measurements. Reapproved 2003.

ASTM G96. Standard Guide for On Line Monitoring of Corrodion in Plant Equipment (Electrical and Electrochemical Methods). Standard G96-90.

Avci, R., B. H. Davis, M. L. Wolfenden, I. B. Beech, K. Lucas, and D. Paul. 2013. Mechanism of MnS-mediated pit initiation and propagation in carbon steel in an anaerobic sulfidogenic media. *Corros. Sci.* 76: 267–274.

Beech, I. B. 2004. Corrosion of technical materials in the presence of biofilms—Current understanding and state-of-the art methods of study. *Int. Biodeter. Biodegr.* 53(3): 177–183.

Beech, I. B., and S. A. Campbell. 2008. Accelerated low water corrosion of carbon steel in the presence of a biofilm harbouring sulphate-reducing and sulphur-oxidising bacteria recovered from a marine sediment. *Electrochim. Acta.* 54(1): 14–21.

Beech, I. B., and C. C. Gaylarde. 1999. Recent advances in the study of biocorrosion: An overview. *Rev. Microbiol.* 30(3): 117–190.

Beech, I. B., and J. Sunner. 2004. Biocorrosion: Towards understanding interactions between biofilms and metals. *Curr. Opin. Biotech.* 15(3): 181–186.

Beech, I. B., C. C. Gaylarde, J. J. Smith, and G. G. Geesey. 1991. Extracellular polysaccharides from Desulfovibrio desulfuricans and Pseudomonas fluorescens in the presence of mild and stainless steel. *Appl. Microbiol. Biotechnol.* 35(1): 65–71.

Beech, I. B., V. Zinkevich, R. Tapper, R. Gubner, and R. Avci. 1999. Study of the interaction of sulphate-reducing bacteria exopolymers with iron using X-ray photoelectron spectroscopy and time-of-flight secondary ionisation mass spectrometry. *J. Microbiol. Meth.* 36: 3–10.

Beese-Vasbender, P. F., J. P. Grote, J. Garrelfs, M. Stratmann, and K. J. Mayrhofer. 2015. Selective microbial electrosynthesis of methane by a pure culture of a marine lithoautotrophic archaeon. *Bioelectrochem.* 102: 50–55.

Bermont-Bouis, D., M. Janvier, P. A. D. Grimont, I. Dupont, and T. Vallaeys. 2007. Both sulfate-reducing bacteria and *Enterobacteriaceae* take part in marine biocorrosion of carbon steel. *J. Appl. Microbiol.* 102: 161–168.

Bhagobaty, R. K. 2014. Culture dependent methods for enumeration of sulphate reducing bacteria (SRB) in the oil and gas industry. *Rev. Env. Sci. Bio/Technol.* 13: 11–16.

Bryant, R. D., W. Jansen, J. Boivin, E. J. Laishley, and J. W. Costerton. 1991. Effect of hydrogenase and mixed sulfate-reducing bacterial populations on the corrosion of steel. *Appl. Environ. Microb.* 57(10): 2804–2809.

Caffrey, S. M., H. S. Park, J. K. Voordouw, Z. He, J. Zhou, and G. Voordouw. 2007. Function of periplasmic hydrogenases in the sulfate-reducing bacterium *Desulfovibrio vulgaris* Hildenborough. *J. Bacteriol.* 189(17): 6159–6167.

Caffrey, S. M., H. S. Park, G. Voordouw, and J. Been. 2008. Gene array analysis of sulfate-reducing bacteria grown on an iron electrode under conditions of cathodic protection. In *Corrosion 2008*. Houston, TX: NACE International.

Callbeck, C. M., X. Dong, I. Chatterjee, A. Agrawal, S. M. Caffrey, C. W. Sensen, and G. Voordouw. 2011. Microbial community succession in a bioreactor modeling a souring low-temperature oil reservoir subjected to nitrate injection. *Appl. Microbiol. Biot.* 91(3): 799–810.

Cavallaro, A. N., M. E. Martinez, H. Ostera, H. Panarello, and R. R. Cordero. 2005. Oilfield reservoir souring during waterflooding: A case study with low sulphate concentration in formation and injection waters. In *SPE International Symposium on Oilfield Chemistry*. Society of Petroleum Engineers.

Cetin, D., and M. L. Aksu. 2009. Corrosion behavior of low-alloy steel in the presence of *Desulfotomaculum* sp. *Corros. Sci.* 51(8): 1584–1588.

Chamritski, I. G., G. R. Burns, B. J. Webster, and N. J. Laycock. 2004. Effect of iron-oxidizing bacteria on pitting of stainless steel. *Corrosion.* 60(7): 658–669.

Chen, J. R., S. D. Chyou, S. I. Lew, C. J. Huang, C. S. Fang, and W. S. Tse. 1988. Investigations of the biological corrosion of condenser tubes by scanning Auger microprobe techniques. *Appl. Surf. Sci.* 33: 212–219.

Cheng, S., J. Tian, S. Chen, Y. Lei, X. Chang, T. Liu, and Y. Yin. 2009. Microbially influenced corrosion of stainless steel by marine bacterium *Vibrio natriegens*:(I) Corrosion behavior. *Mater. Sci. Eng. C.* 29(3): 751–755.

Chongdar, S., G. Gunasekaran, and P. Kumar. 2005. Corrosion inhibition of mild steel by aerobic biofilm. *Electrochimica Acta.* 50(24): 4655–4665.

Chukhrov, F. V., and A. I. Gorshkov. 1981. Iron and manganese oxide minerals in soils. *T. RSE Earth.* 72(3): 195–200.

Dang, H., R. Chen, L. Wang, S. Shao, L. Dai, Y. Ye, L. Guo, G. Huang, and M. G. Klotz. 2011. Molecular characterization of putative biocorroding microbiota with a novel niche detection of Epsilon-and Zetaproteobacteria in Pacific Ocean coastal seawaters. *Environ. Microbiol.* 13(11): 3059–3074.

Daniels, L., N. Belay, B. S. Rajagopal, and P. J. Weimer. 1987. Bacterial methanogenesis and growth from CO_2 with elemental iron as the sole source of electrons. *Science.* 237(4814): 509–511.

Davis, G. D., and C. M. Dacres. Dacco Sci, Inc., 2000a. *Portable, hand-held, in situ electrochemical sensor for evaluating corrosion and adhesion on coated or uncoated metal structures*. U.S. Patent 6,054,038.

Davis, G. D., C. M. Dacres, and L. A. Krebs. 2000b. In situ corrosion sensor for coating testing and screening. *Mater. Performance.* 39(2): 46.

Davis, G. D., L. A. Krebs, and C. M. Dacres, 2002. Coating evaluation and validation of accelerated test conditions using an in-situ corrosion sensor. *J. Coating. Technol. 74*(935): 69–74.

Dickinson, W. H., and Z. Lewandowski. 1996. Manganese biofouling and the corrosion behavior of stainless steel. *Biofouling.* 10(1–3): 79–93.

Dickinson, W. H., F. Caccavo, B. Olesen, and Z. Lewandowski. 1997. Ennoblement of stainless steel by the manganese-depositing bacterium *Leptothrix discophora. Appl. Environ. Microbiol.* 63(7): 2502–2506.

Dinh, H. T., J. Kuever, M. Mußmann, A. W. Hassel, M. Stratmann, and F. Widdel. 2004. Iron corrosion by novel anaerobic microorganisms. *Nature.* 427(6977): 829–832.

Duncan, K. E., L. M. Gieg, V. A. Parisi, R. S. Tanner, S. G. Tringe, J. Bristow, and J. M. Suflita. 2009. Biocorrosive thermophilic microbial communities in Alaskan North Slope oil facilities. *Environ. Sci. Technol.* 43(20): 7977–7984.

Enning, D., H. Venzlaff, J. Garrelfs, H. T. Dinh, V. Meyer, K. Mayrhofer, A. W. Hassel, M. Stratmann, and F. Widdel. 2012. Marine sulphate-reducing bacteria cause serious corrosion of iron under electroconductive biogenic mineral crust. *Environ. Microbiol.* 14(7): 1772–1787.

Enning, D., and J. Garrelfs. 2014. Corrosion of iron by sulfate-reducing bacteria: New views of an old problem. *Appl. Environ. Microbiol.* 80(4): 1226–1236.

Esnault, L., M. Jullien, C. Mustin, O. Bildstein, and M. Libert. 2011. Metallic corrosion processes reactivation sustained by iron-reducing bacteria: Implication on long-term stability of protective layers. *Phys. Chem. Earth. Pt. A/B/C*, 36(17): 1624–1629.

Esquivel, R. G., G. Zavala Olivares, M. J. Gayosso, and A. Gayosso Trejo. 2011. Cathodic protection of XL 52 steel under the influence of sulfate reducing bacteria. *Mater. Corros.* 62: 61–67.

Fardeau, M. L., J. L. Cayol, M. Magot, and B. Ollivier. 1993. H2 oxidation in the presence of thiosulfate, by a Thermoanaerobacter strain isolated from an oil-producing well. *FEMS Microbiol. Letters.* 113(3): 327–332.

Fichter, J., E. J. Summer, C. Janes, and G. Hamblin. 2015. Direct comparison of microbial population analysis using ATP quantification, culture based enumeration in bug bottles, epifluorescence microscopy and shotgun population analysis by sequencing 16S amplicons. In *Corrosion 2015*. Houston, TX: NACE International.

Finkenstadt, V. L., G. L. Côté, and J. L. Willett. 2011. Corrosion protection of low-carbon steel using exopolysaccharide coatings from *Leuconostoc mesenteroides. Biotechnol. Letter.* 33(6): 1093–1100.

Fonseca, I. T., M. J. Feio, A. R. Lino, M. A. Reis, and V. L. Rainha. 1998. The influence of the media on the corrosion of mild steel by *Desulfovibrio desulfuricans* bacteria: An electrochemical study. *Electrochim. Acta.* 43(1): 213–222.

Fredrickson, J. K., and Y. A. Gorby. 1996. Environmental processes mediated by iron-reducing bacteria. *Curr. Opin. Biotech.* 7(3): 287–294.

Gevertz, D., A. J. Telang, G. Voordouw, and G. E. Jenneman. 2000. Isolation and characterization of strains CVO and FWKO B, two novel nitrate-reducing, sulfide-oxidizing bacteria isolated from oilfield brine. *Appl. Environ. Microb.* 66(6): 2491–2501.

Gieg, L. M., I. A. Davidova, K. E. Duncan, and J. M. Suflita. 2010. Methanogenesis, sulfate reduction and crude oil biodegradation in hot Alaskan oilfields. *Environ. Microbiol.* 12(11): 3074–3086.

Gieg, L. M., T. R. Jack, and J. M. Foght. 2011. Biological souring and mitigation in oil reservoirs. *Appl. Microbiol. Biot.* 92(2): 263–282.

Gittel, A., K. B. Sørensen, T. L. Skovhus, K. Ingvorsen, and A. Schramm. 2009. Prokaryotic community structure and sulfate reducer activity in water from high-temperature oil reservoirs with and without nitrate treatment. *Appl. Environ. Microb.* 75(22): 7086–7096.

Glass, E. M., J. Wilkening, A. Wilke, D. Antonopoulos, and F. Meyer. 2010. Using the metagenomics RAST server (MG-RAST) for analyzing shotgun metagenomes. *Cold Spring Harbor Protocols.* 2010(1): doi:10.1101/pdb.prot5368 Cold Spring Harb Protoc2010.

Govil, I. M. 2001. Proton induced X-ray emission-A tool for non-destructive trace element analysis. *Current Science-Bangalore.* 80(12): 1542–1549.

Greene, E. A., and G. Voordouw. 2003. Analysis of environmental microbial communities by reverse sample genome probing. *J. Microbiol. Meth.* 53(2): 211–219.

Grigoryan, A., and G. Voordouw. 2008. Microbiology to help solve our energy needs. *Ann. NY Acad. Sci.* 1125(1): 345–352.

Gu, T., D. Xu, P. Zhang, Y. Li, and A. L. Lindenberger. 2014. Microbiologically influenced corrosion and its impact on metals and other materials. *Microbiology for Minerals, Metals, Materials and Environment.* Boca Raton, FL: CRC Press.

Guan, J., L. P. Xia, L. Y. Wang, J. F. Liu, J. D. Gu, and B. Z. Mu. 2013. Diversity and distribution of sulfate-reducing bacteria in four petroleum reservoirs detected by using 16S rRNA and dsrAB genes. *Int. Biodeter. Biodegr.* 76: 58–66.

Herrera, L. K., and H. A. Videla. Role of iron-reducing bacteria in corrosion and protection of carbon steel. *Int. Biodeterior. Biodegradation.* 63(7): 891–895.

Iken, H., L. Etcheverry, A. Bergel, and R. Basseguy. 2008. Local analysis of oxygen reduction catalysis by scanning vibrating electrode technique: A new approach to the study of biocorrosion. *Electrochim. Acta.* 54(1): 60–65.

Iino, T., K. Ito, S. Wakai, H. Tsurumaru, M. Ohkuma, and S. Harayama. 2015. Iron corrosion induced by nonhydrogenotrophic nitrate-reducing *Prolixibacter* sp. strain MIC1-1. *Appl. Environ. Microbiol.* 81(5): 1839–1846.

Ito, T., J. L. Nielsen, S. Okabe, Y. Watanabe, and P. H. Nielsen. 2002. Phylogenetic identification and substrate uptake patterns of sulfate-reducing bacteria inhabiting an oxic-anoxic sewer biofilm determined by combining microautoradiography and fluorescent in situ hybridization. *Appl. Environ. Microb.* 68(1): 356–364.

Jack, T. R., M. J. Wilmott, and R. L. Sutherby. 1995. *Indicator minerals formed during the external corrosion of line pipe.* No. CONF-950304. Houston, TX: NACE International.

Johansson, L. S., and T. Saastamoinen. 1999. Investigating early stages of biocorrosion with XPS: AISI 304 stainless steel exposed to *Burkholderia* species. *Appl. Surf. Sci.* 144: 244–248.

Johnston, S. L., and G. Voordouw. 2012. Sulfate-reducing bacteria lower sulfur-mediated pitting corrosion under conditions of oxygen ingress. *Environ. Sci. Technol.* 46(16): 9183–9190.

Jurelevicius, D., I. V. D. Weid, E. Korenblum, E. Valoni, M. Penna, and L. Seldin. 2008. Effect of nitrate injection on the bacterial community in a water-oil tank system analyzed by PCR-DGGE. *J. Ind. Microbiol. Biotechnol.* 35(4): 251–255.

Kaster, K. M., A. Grigoriyan, G. Jenneman, and G. Voordouw. 2007. Effect of nitrate and nitrite on sulfide production by two thermophilic, sulfate-reducing enrichments from an oilfield in the North Sea. *Appl. Microbiol. Biotech.* 75(1): 195–203.

Kato, S., 2016. Microbial extracellular electron transfer and its relevance to iron corrosion. *Microbial Biotech.* 9(2): 141–148.

Khatib, Z. I., and J. R. Salanitro. 1997. Reservoir souring: Analysis of surveys and experience in sour waterfloods. In *SPE Annual Technical Conference and Exhibition.* Society of Petroleum Engineers.

Kip, N., and J. A. Van Veen. 2015. The dual role of microbes in corrosion. *The ISME Journal.* 9(3): 542–551.

Kuczynski, J., J. Stombaugh, W. A. Walters, A. González, J. G. Caporaso, and R. Knight. 2012. Using QIIME to analyze 16S rRNA gene sequences from microbial communities. *Curr. Protoc. Bioinfo.* 36: 10.7.1–10.7.20.

Kuehn, M., M. Hausner, H. J. Bungartz, M. Wagner, P. A. Wilderer, and S. Wuertz. 1998. Automated confocal laser scanning microscopy and semiautomated image processing for analysis of biofilms. *Appl. Environ. Microbiol* 64(11): 4115–4127.

Kumaraswamy, R., S. Ebert, M. R. Gray, P. M. Fedorak, and J. M. Foght. 2011. Molecular- and cultivation-based analyses of microbial communities in oilfield water and in microcosms amended with nitrate to control H_2S production. *Appl. Microbiol. Biotechnol.* 89(6): 2027–2038.

Labena, A., M. A. Hegazy, H. Horn, and E. Müller. 2015. The biocidal effect of a novel synthesized gemini surfactant on environmental sulfidogenic bacteria: Planktonic cells and biofilms. *Mater. Sci. Eng.* 47: 367–375.

Landoulsi, J., K. E. Kirat, C. Richard, D. Feron, and S. Pulvin. 2008. Enzymatic approach in microbial-influenced corrosion: A review based on stainless steels in natural waters. *Environ. Sci. Technol.* 42(7): 2233–2242.

Larsen, J., K. Rasmussen, H. Pedersen, K. Sørensen, T. Lundgaard, and T. L. Skovhus. 2010. Consortia of MIC bacteria and archaea causing pitting corrosion in top side oil production facilities. In *Corrosion 2010*. Houston, TX: NACE International.

Larsen, J., K. Sørensen, B. Højris, and T. L. Skovhus. 2009. Significance of troublesome sulfate-reducing prokaryotes (SRP) in oilfield systems. In *Corrosion 2009*. Houston, TX: NACE International.

Larsen, J., S. Zwolle, B. V. Kjellerup, B. Frølund, J. L. Nielsen, and P. H. Nielsen. 2005. Identification of bacteria causing souring and biocorrosion in the Halfdan field by application of new molecular techniques. *Corrosion 2005*. Houston, TX: NACE International.

Lee, M. H. P., S. M. Caffrey, J. K. Voordouw, and G. Voordouw. 2010. Effects of biocides on gene expression in the sulfate-reducing bacterium *Desulfovibrio vulgaris* Hildenborough. *Appl. Microbiol. Biotechnol.* 87(3): 1109–1118.

Lee, N., Nielsen, P. H., Andreasen, K. H., Juretschko, S., Nielsen, J. L., Schleifer, K. H. and Wagner, M. 1999. Combination of fluorescent in situ hybridization and microautoradiography—A new tool for structure-function analyses in microbial ecology. *Appl. Environ. Microbiol.* 65(3): 1289–1297.

Lee, W., and W. G. Characklis. 1993. Corrosion of mild steel under anaerobic biofilm. *Corrosion.* 49(3): 186–199.

Lee, W., Z. Lewandowski, P. H. Nielsen, and W. A. Hamilton. 1995. Role of sulfate-reducing bacteria in corrosion of mild steel: A review. *Biofouling.* 8(3): 165–194.

Lewandowski, Z., and H. Beyenal. 2009. Mechanisms of microbially influenced corrosion. In: *Marine and Industrial Biofouling*: 35–65. Flemming, H.-C., Murthy, P. S., Venkatesan, R. and Cooksey, K. E. (eds.), Springer, London.

Li, S. Y., Y. G. Kim, S. Jung, H. S. Song, and S. M. Lee. 2007. Application of steel thin film electrical resistance sensor for in situ corrosion monitoring. *Sens. Actuator B-Chem.* 120(2): 368–377.

Lin, S., G. Voordouw, and V. Coombe. 2008. Sulfur production associated with souring control by nitrate injection: A potential corrosion risk? In *Corrosion 2008*. NACE International.

Little, B., P. Wagner, and D. Duquette. 1988. Technical note: Microbiologically induced increase in corrosion current density of stainless steel under cathodic protection. *Corrosion.* 44(5): 270–274.

Little, J. B., J. S. Lee, R. I. Ray, S. Austin, and J. C. Biffinger. 2013. Ennoblement due to biofilms: Indicator for potential corrosion and source of electrical energy. *Recent Pat. Mater. Sci.* 6(1): 20–28.

Little, B., P. Wagner, K. Hart, R. Ray, D. Lavoie, K. Nealson, and C. Aguilar. 1998. The role of biomineralization in microbiologically influenced corrosion. *Biodegradation.* 9(1): 1–10.

Magot, M., G. Ravot, X. Campaignolle, B. Ollivier, B. K. C. Patel, M. L. Fardeau, P. Thomas, J. L. Crolet, and J. L. Garcia. 1997. *Dethiosulfovibrio peptidovorans* gen. nov., sp. nov., a new anaerobic, slightly halophilic, thiosulfate-reducing bacterium from corroding offshore oil wells. *Int. J. Syst. Evol. Microbiol.* 47(3): 818–824.

Mand, J., T. R. Jack, G. Voordouw, and H. Hoffmann. 2014. Use of molecular methods (pyrosequencing) for evaluating MIC potential in water systems for oil production in the North Sea. In *SPE International Oilfield Corrosion Conference and Exhibition*. Society of Petroleum Engineers.

McBeth, J. M., B. J. Little, R. I. Ray, K. M. Farrar, and D. Emerson. 2011. Neutrophilic iron-oxidizing "Zetaproteobacteria" and mild steel corrosion in nearshore marine environments. *Appl. Environ. Microbiol.* 77(4): 1405–1412.

Mcinerney, M. J., G. Voordouw, G. E. Jenneman, and K. L. Sublette. 2007. Oilfield microbiology: 898–911. In Hurst, C., Crawford, R., Garland, J., Lipson, D., Mills, A., Stetzenbach, L. (eds.), *Manual of Environmental Microbiology*, Third Edition. Washington, DC.: ASM Press.

Mori, K., H. Tsurumaru, and S. Harayama. 2010. Iron corrosion activity of anaerobic hydrogen-consuming microorganisms isolated from oil facilities. *J. Biosci. Bioeng.* 110(4): 426–430.

Muthukumar, N., A. Rajasekar, S. Ponmariappan, S. Mohanan, S. Maruthamuthu, S. Muralidharan, P. Subramanian, N. Palaniswamy, and M. Raghavan. 2003. Microbiologically influenced corrosion in petroleum product pipelines—A review. *Indian J Exp. Biol.* 41(9): 1012–1022.

NACE TM0194. 2014. Field monitoring of bacterial growth in oil and gas systems. ISBN: 1-57590-192-7.

NACE TM0212. 2012. Standard test method for detection, testing and evaluation of microbially influenced corrosion of internal surfaces of pipelines. ISBN: 1-57590-255-9.

Nguyen, T. M. P., X. Sheng, Y. P. Ting, and S. O. Pehkonen. 2008. Biocorrosion of AISI 304 stainless steel by *Desulfovibrio desulfuricans* in seawater. *Ind. Eng. Chem. Res.* 47(14): 4703–4711.

Oblinger, J. L., and J. A. Koburger. Understanding and teaching the most probable number technique. *J. Milk Food Technol.* 38(9): 540–545.

Okoro, C., S. Smith, L. Chiejina, R. Lumactud, D. An, H. S. Park, J. Voordouw, B. P. Lomans, and G. Voordouw. 2014. Comparison of microbial communities involved in souring and corrosion in offshore and onshore oil production facilities in Nigeria. *J. Ind. Microbiol. Biotechnol.* 41(4): 665–678.

Okoro, C. C., and O. Amund. 2015. Souring and corrosion potentials of onshore and offshore oil-producing facilities in the Nigerian oil-rich Niger Delta. *Petrol. Sci. Technol.* 33(17–18): 1563–1570.

Pang, X., C. Shi, M. Arafin, and R. Zhang. 2015. A highly sensitive amperometric sensor for monitoring sulfide and microbiologically influenced corrosion. In *Corrosion 2015*. Houston, TX: NACE International.

Pankhania, I. P., A. N. Moosavi, and W. A. Hamilton. 1986. Utilization of cathodic hydrogen by *Desulfovibrio vulgaris* (Hildenborough). *Microbiol.* 132(12): 3357–3365.

Papavinasam, S. 2013. *Corrosion Control in the Oil and Gas Industry*. ISBN: 9780123970220, Gulf Professional Publishing Elsevier, Burlington, MA.

Park, H. S., I. Chatterjee, X. Dong, S. H. Wang, C. W. Sensen, S. M. Caffrey, T. R. Jack, J. Boivin, and G. Voordouw. 2011. Effect of sodium bisulfite injection on the microbial community composition in a brackish-water-transporting pipeline. *Appl. Environ. Microbiol.* 77(19): 6908–6917.

Park, H. S., J. Mand, T. R. Jack, and G. Voordouw. 2013. Microbial community composition and MIC in water systems for bitumen production by SAGD. In *Corrosion 2013*. Houston, TX: NACE International. Conference and Expo, Paper 2768.

Pinnock, T., and G. Voordouw. 2016. Use of carbon steel beads to determine microbially influenced pitting corrosion under flow conditions. NACE International Corrosion 2016. Conference & Expo, Vancouver, BC, March 6–10. Paper 7772.

Pollock, S. 2012. Accelerated corrosion of UST equipment (an ethanol hangover). In *23rd National Tank Conference*, March 19–21, 2012, St. Louis, MO.

Poulsen, L. K., G. Ballard, and D. A. Stahl. 1993. Use of rRNA fluorescence in situ hybridization for measuring the activity of single cells in young and established biofilms. *Appl. Environ. Microbiol.* 59(5): 1354–1360.

Prack, E. R. 1993. An introduction to process visualization capabilities and considerations in the environmental scanning electron microscope (ESEM). *Microsc. Res. Tech.* 25(5–6): 487–492.

Ramos-Padrón, E., S. Bordenave, S. Lin, I. M. Bhaskar, X. Dong, C. W. Sensen, J. Fournier, G. Voordouw, and L. M. Gieg. 2010. Carbon and sulfur cycling by microbial communities in a gypsum-treated oil sands tailings pond. *Environ. Sci. Technol.* 45(2): 439–446.

Rozanova, E. P., I. A. Borzenkov, A. L. Tarasov, L. A. Suntsova, C. L. Dong, S. S. Belyaev, and M. V. Ivanov. 2001. Microbiological processes in a high-temperature oilfield. *Microbiol.* 70(1): 102–110.

RP0775, NACE Standard. 2005. Preparation, installation, analysis, and interpretation of corrosion coupons in oilfield operations. Houston, TX: NACE International.

Schloss, P. D., S. L. Westcott, T. Ryabin, J. R. Hall, M. Hartmann, E. B. Hollister, R. A. Lesniewski et al. Introducing mothur: Open-source, platform-independent, community-supported software for describing and comparing microbial communities. *Appl. Environ. Microbiol.* 75(23): 7537–7541.

Sharma, M., D. An, K. Baxter, M. Henderson, L. Edillon, and G. Voordouw. 2016. Understanding the role of microbes in frequent coiled tubing failures. NACE International Corrosion 2016. Conference & Expo, Vancouver, BC, March 6–10. Paper 7815.

Shen, Y., and G. Voordouw. 2015. Primers for dsr genes and most probable number method for detection of sulfate-reducing bacteria in oil reservoirs protocol, part of the series *Springer Protocols Handbooks*, 1–9.

Shen, Y., L. G. Stehmeier, and G. Voordouw. 1998. Identification of hydrocarbon-degrading bacteria in soil by reverse sample genome probing. *Appl. Environ. Microbiol.* 64(2): 637–645.

Sherar, B. W. A., I. M. Power, P. G. Keech, S. Mitlin, G. Southam, and D. W. Shoesmith. 2011. Characterizing the effect of carbon steel exposure in sulfide containing solutions to microbially induced corrosion. *Corros. Sci.* 53(3): 955–960.

Shi, X., R. Avci, and Z. Lewandowski. 2002. Microbially deposited manganese and iron oxides on passive metals—Their chemistry and consequences for material performance. *Corrosion.* 58(9): 728–738.

Skovhus, T. L., B. Højris, A. M. Saunders, T. R. Thomsen, M. Agerbaek, and J. Larsen. 2009. Practical use of new microbiology tools in oil production. *SPE Production & Operations.* 24(1): 180–186.

Soh, J. X. Dong, S. M. Caffrey, G. Voordouw, and C. W. Sensen. 2013. Phoenix 2: A locally installable large-scale 16S rRNA gene sequence analysis pipeline with Web interface. *J. Biotechnol.* 167(4): 393–403.

Sowards, J. W., C. H. D. Williamson, T. S. Weeks, J. D. McColskey, and J. R. Spear. 2014. The effect of *Acetobacter* sp. and a sulfate-reducing bacterial consortium from ethanol fuel environments on fatigue crack propagation in pipeline and storage tank steels. *Corros. Sci.* 79: 128–138.

Starosvetsky, D., R. Armon, J. Yahalom, and J. Starosvetsky. 2001. Pitting corrosion of carbon steel caused by iron bacteria. *Int. Biodeterior. Biodegradation.* 47(2): 79–87.

Stevenson, B. S., H. S. Drilling, P. A. Lawson, K. E. Duncan, V. A. Parisi, and J. M. Suflita. 2011. Microbial communities in bulk fluids and biofilms of an oil facility have similar composition but different structure. *Environ. Microbiol.* 13(4): 1078–1090.

Surman, S. B., J. T. Walker, D. T. Goddard, L. H. G. Morton, C. W. Keevil, W. Weaver, A. Skinner, K. Hanson, D. Caldwell, and J. Kurtz. 1996. Comparison of microscope techniques for the examination of biofilms. *J. Microbiol. Methods.* 25(1): 57–70.

Tan, Y. J., S. Bailey, B. Kinsella, and A. Lowe. 2000. Mapping corrosion kinetics using the wire beam electrode in conjunction with electrochemical noise resistance measurements. *J. Electrochem. Soc.* 147(2): 530–539.

Tan, Y., 2009. Sensing localised corrosion by means of electrochemical noise detection and analysis. *Sens. Actuat. B-Chem.* 139(2): 688–698.

Tan, Y. J. 1998. Wire beam electrode: A new tool for studying localised corrosion and other heterogeneous electrochemical processes. *Corros. Sci.* 41(2): 229–247.

Tardy-Jacquenod, C., P. Caumette, R. Matheron, C. Lanau, O. Arnauld, and M. Magot. 1996. Characterization of sulfate-reducing bacteria isolated from oil-field waters. *Can. J. Microbiol.* 42(3): 259–266.

Telang, A. J., S. Ebert, J. M. Foght, D. Westlake, G. E. Jenneman, D. Gevertz, and G. Voordouw. 1997. Effect of nitrate injection on the microbial community in an oilfield as monitored by reverse sample genome probing. *Appl. Environ. Microbiol.* 63(5): 1785–1793.

Telegdi, J., Z. Keresztes, G. Pálinkás, E. Kálmán, and W. Sand. 1998. Microbially influenced corrosion visualized by atomic force microscopy. *Appl. Phys. A. Mater. Sci. Process.* 66: S639–S642.

Uchiyama, T., K. Ito, K. Mori, H. Tsurumaru, and S. Harayama. 2010. Iron-corroding methanogen isolated from a crude-oil storage tank. *Appl. Environ. Microbiol.* 76(6): 1783–1788.

Ulrich, G. A., L. R. Krumholz, and J. M. Suflita. 1997. A rapid and simple method for estimating sulfate reduction activity and quantifying inorganic sulfides. *Appl. Environ. Microbiol.* 63(4): 1627–1630.

Usher, K. M., A. H. Kaksonen, D. Bouquet, K. Y. Cheng, Y. Geste, P. G. Chapman, and C. D. Johnston. 2015. The role of bacterial communities and carbon dioxide on the corrosion of steel. *Corros. Sci.* 98: 354–365.

Van der Kraan, G. M., M. D. Ridder, B. P. Lomans, and G. Muyzer. 2010. Sampling and nucleic extraction procedures from oil reservoir samples. In *Applied Microbiology and Molecular Biology in Oilfield Systems*: 7–16. Whitby, C. and Skovhus, T. L. (Eds.). Springer, Netherlands.

Videla, H. A., and L. K. Herrera. 2005. Microbiologically influenced corrosion: Looking to the future. *Intl. Microbiol.* 8(3): 169.

Videla, H. A., and W. G. Characklis. 1992. Biofouling and microbially influenced corrosion. *Intl. Biodeterior. Biodegradation.* 29(3): 195–212.

Videla, H. A., F. Bianchi, M. M. S. Freitas, C. G. Canales, and J. F. Wilkes. 1994. Monitoring biocorrosion and biofilms in industrial waters: A practical approach. *ASTM special technical publication* 1232: 128.

Von Der Weid, I., E. Korenblum, D. Jurelevicius, A. S. Rosado, R. Dino, G. V. Sebastian, and L. Seldin. 2008. Molecular diversity of bacterial communities from subseafloor rock samples in a deep-water production basin in Brazil. *J. Microbiol. Biotechnol.* 18(1): 5–14.

Voordouw, G. 1992. Evolution of hydrogenase genes. *Adv. Inorg. Chem.* 38: 397–422.

Voordouw, G. 1995. The genus *Desulfovibrio*: The centennial. *Appl. Environ. Microbiol.* 61(8): 2813.

Voordouw, G. 2000. Microbial communities in oilfields. In *Applied Microbial Systematics*, 315–332. Priest, F. G. and Goodfellow, M. (Eds.). Springer, Netherlands.

Voordouw, G. 2011. Production-related petroleum microbiology: Progress and prospects. *Curr. Opin. Biotechnol.* 22(3): 401–405.

Voordouw, G., and A. J. Telang. 1999. A genome probe survey of the microbial community in oilfields. In *Proceedings of the 8th International Symposium on Microbial Ecology.* Halifax, Canada: Atlantic Canada Society for Microbial Ecology.

Voordouw, G., and M. Nemati. 2000. Identification and characterization of sulfate-reducing bacteria involved in microbially influenced corrosion in oilfields. In *Corrosion 2000.* Houston, TX: NACE International.

Voordouw, G., D. W. S. Westlake, and P. M. Fedorak. 2002. Contacting master solid surface having isolated spots of genomic deoxyribonucleic acid from each different selected standard microorganism affixed thereto with labeled reverse genome probe prepared from sample, detecting hybridization. U.S. Patent 6,376,173, issued April 23, 2002.

Voordouw, G., J. D. Strang, and F. R. Wilson. 1989. Organization of the genes encoding [Fe] hydrogenase in *Desulfovibrio vulgaris* subsp. *oxamicus Monticello*. *J. Bacteriol.* 171(7): 3881–3889.

Voordouw, G., J. K. Voordouw, R. R. Karkhoff-Schweizer, P. M. Fedorak, and D. W. S. Westlake. 1991. Reverse sample genome probing, a new technique for identification of bacteria in environmental samples by DNA hybridization, and its application to the identification of sulfate-reducing bacteria in oilfield samples. *Appl. Environ. Microbiol.* 57(11): 3070–3078.

Voordouw, G., J. K. Voordouw, T. R. Jack, J. Foght, P. M. Fedorak, and D. W. S. Westlake. 1992. Identification of distinct communities of sulfate-reducing bacteria in oilfields by reverse sample genome probing. *Appl. Environ. Microbiol.* 58(11): 3542–3552.

Voordouw, G., P. Menon, T. Pinnock, M. Sharma, Y. Shen, A. Venturelli, J. Voordouw, and A. Sexton. 2016. Use of homogeneously-sized carbon steel ball bearings to study microbially-influenced corrosion in oilfield samples. *Front. Microbiol.* 7: 351.

Voordouw, G., S. M. Armstrong, M. F. Reimer, B. Fouts, A. J. Telang, Y. Shen, and D. Gevertz. 1996. Characterization of 16S rRNA genes from oilfield microbial communities indicates the presence of a variety of sulfate-reducing, fermentative, and sulfide-oxidizing bacteria. *Appl. Environ. Microbiol.* 62(5): 1623–1629.

Voordouw, G., V. Niviere, F. G. Ferris, P. M. Fedorak, and D. W. S. Westlake. 1990. Distribution of hydrogenase genes in *Desulfovibrio* spp. and their use in identification of species from the oilfield environment. *Appl. Environ. Microbiol.* 56(12): 3748–3754.

Voordouw, G., Y. Shen, C. S. Harrington, A. J. Telang, T. R. Jack, and D. W. S. Westlake. 1993. Quantitative reverse sample genome probing of microbial communities and its application to oilfield production waters. *Appl. Environ. Microbiol.* 59(12): 4101–4114.

Wadood, H. Z., A. Rajasekar, Y. P. Ting, and A. N. Sabari. 2015. Role of *Bacillus subtilis* and *Pseudomonas aeruginosa* on corrosion behaviour of stainless steel. *Arab. J. Sci. Eng.* 40(7): 1825–1836.

Walker, J. T., K. Hanson, D. Caldwell, and C. W. Keevil. 1998. Scanning confocal laser microscopy study of biofilm induced corrosion on copper plumbing tubes. *Biofouling.* 12(4): 333–344.

Wang, H., C. Hu, X. Hu, M. Yang, and J. Qu. 2012. Effects of disinfectant and biofilm on the corrosion of cast iron pipes in a reclaimed water distribution system. *Water. Res.* 46(4): 1070–1078.

Wang, W., X. Zhang, and J. Wang. 2009. Heterogeneous electrochemical characteristics of biofilm/metal interface and local electrochemical techniques used for this purpose. *Mater. Corros.* 60(12): 957–962.

Waterman, D. K., S. W. Powell, B. Sweetman, and W. J. Maciejewski. 1997. Corrosion monitoring tool. U.S. Patent 5,627,749.

Wei, L., F. Ma, and G. Zhao. 2010. Composition and dynamics of sulfate-reducing bacteria during the waterflooding process in the oilfield application. *Bioresour. Technol.* 101(8): 2643–2650.

Williamson, C. H., L. A. Jain, B. Mishra, D. L. Olson, and J. R. Spear. 2015. Microbially influenced corrosion communities associated with fuel-grade ethanol environments. *Appl. Microbiol. Biot.* 99(16): 6945–6957.

Xu, D., Y. Li, F. Song, and T. Gu. 2013. Laboratory investigation of microbiologically influenced corrosion of C1018 carbon steel by nitrate reducing bacterium *Bacillus licheniformis*. *Corros. Sci.* 77: 385–390.

Xue, Y., G. Voordouw, and L. M. Gieg. 2015. Laboratory protocols for investigating microbial souring and potential treatments in crude oil reservoirs. In *Hydrocarbon and Lipid Microbiology Protocols, Springer Protocols Handbooks*: 1–28. McGenity, T. J., Timmis, K. N., Nogales Fernandez, B. (eds.). (Heidelberg: Springer-Verlag.

Xue, Y., and G. Voordouw. 2015. Control of microbial sulfide production with biocides and nitrate in oil reservoir simulating bioreactors. *Front. Microbiol.* 6: 1387.

Yang, L., N. Sridhar, O. Pensado, and D. S. Dunn. 2002. An in-situ galvanically coupled multi-electrode array sensor for localized corrosion. *Corrosion* 58(12): 1004–1014.

Yu, L., J. Duan, X. Du, Y. Huang, and B. Hou. 2013. Accelerated anaerobic corrosion of electroactive sulfate-reducing bacteria by electrochemical impedance spectroscopy and chronoamperometry. *Electrochem. Commun.* 26: 101–104.

Yuan, S. J., and S. O. Pehkonen. 2007. Microbiologically influenced corrosion of 304 stainless steel by aerobic *Pseudomonas* NCIMB 2021 bacteria: AFM and XPS study. *Colloids. Surf. B Biointerfaces.* 59(1): 87–99.

Zarasvand, K. A., and V. R. Rai. 2014. Microorganisms: Induction and inhibition of corrosion in metals. *Int. Biodeterior. Biodegradation.* 87: 66–74.

Zaveri, N. A. 2008. *Biocorrosion studies of surface modified bioimplant material in simulated body fluids.* M.S. dissertation, Utah State University, Logan, UT.

Zaveri, N., R. Sun, N. Zufelt, A. Zhou, and Y. Q. Chen. 2007. Evaluation of microbially influenced corrosion with electrochemical noise analysis and signal processing. *Electrochim. Acta.* 52(19): 5795–5807.

Zhang, T., H. H. P. Fang, and B. C. B. Ko. 2003. Methanogen population in a marine biofilm corrosive to mild steel. *Appl. Microbiol. Biot.* 63(1): 101–106.

Zhu, X., J. II Kilbane, A. Ayala, and H. Modi. 2005. Application of quantitative, real-time PCR in monitoring microbiologically influenced corrosion (MIC) in gas pipelines. In *Corrosion 2005.* Houston, TX: NACE International.

Zinkevich, V., I. Bogdarina, H. Kang, M. A. W. Hill, R. Tapper, and I. B. Beech. 1996. Characterisation of exopolymers produced by different isolates of marine sulphate-reducing bacteria. *Int. Biodeterior. Biodegradation.* 37(3): 163–172.

10 Quantification of Sulfate-Reducing Microorganisms by Quantitative PCR

Current Challenges and Developments

Júlia R. de Rezende

CONTENTS

ABSTRACT

Quantifying sulfate-reducing microorganisms (SRMs) in oil reservoirs and production systems is important for understanding, predicting, and mitigating reservoir souring and biocorrosion. Monitoring SRM population size as well as understanding the factors that stimulate or inhibit growth contribute to informed management of microbiological control strategies. In recent years, molecular tools like quantitative PCR (qPCR) have gained preference relative to cultivation-based methods such as most probable number enumeration, due to advantages that include standardization, rapid throughput, and labor costs. However, a challenge facing probe- and primer-based oilfield screening is that tools developed based on "known" genetic diversity are used to screen "unknown" extreme environments, despite awareness that new microbial diversity is continually being discovered. DNA-based qPCR does not necessarily distinguish industrially relevant subpopulations, for example, active vs. dormant vs. dead cells; or sulfidogenic SRMs vs. SRMs that may be fermentative or reduce nitrate in oilfield environments. Distinguishing between these physiological categories influences decisions by operators, since different tools yield results that can over- or underestimate souring and biocorrosion risks. Here we discuss advantages, caveats, and new developments in SRM quantification by qPCR in order to refresh, update, and broaden the range of alternatives for consideration in the oil and gas sector.

10.1 INTRODUCTION

Sulfate-reducing microorganisms are the main cause of reservoir souring and biocorrosion or microbiologically influenced corrosion (MIC) in offshore oilfields. For this reason, controlling SRM populations is a primary goal of most microbiological control methods in reservoirs and associated infrastructure. The need for souring mitigation measures and the effectiveness of different treatment options are often evaluated in terms of SRM quantification in oilfield samples. For this reason, acknowledging technical bias and limitations of each method, incorporating new methods, updating protocols, and developing improvements to SRM quantification are critical for achieving reliable, high-quality microbial monitoring in the oil industry.

For several decades, the standard method to quantify SRMs in oil and gas systems has been most probable number (MPN) enumerations in artificial growth medium (NACE, 2014). Referred to at times in the industry as “bug bottles,” this approach suffers well-known limitations that are widely demonstrated and discussed (e.g., Zhu and Kilbane 2005; Price et al. 2010; Keasler and Chatterjee 2014). In recent years, molecular microbiology and genomics tools have been increasingly employed to address oilfield microbiological challenges (Whitby and Skovhus 2011; Skovhus, Caffrey, and Hubert 2014), especially techniques based on nucleic acid detection, above all quantitative polymerase chain reaction (qPCR) and PCR-enabled high throughput amplicon sequencing. This chapter focuses on qPCR and its advantages, limitations, and recent improvements, in particular for the detection and quantification of SRMs. Other recent literature sources are recommended for more general technical reviews of qPCR and technical recommendations (e.g., Bustin et al. 2009; Ruijter et al. 2009; Brankatschk et al. 2012; Callbeck et al. 2013; Rudkjøbing, Wolff, and Skovhus 2014).

10.2 QUANTIFYING TOTAL BACTERIA AND ARCHAEA BY qPCR

Quantitative PCR has started to gain preference over cultivation-based quantification of SRMs for several reasons. A major caveat of cultivation-based techniques is that they are highly dependent on the growth medium chosen to stimulate relevant microbial populations. For example, the growth medium may stimulate a few species of SRMs that are better adapted to the medium characteristics while many other SRM populations with various physiologies and often depending on the activity of other anaerobes may thrive in field conditions but remain undetected in cultivation-based tests due to the unfavorable growth medium. Most media select for only a small fraction of viable populations, which are very often not the most relevant under environmental conditions. The failure of standard cultivation techniques to adequately survey and quantify microbial populations has been referred to as the "great plate count anomaly" (Staley and Konopka 1985; Epstein 2013). On the other hand, quantitative PCR is based on deoxyribonucleic acid (DNA) extraction from cells, offering a better "snapshot" of the microbial community at the time of sampling, compared to more selective enrichment-based methods, such as most probable number enumerations. However, DNA extraction is not without biases and limitations as discussed below. Other advantages of qPCR over MPN approaches from a practical perspective include that qPCR enables results to be obtained within hours rather than several days or weeks and that the method is less labor intensive and has a higher throughput than MPN. Another significant advantage of qPCR is its flexibility: Several assays can be performed from a single DNA extract within hours. This allows the quantification of different taxonomic groups (e.g., from the domain to strain level) or multiple individual targets from the same original sample (e.g., Bacchetti De Gregoris et al. 2011; Jiménez et al. 2012; Pfeiffer et al. 2014; Ziels et al. 2015; Vigneron et al. 2016).

One of the main requirements for an effective qPCR assay is the design of primers that target a suitable gene or region of a gene with appropriate coverage and specificity. A suitable gene is ideally present in all target organisms and only the target organisms. Once the gene is defined, several tools and databases are available for primer design and *in silico* analyses of coverage and specificity of the primer oligonucleotides (typically 18–20 bases long), for example, ARB (Ludwig et al. 2004); SILVA, in particular the TestPrime tool (Klindworth et al. 2013); probeCheck (Loy et al. 2008); Primer3 (Untergasser et al. 2012); and primer-BLAST (Ye et al. 2012). Online meta-databases, such as omictools.com (Henry et al. 2014), contain links to many helpful resources.

The 16S ribosomal ribonucleic acid (rRNA) gene is widely used as a biomarker for identifying and quantifying bacteria and archaea as it is essential for cell function and is therefore ubiquitous. The 16S rRNA gene contains conserved and variable regions that allow the design of assays targeting different taxonomic levels (Amann, Ludwig, and Schleifer 1995; Klindworth et al. 2013). A well-known issue when using the 16S rRNA gene to quantify bacterial and archaeal populations is that multiple copies of this gene can be present in a single genome or cell, and knowledge of this gene copy number is necessary for an accurate conversion from gene abundance, quantified by qPCR, to cell abundance. The online database rrnDB (Stoddard et al. 2015) is an excellent tool to address this issue. Based on 180 sequenced genomes to date (February 2016), rrnDB shows that archaeal cells have between one and four copies of the 16S rRNA gene per cell. This distribution has a mode and median of one and mean of 1.63 ± 0.88. In the bacterial domain, however, the distribution is much wider. Based on 2,865 sequenced genomes, bacterial cells may have between one and 15 copies of the 16S rRNA gene per cell, with a mode of 2, median of 4 and mean of 4.15 ± 2.75. The distribution of gene copy numbers from known genomes can also be obtained at rrnDB for all taxonomic levels. The difference of 16S rRNA gene copy numbers at the domain level between bacteria and archaea highlights one source of error of qPCR assays: using a single value for the gene copy number to convert gene abundance into cell abundance. This concern is also relevant for the quantification of SRMs in particular. Even though the difference is usually within an order of magnitude, it should be taken into consideration when comparing different populations and communities that are composed of largely distinct groups. Distributions of 16S rRNA gene copy number should also be considered when using high-throughput sequencing data for a semiquantitative assessment of microbial community composition based on the relative abundance of distinct taxa. As an example with some oilfield-relevant genera, *Desulfotomaculum* genomes have three to 10 copies of the 16S rRNA gene (based on six sequenced genomes), *Desulfovibrio* have one to five copies of the gene (n = 14) whereas sulfate-reducing archaea *Archaeoglobus* spp. appear to have only one copy of the gene (n = 6). The 16S rRNA gene copy number bias can be reduced by using a recently developed tool that rapidly incorporates rrnDB information into qPCR and sequencing data (CopyRighter; Angly et al. 2014). Other sources of bias resulting from upstream processing steps (sampling, DNA extraction, PCR, etc.) must still be kept in mind, but this is a valuable contribution to improve data interpretation.

Besides the variation in 16S rRNA gene copy number, the wide distribution of SRMs in several phyla and even in both bacterial and archaeal domains means that prior characterization of the SRM diversity is necessary for the selection of an appropriate assay, or multiple assays should be performed to maximize coverage and specificity. A common solution to the problems of specificity and gene copy number is to instead target protein-coding genes that are associated with sulfate reduction, such as the *aprA* gene or the *dsrAB* gene (e.g., Blazejak and Schippers 2011; Müller et al. 2015). The section below focuses on *dsrAB* gene quantification as it has a larger database of sequences and is more commonly used for PCR assays.

10.3 QUANTIFYING SULFATE-REDUCING MICROORGANISMS BY qPCR

The *dsrAB* gene is a phylogenetic marker for sulfate-reducing microorganisms. This single-copy gene encodes the dissimilatory (bi)sulfite reductase (Dsr), which is a key enzyme in dissimilatory sulfate reduction that catalyzes conversion of sulfite into sulfide. The phylogeny of the *dsrAB* gene broadly correlates with the 16S rRNA gene phylogeny, facilitating studies of SRM diversity. Furthermore, there is a large database of *dsrAB* sequences publicly available, which was recently updated, curated, and annotated (Müller et al. 2015). A comprehensive database provides a good basis for the design of primers specific to each system. Müller and colleagues (2015) recently performed updated coverage and specificity testing for *dsrA*- and *dsrB*-targeted primers. It is recommended that such checks be carried out regularly given the pace at which new microbial diversity is continually being discovered. For example, in the same work, the authors showed that at least 13 family-level lineages of DsrAB have no cultivated representatives, demonstrating the large and growing diversity of SRMs.

Besides constantly expanding the known diversity of *dsrAB*-bearing SRMs, there has been recent evidence of sulfate-reducing microorganisms that may use a different enzymatic pathway. Milucka and colleagues (2012) demonstrated the occurrence of anaerobic oxidation of methane coupled to sulfate reduction and polysulfide disproportionation, which seems to be based on a new, unresolved pathway that does not involve the Dsr enzyme or any of the canonical enzymes for sulfate reduction. This raises important questions: How many SRM populations are being missed by quantifying *dsrA* or *dsrB* gene abundance, and how large are these populations in oil reservoirs and other environments? While these questions remain unresolved at present, they highlight a fundamental limitation of biomarker-based genomics approaches, which is that they rely on the already-known diversity and sequence data to screen unknown environments.

A useful and simple way to assess whether SRM numbers are underestimated by any particular method is to calculate cell-specific sulfate reduction rates (csSRR) and compare them to the range of csSRR observed from pure cultures of SRMs (e.g., Detmers et al. 2001; Sim et al. 2011) or in natural environmental samples (Sahm et al. 1999; Hoehler and Jørgensen 2013) as appropriate. Detmers and colleagues (2001) reported csSRR for over 30 SRMs at exponential growth phases, which ranged from 0.9 to 434 fmol SO_4^{2-} cell^{-1} day^{-1}. One way to calculate csSRR is to divide bulk sulfate reduction rates (SRR) by the estimated number of active SRMs (e.g., de Rezende et al. 2016). Bulk SRR can be measured or derived from measurements of sulfate consumption or sulfide production. Pure cultures of SRMs growing exponentially in artificial media under excess substrate conditions represent ideal circumstances that allow high csSRR. In environmental conditions, the csSRR of SRMs are often tenfold to hundredfold lower (Sahm et al. 1999; Hoehler and Jørgensen 2013). Therefore, if the calculated csSRR for the total SRM community in an environmental sample are orders of magnitude below the pure culture csSRR range, it may be due to less favorable growth conditions. On the other hand, calculated csSRR being orders of magnitude above the csSRR of exponentially growing pure cultures of SRMs suggest that estimated SRM numbers may be underestimated by orders of magnitude. This inference, of course, assumes that bulk SRR are not overestimated.

In the next lines, further challenges associated to qPCR quantification of SRMs are addressed and recent improvements to tackle each challenge are presented.

10.3.1 Potential Overestimation of Living SRM Populations by DNA-Based qPCR

It is well known that usual DNA extraction techniques from bulk environmental samples include DNA from both living and dead cells as extracellular DNA may persist for hours, days, or several years (Willerslev and Cooper 2005; Torti, Lever, and Jørgensen 2015). Microbial abundances estimated by downstream techniques, such as qPCR, may thus overestimate living populations if DNA from dead cells gets extracted intact, such that the fragment can be amplified in the PCR reaction. For example, it has been estimated that soluble DNA, which corresponds to (extracellular) dissolved DNA and DNA adsorbed onto particles may contribute between 10% and 83% of total DNA in various types of sediments (Lever et al. 2015). For the purpose of microbial control, an overestimation based on quantifying DNA from both living and dead cells has the advantage of offering a conservative measure of microbial abundance. But there is a downside in that if dead cells and extracellular DNA represent a significant proportion of the estimated total, despite having no corresponding functional potential in terms of souring or MIC risk, high numbers may lead to a hastened decision to invest in an unnecessary microbial control treatment. Furthermore, high background of DNA not associated to living cells can mislead community characterization efforts that aim to identify the organisms related to the microbiological issue.

An improvement to this aspect is the removal of DNA from dead cells prior to DNA extraction, for example, by pretreating samples with DNA-intercalating dyes, such as ethidium monoazide (EMA) or propidium monoazide (PMA; Nocker, Cheung and Camper 2006). This strategy has been applied for quantifying microorganisms in different environments, such as marine sediments (Leloup et al. 2009) and wastewater treatment (Li et al. 2014) and is widely applied in the food industry (Elizaquível, Aznar, and Sánchez 2014). The success of EMA/PMA treatment has its own limitations as reported elsewhere (e.g., Schmidlin et al. 2010; M. J. Taylor, Bentham and Ross 2014). Issues include the penetration of the dye into living cells and the inefficient inhibition of DNA amplification from dead cells, in particular due to the presence of particles that interfere with the photoactivation of the dye, chemical adsorption, proportion of dead cells, and salt concentration, among others. Technical improvements to EMA/PMA treatment have been reviewed and proposed recently (Fittipaldi et al. 2011; Fittipaldi, Nocker, and Codony 2012; Nkuipou-Kenfack et al. 2013), including the example of distinguishing between live and dead methanogens in environmental samples (Heise et al. 2016). EMA/PMA strategies may thus hold promise for improving oilfield monitoring.

Other protocols have been developed for an improved extraction of DNA, separating nonsoluble DNA, mainly associated to living cells, from soluble DNA (Corinaldesi, Danovaro, and Dell'Anno 2005; Corinaldesi et al. 2014; Lever et al. 2015). Moreover, Lever, and colleagues (2015) thoroughly assessed various aspects of DNA and RNA extraction protocols and proposed a modular method, suggesting adjustments based on sample characteristics, such as sample matrices, mineralogies, and PCR-inhibiting organic matter. Such developments help increasing yield and quality of extracted nucleic acids, targeting living cells, and lead to a more accurate assessment of microbial community and estimation of microbial abundance.

10.3.2 Quantifying Living SRMs: Active and Dormant Subpopulations

It is also important to consider and differentiate the quantitative contribution of active and dormant cells to the total abundance of living microbial populations. It is estimated that dormant subpopulations contribute from 20% to 80% of total cells in the environment (Lennon and Jones 2011). Endospores are particularly resistant and long-living cells among these dormant subpopulations. For example, certain SRM endospores were able to survive repeated autoclaving treatments at temperatures above 120°C and subsequently grow and reduce sulfate at 50°C and 60°C (O'Sullivan et al. 2015). This finding is relevant in the context of souring and MIC as these

SRMs were related to *Desulfotomaculum* spp. isolated from hot subsurface environments, and similar temperatures and conditions may be found in oilfields. It has also been shown that dormant, viable SRM endospores are present in cold seawater (de Rezende et al. 2013), and these endospores could therefore potentially germinate in reservoirs following seawater injection for secondary oil recovery. Dormant SRMs are problematic for long-term souring and MIC control as factors that control shifts between dormancy and activity in the environment are not well understood. It is known that stressful conditions may lead to dormancy rather than death; an example being when low doses of biocide are applied (Keasler et al. 2013). On the other hand, dormant cell activation may simply follow a change in temperature, as observed for thermophilic SRM endospores in cold marine sediments (Hubert et al. 2009; de Rezende et al. 2013), or an increase in the concentration of organic and inorganic compounds that trigger germination (Olguín-Araneda et al. 2015). This underscores the importance of frequent monitoring of microbial communities in oil reservoirs. Quantifying active and dormant populations separately rather than in a bulk total provides a more informative data set of organisms that are immediately problematic and others that are potential hazards, and this knowledge can be applied to further improve souring models and risk management.

Active subpopulations can be addressed by quantifying rRNA or mRNA by reverse-transcription quantitative PCR (RT-qPCR), which gives an estimation of abundance of active cells and activity levels together. The resulting data may not be directly translated to cell numbers as the abundance of transcripts per cell can vary by orders of magnitude according to species and the physiological state of the cell (Fegatella et al. 1998; Klappenbach, Dunbar, and Schmidt 2000; Blazewicz et al. 2013; Lankiewicz, Cottrell, and Kirchman 2016); however, composite indices around functional potential are ultimately of greater relevance for assessing risks, that is, related to souring and MIC in oilfield settings. Once again, targeting 16S rRNA for SRM quantification relies on oligonucleotide probes that cover the known diversity with appropriate specificity as discussed above at the DNA level for the 16S rRNA gene. Targeting the *dsrAB* mRNA is of interest because it relates to the expression of a gene that is directly linked to the production of H_2S rather than the genetic potential that is inferred from DNA-based gene quantification. As such, estimates of mRNA abundance can be more directly related to sulfide production or metabolic rates. There are significant technical challenges associated with mRNA isolation from environmental and industrial samples, including quantitative preservation and extraction of intact mRNA and the difficulty to produce adequate standards for RT-qPCR (Johnson et al. 2005; S. Taylor et al. 2010; Gadkar and Filion 2013). Despite technical challenges, the application of RT-qPCR to quantify *dsrA* mRNA in production water samples has revealed interesting information that complemented DNA-based qPCR and microscopy-based techniques as exemplified by Price and colleagues (2010). The amount of sulfide in a production well in a North Sea field correlated better with *dsrA* mRNA abundance per cell than either *dsrAB* gene abundance or SRM cell abundance. This suggested an increase of activity per cell rather than an increase in cell numbers and revealed that high numbers of *dsrAB* gene or SRM cells may not automatically mean high sulfide production (Price et al. 2010). The combination of techniques therefore provided an insight into the active fraction of the SRM populations.

Combined strategies are also necessary to estimate the abundance of dormant subpopulations. Enzymatic assays, such as ATP/AMP ratio measurement, can also indicate the proportion of dormant and active cells (Keasler et al. 2013). However, both enzymatic assays and standard qPCR assays are unlikely to quantify endospores since both approaches rely on effective and quantitative cell lysis and harsher extraction protocols are necessary to access endospores (Dineen et al. 2010; Wunderlin et al. 2013). To illustrate this bias, studies that quantified endospores in marine sediments using DNA-independent methods have reported abundances of 10^5 to 10^7 endospores per cm^3 of sediment (Fichtel et al. 2007; Lomstein et al. 2012; Langerhuus et al. 2012) whereas a discrepantly low frequency of endospore-related genes has been observed in DNA-based studies of various environments, including sub-seafloor sediments (Filippidou et al. 2015; Kawai et al. 2015). Although there is little quantitative data on the abundance of endospores in oil reservoirs, it has been shown that *Firmicutes*, which encompasses all known endospore-forming taxa, including the sulfate-reducing *Desulfotomaculum* and *Desulfosporosinus* genera, was the most frequently reported phylum in 16S rRNA gene studies of microbial communities in oil reservoirs (Hubert et al. 2012). Therefore, the quantification of endospore-forming SRMs may be a relevant but often overlooked factor to improve long-term microbial monitoring for souring and MIC control. Recent technical advancement to quantify endospore-forming bacteria include improved DNA extraction from bacterial endospores and a recently developed qPCR assay targeting the stage 0 sporulation gene A (*spo0A*), a genetic biomarker for endospore-forming bacteria (Bueche et al. 2013; Wunderlin et al. 2013).

10.3.3 Issues beyond qPCR: Distinguishing Sulfidogenic and Nonsulfidogenic SRMs

Sulfate-reducing microorganisms, as the name suggests, are usually presumed to perform sulfate reduction. However, some are known to gain energy from other metabolisms. A syntrophic, fermentative lifestyle is widely known, but also nitrate reduction and microaerophilia have been reported (Cypionka 2000; Rabus, Hansen, and Widdel 2006; Plugge et al. 2011). All four metabolisms are realistic in oilfield scenarios: Fermentation may occur in reservoir and wells prior to seawater injection/breakthrough; sulfate reduction may be favored in the presence of high concentrations of sulfate in injected seawater; nitrate reduction could follow souring mitigation treatment by nitrate injection; microaerophilia may be encountered in top facilities, corroded pipes, and injection wells. Furthermore the different SRM metabolisms may occur in counterintuitive conditions; for example, the presence of sulfate does not necessarily result in sulfate reduction-based growth even though it is expectedly more favorable (Gieg, Duncan, and Suflita 2008). Versatile SRMs that have been isolated contained the genetic potential for sulfate reduction and switched between metabolisms according to the cultivation conditions (Plugge et al. 2010). Therefore, a SRM quantification based on gene abundance (*dsrAB*, 16S rRNA gene) cannot necessarily distinguish or identify whether the quantified SRMs are sulfidogenic or not in a given system. Surprisingly, the same limitation occurs even with mRNA quantification as it has been shown that a syntrophic member of the *Desulfotomaculum* cluster I that does not perform sulfate reduction not only contains the *dsrAB* gene, but also expresses detectable *dsrAB* mRNA in the presence and absence of sulfate (Imachi et al. 2006). In the context of souring, production of sulfide is the ultimate concern related to SRM activity; thus false positives may lead to overestimation of sulfidogenic populations or may hide the true causes of problems that are being encountered and diagnosed. A better understanding of the factors that stimulate one metabolism or the other and a combination of chemical data and qPCR data will help develop more accurate prediction and troubleshooting tools for souring and MIC issues.

10.4 FINAL CONSIDERATIONS

Despite the caveats discussed here, qPCR remains one of the best tools available for the quantification of total microbial populations and is especially valuable for monitoring trends and relative changes in abundance for similar samples, including at oilfield operations in need of monitoring related to souring and MIC. It is important that operators are aware of limitations and biases associated with all assessment methods being employed and that standard procedures are reviewed regularly to incorporate new developments. As presented here, issues include (i) continually checking and updating primer coverage following discovery of new diversity, (ii) potential false negative results due to dead cells or to neglect of dormant populations, and (iii) potential false positive results due to the nonsulfidogenic metabolism of SRMs.

In fact, when the main concern are the consequences of the activity of sulfidogenic organisms, regardless of their identity, evolution, and genetic potential, quantification approaches that are based on the measurement of sulfate reduction should complement qPCR. MPN enumeration is limited by the small fraction of SRM diversity that is stimulated and detected using traditional cultivation media. However, improving the understanding of SRM physiology and developing better media will enhance the effectiveness of MPN and other physiology-based quantification techniques. Media development may be a challenging, perhaps low-reward, task, but a continuous effort is justified by the potential to test hypotheses, understand switches between dormancy and activation, and better simulate conditions that can generate useful input data for reservoir management models. This may especially pay dividends for a given operation where local SRM diversity is the focus of research attention for improved growth-based assessments. Examples of media improvement in the last years include low-nutrient media; addition of signaling compounds, such as cAMP and homoserine lactones; co-encapsulation into micro-droplets; diffusion chambers; sterilized natural medium; use of radioactive sulfate tracer to score positive MPN tubes; use of surface to support biofilm development; and several others (e.g., Vester and Ingvorsen 1998; Epstein 2013; Tanaka et al. 2014; Lagier et al. 2015).

Ultimately, it is important to acknowledge that all methods are limited but have advantages and disadvantages. Despite some limitations, monitoring trends with qPCR provides valuable information related to souring and corrosion development rapidly and early enough for mitigation measures to be put in place (e.g., Sørensen et al. 2012; Keasler and Chatterjee 2014). Furthermore, quantitative microbial data is essential for understanding complex processes, such as biocorrosion (e.g.,

Vigneron et al. 2016). Being aware of the biases inherent to qPCR or any other chosen method will limit overinterpretation of numbers. If a methodological problem cannot be fixed, then at least an estimation of its effect by testing appropriate controls and spiked samples can yield valuable information. Combining qPCR data with alternative methods, such as cell counts by DAPI staining, fluorescence in situ hybridization (FISH), and physiology-based methods, is highly recommended when qPCR alone fails to adequately explain trends observed in chemical data. A polyphasic approach that includes physicochemical data and microbial quantification by a combination of methods results in a more accurate assessment of the abundance and development of microbial communities. This, in turn, leads to better monitoring data, better reservoir models to predict risks, and improved control of souring and MIC.

REFERENCES

Amann, R I, W Ludwig, and K-H Schleifer. 1995. Phylogenetic identification and in situ detection of individual microbial cells without cultivation. *Microbiological Reviews* 59 (1): 143–69.

Angly, F E, P G Dennis, A Skarshewski, I Vanwonterghem, P Hugenholtz, and G W Tyson. 2014. CopyRighter: A rapid tool for improving the accuracy of microbial community profiles through lineage-specific gene copy number correction. *Microbiome* 2: 11. doi: 10.1186/2049-2618-2-11.

Bacchetti de Gregoris, T, N Aldred, A S Clare, and G J Burgess. 2011. Improvement of phylum- and class-specific primers for real-time PCR quantification of bacterial taxa. *Journal of Microbiological Methods* 86 (3): 351–6.

Blazejak, A, and A Schippers. 2011. Real-time PCR quantification and diversity analysis of the functional genes *aprA* and *dsrA* of sulfate-reducing prokaryotes in marine sediments of the Peru continental margin and the Black Sea. *Frontiers in Microbiology* 2: 253. doi: 10.3389/fmicb.2011.00253.

Blazewicz, S J, R L Barnard, R A Daly, and M K Firestone. 2013. Evaluating rRNA as an indicator of microbial activity in environmental communities: Limitations and uses. *The ISME Journal* 7 (11): 2061–8. doi: 10.1038/ismej.2013.102.

Brankatschk, R, N Bodenhausen, J Zeyer, and H Bürgmann. 2012. Simple absolute quantification method correcting for quantitative PCR efficiency variations for microbial community samples. *Applied and Environmental Microbiology* 78 (12): 4481–9. doi: 10.1128/AEM.07878-11.

Bueche, M, T Wunderlin, L Roussel-Delif, T Junier, L Sauvain, N Jeanneret, and P Junier. 2013. Quantification of endospore-forming *Firmicutes* by quantitative PCR with the functional gene *spo0A*. *Applied and Environmental Microbiology* 79 (17): 5302–12. doi: 10.1128/AEM.01376-13.

Bustin, S A, V Benes, J A Garson, J Hellemans, J Hugget, M Kubista, R Mueller et al. 2009. The MIQE guidelines: Minimum information for publication of quantitative real-time PCR experiments. *Clinical Chemistry* 55 (4): 611–22. doi: 10.1373/clinchem.2008.112797.

Callbeck, C M, A Sherry, C R J Hubert, N D Gray, G Voordouw, and I M Head. 2013. Improving PCR efficiency for accurate quantification of 16S rRNA genes. *Journal of Microbiological Methods* 93 (2): 148–52. doi: 10.1016/j.mimet.2013.03.010.

Corinaldesi, C, R Danovaro, and A Dell'Anno. 2005. Simultaneous recovery of extracellular and intracellular DNA suitable for molecular studies from marine sediments. *Applied and Environmental Microbiology* 71 (1): 46–50. doi: 10.1128/AEM.71.1.46.

Corinaldesi, C, M Tangherlini, G M Luna, and A Dell'Anno. 2014. Extracellular DNA can preserve the genetic signatures of present and past viral infection events in deep hypersaline anoxic basins. *Proceedings of the Royal Society B* 281: 20133299. doi: 10.1098/rspb.2013.3299.

Cypionka, H. 2000. Oxygen respiration by *Desulfovibrio* species. *Annual Review of Microbiology* 54: 827–48. doi: 10.1146/annurev.micro.54.1.827.

de Rezende, J R, C R J Hubert, H Røy, K U Kjeldsen, and B B Jørgensen. 2016. Estimating the abundance of endospores of sulfate-reducing bacteria in environmental samples by inducing germination and exponential growth. *Geomicrobiology Journal*. doi: 10.1080/01490451.2016.1190805.

de Rezende, J R, K U Kjeldsen, C R J Hubert, K Finster, A Loy, and B B Jørgensen. 2013. Dispersal of thermophilic Desulfotomaculum endospores into Baltic Sea sediments over thousands of years. *The ISME Journal* 7 (1): 72–84. doi: 10.1038/ismej.2012.83.

Detmers, J, V Brüchert, K S Habicht, and J Kuever. 2001. Diversity of sulfur isotope fractionations by sulfate reducing prokaryotes. *Applied and Environmental Microbiology* 67 (2): 888–94. doi: 10.1128/AEM.67.2.888-894.2001.

Dineen, S M, R Aranda IV, D L Anders, and J M Robertson. 2010. An evaluation of commercial DNA extraction kits for the isolation of bacterial spore DNA from soil. *Journal of Applied Microbiology* 109 (6): 1886–96. doi: 10.1111/j.1365-2672.2010.04816.10.

Elizaquível, P, R Aznar, and G Sánchez. 2014. Recent developments in the use of viability dyes and quantitative PCR in the food microbiology field. *Journal of Applied Microbiology* 116 (1): 1–13. doi: 10.1111/jam.12365.

Epstein, S S. 2013. The phenomenon of microbial uncultivability. *Current Opinion in Microbiology* 16 (5): 636–42. doi: 10.1016/j.mib.2013.08.003.

Fegatella, F, J Lim, S Kjelleberg, and R Cavicchioli. 1998. Implications of rRNA operon copy number and ribosome content in the marine oligotrophic ultramicrobacterium *Sphingomonas* sp. strain RB2256. *Applied and Environmental Microbiology* 64 (11): 4433–8.

Fichtel, J, J Köster, J Rullkötter, and H Sass. 2007. Spore dipicolinic acid contents used for estimating the number of endospores in sediments. *FEMS Microbiology Ecology* 61 (3): 522–32. doi: 10.1111/j.1574-6941.2007.00354.10.

Filippidou, S, T Junier, T Wunderlin, C-C Lo, P-E Li, P S Chain, and P Junier. 2015. Under-detection of endospore-forming *Firmicutes* in metagenomic data. *Computational and Structural Biotechnology Journal* 13: 299–306. doi: 10.1016/j.csbj.2015.04.002.

Fittipaldi, M, F Codony, B Adrados, A K Camper, and J Morató. 2011. Viable real-time PCR in environmental samples: Can all data be interpreted directly? *Microbial Ecology* 61 (1): 7–12. doi: 10.1007/s00248-010-9719-1.

Fittipaldi, M, A Nocker, and F Codony. 2012. Progress in understanding preferential detection of live cells using viability dyes in combination with DNA amplification. *Journal of Microbiological Methods* 91 (2): 276–89. doi: 10.1016/j.mimet.2012.08.007.

Gadkar, V J, and M Filion. 2013. Quantitative real-time polymerase chain reaction for tracking microbial gene expression in complex environmental matrices. *Current Issues in Molecular Biology* 15 (2): 45–58.

Gieg, L M, K E Duncan, and J M Suflita. 2008. Bioenergy production via microbial conversion of residual oil to natural gas. *Applied and Environmental Microbiology* 74 (10): 3022–9. doi: 10.1128/AEM.00119-08.

Heise, J, M Nega, M Alawi, and D Wagner. 2016. Propidium monoazide treatment to distinguish between live and dead methanogens in pure cultures and environmental samples. *Journal of Microbiological Methods* 121: 11–23. doi: 10.1016/j.mimet.2015.12.002.

Henry, V J, A E Bandrowski, A-S Pepin, B J Gonzalez, and A Desfeux. 2014. OMICtools: An informative directory for multi-omic data analysis. *Database*. Vol. 2014: article ID bau069. doi: 10.1093/database/bau069.

Hoehler, T M, and B B Jørgensen. 2013. Microbial life under extreme energy limitation. *Nature Reviews Microbiology* 11 (2): 83–94. doi: 10.1038/nrmicro2939.

Hubert, C, A Loy, M Nickel, C Arnosti, C Baranyi, V Brüchert, T Ferdelman et al. 2009. A constant flux of diverse thermophilic bacteria into the cold Arctic seabed. *Science* 325 (5947): 1541–4. doi: 10.1126/science.1174012.

Hubert, C, T B P Oldenburg, M Fustic, N D Gray, S R Larter, K Penn, A K Rowan et al. 2012. Massive dominance of *Epsilonproteobacteria* in formation waters from a Canadian oil sands reservoir containing severely biodegraded oil. *Environmental Microbiology* 14 (2): 387–404. doi: 10.1111/j.1462-2920.2011.02521.10.

Imachi, H, Y Sekiguchi, Y Kamagata, A Loy, Y-L Qiu, P Hugenholtz, N Kimura, M Wagner, A Ohashi, and H Harada. 2006. Non-sulfate-reducing, syntrophic bacteria affiliated with *Desulfotomaculum* cluster I are widely distributed in methanogenic environments. *Applied and Environmental Microbiology* 72 (3): 2080–91. doi: 10.1128/AEM.72.3.2080.

Jiménez, N, B E L Morris, M Cai, F Gründger, J Yao, H H Richnow, and M Krüger. 2012. Evidence for in situ methanogenic oil degradation in the Dagang oil field. *Organic Geochemistry* 52: 44–54.

Johnson, D R, P K H Lee, V F Holmes, and L Alvarez-Cohen. 2005. An internal reference technique for accurately quantifying specific mRNAs by real-time PCR with application to the *tceA* reductive dehalogenase gene. *Applied and Environmental Microbiology* 71 (7): 3866–71. doi: 10.1128/AEM.71.7.3866.

Kawai, M, I Uchiyama, H Takami, and F Inagaki. 2015. Low frequency of endospore-specific genes in subseafloor sedimentary metagenomes. *Environmental Microbiology Reports* 7 (2): 341–50. doi: 10.1111/1758-2229.12254.

Keasler, V, B Bennett, C Keller, P Whalen, J Cairns, and R M De Paula. 2013. Expanding the microbial monitoring toolkit: Evaluation of traditional and molecular monitoring methods. *International Biodeterioration & Biodegradation* 81: 51–6. doi: 10.1016/j.ibiod.2012.07.002.

Keasler, V, and I Chatterjee. 2014. Using the power of molecular microbiological methods in oilfield corrosion management to diagnose microbiologically influenced corrosion. In *Applications of Molecular Microbiological Methods*, edited by T L Skovhus, S M Caffrey, and C R J Hubert, 23–31. Norfolk: Caister Academic Press.

Klappenbach, J A, J M Dunbar, and T M Schmidt. 2000. rRNA operon copy number reflects ecological strategies of bacteria. *Applied and Environmental Microbiology* 66 (4): 1328–33. doi: 10.1128/AEM.66.4.1328-1333.2000.

Klindworth, A, E Pruesse, T Schweer, J Peplies, C Quast, M Horn, and F O Glöckner. 2013. Evaluation of general 16S ribosomal RNA gene PCR primers for classical and next-generation sequencing-based diversity studies. *Nucleic Acids Research* 41 (1): 1–11.

Lagier, J-C, P Hugon, S Khelaifia, P-E Fournier, B La Scola, and D Raoult. 2015. The rebirth of culture in microbiology through the example of culturomics to study human gut microbiota. *Clinical Microbiology Reviews* 28 (1): 237–64. doi: 10.1128/CMR.00014-14.

Langerhuus, A T, H Røy, M A Lever, Y Morono, F Inagaki, B B Jørgensen, and B A Lomstein. 2012. Endospore abundance and D:L-amino acid modeling of bacterial turnover in holocene marine sediment (Aarhus Bay). *Geochimica et Cosmochimica Acta* 99: 87–99. doi: 10.1016/j.gca.2012.09.023.

Lankiewicz, T S, M T Cottrell, and D L Kirchman. 2016. Growth rates and rRNA content of four marine bacteria in pure cultures and in the Delaware estuary. *The ISME Journal* 10: 823–32. doi: 10.1038/ismej.2015.156.

Leloup, J, H Fossing, K Kohls, L Holmkvist, C Borowski, and B B Jørgensen. 2009. Sulfate-reducing bacteria in marine sediment (Aarhus Bay, Denmark): Abundance and diversity related to geochemical zonation. *Environmental Microbiology* 11 (5): 1278–91. doi: 10.1111/j.1462-2920.2008.01855.10.

Lennon, J T, and S E Jones. 2011. Microbial seed banks: The ecological and evolutionary implications of dormancy. *Nature Reviews Microbiology* 9 (2): 119–30. doi: 10.1038/nrmicro2504.

Lever, M A, A Torti, P Eickenbusch, A B Michaud, T Šantl-Temkiv, and B B Jørgensen. 2015. A modular method for the extraction of DNA and RNA, and the separation of DNA pools from diverse environmental sample types. *Frontiers in Microbiology* 6: 476. doi: 10.3389/fmicb.2015.00476.

Li, D, T Tong, S Zeng, Y Lin, S Wu, and M He. 2014. Quantification of viable bacteria in wastewater treatment plants by using propidium monoazide combined with quantitative PCR (PMA-qPCR). *Journal of Environmental Sciences* 26 (2): 299–306. doi: 10.1016/S1001-0742(13)60425-8.

Lomstein, B A, A T Langerhuus, S D'Hondt, B B Jørgensen, and A J Spivack. 2012. Endospore abundance, microbial growth and necromass turnover in deep sub-seafloor sediment. *Nature* 484 (7392): 101–4. doi: 10.1038/nature10905.

Loy, A, R Arnold, P Tischler, T Rattei, M Wagner, and M Horn. 2008. probeCheck—A central resource for evaluating oligonucleotide probe coverage and specificity. *Environmental Microbiology* 10 (10): 2894–8.

Ludwig, W, O Strunk, R Westram, L Richter, H Meier, Y A Buchner et al. 2004. ARB: A software environment for sequence data. *Nucleic Acids Research* 32 (4): 1363–71.

Milucka, J, T G Ferdelman, L Polerecky, D Franzke, G Wegener, M Schmid, I Lieberwirth, M Wagner, F Widdel, and M M M Kuypers. 2012. Zero-valent sulphur is a key intermediate in marine methane oxidation. *Nature* 491 (7425): 541–6. doi: 10.1038/nature11656.

Müller, A L, K U Kjeldsen, T Rattei, M Pester, and A Loy. 2015. Phylogenetic and environmental diversity of DsrAB-type dissimilatory (bi)sulfite reductases. *The ISME Journal* 9: 1152–65. doi: 10.1038/ismej.2014.208.

Nkuipou-Kenfack, E, H Engel, S Fakih, and A Nocker. 2013. Improving efficiency of viability-PCR for selective detection of live cells. *Journal of Microbiological Methods* 93 (1): 20–4. doi: 10.1016/j.mimet.2013.01.018.

Nocker, A, C-Y Cheung, and A K Camper. 2006. Comparison of propidium monoazide with ethidium monoazide for differentiation of live vs. dead bacteria by selective removal of DNA from dead cells. *Journal of Microbiological Methods* 67 (2): 310–20. doi: 10.1016/j.mimet.2006.04.015.

O'Sullivan, L A, E G Roussel, A J Weightman, G Webster, C R J Hubert, E Bell, I Head, H Sass, and J R Parkes. 2015. Survival of *Desulfotomaculum* spores from estuarine sediments after serial autoclaving and high-temperature exposure. *The ISME Journal* 9 (4): 922–33. doi: 10.1038/ismej.2014.190.

Olguín-Araneda, V, S Banawas, M R Sarker, and D Paredes-Sabja. 2015. Recent advances in germination of *Clostridium* spores. *Research in Microbiology* 166 (4): 236–43. doi: 10.1016/j.resmic.2014.07.017.

Pfeiffer, S, M Pastar, B Mitter, K Lippert, E Hackl, P Lojan, A Oswald, and A Sessitsch. 2014. Improved group-specific primers based on the full SILVA 16S rRNA gene reference database. *Environmental Microbiology* 16 (8): 2389–407.

Plugge, C M, J C M Scholten, D E Culley, L Nie, F J Brockman, and W Zhang. 2010. Global transcriptomics analysis of the *Desulfovibrio vulgaris* change from syntrophic growth with *Methanosarcina barkeri* to sulfidogenic metabolism. *Microbiology* 156 (9): 2746–56. doi: 10.1099/mic.0.038539-0.

Plugge, C M, W Zhang, J C M Scholten, and A J M Stams. 2011. Metabolic flexibility of sulfate-reducing bacteria. *Frontiers in Microbiology* 2: 81. doi: 10.3389/fmicb.2011.00081.

Price, A, L A Álvarez, C Whitby, and J Larsen. 2010. Detection of SRP activity by quantification of mRNA for the dissimilatory (bi)sulfite reductase gene (*dsrA*) by reverse transcription quantitative PCR. *NACE International Corrosion Conference & Expo*, 10253.

Rabus, R, T A Hansen, and F Widdel. 2006. Dissimilatory sulfate and sulfur-reducing prokaryotes. In *The Prokaryotes*, 3rd ed., edited by M Dworkin, S Falkow, E Rosenberg, K-H Schleifer, and E Stackebrandt, 659–768. New York: Springer-Verlag. doi: 10.1007/0-387-30742-7_22.

Rudkjøbing, V B, T Y Wolff, and T L Skovhus. 2014. Quantitative real-time polymerase chain reaction (qPCR) methods for abundance and activity measures. In *Applications of Molecular Microbiological Methods*, edited by T L Skovhus, S M Caffrey, and C R J Hubert, 193–201. Norfolk: Caister Academic Press.

Ruijter, J M, C Ramakers, W M H Hoogaars, Y Karlen, O Bakker, M J B van den Hoff, and A F M Moorman. 2009. Amplification efficiency: Linking baseline and bias in the analysis of quantitative PCR data. *Nucleic Acids Research* 37 (6): e45. doi: 10.1093/nar/gkp045.

Sahm, K, B J MacGregor, B B Jørgensen, and D A Stahl. 1999. Sulphate reduction and vertical distribution of sulphate-reducing bacteria quantified by rRNA slot-blot hybridization in a coastal marine sediment. *Environmental Microbiology* 1 (1): 65–74.

Schmidlin, M, M Alt, P Brodmann, and C Bagutti. 2010. Insufficient distinction between DNA from viable and nonviable *Staphylococcus aureus* cells in wipe-samples by use of propidium monoazide-PCR. *Applied Biosafety* 15 (4): 180–5.

Sim, M S, S Ono, K Donovan, S P Templer, and T Bosak. 2011. Effect of electron donors on the fractionation of sulfur isotopes by a marine *Desulfovibrio* sp. *Geochimica et Cosmochimica Acta* 75 (15): 4244–59. doi: 10.1016/j.gca.2011.05.021.

Skovhus, T L, S M Caffrey, and C R J Hubert. 2014. *Applications of Molecular Microbiological Methods*. Edited by T L Skovhus, S M Caffrey, and C R J Hubert. Norfolk: Caister Academic Press.

Sørensen, K B, S Juhler, U S Thomsen, and J Larsen. 2012. Cost efficient MIC management system based on molecular microbiological methods. *NACE International Corrosion Conference & Expo*, C2012–0001111.

Staley, J T, and A Konopka. 1985. Measurement of in situ activities of nonphotosynthetic microorganisms in aquatic and terrestrial habitats. *Annual Review of Microbiology* 39 (1): 321–46.

Stoddard, S F, B J Smith, R Hein, B R K Roller, and T M Schmidt. 2015. rrnDB: Improved tools for interpreting rRNA gene abundance in bacteria and archaea and a new foundation for future development. *Nucleic Acids Research* 43 (D1): D593–8.

Tanaka, T, K Kawasaki, S Daimon, W Kitagawa, K Yamamoto, H Tamaki, M Tanaka, C H Nakatsu, and Y Kamagata. 2014. A hidden pitfall in the preparation of agar media undermines microorganism cultivability. *Applied and Environmental Microbiology* 80 (24): 7659–66. doi: 10.1128/AEM.02741-14.

Taylor, M J, R H Bentham, and K E Ross. 2014. Limitations of using propidium monoazide with qPCR to discriminate between live and dead *Legionella* in biofilm samples. *Microbiology Insights* 7: 15–24. doi: 10.4137/MBI.S17723.

Taylor, S, M Wakem, G Dijkman, M Alsarraj, and M Nguyen. 2010. A practical approach to RT-qPCR-publishing data that conform to the MIQE guidelines. *Methods* 50 (4): S1–5. doi: 10.1016/j.ymeth.2010.01.005.

Torti, A, M A Lever, and B B Jørgensen. 2015. Origin, dynamics, and implications of extracellular DNA pools in marine sediments. *Marine Genomics* 24: 185–96. doi: 10.1016/j.margen.2015.08.007.

Untergasser, A, I Cutcutache, T Koressaar, J Ye, B C Faircloth, M Remm, and S G Rozen. 2012. Primer3—New capabilities and interfaces. *Nucleic Acids Research* 40 (15): e115.

Vester, F, and K Ingvorsen. 1998. Improved most-probable-number method to detect sulfate-reducing bacteria with natural media and a radiotracer. *Applied and Environmental Microbiology* 64 (5): 1700–7.

Vigneron, A, E B Alsop, B Chambers, B P Lomans, I M Head, and N Tsesmetzis. 2016. Complementary microorganisms in highly corrosive biofilms from an offshore oil production facility. *Applied and Environmental Microbiology* 82 (8): 2545–2554. doi: 10.1128/AEM.03842-15.

Whitby, C, and T L Skovhus. 2011. *Applied Microbiology and Molecular Biology in Oil Field Systems*. Edited by C Whitby and T L Skovhus. Berlin: Springer.

Willerslev, E, and A Cooper. 2005. Ancient DNA. *Proceedings of the Royal Society of London B* 272 (1558): 3–16. doi: 10.1098/rspb.2004.2813.

Wunderlin, T, T Junier, L Roussel-Delif, N Jeanneret, and P Junier. 2013. Stage 0 sporulation gene A as a molecular marker to study diversity of endosporeforming Firmicutes. *Environmental Microbiology Reports* 5 (6): 911–24. doi: 10.1111/1758-2229.12094.

Ye, J, G Coulouris, I Zaretskaya, I Cutcutache, S Rozen, and T L Madden. 2012. Primer-BLAST: A tool to design target-specific primers for polymerase chain reaction. *BMC Bioinformatics* 13: 134.

Zhu, X, and J J Kilbane II. 2005. Faster and more accurate data collection for microbiologically influenced corrosion. *SPE International Symposium on Oilfield Chemistry*, Houston, TX, SPE 93089.

Ziels, R M, D A C Beck, M Martí, H L Gough, H D Stensel, and B H Svensson. 2015. Monitoring the dynamics of syntrophic β-oxidizing bacteria during anaerobic degradation of oleic acid by quantitative PCR. *FEMS Microbiology Ecology* 91 (4): 1–13.

11 Application of Biocides and Chemical Treatments to Both Combat Microorganisms and Reduce (Bio)Corrosion

Brandon E. L. Morris and Geert M. van der Kraan

CONTENTS

ABSTRACT

The recognition of biocorrosion, or microbially influenced corrosion, as a problem has increased over the last several decades in the oil and gas industry, with an estimated 15–30% of all corrosion incidents being attributed to the metabolic activity of microorganisms. However, corrosion in general still plagues the industry. Exact costs of repairs and lost revenue due to unexpected infrastructure failure or shutdown are not readily available, but are easily estimated to run into the U.S. billion dollar range. Slowing or reducing both abiotic and biotic corrosion processes can greatly extend the longevity of the assets and reduce the incidence of failure in the system, routinely through the use of performance chemicals such as corrosion inhibitors, biocides, or sulfide scavenging chemistries. Corrosion-resistant alloys may also be used to provide longer-term protection for new installations, but their cost is typically prohibitive for large-scale applications or installations. In this chapter, we discuss the performance and functionality of common oilfield chemicals used to combat abiotic and biotic corrosion. Background information is provided on the different chemistries, and comparisons are made on the performance and efficacy of each under different field conditions. The chapter provides several examples of chemical application strategies and ends with suggestions on how to implement meaningful KPIs for monitoring biotic and abiotic corrosion processes in the oil and gas industry.

11.1 INTRODUCTION

The total annual costs of corrosion to the upstream oil and gas industry are hard to estimate but definitively are of the order of billions of USD. These costs can be direct or indirect. Direct costs related to corrosion typically comprise two areas (Sridhar et al. 2001): the design, manufacture, and construction of assets for corrosion resistance and (Nernst 1889) the maintenance and management of the system. Design of upstream oil and gas assets such as platforms, separators, wellheads, tanks, pipes, and valves for corrosion resistance would require the use of alloys to replace traditional carbon steel, such as stainless steels containing corrosion-resistant metals like nickel (Ni), chromium (Cr), or molybdenum (Mo). Cladding or coating of the internal pipeline surfaces with a composite or corrosion-resistance alloy (CRA) may also be considered to enhance surface durability and strength. Operators and corrosion engineers often consider increasing the wall thickness of the pipeline or other assets to provide corrosion allowance or adding cathodic protection systems with each decision during that process significantly impacting the overall lifespan of the asset and performance of the system. Maintenance and management costs include the labor costs for asset inspection, laboratory testing, the initiation of a corrosion-management system, replacement of corroded components, and loss of production. The use of chemicals such as corrosion inhibitors and biocides to reduce corrosion is also typically classified as direct costs. Indirect costs of corrosion may arise after a spill or pipeline rupture and can lead to significant environmental impact, litigation, and legal fallout.

Corrosion is generally considered to be a complex, gradual process that deteriorates metal by oxidation, chemical contamination, or electrolysis but truly affects every material within the upstream process, including cement, pipeline coatings, and paint as well as the overall system integrity. For the context of this chapter, we mainly focus on corrosion of metals in general and iron (Fe) in particular. Chemical, biological, and operational parameters all contribute to the overall rates of corrosion seen in the field with each playing a unique and complex role in the process. In a pipeline, the internal corrosion rates are inextricably tied to the composition and partial pressure of gasses; the water chemistry, temperature, and operation flow velocities; as well as the nature of the oil being transported. High corrosion rates have been reported in the presence of sulfide (H_2S), carbon dioxide (CO_2), oxygen (O_2), and elevated chloride concentrations as well as large sulfur or polysulfide concentrations (Sridhar et al. 2001). Other factors, such as organic acid content, the pH of the system, the presence of solids, and microbial biomass, are also significant contributors to corrosion in the upstream sector. The presence and combination of each of the factors above strongly affect the corrosion rates and have a synergistic effect on the overall operational efficiency or detriment of the asset.

Electrochemical mechanisms are the main drivers of metal and alloy corrosion in aqueous solutions (Nernst 1889). It is generally accepted that four fundamental elements, referred to as ACME, must be present for corrosion to occur: an anode, a cathode, a metallic conductor, and an electrolytic conductor (Papavinasam 2014). Each element contributes separately to overall corrosion severity and plays a crucial role in the process. At the anodic site, metal ions are solubilized after losing electrons and carried away from the material surface, thereby leading to an oxidized (i.e., electron deficient) state of the metal at this location on the object. This process of gradual metal ion dissolution and deterioration is general corrosion. The electrolytic conductor, or electrolyte, is often a liquid that allows the metal ions to move away from the anode with other charged species in solution. The conductivity of electrolytes will be enhanced as concentrations of cations and anions in solution increase. The anions in the electrolyte will move toward the anode where they lose electrons, and the cations in the electrolyte will move toward the cathode where they gain electrons. At the cathode, positively charged species in the electrolyte will be deposited after being reduced (i.e., gaining electrons). These ions deposited at the cathode may be metal from the anodic site or other species, such as hydrogen or oxygen ions in solution, depending on the pH of the system (Papavinasam 2014). The last element, the metallic conductor, carries electrons left by the metal ions at the anodic site back to the cathode.

All four of the ACME elements must be present for corrosion to occur. However, since the anode, cathode, and metallic conductor can all occur simultaneously on the metal surface, submersion of the object into an electrolyte may be all that is needed to start the corrosion process (Enning and Garrelfs 2014). The tendency to corrode and the rate of corrosion will be partially driven by the Gibbs free energy change of the system (ΔG) with a negative overall value for the process leading to more stable end products and spontaneous reactions. The propensity of a specific metal to corrode is referred to as its standard redox potential (E_o) and is measured experimentally in comparison to a reference electrode (Papavinasam 2014). In a seawater environment, graphite and gold (Au) are the least reactive and magnesium (Mg) and aluminum (Al) are among the most reactive (RS-TR-67-11, 2011). Knowledge of the standard redox potential of a metal is valuable when comparing the different tendencies of metals to corrode in the same electrolyte but cannot be used to determine the rate of corrosion in that environment. Rates are proportional to the speed of electron transfer between the electrode and the electrolyte with the slowest ACME process being the limiting factor for corrosion rates in the system (Papavinasam 2014). The process described above, whereby all of the ACME elements are present, would lead to a uniform thinning of the metal over time. However, in the real world, localized corrosion driven by surface anomalies, pipeline velocities, headspace gas compositions, the presence of passive or conductive films on the metal surface, metallurgical compositions, and microorganisms can change the electrochemical nature of the ACME elements and alter the location and/or severity of the corrosion process (Beech et al. 2010). Therefore, local monitoring protocols and performance indicators should be established for assets in different locations and under varying service conditions.

11.2 THE ROLE OF MICROORGANISMS IN THE CORROSION PROCESS

In recent years, it has become clear that microorganisms cannot only accelerate the electrochemical reactions associated with corrosion but may help to inhibit the corrosion process as well (Kip and van Veen 2015). These processes are termed microbiologically influenced corrosion (MIC) and microbiologically influenced corrosion inhibition (MICI). Several microbiological processes that significantly increase the rate of corrosion are hydrogen consumption, which increases the rate of cathodic reactions (Videla and Herrera 2005), the release of acidic metabolites, certain enzymatic cellular processes, manganese oxidation, and iron oxidation. Groups of microorganisms associated with increased corrosion rates and threats are sulfate-reducing bacteria and archaea, sulfur-oxidizing bacteria, iron-reducing and iron-oxidizing bacteria, manganese-oxidizing bacteria, and microorganisms capable of excreting acidic metabolic end products into the environment, such as organic acids. In addition, there is increasing evidence that methanogens also play a role in microbiologically inducted corrosion (Uchiyama et al. 2010). The biggest impact on corrosion rates often comes from sulfate-reducing microorganisms that are capable of rapid iron oxidation (Emerson et al. 2010; Enning and Garrelfs 2014). For iron corrosion in particular, the microbially catalyzed processes can be broken down into chemical MIC (CMIC) and electrical MIC (EMIC) that differ in the sources of electrons for the process and overall impact on the system (Enning et al. 2012).

Work has also been performed to show that microorganisms can play a role in protecting a metal surface against abiotic and biotic corrosion (Potekhina et al. 1999; Videla and Herrera 2005). Mainly, the process of using biology to protect a system from corrosion involves controlling the microbial ecology of that system in order to displace problematic microorganisms with beneficial or nontroublesome microorganisms in a process known as biocompetitive exclusion (Videla and Herrera 2005). Although still in the early days and not yet demonstrated for mixed-species biofilms outside of the lab, promising results have been shown by using microorganisms to remove corrosive compounds from a system, such as through oxygen consumption (Jayaraman et al. 1997) or by the use of sulfate-reducing bacteria that generate antimicrobial peptides that can be released in situ (Jayaraman et al. 1999). Future research in this area will need to focus on the complex signaling mechanisms that play a role in mixed-species biofilms in order to control the process to the extent needed in the field. However, one of the most promising areas of research in our opinion is using microorganisms to catalyze the process of metal-sulfide, -oxide, or -carbonate precipitation in a system to form a low-energy, stable protective barrier on the metal surface. It must be said though that any film laid down can be dislodged or disrupted through normal activities, such as solids production from the well or pigging. For iron in particular, typical forms of iron sulfides seen in the upstream oil and gas industry are shown in Table 11.1.

These iron sulfide crusts may offer protection to metal surfaces in the field if the films are thin, uniform, and other corrosion-inhibiting chemistries are used but can become problematic as the layer grows and cracks. A thick corrosion crust will, however, facilitate the corrosion process if the film is conductive, such as in the case of EMIC (Enning et al. 2012). This conductive crust allows the EMIC organisms to catalyze the corrosion process even when not in direct contact with the metal surface. Therefore, it is important to have a well-designed microbial control program from the onset of operations to help prevent excessive microbial growth, accumulation of bacterially produced extracellular polymeric substances that help form robust biofilms (Beech et al. 2010), and the onset of corrosion processes due to unstable water chemistry. For these programs, biocidal actives play an important role in the overall success of the microbial control program. Some of the most often used biocides in the upstream oil and gas industry and their effective biological and chemical ranges are described in Table 11.2.

TABLE 11.1
Types of Iron Sulfides Seen in Upstream Oil Facilities

Sulfide	Chemical Formula	Stoichiometry
Greigite	Fe_3S_4	–
Pyrite	FeS_2	S or Fe deficient
Marcasite	FeS_2	S deficient, instable
Mackinawite	$Fe_{1+x}S$	$x = 0.057 - 0.064$
Pyrrhotite	$Fe_{1-x}S$	$x = 0 - 0.14$
Smythite	$Fe_{3+x}S_4$	$x = 0 - 0.25$

Source: N. Sridhar, D. S. Dunn, and M. Seth, M., *Corrosion*, 57, 598–613, 2001.

TABLE 11.2
Overview of Typical Nonoxidizing Biocides Used in the Oil and Gas Industry

Active	Efficacy	Effective pH Range
Glutaraldehyde	B, F, A, biofilm	up to 9.5
THPS[a]	B, F, A, biofilm	3–11
DBNPA[b]	B, F, A, biofilm	up to 8.5
DiDAC[c]	B, F, A, biofilm	3–11
CMIT/MIT[d]	B, F, A	2–9
EBCO[e]	B	7–11

Note: B = bacteria; F = fungi; A = algae.
[a] Tetrakis(hydroxymethyl) phosphonium sulfate
[b] 2,2-Dibromo-3-nitrilopropionamide
[c] N, N Didecyl-N,N-dimethylammoniumchloride
[d] 5-Chloro-2-methyl-4-isothiazolin-3-one/2-methyl-4-isothiazolin-3-one
[e] 5-Ethyl-3,7-dioxa-1-azabicyclo(3,3,0)octane

11.3 THE USE OF BIOCIDES TO MITIGATE MICROBIOLOGICALLY INDUCED CORROSION

Upstream oil and gas operations exposed to water or multiphase fluids are typically treated with functional chemicals to mitigate corrosion and biocorrosion, and chemical mitigation strategies are defined around two classes of chemicals: biocides (microbicides) and corrosion inhibitors (CI), often applied in parallel. The first class comprises organic and inorganic molecules that have the ability to disrupt cellular physiology by disrupting the membrane of the cell, interfering with cellular metabolic processes, or by interrupting the ability of the cell to generate energy in the form of ATP (Lee et al. 2010). This class of chemicals limits the proliferation of unwanted microorganisms in industrial systems in order to keep microbial issues, such as biocorrosion, in check. The other class of chemicals used during a corrosion mitigation strategy comprises organic, surfactant-like molecules that are used in small concentrations and act to form an anticorrosive protective layer on metal surfaces. These molecules consist of a surface-active head group, typically including an amine functionality and a hydrophobic aliphatic tail. After the film-forming process is completed, the aliphatic tails act to repel hydrophilic molecules and form a double protective layer on the metal surface. This layered structure prevents water intrusion and inhibits the formation of anodic and cathodic zones, thereby inhibiting the ACME process and corrosion. However, the division between chemical corrosion inhibitors and biocidal actives is not absolute, and many corrosion inhibitors also have intrinsic biocidal properties or will be combined with a biocidal active during application to the system. A well-known example of molecules with dual functionality are quaternary ammonium salts. These molecules are typically applied during a corrosion inhibitor program but also have biocidal properties and are used in formulations containing a mixture of other biocidal actives and quaternary ammonium compounds within microbial control programs.

11.4 BIOCIDE FUNCTIONALITIES AND THEIR PERFORMANCE

In the section below, biocidal actives commonly used in the petroleum industry are described, and we discuss their chemical properties, modes of action, and performance characteristics for each molecule type. An example structure for each of the different classes is shown in Figure 11.1.

In Section 11.5, we provide a short introduction to the application of biocides and considerations that are taken into account during application.

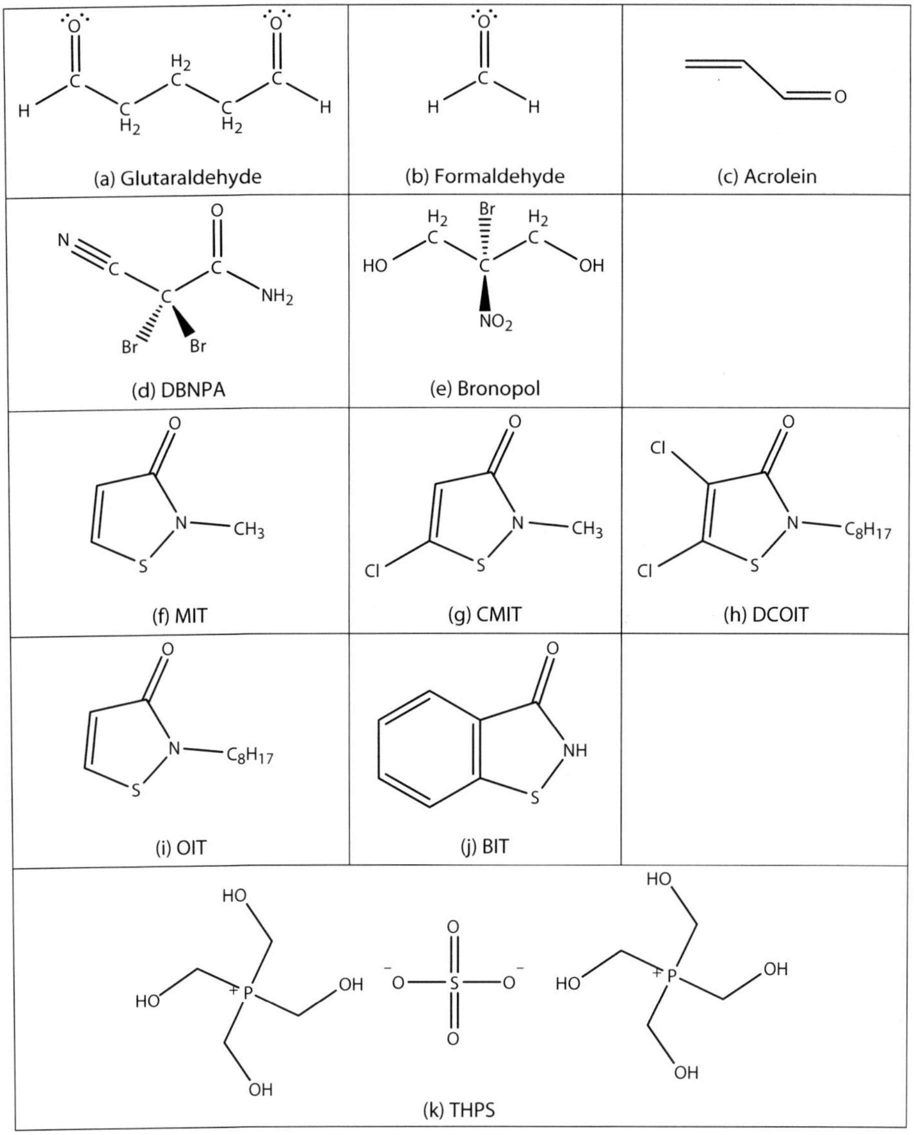

FIGURE 11.1 Overview of typical biocidal actives used in the oil and gas industry.

In general, biocides are functional chemicals applied to reduce and/or diminish the proliferation of microorganisms that thrive in industrial systems where water is present by inhibiting their metabolic activities that, over time, lead to system deterioration, reduction in oil quality, and costly repairs. There are many options in today's markets, yet biocides can typically be divided into two subclasses: oxidizing and nonoxidizing chemistries. Due to the corrosive nature of many of the oxidizing biocides, this section focuses more on the organic nonoxidizing biocides. We have, however, provided an overview of oxidizing biocides and their modes of action below.

11.4.1 Oxidizing Biocides

Oxidizing biocides are typically composed of inorganic compounds. Due to their very reactive nature, they are short-lived in many industrial facilities and their activity is quick (minutes to hours). Oxidizing chemistries can be a cost-effective way of treating large volumes of water if the system has been engineered to be corrosion resistant before application. The molecules have very low environmental impact and persistence due to the high reactivity of these species. Generally, oxidizing chemistries function by disruption of the bacterial cell membrane and inactivation of proteins important to the life of the cell (Finnegan et al. 2010). However, they are generally considered to be ineffective against biofilms as the reactive species only treat the surface of the biofilm and are unable to penetrate the biofilm matrix.

In the petroleum industry, the most often applied type of oxidizing chemistry is based upon chlorine (Cl_2) due to high solubilities in water (up to 7100 mg/l) under standard conditions (20°C and pressures of 1 atm). In contact with water, the chlorine molecule will hydrolyze completely to form hypochlorous acid (HClO) (Wallen and Henrikson 1989). At pH values above 8.5, the hypochlorous acid deprotonates and produces the less biocidal hypochlorite (ClO^-). Therefore, the efficacy of chlorine chemistries can vary greatly with pH but also with temperature and the amount of light in the system (Keevil et al. 1990). Using electrochlorination, hypochlorite can be generated at the site of treatment for easy dosing and application. If electrochlorination is not an option, other forms of oxidizing biocides also applied can vary from sodium hypochlorite (NaClO), chlorine dioxide (ClO_2), bromide and bromine chloride (BrCl), various peroxides, and combinations of oxidizing chemistries herein. However, the majority of applications in oil and gas are treated with chlorine when an oxidizing chemistry is preferred.

Most often, operators would choose to apply an oxidizing chemistry when there is limited time to treat the water. As a rule of thumb, chlorination processes should be performed using a minimum contact time of 15 minutes with concentrations of free chlorine at values of 0.2 mg/L, up to 0.5 mg/L (Flynn 2009). Water chemistry may make it necessary to use high concentrations of chlorine (i.e., higher chlorine demands), but the concentrations above are generally recommended. Other parameters that may affect chlorine demand are pH; the presence of ammonia, sulfides, nitrites, or hydrocarbons; and the presence of reducing agents, such as ammonium bisulfite. In general, chlorine demand is defined as the difference between the amount added to water and the quantity of free and combined available chlorine that remains at the end of a specified contact time. Unlike nonoxidizing biocides, chlorine will be generally applied as a low continuous dose (Palmer and King 2008; Flynn 2009). This holds true for industrial cooling water systems as well as for petroleum systems. Dosage automation is possible because oxidizers can be easily monitored using online systems. Oxidizing agents are frequently applied in seawater injection systems either with or without other nonoxidizing biocides (Kelland 2009). The application depends on which part of the system is under treatment. An oxidizer, such as hypochlorite or chlorine, may be dosed at the seawater lift pumps, where the seawater is pretreated (and flocculated) before a first filtration takes place. Chlorine will, however, be removed in a deaeration tower. So downstream of the tower, a secondary treatment point may be included. It is recommended that oxidizer dosing be alternated with dosing of other chemistries, such as oxygen scavengers, in order to avoid deactivation. Typically, the application point remains upstream of the tower, and therefore, chlorine is exposed to the downstream pipework.

Oxidizing chemistries are also used frequently in the hydraulic fracturing industries. In general, oxidizing agents are either applied to break the polymers into smaller molecular weights that can move downhole to permit gas flow and the start of production or for biocidal purposes. In the first case, oxidizing agents like persulfate salts (along with cellulases) are used (Barati and Liang 2014). For biocide treatment of fracturing water, oxidizing chemistries, such as chlorine, chlorine dioxide, and chlorite, are often used successfully (Malcolm et al. 2015). The case study outlines the use of chlorine dioxide due to its nonpersistence in flowback waters. However, this study describes the use of ClO_2 combined with low amounts of Bronopol to achieve microbial control. In the paper, the authors discuss clear reasons for using ClO_2 over other oxidizing chemistries. Hartwick et al. (1997) show that the application of peroxide and ozone-based biocides leads to oxygenic waters that require an additional oxygen scavenger, shifting the cost of chemicals from biocides to oxygen scavengers. Bleach is frequently applied in the petroleum industry; however, there are some concerns over the high pH of the water when bleach is applied, causing an interference with other fracturing fluids. A drawback of using Cl_2 is the formation of short halogenated hydrocarbons that may present toxicity concerns and health hazards (Kahrilas et al. 2014). Several publications are available that provide a comprehensive overview of all biocides used in the hydraulic fracturing industry (Finnegan et al. 2010; Kahrilas et al. 2014). Several reviews also discuss the shortcomings of oxidizing agents (Flynn 2009). From the literature, it can be concluded that oxidizing agents can be used for water pretreatment but are less effective when used for downhole operations and, in that application, should be combined with a nonoxidizing biocide for the best microbial control outcomes. The concentration of residual chlorine must be carefully monitored as it can induce corrosion to metal as four chlorine molecules have the same effect as one oxygen molecule (Palmer and King 2008). Unfortunately, chlorine is less effective against SRBs that typically occur later in a seawater injection system, and therefore, a nonorganic biocide is still required to maintain control (Boivin 1995; Jayaraman et al. 1999).

11.4.2 Common Nonoxidizing Biocides for the Oil and Gas Industry

Nonoxidizing biocides are, in general, short organic molecules. They are generally considered to be less reactive than oxidizing biocides and, therefore, do not have the disadvantages of oxidizing biocides, but the lower reactivity allows for better penetration into biofilms and the water injection systems. Reactive properties of the nonoxidizing chemistries should be taken into account when choosing the right active for the system and application. For example, isothiazolones are known for their slow mode of action, which makes them suitable as long-term preservatives. A molecule like 2,2-Dibromo-3-nitrilopropionamide (DBNPA) is far more reactive and should, therefore, be applied in systems where the contact time is short or a quick reduction in cells is desired. In the section below, the following biocidal actives will be reviewed in extensive detail (Figure 11.1).

- Aldehydes (specifically glutaraldehyde)
- THPS (tetrakis(hydroxymethyl)phosphonium sulfate)
- 2,2-Dibromo-3-nitrilopropionamide (DBNPA)
- 2-bromo-2-nitropropane-1,3-diol (Bronopol)
- Isothiazolones (e.g., CMIT/MIT)
- Didecyldimethylammoniumchloride (DDAC)
- Alkyldimethylbenzylammonium chloride (ADBAC)
- Tributyltetradecylphosphonium chloride (TTPC)

Aldehydes belong to the class of electrophilic active agents that have been in use for over 40 years. The electron-deficient oxygen atoms allow the carbon atoms to react with nucleophilic molecules and cellular moieties, such as amino-, amide-, or thiol-functionalities found in many cellular components (Paulus 2005). Aldehydes quickly react with these cellular features, disabling their function and thereby disrupting the overall metabolism of the cell. The most commonly used aldehydes in the petroleum industry are glutaraldehyde, formaldehyde, and acrolein. Glutaraldehyde has two functional groups, allowing the molecule to help catalyze cross-linking reactions with two separated nucleophilic entities in or on a cell (Gorman and Scott 1980). In the following section, we will focus on glutaraldehyde. Formaldehyde and acrolein are also mentioned as they are still used in the oil industry, but the molecules are slowly being phased out due to severe environmental, health, and safety (EH&S) issues and intrinsic severe handling risks.

11.4.3 Glutaraldehyde ($C_5H_8O_2$)

Glutaraldehyde (1,5-Pentanedial) is a colorless liquid that contains two aldehyde functionalities. The molecule consists of five carbon atoms and can react irreversibly with primary and secondary amines (Boivin 1995; Paulus 2005). Protonation of amine groups will take place in environments where the pH is low, impacting the reactivity of glutaraldehyde versus systems that are above neutral pH. The biocidal effect of glutaraldehyde is based on its reactivity with cellular nucleophilic substances causing cross-linkages that prevent cellular metabolism. The reactivity of the carbonyl groups is increased with amino groups if they are not protonated; therefore, the molecule has improved efficacy at pH values around 7.5 to 8.4 that are often found in seawater. In addition to the reaction with amino groups, glutaraldehyde can also react with thiol (R-SH) and hydroxy groups (R-OH). This reaction disrupts the tertiary structures of proteins and causes them to lose their catalytic functions. Besides good biocidal properties against planktonic cells, glutaraldehyde has also been shown to be effective against spores. Parameters that should be considered when using glutaraldehyde are system pH, contact time, and system temperature. In the petroleum industry, glutaraldehyde can be used during water flooding, in produced water or drilling mud, in completion and packer fluids, and also frequently in the water treatment industry.

The molecule is versatile and is largely unaffected by salt. The reactivity of the molecule does, however, increase above pH 8. This is an advantage as well as a disadvantage as the biocidal efficacy of glutaraldehyde increases but will, as a consequence, have a decreased half-life. Glutaraldehyde can sustain temperatures up to around 85°C depending on the pH of the environment. In injection and production water, glutaraldehyde can be dosed from 50 ppm up to 2500 ppm active on a batch fashion, typically for 4 to 6 hours. At elevated pH dosage can be lowered from 25 to 500 ppm (e.g., for drilling muds). Glutaraldehyde will react with sulfides (H_2S, HS^-) but to a far lesser extent than other aldehydes. Ammonia interferes with the molecule as do oxygen scavengers like HSO_3^-, and the molecule can be combined with other nonoxidizing biocides for specialized applications. The molecule will also retain activity in the presence of an oil phase due to its high water solubility.

11.4.4 Formaldehyde (H_2CO) and Acrolein (C_3H_4O)

Formaldehyde is the smallest aldehyde, with a hydrogen-only containing R-group. It is a flammable gas in its pure form, but when dissolved in water, it becomes formalin and is generally offered on the market as a 37% solution that also contains methanol as a stabilizing agent. The efficacy of the molecule is significant against bacteria, spores, and viruses. The O-atoms in the molecule are electron-deficient because there is no "softening" from an extended R-group. As compared to formaldehyde, acrolein or 2-propenal contains a molecule with two functional groups, an aldehyde group and a vinyl double bond. This dual functionality makes it a good biocide that can act against nucleophiles. The double bond in both molecules is effectively attracted toward thiol groups (R-SH) that contribute to many cellular functions. Acrolein is still currently offered on the market as a colorless to straw-colored liquid that gives off a lacramating odor (Paulus 2005). Polymerization can occur if the molecule is not stabilized, but commercial formulations are typically stabilized with hydroquinone. The aldehyde group can react with amino groups similar to glutaraldehyde, and the double bond is known for its reactivity with nucleophilic sulfur groups.

Formaldehyde is also widely used for biocidal applications in all aspects of the oil industry (Penkala et al. 2004). It will, however, react with sulfide scavengers and oxygen scavengers, removing the molecule for biocidal efficacy. Acrolein will react with similar molecules but also with other sulfur groups. Due to the severe reactivity of both molecules, they are nonpersistent and degrade quickly in the environment. Despite good performance as biocides, health and environmental concerns are leading to both molecules being slowly phased out of the market. Formaldehyde is acutely toxic to humans via all routes of exposure (dermal, oral, and inhalation). When formaldehyde is in contact with skin, it will cause burns and allergic reactions. At any time, direct contact with the chemical needs to be avoided. The use of formaldehyde has come under more and more regulatory pressure and is generally not supported in several regions. As an example, the European reclassification of formaldehyde as a class 1B carcinogen led to significant decrease in its uses. Acrolein has also seen increased pressure from regulators because of high toxicity and the need for special engineering controls to avoid any contact with the substance (especially offshore). Like formaldehyde, all routes of exposure induce an acute toxicity, and acrolein imposes an acute threat for all aquatic life. However, despite the severe environmental profile of the active, it is currently still in use offshore within several regions of the North Sea sector.

11.4.5 Tetrakis Hydroxymethyl Phosphonium Sulfate

Tetrakis hydroxymethyl phosphonium sulfate (THPS) is a quaternary phosphonium compound $[P(CH_2OH)_4]_2SO_4$. In its pure form, it is a slightly waxy solid that mixes well with water and provides an easy-to-use biocide for injection systems. The molecule has a relatively fast mode of action and works well against various organism types, including higher organisms such as fungi. The phosphine, THP, is responsible for its biocidal properties as it interferes with disulfide linkages in proteins, causing them to lose catalysis due to the breakdown of tertiary structure. In addition, the loss of free thiol groups leads to cell destabilization. THPS is an all-around biocide, which is applied in topside water and injection water as well as in three-phase production systems and pipelines. THPS can also be applied in water used for hydrotesting.

The active is effective in reducing the growth and proliferation of many organism types and may also function to keep reservoirs from souring by severely reducing the amount of microbiological activity and through sulfide-scavenging activity. Under continuous dosing, the molecule is applied at concentrations of 10 to 75 ppm. When applied in a batched fashion, typical concentrations are higher and may range from 70 to 300 ppm active ingredient in the product. The molecule has some drawbacks, such as pH sensitivity and reactivity with oxidizers and sulfites. Temperature effects also influence the degradation of the active, similar to other biocides with aldehyde functionalities. However, THPS is known to remain stable up to temperatures exceeding 120°C and may be combined with other organic biocides for combined functionalities or biocidal applications. Weekly batch doses of both THPS and glutaraldehyde have been shown to successfully control severe microbial induced corrosion (Enning et al. 2016).

11.4.6 Bronopol (2-Bromo-2-Nitropropane-1,3-Diol)

This molecule contains a severely electron-deficient carbon at its center, which is caused by the combination of the bromine group and the hydroxyl (-OH) R-groups (Paulus 2005). In its pure form, bronopol is a solid white powder that is miscible with water up to 280 grams per liter. The biocidal efficacy of bronopol is different under oxic and anoxic conditions. In oxygen-saturated water, its target is the amino acid cysteine; the reaction with this amino acid forms oxidizing agents by converting the cysteine to cystine (holding a disulfide bond), which can interfere with various cellular processes. When no oxygen is present, the molecule reacts with various nucleophilic components present inside a cellular body. Typically, this active is applied for longer-term applications as the rate of reactive is not quick, and longer contact times are needed for effective treatment of a system. The molecule is mainly used in the tight and shale gas industry due to its rapid degradation in slightly alkaline environments. Disadvantages of the molecule are extreme sensitivity toward sulfides, and its instability in heated environments above 40°C. Significant performance with bronopol is best achieved when combined with other biocidal actives.

11.4.7 DBNPA (2,2-Dibromo-3-Nitrilopropionamide)

This active is typically used for treating water-containing topside systems, both the fluids and the facilities themselves. DBNPA can also be applied as one of two biocides approved for use in seawater sulfate removal units to avoid biofilm formation on the nanofiltration membranes (Exner et al. 1973). Due to its rapid mode of action, a quick cell reduction can be observed, and dosage levels are generally low (<50 ppm). Although this active is a solid white powder similar to bronopol, it is not very soluble in water, and commercial formulations need to contain solvents. Common DBNPA formulations on the market are sold as 5%–20% active materials, and the molecule is mainly effective against bacteria. The majority of its efficacy comes from interfering with thiol and amine groups within cellular components, mainly enzymes, and thereby disrupting normal cellular function. Despite similarities to oxidizing agents, DBNPA is still categorized as a nonoxidizing biocide, due to the absence of a halogen release upon reaction.

Major uses for DBNPA are found in the shale and tight gas industries but also may be used in various decontamination operations of industrial water-holding systems, such as hydrotesting or for cleaning topside production facilities. The active is rapidly degrading in water systems. Like many other biocides, however, the active is not compatible with sulfites and sulfides, and alkaline environments will contribute to the rate of degradation.

11.4.8 Isothiazolones

The group of isothiazolone biocides mainly comprise cyclic structures containing a nitrogen–sulfur link. They may also contain chlorine atoms substituted for an (-H) group. The nitrogen–sulfur is the cause of the main antimicrobial effect as this link will break open when it interacts with thiol groups contained in cellular components (Williams 2007). The earlier described chlorine side groups will make the nitrogen–sulfur link more reactive as these groups draw electrons from the ring. Isothiazolones are widely used in the preservation of fuel due to their partitioning coefficient. Due to their extreme H_2S sensitivity, isothiazolone use in the oil and gas industry is very low. The most commonly applied isothiazolones are CMIT (A), MIT

(B), DCOIT (C), OIT (D), and BIT (E). The combination of CMIT/MIT (3:1) is also widely used.

- CMIT 5-Chloro-2-methyl-4-isothiazolin-3-one (A)
- MIT 2-methyl-4-isothiazolin-3-one (B)
- DCOIT 4,5-dichloro-2-n-octyl-4-isothiazolin-3-one (C)
- OIT 2-n-octyl-4-isothiazolin-3-one (D)
- BIT 1,2-benzisothiazolin-3-one (E)

11.4.9 CMIT/MIT

As mentioned previously, the vast majority of fuel preservation is performed by applying a combination of 5-Chloro-2-methyl-4-isothiazolin-3-one (CMIT) and 2-methyl-4-isothiazolin-3-one (MIT). This CMIT/MIT combination is also the other approved biocide in seawater sulfate removal units, next to DBNPA. The manufacturing process of CMIT/MIT dictates the ratio of these molecules to be 3:1 (CMIT:MIT), and both molecules are electrophilic; CMIT has an additional -Cl (electron withdrawing) atom in the 5 position, which make this active highly reactive. Most concentrated formulations containing CMIT/MIT are 14% and have an acidic pH with the liquid having a light or colorless tint. Their main biocidal efficacy can be attributed to their ability to interfere with enzymatic reactions involved in vital cellular processes, such as the Krebs cycle. In a later stage of biocidal treatment with this combination, free radicals are formed that also contribute to a gradual increased disruption of cells. This makes these molecules highly effective as biocides and leads to lower dosage levels (5–25 ppm active range). The slow mode of action, however, makes them typically unsuitable for waterflooding. Compatibility wise, CMIT/MIT is not compatible with amines, thiols, or mercaptans and is stable up to temperatures of 55°C.

11.4.10 Quaternary Ammonium and Phosphonium Compounds

These actives are known as quats for short and comprise a molecule containing a central quaternary nitrogen atom that carries a permanent positive charge and is bonded to four carbon-containing "R" substituents. The substituents are typically alkyl side chains but may also consist of aromatic hydrocarbons as in the case of alkyl dimethyl benzyl ammonium chloride (ADBAC). Biocidal efficacy of the molecule is affected by side chain length with an optimum around 14 to 16 carbons. Quats are lytic biocides, known to be membrane active and function by dissolution and disruption of the bacterial cell wall as well as binding to anionic functional groups on the membrane surface. These molecules are relatively stable and can provide the system with a combined effect of reducing bacterial numbers through biocidal activity and forming a protective film on the metal surface to combat corrosion caused by MIC (described in more detail below). In addition to ADBAC, other common quaternary ammonium compounds are didecyl dimethyl ammonium chloride (DiDAC, Figure 11.2) and tributyl tetradecyl phosphonium chloride (TTPC, Figure 11.3). Each of these molecules has advantages and disadvantages, but the most widely used quat today is DiDAC. In addition to having the best environmental profile of the group, the molecule also performs extremely

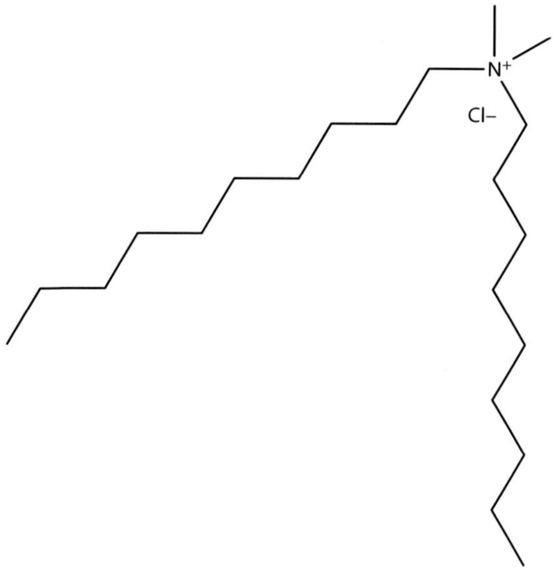

FIGURE 11.2 N, N Didecyl-N,N-dimethylammoniumchloride (DiDAC).

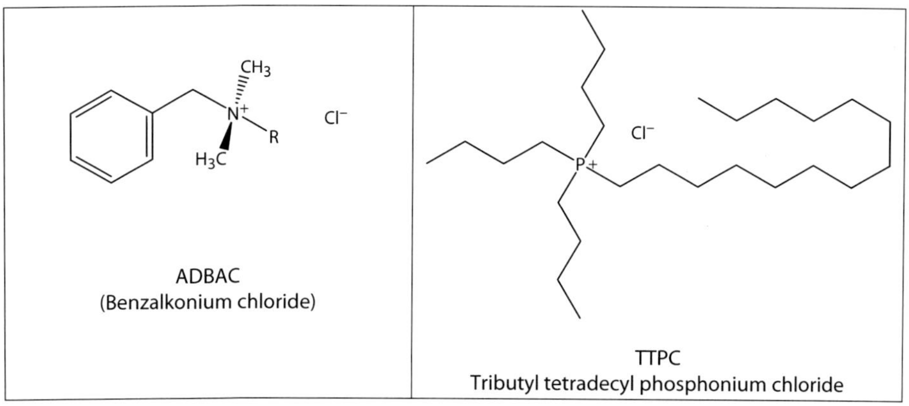

FIGURE 11.3 Overview of typical quaternary ammonium compounds used in the oil and gas industry. N, N Didecyl-N,N-dimethylammoniumchloride (DiDAC, as shown in Figure 11.2) is also a very effective molecule for both biocidal efficacy and corrosion inhibition.

well in a soured system or within topside pipelines and infrastructure. When combined with another biocide, such as glutaraldehyde, the quats provide the system with long-lasting protection while the glutaraldehyde will provide a quick to short-term efficacy.

11.5 APPLICATION OF BIOCIDES

Before the application of biocides is considered, consensus among system engineers and operators should be reached on what the objective of the treatment is and what other chemistries will be dosed into the system during treatment (i.e., oxygen and hydrogen sulfide scavengers, scale inhibitors, dispersants, water clarifiers, de-emulsifiers, etc.). As can be concluded from the preceding paragraphs, there is a variety of biocide choices from which an operator can choose, each with its own strengths, weaknesses, and ideal applications. Does the treatment need to be quick or is a longer-term protection needed or does the system require dynamic treatment in which the water is constantly being replenished, such as in an oil separator or pipeline, or is the system static as in the case of a long-term storage tank? It would also be important to consider water recycling in the system due to the formation of "incubation loops" where microorganisms in the water are being moved from one place to another while allowing them to proliferate and avoid washout. For example, water returned from an oil storage tank to the separator tank might experiment consistent increases in cellular numbers over time. In order to set up a good microbial control program, these aspects need to be considered before a single biocide or combination is selected. Other system properties to consider are the temperature of the water in the system, the system pH, the amount of water to be treated, and the contact time within the system. Depending on the system, it may also be advisable to include multiple dosing points along the water flow path.

Nonoxidizing biocides are typically added to a system via slug dose once or twice per week, depending on the application, for several hours. However, oxidizing biocides are dosed continuously based upon the system chlorine demand as discussed previously. How and when to apply biocides will depend partly on the bacteriological activity of the system and the lifespan and efficacy of the biocidal active used for treatment. When contamination is significant or the system contains corrosion products, larger batch doses for longer periods of time should be considered, especially if biofilms are present in the system. In a biofilm, organisms are more resistant to biocides and require longer exposure time to the active to be effectively removed (Jayaraman et al. 1999). This can be due to several reasons: cells found in a biofilm are typically less active than in the planktonic phase, so they may be less prone to disruption by a biocide. Also, diffusion limitation through the film itself plays a role as cells buried deeper in the biofilm will be exposed to less active if the formulation cannot penetrate the EPS and cellular matrix of the biofilm. Therefore, typical recommended dosages during biocide slug dosages can vary. As a guideline, the following parameters can be considered. Application rates can vary from 50 to 1000 ppm under normal circumstances (lower in the rare case of isothiazolones). Dosing takes place in a weekly or twice weekly batch fashion, ranging from 1 to 5 hours per dose depending on the volume of water being treated and capacity of the pumps to supply the biocidal formulation. Batches may occur up to three times per week in severe cases but typically occur once or twice per week. Choices on the ppm level applied depend on the level of microorganisms and biological activity in the system as well as the goal and production needs at the time of treatment. If the goal is to prevent further proliferation of microorganisms in the system and keep cellular numbers from significantly increasing, treatment can be performed using the bacteriostatic concentration of a formulation. This is the concentration where cells exposed to the active go into a state of metabolic arrest. If the goal of treatment is to disinfect a system, the biocide should be dosed at the bactericidal concentration that will destroy cells in the system and reduce overall numbers. The latter goal is most often the treatment scenario considered.

11.6 THE USE OF CORROSION INHIBITORS TO MITIGATE CORROSION

Areas of upstream oil and gas operations that are exposed to water are typically treated with functional chemicals to mitigate corrosion and biocorrosion. Most often, the classes of chemical used are known as corrosion inhibitors (CIs). These are organic, surfactant-like molecules that are used in small concentrations and act to form an anticorrosive protective layer on the metal surface. These molecules consist of a surface-active head group, typically including an amine functionality, and a hydrophobic aliphatic tail.

After the film-forming process is completed, the aliphatic tails act to repel hydrophilic molecules and form a double protective layer on the metal surface. This layered structure prevents water intrusion and inhibits the formation of anodic and cathodic zones, thereby inhibiting the ACME process and corrosion.

Traditionally, corrosion inhibitors were manufactured to be highly oil soluble. However, this feature tended to increase the environmental impact of the molecules and led to elevated bioaccumulation and toxicity profiles. Looking at the current situation, the molecules and technological developments made in the corrosion-inhibitor industry have matured over the last 20 to 30 years and now offer better corrosion protection versus older technologies while also providing a favorable environmental footprint. In the oil and gas industry in particular, corrosion inhibitors are most often imidazolines (Figure 11.4).

The amine groups within the head group act to form a protective film on the metal surface that will become stable and persist with continued use of the chemistry. Corrosion inhibitors containing amines represent around 39% of the current CI market and consist of molecules like amidoamines, imidazolines, imidazoles, fatty akylamines, diamines, and quaternary amines. The efficacy and performance of each of the different compounds depends largely on the R groups attached to the head of the molecule, the water chemistry, the presence and composition of an oil phase, the salinity of the system, and the presence of H_2S or saturated CO_2. Each of these factors combined will help to determine the product used in the system and the dosing rates and strategies employed for optimum protection of the assets.

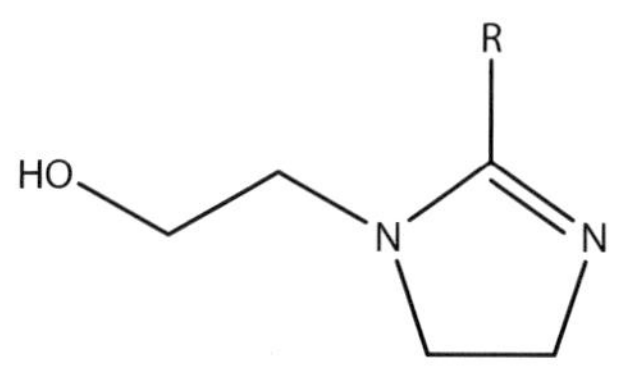

FIGURE 11.4 An example structure of the imidazoline family. In this case, hydroxyethyl imidazoline is shown with the R-group being representative of an alkyl chain derivative.

11.7 CORROSION INHIBITOR FUNCTIONALITIES AND THEIR EFFECTS ON PERFORMANCE

In the section below, we would like to describe the chemical functionalities found in modern corrosion inhibiting actives, and describe typical performance characteristics for each molecule type. An example structure for each of the different classes is shown in Figure 11.5.

11.7.1 Imidazolines

Corrosion inhibitors in this class are the most widely used and often among the most effective offerings available today. These molecules are generally represented as shown in Figure 11.5a. The functionality is based on two parts: (i) a hydrophobic chain that is oriented away from the metal surface and (ii) a nitrogen-containing, ring-based head group that is able to adsorb to the metal surface through electron sharing (Figure 11.6). The chain length of the tail affects both the performance of the molecule and the environmental profile of the molecule. Balancing the two aspects of the molecule would depend on the regulatory structure within the end country of use and the desired performance profile of the targeted application. However, research has shown that molecules with chain lengths of less than 12 carbons do not provide protection as corrosion inhibitors (Zuo 2007). In general, these molecules are formed by reacting carboxylic acids with primary amines. The carboxylic group can be derived from a low-cost natural source, such as crude or refined tall oil (Cutler et al. 1967).

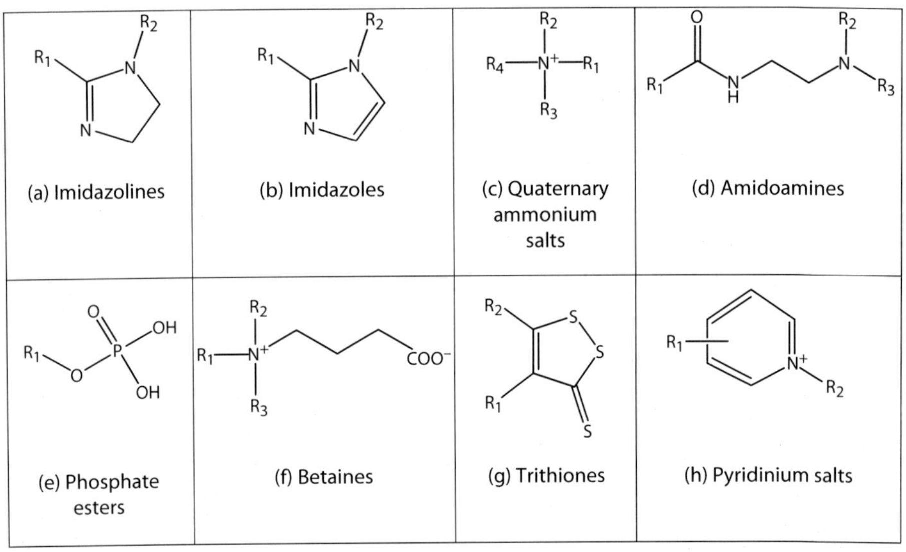

FIGURE 11.5 Overview of typical corrosion inhibitors used in the oil and gas industry.

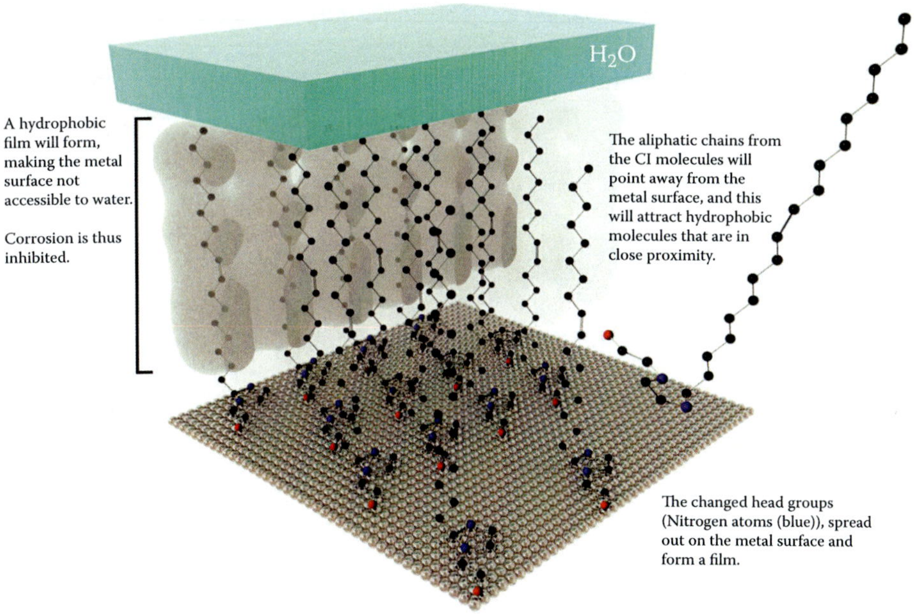

FIGURE 11.6 An artistic rendition of the mode of action for organic corrosion inhibitors.

11.7.2 Fatty Amine Adducts

This class of corrosion inhibitors is widely applied across the oil and gas industry, and use is particularly broad in the Middle East. These molecules are similar to quaternary ammonium ions (described below) but have often had one or two of the hydrogens of the ammonium ion replaced with fatty acids ranging in length from 18 to 20 carbons. The charge and protection profile of the molecule can vary with pH, but these molecules are known to be effective within typical seawater ranges. These adducts are used for long-term preservation of structures that will be exposed to prolonged seawater exposure, such as injectors or pipelines used for seawater transport. Depending on the fatty acid adduct, these molecules can also have biocidal properties that offer additional protection against biofilm-mediated MIC.

11.7.3 Alkoxylated Amines and Amides

In many corrosion-inhibitor systems, surfactants are required to perform several functions. First, the surfactant is often needed to solubilize the corrosion-inhibiting active ingredient. Second, the surfactant can act to clean the surface of the metal before treatment in order to form a more uniform and therefore more protective film. Last, the surfactant can act to improve the penetration of the active ingredient into pores on the material and the biological or chemical abnormalities on the metal surface. By adding ethoxylated, propoxylated alkylphenol amines and amides to the head group, the molecule will become water soluble and therefore will not require additional surfactants, often decreasing the direct costs of the formulation and enhancing the biodegradability and environmental profile of the product. The oxygen atom in the head groups maintains adsorption of the molecule to the metal surface.

11.7.4 Phosphate Esters

Phosphate esters, as shown generally in Figure 11.5e, are widely used in the oil and gas industry and offer some of the best toxicity profiles for these classes of compounds. In general, this class of molecule offers protection at moderate temperatures and pH ranges and can even offer protection in the presence of trace amounts of oxygen. Ethoxylated tridecylalcohol phosphate esters may also be used to prevent calcium precipitation during well stimulation by forming stable calcium salts and during routine oil and gas production scenarios. The phosphate esters are also used in refineries, at higher temperatures, to complex with soluble iron. These molecules therefore have wide uses in upstream, midstream, and downstream oil and gas applications and are widely praised for their excellent film-forming abilities and increased performance against pitting corrosion. They may also be used in the presence of harder brines.

11.7.5 Nitrogen Quaternaries

This family of molecules is also often applied as film-forming corrosion inhibitors, and certain compounds within the family are also known to have biocidal properties as well. General structural aspects of the family are shown in Figure 11.5c. The term "quaternary ammonium" applies to positively charged molecules that have had the hydrogen atoms of the ammonium parent ion replaced with alkyl or aryl groups. The quaternary ammonium compounds have a permanent charge, independent of the pH in the system being treated. For transportation and stabilization of the compounds, they are prepared as salts typically with a chloride ion. Synthesis of these compounds is carried out by a process known as quaternarization (Garverick 1994). Under certain conditions, these molecules may also have uses as deemulsifiers. One of the most widely used quaternary ammonium compounds in the oil and gas industry is N, N Didecyl-N,N-dimethylammoniumchloride, also known as DiDAC, shown in Figure 11.2.

11.8 CORROSION INHIBITOR USES AND APPLICATIONS

For general uses today, the available corrosion inhibitor actives can adequately serve the industry needs for typical waterflood and hydraulic fracturing applications. However, new chemistries will be needed as the industry moves to nontraditional operations in high-temperature and high-pressure ultradeep reservoirs. The current offerings are able to provide protection again aquatic corrosion, CO_2-associated corrosion, and H_2S-associated corrosion; however, misapplication or selection of the wrong actives may, under certain circumstances, lead to an increase in microbial numbers and MIC pitting rates (Harris et al. 2010).

Typically, corrosion inhibitors applied at ranges of 10 to 100 ppm active ingredient are effective at inhibiting the corrosion process. However, the amount of chemistry used may vary depending on the temperature and flow velocity of the system with increased dosing rates required for increases in both of those parameters. In certain extreme cases, concentrations up to 300 ppm active have been used in the field. Formulations are typically provided as 20% to 30% active ingredient preparations and are diluted and applied to the system on-site at the point of dosing. At the onset of pipeline operations, it is important to have identified key performance indicators that will be monitored over the lifespan of the asset as well as having a protocol and frequency for pipeline treatment with corrosion inhibitors or biocidal actives. Unfortunately, it is rare for systems to be treated proactively with corrosion inhibitors and biocides, and the onset of functional chemistry dosing begins after risk levels begin to rise. This reactive approach to dosing can be effective if KPIs have been monitored and thorough documentation of system status and maintenance has been maintained. However, corrosion inhibitors are far less effective after the onset of corrosion processes due to surface anomalies that develop on the interior walls of pipes and result in the inability of the chemical active to form a uniform protective film. Therefore, we would recommend a proactive approach and the implementation of MIC and microbial control programs as early as possible. The design of the program, actives selected, and dosing frequency should be tailored to the unique aspects of the system being treated. However, some of important considerations for the application of corrosion inhibitors are the following:

- The pour point of the formulation. For outdoor use, a lower minimum of –30°C is typically required.
- Solubility of the formulation and overall protection and versatility of the protective film formed on the metal surface.
- The performance of the formulation (typically aimed for is 0.1–0.3 mm/year). Typical performance cutoffs and ratings in the oil and gas industry are shown in Table 11.3.

TABLE 11.3
Typical Performance Ratings for Corrosion-Inhibiting Chemistries

Relative Corrosion Resistance	mpy	mm/yr
Outstanding	<1	<0.02
Excellent	1–5	0.02–0.1
Good	5–20	0.1–0.5
Fair	20–50	0.5–1.0
Poor	50–200	1–5
Unacceptable	200+	5+

- The tendency to form emulsions. When applied in batch mode, emulsions can appear between the oil and water phases and the need to use deemulsifiers. In general, emulsions caused by the use of performance chemistries should always be avoided.

The performance of a corrosion inhibitor is calculated by determining the percentage of inhibition efficiency, which is expressed using the following equation:

$$[1.00 - (\text{Weight loss (with inhibitor)}/\text{Weight loss (without inhibitor)})] \times 100 \quad (11.1)$$

This equation is used to describe the amount of metal surface covered by a selected formulation. Currently, a performance of 90% or above is expected. The formulation and presence or absence of a surfactant can greatly increase the effective coverage of a chemical by helping to clean the metal surface during dosing and to get better coverage into surface anomalies and crevices in the pipeline. Application of the corrosion inhibitors involves a process of (i) preparing the formulation and diluting the concentrated formulation down to use concentrations. During this step, it is recommended to minimize oxygen ingress as much as possible. (ii) Depending on the status of the system, often a series of brush pigs or foam pigs are used to clean the interior surface of the pipeline of chemical debris, waxes, and microbial biofilms. (iii) The dosing of the corrosion inhibitor begins, and (iv) the system is monitored for efficacy by routinely pulling coupons from the interior pipeline surface. Areas of injection in the pipeline should be selected to ensure proper mixing of the formulation, for example, in areas that do not have laminar flow. However, it may also be necessary to dose areas of increased corrosion with an additional dose as needed. Our findings indicate that low, continuous dosage of corrosion-inhibiting chemistries is often the most effective dosing strategy versus a weekly batch dose. This is especially the case with systems under high flowing conditions.

11.9 FUTURE PERSPECTIVES AND THE DEVELOPMENT OF KPIs

Our understanding of MIC and the different process therein continues to expand (Enning and Garrelfs 2014). Operators are increasingly becoming aware of the role that microorganisms play in upstream oil and gas systems and beginning to understand the benefits of targeted microbial control and functional chemistry applications. However, it is still difficult to definitively determine the overall contributions of different corrosion processes within a system experiencing corrosion. Each aspect of the system, including water chemistry, sulfide concentrations, flow rates, the presence of headspace gases, and microbial populations, contributes to the overall severity of corrosion problems. Therefore, our suggestion for the industry would be to define international standards that allow field operators to evaluate, reduce, and mitigate general corrosion as well as MIC. We are certain that targeted and well-defined chemical applications and microbial control programs are also part of the solution. These programs are better prepared for success when started early in the life of the asset and defined around meaningful KPIs. For the most recent developments regarding methods to monitor and track MIC, please see Chapter 3, Section 3.1

in this book. Meaningful monitoring programs are routinely comprised of bug bottle testing and coupon monitoring as well as DNA-based technologies (Keasler et al. 2010; Baldoni-Andrey et al. 2015). When a robust baseline is established, it is easier to determine when microbial processes or operational changes are leading to increased corrosion rates in the system. This baseline can only be established when operators shift from reactive to proactive strategies to combat corrosion. Several operators have now deployed ways to assess MIC threat. An example of one operator strategy includes a pyramid of detection systems, such as (i) SRB bug bottles that indicate an increase in microbial populations relative to a baseline below 10^3 colony-forming units (CFUs), (ii) coupons pulled from all assets monthly and visually and optically assessed for corrosion rates and the presence of pitting corrosion using surface analytical techniques, and (iii) quantitative PCR (qPCR) used in combination with high-throughput next-generation sequencing (NGS) technologies to profile the populations in biofilms and more accurately determine the bacterial numbers in a system. When two of the three methods detect an increased corrosion threat in the system, a proactive regime combining a suite of biocidal actives and corrosion-inhibiting chemistries are dosed until the populations fall below baseline levels. Industry experiences have now shown that bug bottles alone do not accurately assess the threats of MIC in the field. Particular areas of concern for increased MIC threat are upstream water injection systems, including floodwater and produced water subsea injection lines (Korenblum et al. 2010). Operators have also seen increased corrosion in midstream crude transmission lines.

Increasingly, novel DNA-based sequencing technologies are being used to form the basis of MIC monitoring programs and used as part of KPI guidelines. Third-party labs and consultancy companies are now offering tailored MIC monitoring services in collaboration with oil operators. As we continue to learn more about corrosion and the triggers to corrosion in a particular system, we can move from being reactionary to being able to predict the occurrence and severity of corrosion processes. Then, with predictions in hand, the functional chemistries used in the oil and gas arena can be used in the most efficient and cost-effective manner.

REFERENCES

Army Missile Command Report RS-TR-67-11. 2011. Practical Galvanic Series.

Baldoni-Andrey, P., Lesage, N., Pedenaud, P., and Jacob, M. 2015, May. Detection and Monitoring of Biofilm Growth in the Seawater Sulfate Removal Units. In *SPE Produced Water Handling and Management Symposium*. Society of Petroleum Engineers.

Barati, R., and Liang, J. T. 2014. A review of fracturing fluid systems used for hydraulic fracturing of oil and gas wells. *J Appl Polym Sci.*, 131(16). doi: 10.1002/app.40735.

Beech, I. B., Sunner, J. A., and Hiraoka, K. 2010. Microbe-surface interactions in biofouling and biocorrosion processes. *Int Microbiol.*, 8: 157–168.

Boivin, J. 1995. Oil industry biocides. *Mater Perform.*, 34: 65–68.

Cutler, R. A., Cimijotti, E. B., Okolowich, T. J., and Wetterau, W. F. 1967. Alkylbenzyldimethylammonium chlorides. A comparative study of the odd and even chain homologues of 12 different quaternary ammonium compound type antimicrobial agents. *Soap Chem Spec.*, 84–92.

Emerson, D., Fleming, E. J., and McBeth, J. M. 2010. Iron-oxidizing bacteria: An environmental and genomic perspective. *Annu Rev Microbiol.*, 64: 561–583.

Enning, D., Venzlaff, H., Garrelfs, J., Dinh, H. T., Meyer, V., Mayrhofer, K. et al. 2012. Marine sulfate-reducing bacteria cause serious corrosion of iron under electroconductive biogenic mineral crust. *Environ Microbiol.*, 14: 1772–1787.

Enning, D., and Garrelfs, J. 2014. Corrosion of iron by sulfate-reducing bacteria: New views of an old problem. *Appl Environ Microbiol.*, 80: 1226–1236.

Enning, D., Smith, R., and Stolle, J. 2016, June. Evaluating the Efficacy of Weekly THPS and Glutaraldehyde Batch Treatment to Control Severe Microbial Corrosion in a Simulated Seawater Injection System. In *CORROSION 2016*. NACE International.

Exner, J. H., Burk, G. A., and Kyriacou, D. (1973). Rates and products of decomposition of 2, 2-dibromo-3-nitrilopropionamide. *J Agric Food Chem.*, 21(5): 838–842.

Finnegan, M., Linley, E., Denyer, S. P., McDonnell, G., Simons, C., and Maillard, J. Y. 2010. Mode of action of hydrogen peroxide and other oxidizing agents: Differences between liquid and gas forms. *J Antimicrob Chemother.*, 65(10): 2108–2115.

Flynn, D. 2009. *The Nalco Water Handbook.* 3rd ed. New York: McGraw-Hill.

Garverick, L. (Ed.). 1994. *Corrosion in the Petrochemical Industry*. ASM International.

Gorman, S. P., Scott, E. M., and Russell, A. D. 1980. Antimicrobial activity, uses and mechanism of action of glutaraldehyde. *J Appl Bacteriol.*, 48(2): 161–190.

Harris, J. B., Webb, R., and Jenneman, G. 2010, January. Evaluating corrosion inhibitors as a means to control MIC in produced water. In *CORROSION 2010*. NACE International.

Hartwick, D., Chalut, J., and VanDoorn, E. 1997. *The performance of ozone as a cooling water treatment in water conservation efforts* (No. CONF-970332). NACE International, Houston, TX (USA).

Jayaraman, A., Cheng, E. T., Earthman, J. C., and Wood, T. K. 1997. Axenic aerobic biofilms inhibit corrosion of SAE 1018 steel through oxygen depletion. *Appl Microbiol Biotechnol.*, 48: 11–17.

Jayaraman, A., Hallock, P. J., Carson, R. M., Lee, C. C., Mansfeld, F. B., and Wood, T. K. 1999. Inhibiting sulfate-reducing bacteria in biofilms on steel with antimicrobial peptides generated in situ. *Appl Microbiol Biotechnol.*, 52: 267–275.

Kahrilas, G. A., Blotevogel, J., Stewart, P. S., and Borch, T. 2014. Biocides in hydraulic fracturing fluids: A critical review of their usage, mobility, degradation, and toxicity. *Environ Sci Technol.*, 49(1): 16–32.

Keasler, V., Bennett, B., and McGinley, H. 2010, January. Analysis of bacterial kill versus corrosion from use of common oilfield biocides. In *2010 8th International Pipeline Conference* (pp. 935–944). American Society of Mechanical Engineers.

Keevil, C. W., Mackerness, C. W., and Colbourne, J. S. 1990. Biocide treatment of biofilms. *International Biodeterioration*, 26: 169–179.

Kelland, M. A. 2009. *Production Chemicals for the Oil and Gas Industry.* Boca Raton, FL: CRC Press.

Kip, N., and van Veen, J. A. 2015. The dual role of microbes in corrosion. *ISME J*, 9: 542–551.

Korenblum, E., Valoni, É., Penna, M., and Seldin, L. 2010. Bacterial diversity in water injection systems of Brazilian offshore oil platforms. *Appl Microbiol Biotechnol.*, 85(3): 791–800.

Lee, M. H. P., Caffrey, S. M., Voordouw, J. K., and Voordouw, G. 2010. Effects of biocides on gene expression in the sulfate-reducing bacterium *Desulfovibrio vulgaris* Hildenborough. *Appl Microbiol Biotechnol.*, 87(3), 1109–1118.

Malcolm, A. C., Dempster, S. P., Chandler, B., and Setiawan, S. 2015, December. Evaluation of Chlorine Dioxide as an Oxidative Biocide for Hydraulic Fracturing in Western Canada. *In SPE/CSUR Unconventional Resources Conference*. Society of Petroleum Engineers.

Nernst, W. H. 1889. Die elektromotorische Wirksamkeit der Ionen. *Zeitschrift fuer physikalischer Chemi. Stoechiometrie und Verwandtschaftslehre*, 4: 129–181.

Palmer, A. C., and King, R. A. 2008. *Subsea Pipeline Engineering*. 2nd ed. Tulsa, OK: PennWell Corp.

Papavinasam, S. 2014. *Corrosion Control in the Oil and Gas Industry*. London: Elsevier.

Paulus, W. 2005. *Directory of Microbicides for the Protection of Materials: A Handbook*. Dordrecht: Kluwer Academic Publishers.

Penkala, J., Law, M. D., Horaska, D. D., and Dickinson, A. L. 2004, January. Acrolein 2-propenal: A versatile microbiocide for control of bacteria in oilfield systems. In *CORROSION 2004*. NACE International.

Potekhina, J. S., Sherisheva, N. G., Povetkina, L. P., Pospelov, A. P., Rakitina, T. A., Warnecke, F. et al. 1999. Role of microorganisms in corrosion inhibition of metals in aquatic habitats. Applied Microbiology and Biotechnology, 52: 639–646.

Sridhar, N., Dunn, D. S., and Seth, M., 2001. Application of a general reactive transport model to predict environment under disbonded coatings. *Corrosion*, 57: 598–613.

Uchiyama, T., Ito, K., Mori, K., Tsurumaru, H., and Harayama, S. 2010. Iron-corroding methanogen isolated from a crude-oil storage tank. *Appl Environ Microbiol.*, 76: 1783–1788.

Videla, H. A., and Herrera, L. K. 2005. Microbiologically influenced corrosion: Looking to the future. *Int Microbiol.*, 8: 169–180.

Wallen, B., and Henrikson, S. 1989. Effect of chlorination on stainless steels in seawater. *Materials and Corrosion*, 40(10): 602–615.

Williams, T. M. 2007. The mechanism of action of isothiazolone biocides. *PowerPlant Chemistry*, 9(1): 14.

Zuo, R. 2007. Biofilms: Strategies for metal corrosion inhibition employing microorganisms. *Appl Microbiol Biotechnol.*, 76(6): 1245–1253.

12 MIC Mitigation
Coatings and Cathodic Protection

Sandra L. Wilson and Thomas R. Jack

CONTENTS

ABSTRACT

Failure statistics show that external corrosion has traditionally been a significant cause of failure in oil and gas operations. While nonmetallic lines, which are not susceptible to corrosion, are being introduced in certain applications, carbon steel makes up most of our existing infrastructure and remains the material of choice for many new facilities. Prevention of corrosion for buried and immersed steel structures is achieved through the use of external coatings and cathodic protection; however, certain coating failure modes can compromise the level of protection achieved. Microorganisms have long been associated with external corrosion problems through coating degradation, increased current demand for cathodic protection, and the enhanced corrosion of susceptible steel. Historically, the identification of microbiologically influenced corrosion (MIC) has emphasized the detection of sulfate-reducing bacteria, but more extensive microbial analysis, including the use of emerging molecular methods, shows that the microbial communities associated with exposed steel surfaces are complex and highly variable. Knowledge of the type of coating and its CP compatibility can be of value in predicting the probability of MIC occurring on a given facility.

12.1 INTRODUCTION

Failure statistics (AER Report 2013-B 2013) for small-diameter lines in upstream oil and gas production reveal a history of frequent failures due to corrosion. This problem is being addressed in part by a move to the use of nonmetallics in place of steel; however, nonmetallics cannot be used everywhere. Steel remains a material of choice where cost-effective structural strength is needed. In these facilities, corrosion, including microbiologically influenced corrosion and environmentally assisted cracking (EAC), remain a significant concern. Coatings and cathodic protection (CP) are primary methods of MIC mitigation in upstream production and transportation systems.

12.2 COATINGS

In general, protective coatings work by isolating the surface of the steel from the local environment. In this role, they are referred to as "barrier coatings." Exclusion of water from the metal surface precludes both microbial activity and corrosion. A properly selected, applied, intact coating will prevent MIC without the addition of any other form of protection. Unfortunately, errors in coating selection or application, physical damage, or exposure of susceptible coatings to harmful operating or in-service conditions can lead to coating failure (Norsworthy 2009). Biodegradation can be a factor. When coatings fail, corrosion and environmentally assisted cracking of the underlying steel can ensue.

Coatings have been used over many decades in the oil and gas industry, but the materials employed have changed over time. In barrier coating systems, a water-impermeable component such as wax, asphalt, coal tar, or a suitable polymer (polyethylene [PE], polyvinyl chloride [PVC], polypropylene, epoxy, polyurethane [PU]) blocks water access to the underlying steel surface. While this is the key component of any coating system, other components may be added to protect the impermeable layer from sunlight or physical damage during handling and installation, to optimize physical properties such as temperature tolerance or elasticity, and to ensure tight adhesion to the steel surface.

A good overview of organic coatings has been published by Khaladkar (2011). Use of paints in corrosion protection has been reviewed by Hare (2011). An updated general text on the use of coatings in corrosion protection is available (Munger and Vincent 2014) while Norsworthy (2011) has published a review that includes a discussion of the advantages and disadvantages of various types of coating.

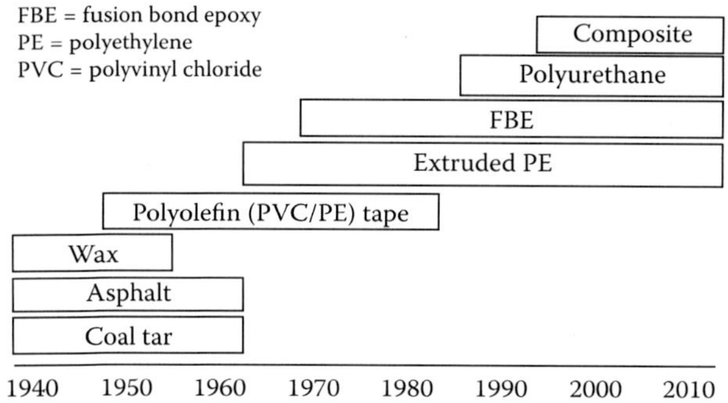

FIGURE 12.1 Coating systems used to protect the external surfaces of pipelines.

Wax, asphalt, and coal tar enamels were popular for field application until the 1960s (Figure 12.1), when they were displaced by newer coating systems that avoided worker exposure to potentially hazardous hot materials and were quicker to apply in the field or could be applied under controlled conditions to the joints of pipe in a coating mill prior to field use. An understanding of older coating systems and their failure modes remains relevant today. Most of the materials noted in Figure 12.1 are still in use, and the early coating systems can still be found in older parts of the oil and gas industry infrastructure. Knowing the type of coating and its failure modes is essential to managing the integrity of any operating system.

Asphalt, coal tar, polyolefin, epoxy, urethane, and composite coating products have all been used to cover welded joints and for external repair work in the field (Papavinasam and Revie 2002; Papavinasam 2014). A wide variety of commercial products is available to coat the internal surfaces of tanks, piping, and other operating units (Papavinasam 2014).

Coating degradation falls within a broad definition of MIC, which considers not only the role of microorganisms in the corrosion or cracking process itself, but also the role of microorganisms in establishing conditions that create the opportunity for corrosion. Microbial degradation can be an important factor in the loss of coating protection, particularly for coatings containing asphalt or polyurethane or for polyolefin tape wrap coating systems (Harris 1963; Norman and Argent 2007). Table 12.1 summarizes the potential consequences of biodegradation in terms of coating performance.

12.2.1 Early Coating Systems

Coal tar and asphalt can be degraded over time by photooxidation, water leaching, and microbial oxidation of lower molecular weight hydrocarbon components (Traxler 1962, 1966, 1968). These processes are supported by exposure to sunlight, oxygen, and wet soil environments. Harris (1963, 1966) isolated asphalt-degrading bacteria from the soil surrounding a buried pipeline. Unlike wax and asphalt, coal tar contains "toxic" components, such as phenols, at levels that can discourage microbial growth (Kulman 1958; Phillips and Traxler 1963; Pankhurst 1973; Pankhurst

TABLE 12.1
Effect of Biodegradation on Some Susceptible Coatings

Coating	Principal Components	Biodegradation	Potential Consequences
Asphalt	Mixture of hydrocarbons	Loss of lower molecular weight hydrocarbons	Embrittlement → cracking, increased permeability (Figure 12.8)
PE tape	Polyethylene polymer	*Polymer resists attack*	*None*
	Primer/adhesive	Loss of susceptible components	Loss of adhesion → coating disbondment (Figure 12.3)
PVC tape	PVC polymer	*Polymer resists attack*	*None*
	Plasticizer	Loss of susceptible plasticizers	Embrittlement → tape cracks/falls apart (Figure 12.4)
	Primer/adhesive	Loss of susceptible components	Loss of adhesion → coating disbondment (Figure 12.3)
Polyurethane	Polyurethane	Cleavage of the polymer backbone	Surface discoloration, chalking Loss of coating integrity

PE = polyethylene; PVC = polyvinyl chloride.

1976). Because the hydrocarbon substrates preferred by microorganisms are also relatively water-soluble, the gross effect of water leaching and biodegradation can be similar. Prolonged water leaching at 60°C in the laboratory results in asphalt with comparable barrier qualities to a degraded field sample after many years of service on a buried pipeline (Jack et al. 2002). In service, all of these processes cause the coating to harden, crack, and become more permeable and less of a barrier to water over time. This results in an increase in current demand for buried structures under CP. For one facility coated with asphalt enamel, the current density required to maintain adequate CP potentials increased from <0.1 to >1.4 mA/m^2 over a 20-year period (Jack 2001).

12.2.2 Polyolefin Coating Systems

Carbon–carbon bonds make up the backbone of polyolefin polymers. In general, these polymers resist biodegradation because the polymer chains are too large to fit into microbial cells and form a relatively rigid hydrophobic mass that excludes water and presents few terminal ends for microbial attack. Pigments are often added to block the penetration of ultraviolet light, which can break the chains and introduce oxygen functions that increase susceptibility to biodegradation (Albertsson and Ranby 1975; Albertsson et al. 1987). While high molecular weight polyolefin polymers tend to resist biodegradation, other components of a polyolefin coating system may be susceptible to attack. Figure 12.2 shows that the microbial reduction of sulfate, a common ion in groundwater, to sulfide by sulfate-reducing microorganisms

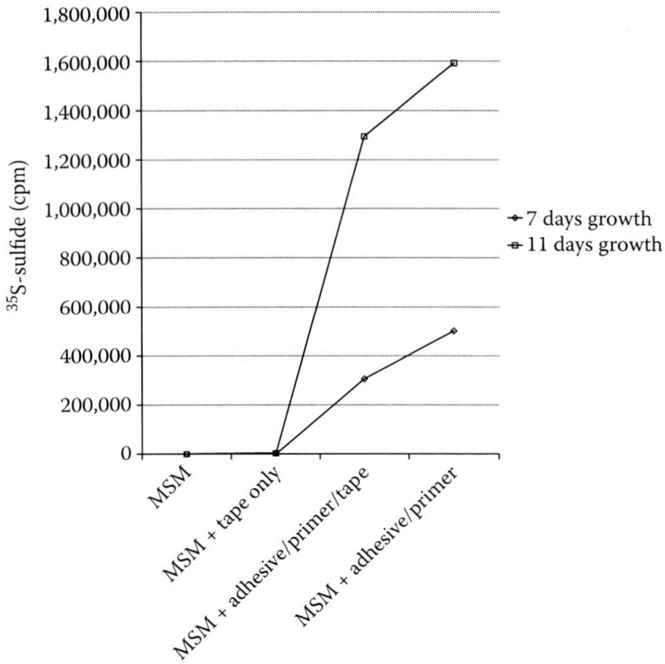

FIGURE 12.2 Reduction of ^{35}S-sulfate to sulfide by laboratory cultures of *Desulfovibrio vulgaris* var. *vulgaris* NCIB 8303 in a mineral salts medium (MSM) with hydrogen present (Jack et al. 1986). (Adapted from S. Papavinasam, *Corrosion Control in the Oil and Gas Industry*, Amsterdam: Elsevier, 2014.)

is sustained by the adhesive/primer components of a disbonded tape coating sample taken from a field corrosion site.

12.2.2.1 Polyethylene

Degradation of the adhesive/primer used to glue field-wrapped PE tape coating onto a pipe surface can result in the formation of disbonded areas of coating that allow water ingress and corrosion of the underlying steel surface (Figure 12.3). This remains a significant issue for early tape wrap coating systems that are still in service. Modern field-applied PE coatings show much better adhesion to the pipe surface than the early tape coating systems and are widely used, particularly in the form of hand wrap tapes used to repair missing or damaged coating in existing operations and as heat shrink sleeves that are used to cover welded joints on lines made up of mill-coated pipe sections. Regardless of improvements, a poor application/installation that allows the ingress of groundwater under the coating can still allow MIC in the field (Li et al. 2000; Li et al. 2001).

12.2.2.2 Polyvinyl Chloride

Polyvinyl chloride $(\text{-}CH_2\text{-}CHCl\text{-})_n$ has also been used to make tape coatings, but for this application, the pure polymer must be supplemented with a

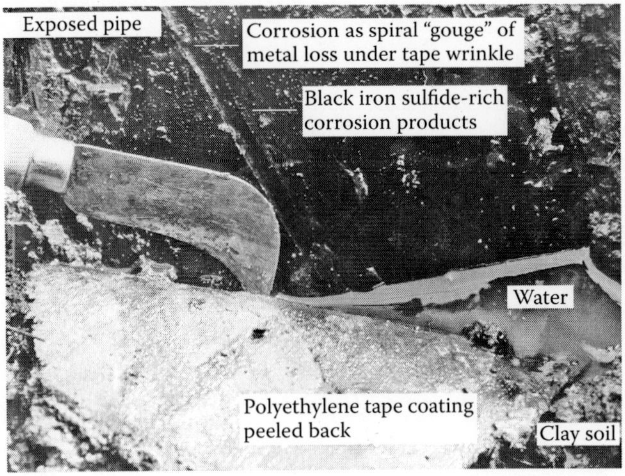

FIGURE 12.3 An iron sulfide/SRB MIC scenario under disbonded PE tape coating on a buried pipeline where corrosion damage has occurred under a wrinkle in the tape that had allowed water to access the steel surface under the coating.

plasticizer (20%–35%) to allow the tape to stretch and conform to metal structures. Plasticizers are typically small organic molecules that fit into the polymer matrix, allowing the polymer chains to slide past one another when the material is stretched. Phthalates are typically used in this role. Unfortunately, phthalates and many other plasticizers are susceptible to microbial attack under a wide range of environmental conditions (Brown 1945; Cavett and Woodrow 1968; Pankhurst and Davies 1968; Pankhurst et al. 1972; Cadmus 1977; Sugatt et al. 1984; Qiu et al. 2004). Biocides can be added to protect coating performance (Lasman and Scullin 1977). Webb et al. (2000) investigated the colonization of plasticized PVC on prolonged exposure to the atmosphere. *Aureobasidium pullullans* (*A. pullullans*) was the first to establish itself after 25 weeks exposure, followed by yeasts and yeast fungi that came to dominate the long-term microbial population on the coating surface. In vitro tests showed that *A. pullullans* taken from a field exposure test could grow on plasticized PVC, produce an extracellular esterase enzyme, degrade the dioctylphthalate plasticizer, and cause measurable weight loss. Microbial communities responsible for phthalate degradation have been described and degradation pathways proposed (Gu et al. 2011). Most of this work has been done under aerobic conditions; however, published studies also confirm anaerobic biodegradation. Benckiser and Ottow (1982) identified a strain of denitrifying *Pseudomonas pseudoalcaligenes* that could use one butyl group from di-n-butylphthalate as its sole source of carbon when nitrate was present, and Qiu et al. (2004) described methanogenic enrichment cultures and cocultures able to degrade phthalate isomers. Loss of plasticizer through water leaching, crystallization, and microbial degradation leaves a material that is brittle and subject to shrinkage and fracture (Figure 12.4).

FIGURE 12.4 PVC tape coating that has lost plasticizer and suffered brittle failure in the field.

12.2.3 Polyurethane

Polyurethanes are a broad class of materials in which units in the backbone of the polymer are held together by carbamate bonds (Figure 12.5). Intervening units (R in Figure 12.5) are commonly polyols, which can be made up of multiple units held together by ester or ether linkages. Selection of different components (R and R' in Figure 12.5) and cross-linking between polymer chains can give significant differences in the behavior of the material. A very wide range of stiffness, hardness, and density can be achieved. In the oil and gas sector, high-density solid PU elastomers are used to make flexible or semirigid coatings and adhesives while low-density rigid foams are used for thermal insulation.

PU paints and coatings are susceptible to biodegradation. Enzymes commonly found in microbial cells can cleave carbamate and ester linkages in the polymer chain or matrix, resulting in a potential loss of product performance. Gu and Mitchell (2004) isolated a bacterium from soil identified as a *Rhodococcus* species that produced esterase when growing on polyester–polyurethane, as the sole source of carbon and energy. A brief review of fungal attack on urethane coatings under aerobic conditions can be found in Little and Ray (2001). Examples cited include one where the growth of *Aureobasidium* sp. on and through a polyester-based urethane paint led to disbondment of the coating (Little and Ray 2001).

Products containing ether linkages in place of ester linkages are more resistant. This has been observed, for example, for fungal attack by species of *Pestalototiopsis*

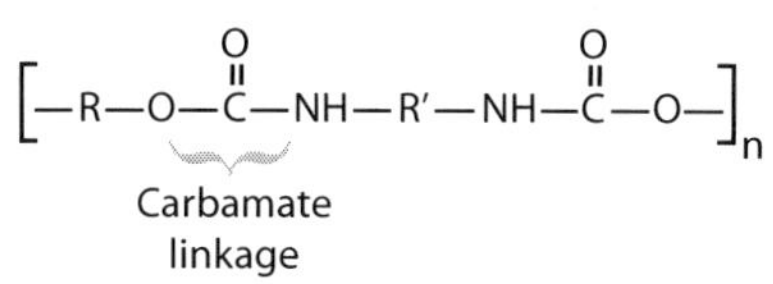

FIGURE 12.5 Generic structure for polyurethanes where R may be (—CH_2CH_2—) or may be a more complex chain incorporating ester or ether linkages.

(Darby and Kaplan 1968). Pankhurst et al. (1972) found that the growth of selected bacteria, fungi, and yeasts was minimal when ether linkages were present. In test products where a spacer with neither ester nor ether linkages was present (as R, Figure 12.5), an enzyme-containing commercial detergent was found to degrade the carbamate linkages in the PU chain (Huang et al. 1978). While the chemical bonds that link polymeric PU together are susceptible to enzymatic cleavage, high-density coatings and closed cell foams restrict the free access of water into the polymer matrix. This constrains microbial activity and enzymatic hydrolysis. In practice, PU coatings and foams are likely to suffer superficial damage through exposure to sunlight, water, and biological activity that leads to discoloration and "chalking" of the surface without seriously compromising the bulk properties of a resistant material. Biocides can be added to susceptible formulations to discourage biodegradation when necessary.

12.2.4 Modern Coating Systems

A wide range of products are now commercially available for a host of applications in the upstream oil and gas industry. These include sophisticated coating systems that combine two or more materials to optimize protection (e.g., high-performance composite coating or HPCC). These products often include epoxy components that intrinsically resist biodegradation. In addition to the development of new coating systems, new strategies are being adopted to make coatings antimicrobial. Recent examples include the use of niobium oxide as a base coat (Cordeiro Antunes de Araujo et al. 2013) or additives, such as molybdenum trioxide (Lackner and Guggenbichler 2013) or bioactive spores of *Paenibacillus polymyxa* (Akid et al. 2008).

12.2.5 Assessing Coating Performance

Early studies used microbial growth on coating materials, the appearance of metabolites such as carbon dioxide, and visual inspection to assess susceptibility to biodegradation. Coating samples were exposed to various aqueous test solutions inoculated with fungi or bacteria under aerobic or anaerobic conditions. In general, aerobic conditions fostered more growth. While growth tests can indicate the presence of coating components with a biocidal or biostatic effect, a visible surge of microbial growth in a bottle test can be fostered by trace quantities of a susceptible coating component or contaminant, the loss of which does not necessarily compromise the performance of the bulk material. To achieve more relevant information, longer term exposure of coated specimens to biologically active environments is carried out to assess the potential impact of biodegradation on physical properties, such as tensile strength, coating adhesion, and insulating ability (Yabannavar and Bartha 1994). Today, batteries of standardized tests are available for characterization of coating performance and selection (Papavinasam and Revie 2006; Thomson and Campion 2011; Papavinasam 2014). The challenge of using short-term accelerated testing to predict long-term field performance has been addressed by Papavinasam and Revie (2006). While it has been noted that exposure to biologically active soil prior to testing can influence results (Jack et al. 1996), these tests are routinely done on pristine coating samples. Specific standard methods and practices to assess the susceptibility of polymers to biodegradation (Gu and Wang 2004; Gu et al. 2011) are not widely used in the oil and gas sector.

Techniques such as electrical impedance spectroscopy can be used to nondestructively assess coating system performance in lab (Mitton et al. 1993; Zhong et al. 2008; Gu et al. 2011) and field (Jack et al. 2002; King et al. 2002). This technique can characterize the resistivity of an intact barrier coating on steel; the onset of water, oxygen, and ion penetration; and the loss of coating adhesion to the underlying steel surface (Mitton et al. 1993). Areas of coating that no longer act as effective barriers and areas of missing coating can be detected in the field by measuring current loss from buried steel structures under CP using aboveground survey techniques, such as direct current voltage gradient (DCVG) (NACE TM0109-2009), alternating current voltage gradient (ACVG), and pipeline current mapping (Ruschau and Kowalski 2006). Current loss can also be mapped in underground pipelines using in-line inspection tools (Janda and Williams 2014; Williams and Janda 2014).

12.3 CATHODIC PROTECTION

Corrosion involves the conversion of metallic iron to a soluble form through the loss of electrons. Equation 12.1 represents the process under anaerobic conditions where the soluble form of iron produced is the ferrous ion, Fe^{2+}. Under aerobic conditions, this may be further oxidized to ferric ion, Fe^{3+}.

$$Fe \leftrightarrow Fe^{2+} + 2 \text{ electrons} \tag{12.1}$$

The electrons given up in this process are picked up off a complementary cathodic area on the adjacent metal surface by electron-accepting chemical reactions. The cathodic reactions commonly encountered involve the reduction of protons or water (Equation 12.2) under anaerobic conditions or the reduction of oxygen (Equation 12.3) in aerobic ones.

$$2H^+ + 2 \text{ electrons} \rightarrow H_2 \text{ or } 2H_2O + 2 \text{ electrons} \rightarrow H_2 + 2OH^- \tag{12.2}$$

$$O_2 + 4H^+ + 4 \text{ electrons} \rightarrow 2H_2O \tag{12.3}$$

To stay the corrosion process, the equilibrium point for Equation 12.1 can be shifted to favor metallic iron by loading the metal surface with electrons. Where a sufficiently negative potential is achieved, corrosion effectively stops. The cathodic protection potential required to do this can be identified theoretically from Pourbaix diagrams (Pourbaix 1974). These diagrams are based on known chemical equilibrium constants for all the possible chemical reactions for iron in a given environment and define the most stable form of iron under various conditions (Verink 2011). The axes of the diagram are oxidation reduction potential (redox potential, Eh) and pH, both of which are, in principle, measurable quantities in any environment. Figure 12.6a presents a simple Pourbaix diagram for iron in the presence of water. Inclusion of sulfur in the system alters the diagram by introducing domains where iron sulfides

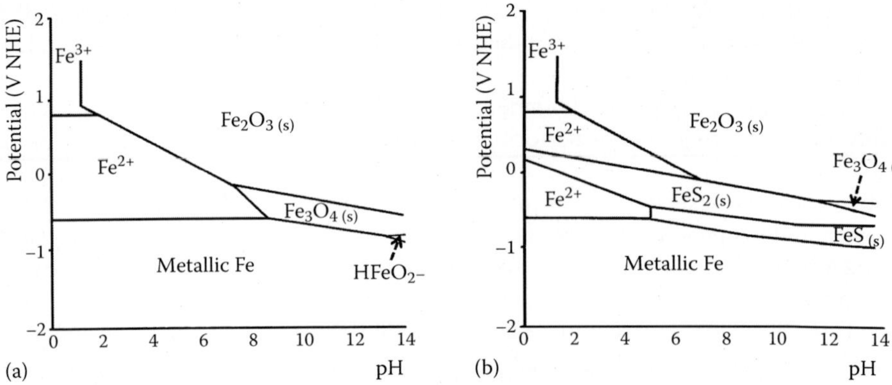

FIGURE 12.6 Simplified Pourbaix diagrams for (a) iron in an aqueous system and (b) for iron in an aqueous environment with sulfide present. Note that the potential is stated relative to a normal hydrogen reference electrode (NHE). (Based on E. D. Verink, Jr., Simplified procedure for constructing Pourbaix diagrams. In *Uhlig's Corrosion Handbook*, ed. R. W. Revie. 93–101. New Jersey: John Wiley & Sons, Inc., 2011; B. Little, R. Ray, and S. Dexter, Introduction. In *A Practical Manual on Microbiologically Influenced Corrosion*, Volume 2, ed. J. G. Stoecker II. 1.1–1.9. Houston: NACE, 2001.)

are the most stable form for iron (Figure 12.6b). Comparison of Figure 12.6a and 12.6b shows that the cathodic protection potential for the boundary between metallic iron and its potential corrosion products at pH 7 drops by about 100 mV if sulfide is present. This indicates that a more negative CP potential will be needed where sulfate-reducing microorganisms are active.

While setting CP potential requirements based on a Pourbaix diagram is theoretically sound, this is not as crisp or straightforward as the simple diagrams in Figure 12.6 might imply. The position of the boundary between two domains in the diagram represents an equilibrium point for the relevant chemical reaction and will shift with changes in the concentration of chemical species involved. Stable species may take different forms depending on the situation. Jing Ning et al. (2014) have prepared separate Pourbaix diagrams for different metastable and stable forms of iron sulfide (mackinawite, greigite, and pyrite) to reflect corrosion products actually observed. The chemistry in a corrosion site may be more complex in other ways. For example, the formation of acetic acid by anaerobic acetogenic bacteria could result in the formation of stable water-soluble iron complexes that are simply not considered in the construction of simple Pourbaix diagrams. In all cases, Pourbaix diagrams present a theoretical picture that assumes the entire system has come to thermodynamic equilibrium. In reality, this may not be so in a system where specific chemical reactions are kinetically constrained or in an open system where corrosion products may be lost through volatilization or diffusion away from the metal surface. Field tests or laboratory experiments must always be used to confirm that predictions made using a Pourbaix diagram actually happen on a meaningful time scale under conditions of interest.

12.3.1 Protective Potentials

As early as 1952, Wormwell and Farrar demonstrated that corrosion of steel in an active culture of sulfate-reducing bacteria (SRB) could be controlled by imposing a potential of –950 mV (CSE) on the metal surface. Many studies over many years generally confirmed this recommendation (Booth and Tiller 1968; Shirouzu et al. 1974; Fischer 1981; Barlo and Berry 1984; Pritula et al. 1987; Guezennec et al. 1990; Kajiyama and Okamura 1999), but more recent reports have suggested that a CP potential of –950 mV (CSE) may not be enough to arrest corrosion in the presence of active SRB (Esquivel et al. 2006; Romero et al. 2008). The exact value needed continues to be a topic of discussion, but NACE Standard SPO169-2013 (2013) recommends a protective potential of –850 mV (CSE) or lower (or a cathodic polarization of 100 mV) in the absence of microbial activity and –950 mV (CSE) or lower (or as much as 300 mV cathodic polarization) where active acid-producing bacteria or SRB are present. The standard suggests a benchmark for effective corrosion control of 0.025 mm/y or less. Driving the CP protection potential lower than about –1050 mV (CSE) to reduce corrosion rates further risks damage to coatings and susceptible steels through the generation of very high pH and/or excessive levels of cathodic hydrogen at the metal surface (Papavinasam 2014).

12.3.2 Increase in pH

Electrons supplied by CP drive cathodic reactions on the metal surface that consume protons or produce hydroxyl ions (Equations 12.2 and 12.3). These processes will increase the pH of the aqueous phase in contact with the metal. This can have beneficial consequences. Where the solution in contact with the metal surface is near saturation with respect to scaling, a shift to higher pH causes the precipitation of calcium carbonate or calcium magnesium carbonate minerals that seal off the metal surface, precluding corrosion and reducing CP current loss. However, the integrity of these protective scales is subject to attack by organic acids where these are produced by associated microbial communities. Guezennec (1994) illustrated this effect in the laboratory using pure cultures of an acid-producing, slime-forming microorganism, *Vibrio natriegens*.

In the absence of effective scale formation, the increase in local pH can discourage microbial activity according to the tolerance of the specific organisms involved. Miyanaga et al. (2007) applied CP to an artificial biofilm of slime-forming *Pseudomonas aeruginosa* PAO1 growing on carbon steel in an aqueous medium in the laboratory and noted that cathodic reactions increased the pH not only on the metal surface, but through the biofilm as well. Where the profile exceeded pH 9, viable cell counts in the biofilm dropped dramatically, leading the authors to suggest pH elevation can discourage activity on protected surfaces. Observation of sulfide deposits under marine macro-fouling on immersed offshore structures protected by CP led to a laboratory study of biofilm formation on steel coupons exposed to estuarine waters of Aberdeen Harbor with and without CP at –950 mV (CSE) (Maxwell 1986). Application of CP did reduce aerobic and anaerobic microbial activity and the number of viable SRB but did not prevent biofilm formation. While surface pH measurements were not reported, carbonate scale was seen to be forming on the steel surface under CP. Pritula et al. (1987) and Esquivel et al. (2006) reported that alkalinity generated by an applied potential of –950 mV (CSE) did not stop SRB growth.

12.3.3 Promotion of Hydrogen-Utilizing Microorganisms

Where conditions allow microbial growth, the application of CP to a metal surface can promote microorganisms able to use reducing power from the CP system to support their core metabolism. This may involve either the use of hydrogen formed by reduction of protons or water on the protected steel surface (Equation 12.2) or the direct uptake of electrons by microbial cells. Hardy (1983) showed that SRB were able to use cathodic hydrogen to reduce sulfate to sulfide (Equation 12.4).

$$4H_2 + SO_4^{2-} \rightarrow S^{2-} + 4H_2O \tag{12.4}$$

Equation 12.4 is the basis for the MIC mechanism originally described by von Wolzogen Kühr and van der Vlugt (1934) in which accelerated corrosion rates were attributed to the use of cathodic hydrogen formed by the anaerobic corrosion of steel (Equations 12.1 and 12.2) in the reduction of sulfate by SRB. A more modern view proposed in 1971 by King and Miller suggested that electrically conductive iron sulfides formed as corrosion products in SRB/iron sulfide corrosion scenarios could link the steel surface with SRB embedded in the iron sulfide matrix. More recently, the direct transfer of electrons from conductive deposits to SRB has been identified as a key process in marine systems showing very high corrosion rates (Enning and Garrelfs 2014).

Regardless of mechanism, the inference is that metal surfaces under cathodic protection can promote the growth of microorganisms able to use the cathodic hydrogen or electrons provided in endless supply by an effective CP system. Guezennec et al. (1990) showed that CP promoted the growth of hydrogen-utilizing microorganisms in marine sediments adjacent to protected pipe surfaces. Kajiyama and Okamura (1999) investigated the effect of various CP potentials on SRB living in clay soil around steel test coupons. Bacterial numbers and iron sulfide deposited through microbial activity reached maximum levels with a CP potential of –970 mV (CSE). Bacterial numbers in soil were a hundred times higher at this potential than at either –750 mV or –1300 mV (CSE) while iron sulfides in the soil were 10 times higher. In addition to the reduction of sulfate to sulfide, hydrogen can be used to reduce carbon dioxide to methane (Equation 12.5) or acetic acid (Equation 12.6) by methanogenic or acetogenic microorganisms, respectively (Rajagopal and LeGall 1989). Daniels et al. first demonstrated the growth of methanogens using elemental iron as the sole source of electrons for reduction of CO_2 in 1987 (Daniels et al. 1987). The potential importance of methanogens and acetogens in MIC in oilfield operations has been recognized in recent years through the use of modern molecular methods based on the analysis of extracted DNA and RNA (Larsen et al. 2010; Larsen et al. 2013; Mand et al. 2014).

$$4H_2 + CO_2 \rightarrow CH_4 + 2H_2O \tag{12.5}$$

$$4H_2 + 2CO_2 \rightarrow HOOCCH_3 + 2H_2O \tag{12.6}$$

Driving the growth of sulfate-reducing, methanogenic, and acetogenic organisms through the application of CP will not have any immediate consequence in terms of corrosion provided that effective protection potentials are maintained; however, an elevated population of active organisms will be in place should CP fail. Acetic acid produced by the reduction of carbon dioxide (Equation 12.6) on protected surfaces may feed other microorganisms in the community, perhaps at a downstream location in a flowing system where effective CP is not in place (Mand et al. 2014).

12.3.4 CP in the Presence of Microbial Activity

Microbial activity can compromise the maintenance of protective CP potentials. The formation of electrically conducting iron sulfide corrosion products will drain electrons from the metal surface and increase the current density required to sustain an adequate potential (Jack 2002; King 2008). Where this increase cannot be met, effective protection may be lost. Microbial activity can also affect the life and current output of sacrificial anodes used in CP applications. SRB activity has been reported to reduce the current output of magnesium anodes (Schick 1990), and King (2008) noted that the mass of sacrificial anodes exposed to biogenic sulfide over a 25-year service life was reduced by 25% for aluminum and magnesium and 14% for zinc.

12.4 COATING FAILURE MODES AND MIC

An intact, properly selected and applied coating or the sustained application of an effective CP potential to exposed steel can prevent corrosion and EAC. Where coating defects or failure allow CP to reach the underlying steel, protection can be sustained; however, problems occur when a coating fails in such a way as to allow water but not CP to reach the metal surface. Figure 12.7 illustrates coating failure scenarios for a "CP-compatible" coating and for a "shielding disbondment."

12.4.1 CP-Compatible Coatings

In the first failure mode, application defects, such as pinholes, allow CP to reach the base metal. CP will then drive cathodic reactions (Equations 12.2 and 12.3) on the steel surface, creating a highly basic local environment. This, in turn, can help to disbond adjacent intact coating through a process referred to as "cathodic disbondment." Thus, pinholes in an FBE coating can trigger the formation of blisters containing a high pH electrolyte. Water trapped under CP-compatible coatings typically has a pH of 9 or more (Jack et al. 2002). Norsworthy has reported a pH of 12 under an

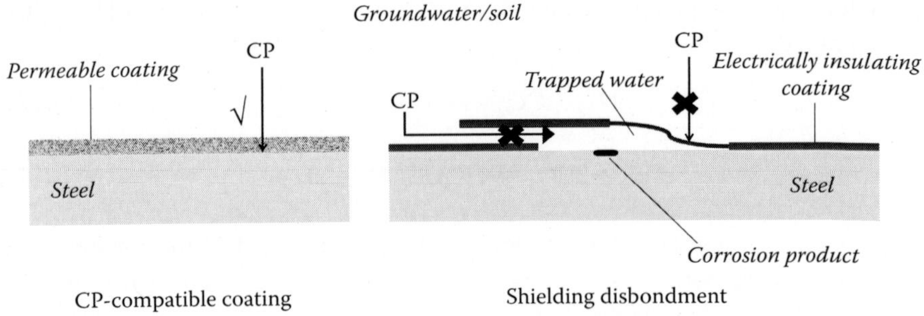

FIGURE 12.7 CP-compatible coatings allow CP to reach underlying steel. Shielding disbondments formed when an electrically insulating coating separates from the steel surface and blocks CP but allows water access to the steel surface. (Adapted from T. R. Jack, MIC in underground environments: External corrosion in the gas pipeline industry. In *Microbiologically Influenced Corrosion* Volume 2, ed. J. G. Stoecker II. 6.1–6.12. Houston: NACE, 2001.)

FBE blister (Norsworthy 2011). This environment does not favor microbial activity, and little corrosion is seen.

Asphalt and coal tar–type coatings that become more permeable over time also allow penetration of CP through the coating and the generation of a high pH environment at the interface between the coating and the protected steel surface. This can selectively draw certain ions and gases, such as carbon dioxide, through the coating to form distinctive deposits in the coating and at the interface. In Figure 12.8, sodium carbonate and bicarbonate crystals have formed under a permeable asphalt coating on a CP-protected gas transmission line (Krist et al. 2000). At a site on another facility, white calcium carbonate deposits were observed in the soil in contact with a permeable asphalt coating under CP (Jack et al. 2002). Under these conditions, corrosion (including MIC) is not an issue.

FIGURE 12.8 Crystals of nahcolite ($NaHCO_3$) and thermonatrite ($Na_2CO_3 \cdot H_2O$) formed under a permeable asphalt coating on a CP-protected gas transmission line.

12.4.2 Shielding Disbondments

Corrosion and EAC have been observed extensively under shielding disbondments. Shielding disbondments occur when an electrically insulating coating, such as a polyolefin tape, pulls away from the pipe surface to create an enclave that allows water access but constrains penetration of effective CP potentials (Figure 12.7). The extent of CP penetration depends on various factors, including the conductivity of the trapped water. In the absence of CP, the water trapped under a shielding disbondment remains at near neutral pH, and the steel surface remains at a freely corroding potential (Kuang and Cheng 2015). Under these conditions, MIC can be expected. Figure 12.3 shows an MIC scenario involving SRB and iron sulfide under disbonded polyolefin tape on a buried pipeline (Jack et al. 1986). This occurred despite adequate close interval survey pipe to soil CP potentials being measured previous to excavation on this section of line (Worthingham et al. 1986). This scenario has also been reported under improperly installed polyethylene shrink sleeves by Korea Gas (Li et al. 2000; Li et al. 2001).

12.4.2.1 Hydrogen Damage

It has been suggested that CP shielding can be mitigated by more negative CP potentials (Kuang and Cheng 2015); however, application of very negative CP potentials to bring SRB/iron sulfide corrosion scenarios under control may have unintended consequences as shown in laboratory work using a Devanathan cell to estimate the flux of cathodic hydrogen into steel under a layer of abiotic synthetic iron sulfide/iron carbonate corrosion products (Jack et al. 2005). The presence of iron sulfide in contact with the charging steel surface appeared to enhance hydrogen entry into the steel matrix. Levels of hydrogen in the steel coupon at –1.2 V (SCE) were five times higher than seen in control experiments without corrosion products and exceeded the critical level of hydrogen needed to initiate midwall delamination in a susceptible X-70 steel that had suffered hydrogen-induced cracking (HIC) problems in the field. Hydrogen sulfide has long been associated with enhanced entry of hydrogen into steel, but attempts to demonstrate embrittlement of steel due to hydrogen influx in active SRB cultures in the absence of iron sulfide–rich corrosion products have not always been successful although this has been reported in some cases (Raman et al. 2005; Javaherdashti et al. 2006). Further work is needed to establish the potential importance of microbes in hydrogen embrittlement issues related to CP overprotection and the promotion of crack tip growth in EAC.

12.4.2.2 Microbial Communities

Molecular methods are providing new insights into the nature of the microbial communities found under disbonded coatings on buried piping systems (Jack et al. 2014). Extraction and analysis of DNA from samples of corrosion products/deposits, water trapped under the disbonded coating, and adjacent soil (Figure 12.7) have been used to document the composition of microbial communities found in an anaerobic corrosion site (Figure 12.9). The microbial community outside the coating had a common soil organism (J) *Comamonadaceae-Acidovorax* as its major component, but this community was different from the community found in trapped water under a disbonded coating. This was dominated by anaerobic methanogenic microorganisms (A to E). A deposit taken from one spot on the surface of the pipe, corrosion product 1, showed a population similarly rich in methanogens; however, a very different community dominated by (K) *Delftia* was found in an adjacent surface deposit, corrosion product 2. Apparently, microbial populations can change significantly depending on where a sample is taken and can be expected to vary even over short distances on the pipe surface.

Species of *Methanobacterium* (A), corrosion product 1 (Figure 12.9), use hydrogen to reduce carbon dioxide to methane (Equation 12.5) and may be linked to active MIC at this site. *Delftia* (K), the dominant organism in corrosion product 2, has not been previously associated with corrosion of steel, but species of *Delftia* have been identified as agents or inhibitors of corrosion on NiZn alloy coatings (San et al. 2012a), NiCu alloys (San et al. 2012b), and copper (Garcia et al. 2012). As better understanding of the microbial populations associated with MIC emerges through the use of molecular methods, new mechanisms and detection tools can be developed to identify and monitor MIC in production facilities. A revolution in the recognition of MIC and predictive modeling of MIC failures has begun (Larsen et al. 2013).

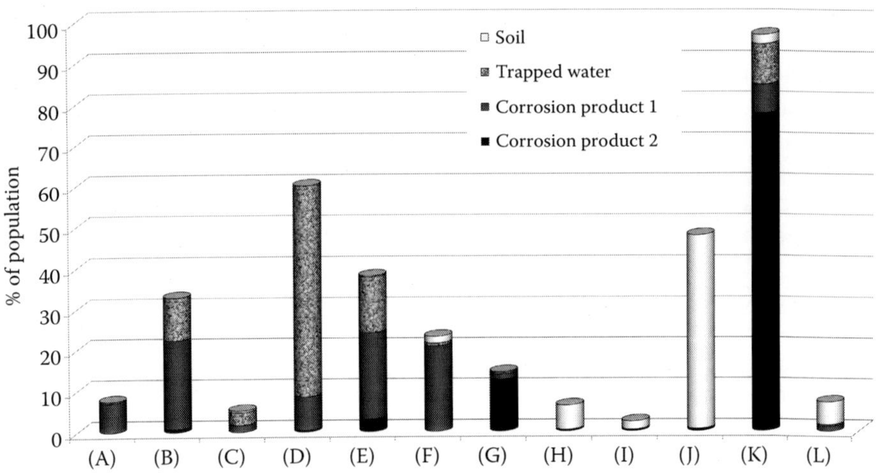

FIGURE 12.9 Composition of microbial communities associated with a disbonded polyethylene tape coating where (A) *Methanobacteriaceae-Methanobacterium*, (B) *Candidatus-Methanoregula*, (C) *Methanomicrobiales-Methanolinea*, (D) *Methanospirillaceae-Methanospirillum*, (E) *Methanosaetaceae-Methanosaeta*, (F) *Rhizobiales-Rhizobiaceae-Rhizobium*, (G) *Sphingomonadaceae-Sphingomonas*, (H) *Burkholderiaceae-Ralstonia*, (I) *Comamonadaceae*, (J) *Comamonadaceae-Acidovorax*, (K) *Comamonadaceae-Delftia*, (L) *Oxalobacteraceae*. (Adapted from T. R. Jack, G. Voordouw, A. Agrawal, J. K. Voordouw, and T. Place, Paper No. 3911, Microbial Communities Associated with External Corrosion on an Operating Pipeline System, Corrosion2014, Houston, TX: NACE, 2014.)

12.5 MANAGEMENT OF MIC

Properly selected and applied coatings can prevent MIC. CP works anywhere a potential that ensures the immunity of iron from corrosion is achieved and consistently maintained. Use of CP with modern coatings that cannot form shielding disbondments will protect steel surfaces exposed by coating gaps, defects, and damage over the life of a facility. Corrosion occurs where adequate CP potentials have not been maintained or where special circumstances, such as shielding coating disbondments, leave the underlying steel unprotected.

MIC is not confined to any particular coating system but is of special concern for coating systems that can form shielding disbondments. Aboveground nondestructive testing (NDT) methods, including visual inspection, can locate external coating issues and potential trouble spots for further investigation and remedial action. Areas of missing or permeable coating on underground facilities can be located using traditional surface survey or ILI methods based on current loss from a CP-protected pipeline as noted previously; however, finding shielding disbondments is more challenging. Relatively new ILI tools based on electromagnetic acoustic transducer (EMATS) technology can be used (Norsworthy 2012); however, these tools are not applicable in upstream production systems where small pipe diameters, sharp bends or elbows, unsuitable valves, complex geometries, and other factors preclude their use. Methodology developed for external corrosion assessment on lines unsuited to ILI may not be appropriate for coatings that cause shielding (ANSI/NACE SP0502-2010). Replacing steel with nonmetallic materials, scheduled shut downs for excavation and inspection, and other measures may be needed to manage the threat of MIC in these situations.

REFERENCES

AER Report 2013-B. 2013. Pipeline Performance in Alberta, 1990–2012, Alberta Energy Regulator, Calgary, Alberta.

Akid, R., H. Wang, T. J. Smith, D. Greenfield, and J. C. Earthman. 2008. Biological functionalization of a sol-gel coating for the mitigation of microbial-induced corrosion. *Adv. Funct. Mater.* 18: 203–211.

Albertsson, A.-C., and B. Ranby. 1975. Biodegradation of synthetic polymers: The C-14 method applied to polyethylenes. In *Proceedings of the Third International Biodegradation Symposium*, ed. J. M. Sharpley, and A. M. Kaplan. 743–751. London: Applied Science Publishers.

Albertsson, A.-C., S. O. Andersson, and S. Karlsson. 1987. The mechanism of biodegradation of polyethylene. *Polym. Degrad. Stab.* 18: 73–87.

ANSI/NACE SP0502-2010, Item No. 21097-SG. Pipeline external corrosion direct assessment methodology. NACE, Houston, TX.

Barlo, T. J., and W. E. Berry. 1984. An assessment of the current criteria for cathodic protection of buried steel pipelines. *Mater. Perform.* 23(9): 9–16.

Benckiser, G., and J. C. G. Ottow. 1982. Metabolism of the plasticizer di-n-butylphthalate by *Pseudomonas pseudoalcaligenes* under anaerobic conditions, with nitrate as the only electron acceptor. *Appl. Environ. Microbiol.* 44: 576–578.

Booth, G. H., and A. K. Tiller. 1968. Cathodic characteristics of mild steel in suspensions of sulphate-reducing bacteria. *Corros. Sci.* 8: 583–600.

Brown, A. E. 1945. Report No. 6067. The problem of fungal growth on synthetic resins, plastics and plasticizers. U.S. Office of Scientific Research and Development, Washington, DC.

Cadmus, E. L. 1977. Microbial deterioration of cable coatings. *Wire J.* May: 94–97.

Cavett, J. J., and M. N. Woodrow. 1968. A rapid method for determining degradation of plasticized PVC by micro-organisms. In *Biodeterioration of Materials: Microbiological and allied aspects*, ed. A. H. Walters, and J. J. Elphick. 162–169. London: Elsevier.

Cordeiro Antunes de Araujo, L., L. Reznik, M. Lutterbach, L. de Carvalho, and E. Servulo. 2013. Paper No. 2675. Biocorrosion of AISI 1020 carbon steel protected with the base coat of niobium oxide. NACE 2013. Houston, TX: NACE.

Daniels, L., N. Belay, B. S. Rajagopal, and P. J. Weimer. 1987. Bacterial methanogenesis and growth from CO_2 with elemental iron as the sole source of electrons. *Science* 237: 509–511.

Darby, R. T., and A. M. Kaplan. 1968. Fungal susceptibility of polyurethanes. *Appl. Microbiol.* 16: 900–905.

Enning, D., and J. Garrelfs. 2014. Corrosion of iron by sulfate-reducing bacteria: New views of an old problem. *Appl. Environ. Microbiol.* 80: 1226–1236.

Esquivel, R. G., G. Z. Olivares, M. J. H. Gayosso, A. G. Trejo, and C. C. Gurrión. 2006. Paper No. 06075. Influence of sulfate reducing bacteria on the cathodic protection potential of XL52 steel. Corrosion2006. Houston, TX: NACE.

Fischer, K. P. 1981. Cathodic protection in saline mud containing sulfate reducing bacteria. *Mater. Perform.* 20 (October): 41–46.

Garcia F., A. L. R. Lopez, J. C. Guillén, L. H. Sandoval, C. R. González, and V. Castaño. 2012. Corrosion inhibition in copper by isolated bacteria. *Anti-Corros. Method M.* 59(1): 10–17.

Gu, J.-D., T. E. Ford, D. B. Mitton, and R. Mitchell. 2011. Microbiological degradation of polymeric materials. In *Uhlig's Corrosion Handbook*, 3rd edition, ed. R. W. Revie. 421–438. New Jersey: John Wiley & Sons, Inc.

Gu, J.-D., and R. Mitchell. 2004. Paper No. 04584. Degradation of polyurethane by bacterium isolated from soil. Corrosion2004. Houston, TX: NACE.

Gu, J.-D., and Y. Wang. 2004. Paper No. 04586. Degradation of the plasticizers ortho-dimethyl phthalate ester by environmental bacteria. Corrosion2004. Houston, TX: NACE.

Guezennec, J. G. 1994. Cathodic protection and microbially induced corrosion. *Int. Biodeterior. Biodegrad.* 34: 275–288.

Guezennec, J., N. J. W. Dowling, M. Conte, E. Antoine, and L. Fiksdal. 1990. Cathodic protection in marine sediments and the aerated seawater column. In *Microbially Influenced Corrosion and Biodeterioration*, ed. N. C. Dowling, M. W. Mittelman, and J. C. Danko. 6.43–6.50. Houston, TX: National Association of Engineers.

Hardy, J. A. 1983. Utilisation of cathodic hydrogen by sulphate-reducing bacteria. *Brit. Corros. J.* 18: 190–193.

Hare, C. H. 2011. Corrosion control of steel by organic coatings. In *Uhlig's Corrosion Handbook*, 3rd edition, ed. R. W. Revie. 971–983. New Jersey: John Wiley & Sons, Inc.

Harris, J. O. 1963. Microbiological Studies Reveal Significant Factors in Oil and Gas Pipeline Backfilled Ditches. Technical Bulletin 135, Kansas State University, Manhattan, Kansas.

Harris, J. O. 1966. Asphalt oxidizing bacteria of the soil. *Ind. Eng. Chem.* 58(6): 65–69.

Huang, S. J., M. Bitritto, K. W. Leong, J. Pavlisko, M. Roby, and J. R. Knox. 1978. The effects of some structural variations on the biodegradability of step-growth polymers. *Adv. Chem. Ser.* 169: 205–214.

Jack, T. R. 2001. MIC in underground environments: External corrosion in the gas pipeline industry. In *Microbiologically Influenced Corrosion*, Volume 2, ed. J. G. Stoecker II. 6.1–6.12. Houston, TX: NACE.

Jack, T. R. 2002. Biological corrosion failures. In *ASM Handbook, Failure Analysis and Prevention*, Volume 11, ed. W. T. Becker, and R. J. Shipley. 881–898. Materials Park, OH: ASM International.

Jack, T. R., M. M. Francis, and R. G. Worthingham. 1986. External corrosion of line pipe. Part II: Laboratory study of cathodic protection in the presence of sulfate-reducing bacteria. In *Biologically Induced Corrosion*, ed. S.C. Dexter. 339–350. Houston, TX: NACE.

Jack, T. R., K. Ikeda-Cameron, P. Crosthwaite, and F. King. 2005. Mitigation of MIC using CP under disbonded coating. Final report prepared for the Gas Research Institute, Contract No. GRI-8616. Calgary, Alberta: NOVA Research & Technology Corporation.

Jack, T. R., F. King, Y. Cheng, and R. Worthingham. 2002. IPC02-27267. Permeable coatings and CP compatibility. In *Proceedings of the International Pipeline Conference*. September 29–October 3, 2002, Calgary, Alberta. New York: ASME.

Jack, T. R., G. Van Boven, M. J. Wilmott, and R. G. Worthingham. 1996. Evaluating performance of coatings exposed to biologically active soils. *Mater. Perform.* 35: 39–45.

Jack, T. R., G. Voordouw, A. Agrawal, J. K. Voordouw, and T. Place. 2014. Paper No. 3911. Microbial Communities Associated with External Corrosion on an Operating Pipeline System. Corrosion2014. Houston, TX: NACE.

Janda, D., and D. Williams. 2014. Lessons learned: Monitoring cathodic protection current from inside the pipe. *Mater. Perform.* 53(1): 42–45.

Javaherdashti, R., R. K. S. Raman, C. Panter, and E. V. Pereloma. 2006. Microbiologically assisted stress corrosion cracking of carbon steel in mixed and pure cultures of sulfate reducing bacteria. *Int. Biodeter. Biodegr.* 58(1): 27–35.

Kajiyama, F., and K. Okamura. 1999. Evaluating cathodic protection reliability on steel pipe in microbially active soils. *Corrosion* 55(1): 74–80.

Khaladkar, P. R. 2011. Using plastics, elastomers, and composites for corrosion control. In *Uhlig's Corrosion Handbook*, 3rd edition, ed. R. W. Revie. 915–970. Hoboken, NJ: John Wiley & Sons, Inc.

King, R. A. 2008. Microbiologically-induced corrosion in the oil industry and the impact of mitigation programmes. *J. Pipeline Eng.* 7(2): 137–143.

King, F., Y. Cheng, L. Gray, B. Drader, and R. Sutherby. 2002. IPC02-27100. Field assessment of FBE-coated pipelines and the implications for stress corrosion cracking. *Proceedings of the International Pipeline Conference*, Calgary, Alberta, September 29–October 3, 2002. New York: ASME.

King, R. A., and J. D. A. Miller. 1971. Corrosion by sulphate-reducing bacteria. *Nature* 233: 491–492.

Krist, K., R. R. Fessler, T. R. Jack, and B. Erno. 2000. Paper No. 00362. Generation of near neutral pH and high pH SCC environments on buried pipelines. Corrosion2000. Houston, TX: NACE.

Kuang, D., and Y. F. Cheng. 2015. Probing potential and solution pH under disbonded coating on pipelines. *Mater. Perform.* 54(5): 40–45.

Kulman, F. E. 1958. Microbiological deterioration of buried pipe and cable coatings. *Corrosion* 14: 23–32.

Lackner, M., and J. P. Guggenbichler. 2013. Prevention of biofouling in hydrocarbons by antimicrobial vessel and pipeline coating for cost savings and an increase in safety and reliability. *Iran. J. Oil and Gas Sci. Technol.* 2: 1–7.

Larsen, J., K. Rasmussen, H. Pedersen, K. Sørensen, T. Lundgaard, and T. L. Skovhus. 2010. Paper No. 10252. Consortia of MIC bacteria and archaea causing pitting corrosion in top side oil production facilities. Corrosion2010. Houston, TX: NACE.

Larsen, J., S. Juhler, K. B. Sørensen, and D. S. Pedersen, 2013. Paper No. 02029. The application of molecular methods for early warning of MIC in pipelines. Corrosion2013. Houston, TX: NACE.

Lasman, H. R., and J. P. Scullin. 1977. Miscellaneous modifying additives. In *Encyclopedia of PVC*, Volume 2, ed. L. I. Nass. 801–845. New York: Marcel Dekker.

Li, S. Y., Y. G. Kim, K. S. Jeon, and Y. T. Kho. 2000. Microbiologically influenced corrosion of underground pipelines under the disbonded coatings. *Met. Mater.* 6(3): 281–286.

Li, S. Y., Y. G. Kim, K. S. Jeon, Y. T. Kho, and T. Kang. 2001. Microbiologically influenced corrosion of carbon steel exposed to anaerobic soil. *Corrosion* 57(9): 815–828.

Little, B., and R. Ray. 2001. Fungal-influenced corrosion of metals in humid environments. In *A Practical Manual on Microbiologically Influenced Corrosion*, Volume 2, ed. J. G. Stoecker II. 2.1–2.6. Houston, TX: NACE.

Little, B., R. Ray, and S. Dexter. 2001. Introduction. In *A Practical Manual on Microbiologically Influenced Corrosion*, Volume 2, ed. J. G. Stoecker II. 1.1–1.9. Houston, TX: NACE.

Mand, J., H. S. Park, T. R. Jack, and G. Voordouw. 2014. Article 268. The role of acetogens in microbially influenced corrosion of steel. *Front. Microbiol.* 5: 1–14.

Maxwell, S. 1986. Effect of cathodic protection on the activity of microbial biofilms. *Mater. Perform.* 25 (November): 53–56.

Miyanaga, K., R. Terashi, H. Kawai, H. Unno, and Y. Tanji. 2007. Biocidal effect of cathodic protection on bacterial viability in biofilm attached to carbon steel. *Biotechnol. Bioeng.* 97(4): 850–857.

Mitton, B., T. E. Ford, E. LaPointe, and R. Mitchell. 1993. Paper No. 296. Biodegradation of complex materials. Corrosion93. Houston, TX: NACE.

Munger, C. G., and L. D. Vincent. 2014. *Corrosion Prevention by Protective Coatings*, 3rd edition. Houston, TX: NACE.

NACE SPO169-2013. Control of External Corrosion on Underground or Submerged Metallic Piping Systems. Houston, TX: NACE.

NACE TM0109-2009. Aboveground Survey Techniques for the Evaluation of Underground Pipeline Coating Condition. Houston, TX: NACE.

Ning, J., Y. Zheng, D. Young, B. Brown, and S. Nešić. 2014. Thermodynamic sulfide corrosion of mild steel. *Corrosion* 70(4): 375–389.

Norman, D., and C. Argent. 2007. Pipeline coatings, external corrosion and direct assessment. NACE paper 07154. Corrosion2007. NACE International.

Norsworthy, R. 2009. Coatings used in conjunction with cathodic protection—Shielding vs. non-shielding pipeline coatings. NACE paper 4017.

Norsworthy, R. 2011. Selection and use of coatings for underground or submersion service. In *Uhlig's Corrosion Handbook*, 3rd edition, ed. R. W. Revie. 985–1000. Hoboken, NJ: John Wiley & Sons, Inc.

Norsworthy, R. 2012. Paper No. C2012-0001673. Importance of locating disbonded coatings with electro-magnetic acoustic transducer technology. NACE 2012. Houston, TX: NACE.

Pankhurst, E. S. 1973. Protective coatings and wrappings for buried pipes: Microbiological aspects. *J. Oil and Gas Chem. Assoc.* 56: 373–381.

Pankhurst, E. S. 1976. Microbiology and the gas industry. *Gas and Engineering Management. (January)* 16(1): 3–16.

Pankhurst, E. S., and J. M. Davies. 1968. Investigation into the effect of microorganisms on PVC pressure-sensitive adhesive tape and its constituents. In *Biodeterioration of Materials*, ed. A. H. Walters, and J. J. Elphick. 302–316. London: Elsevier.

Pankhurst, E. S., M. J. Davies, and H. M. Blake. 1972. The ability of polymers or materials containing polymers to provide a source of carbon for selected microorganisms. In *The Proceedings of the Second International Biodeterioration Symposium*, ed. A. H. Walters, and E. Hueck van Plas. 76–90. London: Applied Science Publishers.

Papavinasam, S. 2014. *Corrosion Control in the Oil and Gas Industry*. Amsterdam: Elsevier.

Papavinasam, S., and R. W. Revie. 2002. Coating Gap Analysis. GRI-8334/CANMET MTL—2002-46 (CF). Chicago, IL: Gas Research Institute.

Papavinasam, S., and R. W. Revie. 2006. Paper No. 06047. Protective Pipeline Coating Evaluation. Corrosion2006. Houston, TX: NACE.

Phillips, U. A., and R. W. Traxler. 1963. Microbial degradation of asphalt. *Appl. Microbiol.* 11: 235–238.

Pourbaix, M. 1974. *Atlas of Electrochemical Equilibria in Aqueous Solutions*. Houston, TX: NACE.

Pritula, V. V. 1987. Protection of metals. (English Translation). *Zaschita Metallov* 23 (1): 133–135.

Qiu, Y.-L., Y. Sekiguchi, H. Imachi, Y. Kamagata, I.-C. Tseng, S.-S. Cheng, A. Ohashi, and H. Harada. 2004. Identification and isolation of anaerobic, syntrophic phthalate isomer-degrading microbes from methanogenic sludges treating wastewater from terephthalate manufacturing. *Appl. Environ. Microbiol.* 70(3): 1617–1626.

Rajagopal, B. S., and J. LeGall. 1989. Utilization of cathodic hydrogen by hydrogen-oxidizing bacteria. *Appl. Microbiol. Biotechnol.* 31: 406–412.

Raman, R. K. S., R. Javaherdashti, C. Panter, and E. V. Pereloma. 2005. Hydrogen embrittlement of a low carbon steel during slow strain testing in chloride solutions containing sulphate reducing bacteria. *Mater. Sci. and Technol.* 21(9): 1094–1098.

Romero, M., O. Rincón, M. Sanz, B. Rincón, L. Ocando, and W. Campos. 2008. Paper No. 08504. Evaluation of the cathodic protection criteria in the presence of SRB. Corrosion2008. Houston, TX: NACE.

Ruschau, G. R., and A. R. Kowalski. 2006. Paper No. 06191. Field investigation of above ground techniques for detecting coating anomalies. NACE 2006. Houston, TX: NACE.

San, N. O., H. Nazir, and G. Dönmez. 2012a. Microbiologically influenced corrosion of NiZn alloy coatings by *Delftia acidovorans* bacterium. *Corros. Sci.* 64: 198–203.

San, N. O., H. Nazir, and G. Dönmez. 2012b. Microbiologically influenced corrosion failure analysis of nickel-copper alloy coatings by *Aeromonas salmonicida* and *Delftia acidovorans* bacterium isolated from pipe system. *Eng. Fail. Anal.* 25: 63–70.

Schick, G. 1990. Corrosion control in the presence of sulfate-reducing bacteria. *Mater. Perform.* 29(July): 26–32.

Shirouzu, T., T. Umino, and K. Muto. 1974. Effect of sulfate-reducing bacteria on corrosion behavior and cathodic protection of mild steel. *Corros. Eng. Tokyo* 23: 393–398.

Sugatt, R. H., D. P. O'Grady, S. Banerjee, P. H. Howard, and W. E. Gledhill. 1984. Shake flask biodegradation of 14 commercial phthalate esters. *Appl. Environ. Microbiol.* 47: 601–606.

Thomson, B., and R. P. Campion. 2011. Testing of polymeric materials for corrosion control. In *Uhlig's Corrosion Handbook*, ed. R. W. Revie. 1107–1116. New Jersey: John Wiley & Sons, Inc.

Traxler, R. W. 1962. Microbial degradation of asphalt. *Biotechnol. Bioeng.* 4: 369–376.

Traxler, R. W. 1966. Bitumen attack by microorganisms. *Ind. Eng. Chem.* 58(6): 59–64.

Traxler, R. W. 1968. Attaque microbienne des bitumes petroliers. *Revue de l'Institute Francais de Petrole*. 23: 140–154.

Verink, E. D. Jr. 2011. Simplified procedure for constructing Pourbaix diagrams. In *Uhlig's Corrosion Handbook*, ed. R. W. Revie. 93–101. New Jersey: John Wiley & Sons, Inc.

von Wolzogen Kühr, C. A. H., and L. S. van der Vlugt. 1934. The graphitization of cast iron as an electrobiochemical process in anaerobic soil. *Water* (The Hague) 18: 147–165.

Webb, J. S., M. Nixon, I. M. Eastwood, M. Greenhalgh, G. D. Robson, and P. S. Handley. 2000. Fungal colonization and biodeterioration of plasticized polyvinyl chloride. *Appl. Environ. Microbiol.* 66: 3194–3200.

Williams, D., and D. Janda. 2014. Paper No. C2014-3821. Developing a standardized process for cathodic protection current measurement on in-service pipelines—Process and procedures for a new technology. Corrosion 2014. Houston, TX: NACE.

Wormwell, F., and T. W. Farrer. 1952. Electrochemical studies of anaerobic corrosion in presence of sulphate-reducing bacteria. *Chem. Ind.* 5: 108–109.

Worthingham, R. G., T. R. Jack, and V. Ward. 1986. External corrosion of line pipe. Part 1: Identification of bacterial corrosion in the field. In *Biologically Induced Corrosion*, ed. S.C. Dexter. 330–339. Houston, TX: National Association of Engineers.

Yabannavar, A. V., and R. Bartha. 1994. Methods for assessment of biodegradability of plastic films in soil. *Appl. Environ. Microbiol.* 60: 3608–3614.

Zhong, C., X. Tang, and Y. F. Cheng. 2008. Corrosion of steel under the defected coating studied by localized electrochemical impedance spectroscopy. *Electrochimica Acta*, 53: 4740–4747.

13 MIC Monitoring

Developments, Tools, Systematics, and Feedback Decision Loops in Offshore Production Systems

Renato De Paula and Victor Keasler

CONTENTS

ABSTRACT

Identifying the right tools for monitoring microbial risk in a production system is key to managing and controlling the risks of MIC-induced failures. Frequently, inadequate monitoring tools are used to perform biomonitoring programs in the field, and failure to connect the obtained data to the producers output dramatically limits the usefulness of these tools. The purpose of this chapter is to discuss recent developments in microbial monitoring, best practices for analyzing results in the context of operational data, and future developments to improve MIC monitoring practices.

The last 10 years have seen a major change in the accepted practices for microbial monitoring. Traditional culture-based analyses are beginning to be replaced by culture-independent methods that overcome some of the major challenges historically encountered, such as the inability to detect organisms that do not grow in culture bottles. Despite a steady increase in the use of new culture-independent techniques in the field, the industry has not yet come to full adoption of these methods as standards. Culture-independent methods include analysis of molecules involved in energy metabolism, such as adenosine mono and triphosphate or specific metabolic enzymes within a microbial cell as well as DNA-based methods, such as quantitative polymerase chain reaction (qPCR), fluorescence in situ hybridization (FISH), and DNA sequencing, among others.

Recent publications have highlighted that microbial numbers alone are insufficient to identify the risk microbes pose in a given system (Geesey 1994; Wei et al. 2007; De Paula et al. 2012; Keasler et al. 2013). Instead, it is critical to compare collected microbial monitoring data to operational outputs from the asset, such as corrosion rates, hydrogen sulfide generation, iron sulfide deposition, fluid injectivity, and oil and water quality. When these correlations are available, specific key performance indicators (KPIs) and intelligent decisions can be made about the immediate and/or potential risk that a microbial population presents in a given system. The holistic data sets also provide insight into specific locations in the system where risks exist and where remediation strategies should be implemented.

Despite significant progress, there is still significant room for development in the area of microbial monitoring and MIC control. As the industry moves to greater levels of automation, the ability to identify and control microbial threats must move from sample collection and analysis to tools that can provide real-time monitoring and response to increased microbial risks. This next generation of development will continue to advance the industry to new levels of microbial risk prediction and control.

13.1 INTRODUCTION

This chapter of the book covers a brief review of monitoring technologies used in the oilfield in the past and present, and it discusses the next generation of monitoring that is coming in the future (Figure 13.1). It is meant to outline advantages and

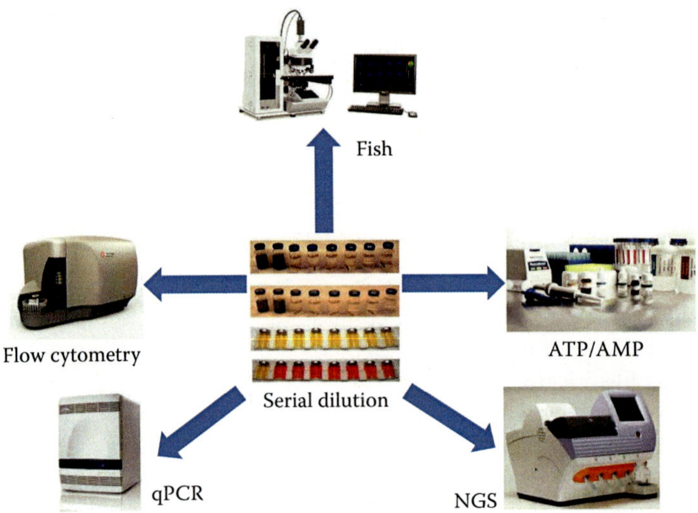

FIGURE 13.1 Expansion of oilfield-monitoring techniques in the past few years.

limitations of monitoring technologies and highlight problems that are still under investigation. In addition, it discusses the current best-in-class methods for monitoring in an offshore production system for microbial risks. This chapter provides a holistic view of the system, including microbial data, corrosion results, environmental and operations feedback, and an action plan that has defined decision points for changes and optimization to the system.

13.2 THE EARLY DAYS OF MONITORING

It has been known for decades that microbes can cause operational challenges in oil and gas systems. They can increase the localized corrosion rate (known as microbiologically influenced corrosion), they can produce hydrogen sulfide (H_2S) as a by-product of their metabolism, and they can negatively influence processes, such as heat transfer efficiency and membrane functionality, through the buildup of biofilm on surfaces (Whitby and Skovhus 2009).

Detection and control of these problematic microbes was initially targeted toward sulfate-reducing bacteria (SRB). For this purpose, a standard media (Postgate 1984) was adopted in the industry. This media has been used for many years for detection of microbes that utilize sulfate as part of their metabolism. This method uses a simple indicator of positive growth as the bottles change from clear/turbid to black based on the production of sulfide and its reaction with iron in the media.

Expansion of culture-based detection methods has included several different selective media types. The most commonly used is phenol red dextrose for the detection of fermentative microbes (Forbes et al. 2007). As the pH drops below a certain level in the culture bottle, the media has a color indicator that changes from red to yellow to confirm the production of acid. Other culture media for nitrate-utilizing microbes, thiosulfate-reducing bacteria, and even methanogens have been proposed and documented in NACE Standard TM01948.

One of the most significant advantages of the culture techniques (including serial dilution, dip slides, plate counting, etc.) is their ease of use, storage, and data interpretation. Additionally, many of these culturing techniques are set as standards for use in the field (ASTM, NACE standards). Given that many of the world's oil and gas fields are in remote areas (both onshore and offshore), complex testing methods present logistical challenges to be performed. In addition, the test method needs to be simple enough to be carried out and interpreted with minimal technical training. In this sense, culture-based methods, such as serial dilution, meet the criteria of simplicity (visual color change in the bottles). Nevertheless, only a small fraction of the operators have adopted the use of the most probable number (MPN) method, where multiple replicates of each sample are inoculated to provide statistically significant results.

The major drawbacks of the serial dilution method for microbial enumeration are the long duration required for culturing and the risk of false negatives. According to NACE Standard TM0194, SRB bottles need to be incubated 28 days while APB bottles need to be incubated for 14 days. Any corrective activity that might be employed to reverse a bacterial upset condition is thus delayed by the 2- to 4-week culture period. Although the time delay is not optimal, the even more challenging problem with the serial dilution method is the rate of false negatives. This is due to the fact that there are numerous different microbes present in oilfields that simply cannot be successfully grown in a small number of culture media. In 2014, Geissler et al. suggested that over 900 distinct genera of microbes have been identified in approximately 7,000 oilfield samples. Although this is unlikely to be an exhaustive list of the microbes present in oilfield systems, it highlights that all microorganisms from these extremely diverse groups of microbes are unlikely to be identified by growth in a handful of different media types. The diverse microbial consortia that has been documented is not unexpected and is consistent with other papers in the literature, showing that serial dilution is only able to detect <1% of the actual microbial population present in the sample (Vazquez-Duhalt and Quintero-Ramirez 2004). Aside from inadequate nutritional supplementation from the culture media, the culture bottles are rarely incubated in the correct conditions mimicking the system. In most instances, samples collected from vessels with different temperatures are generally incubated at a single temperature. This may skew the results and prevent the assessment of the real point of contamination in the system.

Moreover, serial dilution tests present inconveniences, such as storing hundreds of bottles in limited spaces on offshore platforms, and minor hazards, such as using sharp needles to transfer inoculant between bottles. All these challenges led to the development of new enumeration tests aimed to overcome the limitations associated with timing and/or false negatives. These include other culture-based techniques, such as biological activity reaction test (BART) bottles and enzymatic-based tests such as adenosine-5-phosphosulfate (APS) reductase measurement, hydrogenase measurement, and RapidCheck and adenosine triphosphate/monophosphate (ATP/AMP) measurements. Each of these tests has their own advantages and disadvantages, and many of them provide dramatic improvements in terms of time to results as well as minimizing the risk of false negative readings. The expansion of monitoring techniques from simple serial dilution to a plethora of different methods has been remarkable in the past 5 years or so, and it highlights the industry's need for better ways to more accurately assess microbial risks (Figure 13.1).

The frequency of use and acceptance of each of these technologies has been at the discretion of the operator and/or service company and has also been influenced by geographical location of the oil and gas asset where the testing is being performed. It is important to note that although several of the technologies discussed earlier are significant improvements over the conventional serial dilution methods, there are still opportunities for improvement as enumeration methods developed in the medical and academic industries have become more cost-effective and widely available for use.

Other nonmicrobial tests were often employed in conjunction with the enumeration of microbes in order to achieve a more comprehensive picture of the risk in the system. This included testing, such as sulfide and H_2S levels as an indicator of SRB activity or water analysis to understand the change in sulfate levels through a system or process. These test methods combined with a review of the environmental aspect of the asset (temperature, pressure, etc.) provided an assessment of the potential risk. However, it is important to note that most (if not all) of the testing described so far has been focused on an analysis of the water. Although microbes are found in the water, testing of only water samples leaves a large potential that the true risk will not be identified. Microbial populations, whether in oil and gas systems or naturally found in the environment, are often found living symbiotically together in what is termed a biofilm (Lenhart et al. 2014). This mixed community of microbes could have only a couple or potentially dozens of microbial species, all living in the same environment and feeding off of the nutrients and by-products produced by other microbes. Not only do the microbes feed off of each other, but the corrosion mechanisms may be exacerbated by the presence of multiple different microbial species living together (Mand et al. 2014). Such biofilms can present a very significant risk to the oil and gas system and are something that must be monitored to ensure that MIC, biotic H_2S, and/or biofouling are under control. Biofilm formation is common in the field, where improper "operation hygiene" (comingling water sources, flushing the system with untreated water, etc.) and the presence of poorly engineered locations with low velocity/stagnant water occur.

13.3 MONITORING SHIFT IN THE LAST 10 YEARS

As microbial monitoring and risk assessment has evolved, there are two major trends that have emerged. The first is the adoption of DNA-based monitoring strategies within the oil and gas industry. These technologies have been proven in other industries and are now becoming more commonplace in oil and gas operations. The second major trend is around sessile microbial monitoring and involves a shift away from monitoring water samples to instead monitoring biofilms. This can be performed in numerous different ways, and it provides a unique insight that was missed when only the fluid samples were tested. Each of these trends is discussed in more detail below.

The early usage of molecular (or DNA-based) monitoring in oil and gas applications was focused on speciation through denaturing gradient gel electrophoresis (DGGE) (Wang et al. 2008) and single pass sequencing as well as fluorescence in situ hybridization (FISH) (Zeng et al. 2006). Although well established in academic labs for many years, their application in oil and gas was not without challenge. For liquid samples, they often contain hydrocarbons, elevated levels of total dissolved solids or total suspended solids, and/or production chemicals that are used in almost every oil and gas system. Solid samples are even more challenging as they may contain paraffin, asphaltenes, sand, iron sulfide, and corrosion products in addition to the microbes themselves. In spite of the challenges, DNA extraction methods have been developed to effectively work with even the most challenging samples. DGGE, for the first time, provided insight into what microbes were present at the genus and sometimes even the species level. Because the primer sets were designed to a conserved region, it was able to detect most microbes without having to know exactly what you were looking for ahead of time. FISH, on the other hand, gave the ability to enumerate general or specific microbes, assuming the user had designed a probe for the organism types they wanted to explore. Contrary to DGGE, FISH is significantly more costly. As the technologies were used over a period of a few years, adoption of FISH significantly dropped while DGGE remained a viable technology for genus/species identification.

As technology evolved, quantitative PCR (qPCR) has replaced FISH as the method of choice for specific enumeration while high-throughput sequencing began to replace DGGE and single pass sequencing. qPCR provides a faster and more accurate way to determine the absolute number of organisms in a given sample when compared to FISH. However, species-specific qPCR assays are frequently challenged to cover and detect all genera and species from the single family. However, when used together, qPCR and high-throughput sequencing could provide enumeration and identification (at the genus or even species level) that far surpasses both FISH and DGGE. Continual development of the qPCR and sequencing platforms by companies such as Life Technologies, Roche, and Illumina have brought the technology to a place where it has become affordable enough to be used in oil and gas systems on at least a periodic basis. It became clear that the industry was adopting these new molecular-based technologies as two NACE Standards now call them out as part of a monitoring toolkit: NACE TM0212 (2012) and NACE TM0194 (2014).

The use of qPCR for enumeration of organisms belonging to specific metabolic classes (e.g., total sulfate reducing bacteria, etc.) still faces several hurdles due to the fact that dozens of combinations of primer sets must be used to achieve full coverage, and common targets such as the *aps* (Filion et al. 2012) or *dsrA/dsrB* (Spence et al. 2008) genes failed to provide such comprehensive coverage. Instead, semiquantitative information about specific genera can be obtained by extrapolation of numbers of DNA reads from DNA sequencing from the absolute numbers obtained by qPCR for total bacteria or archaea (Kunin et al. 2008).

When used together, qPCR and DNA sequencing provide a different insight into the microbial risk. Both methods overcome the challenge of false negatives that existed with culture-based methods. qPCR can be set up to be very general (detecting total bacteria or total archaea numbers) or very specific (enumerating a particular genera or species) while DNA sequencing does not require prior knowledge of what organisms are present as the primers are designed generally enough to identify most organisms. At the time this chapter was written, qPCR has been focusing more on classes of microbes (SRB, APB, etc.) than on specific genera or species, but it is expected that insight gained going forward will identify high-risk species that should specifically be monitored. In terms of high-throughput DNA sequencing, the identification of organisms at the genus or species level provides insight that can then be gained with respect to nutrient sources used by the microbes, the potential corrosion mechanism they could be accelerating, and even where the organisms might have originated from (reservoir versus topside contamination). Additionally, the knowledge of the microbial population in an oil/gas production facility can help to customize qPCR assays aimed to enumerate the specific problematic organisms on a routine basis. In spite of the positive attributes of qPCR and sequencing, it is still not the major method used for sample analysis in the oilfield. The main reasons are that the costs are still substantially higher than serial dilution and the technologies are currently mostly lab-based and require specialized personnel. Both qPCR and sequencing techniques are moving to the field. Nonetheless, a few more years of development will be required for adoption of these techniques as field methods.

In addition to the detection methods that are being developed for microbes, it is equally critical that the analysis is actually performed on biofilms. This can be done through the analysis of biostuds, corrosion coupons, scale coupons, extended analysis coupons, and even pig debris in certain circumstances. There have been several papers published recently that address the importance of sessile analysis and even provide guidance for how that testing should be performed (Maxwell et al. 2004; Little and Lee 2007). In spite of the value that is gained through sessile monitoring, it continues to be the exception in the industry and not the rule. That is mainly due to (a) logistics to pull coupons for analysis, (b) older oil/gas systems that were simply not engineered with biofilm monitoring sample locations, or (c) the high-risk locations not being accessible. For systems where sessile testing can be performed, there is still a key challenge around placement of coupons in the system. If they are not placed correctly, the data obtained from them may be inaccurate or misleading. This would include issues such as a coupon being placed into an oil phase instead of the water phase, a sidestream device that is experiencing a flow rate or sheer that is much higher than the actual system, or a coupon that is placed upstream of the biocide treatment, among others.

13.4 MONITORING LOOKING FORWARD

Given the rapid development of microbial monitoring tools over the past 10 years, it would not be surprising to see even bigger changes over the next decade as technology continues to evolve. The oil and gas industry appears to be catching up to the forefront of microbial enumeration, speciation, and monitoring technology, and many companies and independent labs continue to look to leverage developments from academics as well as the medical community.

The transfer of sophisticated molecular methods from their lab-based origins to field service practical is a significant challenge. This will happen over the next 2–5 years as qPCR and DNA sequencing technologies continue to advance to the point that both can be performed in the field and decisions made in real time. For that, both methods must be portable and affordable, and the tests must be simple enough to be performed and interpreted by a field worker. One big challenge that remains even as qPCR and DNA sequencing move to the field is the interpretation of data and the creation of an actionable response from it. Recent publications have come to differing conclusions on the ability to link a particular microbial number to an overall risk (De Paula 2015; Eckert and Skovhus 2015). Key questions that have to be answered include the following: (a) Is there a certain microbe number that can always be considered high risk, (b) are their certain genera or species that are always associated with elevated pitting corrosion rates, and (c) can primary biofilm-forming organisms be identified that can give an early warning to upcoming MIC risks? Continued work in this area will be necessary to better define microbial risks in terms of organism number and/or species.

Even with qPCR and microbial speciation moving to a field-based assay, it still requires manual intervention to collect and analyze samples. The most important accomplishment will still performing the tests in situ, on biofilms, where the results can be analyzed and acted upon automatically without the need to collect a sample. The ability to monitor without manual intervention is critical as it minimizes safety risks to the person collecting the sample and makes it easier for otherwise challenging sample points to be monitored routinely. This might include sample points that are in a high-pressure area or are subsea and inaccessible except by dive teams or an unmanned vehicle. Unfortunately, at this time, there are no current commercially available real-time, online systems that are able to differentiate biomass/biofilm from scale or corrosion by-products. Moreover, qPCR and/or speciation have yet to be evaluated as a viable technology to be used for in situ monitoring, and other new technologies, not yet evaluated, may fulfill this gap.

13.5 CURRENT BEST-IN-CLASS MIC MONITORING FOR AN OFFSHORE PRODUCTION SYSTEM

Based on the currently available technology, there is a best-in-class program that can be put in place for MIC monitoring on an offshore production system. This section outlines that recommended monitoring program, taking into account microbial detection, environmental and operational parameters, and corrosion evaluation. Since the focus will be on MIC, we do not discuss other corrosion mechanisms, such as oxygen corrosion, CO_2 corrosion, H_2S corrosion, under-deposit corrosion, or any other nonmicrobial mechanism.

Microbial monitoring results can be used to assess key parameters of a well-functioning operation and help to make decisions on how changes should be implemented in the field. The process is usually followed by a set of information that is highlighted in Figure 13.2. The first part of a successful microbial corrosion–monitoring program requires key performance indicators (KPIs) to be well defined.

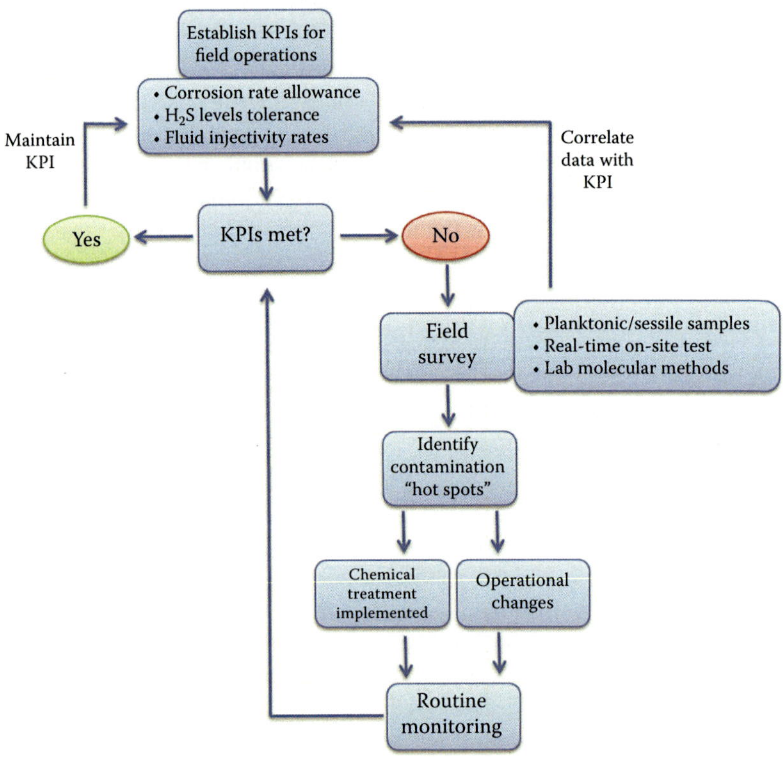

FIGURE 13.2 Decision tree for management of microbial-related issues in the field.

This includes answering questions such as what is the maximum acceptable corrosion rate, what level of H_2S can be tolerated, what is the expected lifetime of the system, and how are corrosion rate and/or H_2S levels going to be monitored.

With KPIs defined, the next step is to perform a survey of the system to identify areas of high, medium, and low risk. This should start at the front end of the system by analyzing a sample from the wellhead or manifold if it is available, and samples should then be taken at points throughout the entire water, oil, and gas systems. Each of these samples should be analyzed for abundance of planktonic microbes as an indication of high and low risk potential. In general, the highest microbial populations will follow the water in the system, but it is important to look for other areas where microbial populations are elevated. In addition to the planktonic sampling, it is critical that biofilm samples are analyzed at several locations throughout the system. This can consist of biostuds from a Robbin's device, corrosion coupons, scale coupons, or other devices that allow for biofilm deposition to be measured. Pig debris can also provide some insight into the sessile bacterial population of flowlines.

There are several methods by which to analyze the planktonic and biofilm samples to gauge risk. Optimally, this would include an onsite evaluation, such as microbial activity by ATP and a deeper analysis into the microbes present using qPCR and speciation via DNA sequencing. These two methods combined together provide insight into microbial activity in the system as well as insight into where the microbes originated from and the potential corrosion mechanisms at play. Unfortunately, there is not an industry standard that defines an acceptable level of microbes. This is going to be completely system-dependent and requires that the microbial numbers are correlated to the corrosion rates observed in the system. In certain systems, it may be observed that 100,000 microbes per gram or per cm^2 may still keep the system within corrosion allowances, but that may not be the case in other systems (De Paula et al. 2014).

In addition to monitoring of the microbes, it is important to review the environmental and operations parameters of the system as well. This starts with a water analysis to identify anything that could boost microbial growth, such as elevated sulfate levels or high levels or organic acids. It can also be used to identify waters that might not be as supportive of widespread microbial growth, such as extremely high levels of chlorides. In addition, the temperatures and pressures should be mapped across the system as this will also provide insight into areas of high and low risk. If the downhole temperatures are particularly high, it is not uncommon to see a significant increase in microbial numbers and activity as the temperature becomes more moderate across the system. This is often seen in conjunction with larger sessile populations, which can increase the corrosion risk. Finally, it is important to identify areas of potential stagnant or low flow as these will be high risk areas. This can include vessels, drains, and sumps as well as pipelines and flowlines.

Once all of the information has been gathered, the next step is to build a heat map of the system with high, medium, and low risk areas identified. Monitoring points are critical for high and medium risk areas and should be considered for at least a few of the low risk areas as well. Although water samples can provide insight into planktonic populations, these high and medium risk areas require the ability to monitor sessile bacterial population.

Once monitoring points have been established, the next question that must be addressed is monitoring frequency and how the coupons should be analyzed. There is no industry-wide standard for how often biofilm samples should be analyzed, but a good rule of thumb for offshore oil and gas production systems would be to monitor monthly with field tests, such as ATP for systems experiencing active microbial corrosion, and quarterly for systems that are not experiencing active microbial corrosion. Similarly, there is no industry standard for how long coupons should be left in a system before they are removed for analysis, but at least one paper did address this in looking at microbial growth at 30, 60, and 90 days from a Robbin's device in a produced water system (De Paula et al. 2014). That paper showed that biofilm growth continues for at least 90 days although the testing did not go any longer than that, so it is unclear if the biofilm had reached a steady-state level or if they would have continued to increase in numbers. With that in mind, a conservative approach would be to leave coupons in the system for at least 90–120 days before removing them for analysis. When the coupons are removed, they can be tested for biofilm (through a test such as ATP, qPCR, or speciation), and they should ideally also be evaluated for metal loss. The most commonly performed corrosion test is simply weight loss, but this does not accurately provide insight into the pitting corrosion rate, which is what microbes would accelerate. In order to address this, some sort of pitting rate analysis, such as white light interferometry, should be performed to identify the deepest pits as well as the total number of pits. By testing for both microbial number/activity and pitting corrosion rate on a single coupon, it provides a stronger correlation to drive decisions to microbial management.

The final part of a best-in-class monitoring program is to have a decision tree in place of how to respond to the microbial and corrosion data points that are obtained. Is there a biocide that has been qualified for the system to be used in case the testing results show out-of-spec microbial numbers and pitting corrosion rates? And equally important is to define how many out-of-spec points are required before a biocide program is either started, modified, or changed. For example, is a single out-of-spec coupon that is pulled once per quarter enough information to drive a change to the biocide program? That will be operator-dependent and will be heavily influenced by the risk that can be tolerated in the system.

13.6 SUMMARY

Microbial monitoring has progressed significantly over the past 10 years as traditional culture-based methods have given way to cutting-edge molecular testing. The insight gained through these new technologies is an order of magnitude better than historical methods, but the methods of microbial monitoring are only as good as the total MIC monitoring program. If sessile samples are not available or simply not utilized, then the decisions can be extremely misleading. If microbial data is not correlated to pitting corrosion rate or overlaid with environmental and operations parameters, then the true risk of the system is not understood. It is only when a holistic program is in place that microbial risks can be truly averted. This must combine microbial testing of biofilm samples, pitting corrosion analysis from properly installed coupons, and KPIs and decisions points in place to drive operational changes based on the results obtained.

REFERENCES

De Paula, R. M. 2015. Microorganisms and MIC: How many are too many? Presented at the 21st Reservoir Microbiology Forum, London.

De Paula, R. M., Keasler, V. V., Bennett, B., Clark, J. C., and Cloud, R. 2014. On-site evaluation of microbiologically induced corrosion and the effects of continuous low dosage corrosion inhibitor application, *Corrosion 2014*, C2014-4066. NACE International, Houston, TX.

De Paula, R., Keasler, V., Bennett, B., Keller, C., Adams, R. C., Vaksman, Z. and Kaplan, H. 2012. Optimization of a microbial control program in an aging Gulf of Mexico asset to minimize the risk of corrosion, *NACE Corrosion 2012*, C2012-0001195.

Eckert, R. B., and Skovhus, T. L. 2015. Advances in the application of molecular microbiological methods in the oil and gas industry and links to microbiologically influenced corrosion. Presented at ISMOS-5 Conference, Stavanger, Norway.

Filion, M. 2012. *Quantitative Real-Time PCR in Applied Microbiology*. Horizon Scientific Press, Poole, UK.

Forbes, B. A., Sahm, D. F., and Weissfeld, A. S. 2007. *Bailey & Scott's Diagnostic Microbiology*, 12th edition. St. Louis, MO: Mosby, Inc.

Geesey, G. G. 1994. *Biofouling and Biocorrosion in Industrial Water Systems*. CRC Press, Boca Raton, FL, USA.

Geissler, B., De Paula, R. M. Keller-Schultz, C., Lilley, J., and Keasler, V. 2014. Data mining to prevent microbiologically influenced corrosion? *NACE Corrosion 2014*, C2014-4080. NACE International, Houston, TX.

Keasler, K., Bennett, B., Keller, C., Whalen, P., Cairns, J., and De Paula, R. M. 2013. Expanding the microbial monitoring toolkit: Evaluation of traditional and molecular monitoring methods, *Int Biodeterior Biodegradation* 81: 51–56.

Kunin, V., Copeland, A., Lapidus, A., Mavromatis, K., and Hugenholtz, P. 2008. A bioinformatician's guide to metagenomics, *Microbiol Mol Biol Rev* 72: 557–578.

Lenhart, T. R., Duncan, K. E., Beech, I. B., Sunner, J. A., Smith, W., Bonifay, V., Biri, B., and Suflita, J. M. 2014. Identification and characterization of microbial biofilm communities associated with corroded oil pipeline surfaces, *Biofouling* 30: 823–835.

Little, B. J., and Lee, J. S. 2007. *Microbiologically Influenced Corrosion*. John Wiley & Sons, Hoboken, NJ, USA.

Mand, J., Park, H. S., Jack, T. R., and Voordouw, G. 2014. The role of acetogens in microbially influenced corrosion of steel, *Front Microbiol* 5: 26.

Maxwell, S., Devine, C., and Rooney, F. 2004. Monitoring and control of bacterial biofilms in oilfield water handling systems, Paper 04752 Corrosion 2004. NACE International, Houston, TX, USA.

NACE Standard TM0194. 2014. Field monitoring of bacterial growth in oil and gas systems. NACE International, Houston, TX.

Postgate, J. R. 1984. *The Sulphate-Reducing Bacteria*, 2nd ed. London: Cambridge Press.

Spence, C., Whitehead, T. R., and Cotta, M. A. 2008. Development and comparison of SYBR Green quantitative real-time PCR assays for detection and enumeration of sulfate-reducing bacteria in stored swine manure, *J Appl Microbiol* 105: 2143–2152.

Vazquez-Duhalt, R., and Quintero-Ramirez, R. 2004. *Petroleum Biotechnology: Development and Perspectives*. Elsevier, Amsterdam, the Netherlands.

Wang, J., Ma, T., Zhao, L., Lv, J., Li, G., Liang, F., and Liu, R. 2008. PCR–DGGE method for analyzing the bacterial community in a high temperature petroleum reservoir, *World J Microbiol Biotechnol* 24: 1981–1987.

Wei, L., Ma, F., Wang, J. H., and Zhao, L. J. 2007. Rapid quantitative detection of sulfate reducing bacteria in oil field, *Huan Jing Ke Xue* 28(2): 441–444.

Whitby, C., and Skovhus, T. L. 2009. *Applied Microbiology and Molecular Biology in Oilfield Systems*. Springer Science & Business Media, Berlin, Germany.

Zeng, J. H., Wu, X. L., Zhao, G. F., and Qian, Y. 2006. Detection of SRPs in injection water of Shenli oil field by FISH, *Huan Jing Ke Xue* 27: 972–976.

14 Review of Current Models for MIC Management

Erlend Stokstad Andersen, Torben Lund Skovhus, and Elizabeth Hillier

CONTENTS

ABSTRACT

Operating offshore oil and gas facilities is often associated with the threat of emerging corrosion-related issues and the subsequent risk of failures. In order to manage this threat, operators commonly use models to support decision making. In the case of managing the threat of microbiologically influenced corrosion (MIC) and establishing the potential rate of degradation, several models have been developed in the last 15–20 years. However, there are few available methods to systematically map the threat of MIC. This is for many reasons: the highly unpredictable nature of microorganisms, disagreement among scientists and operators regarding which groups of microorganisms influence corrosion, and limited knowledge about which (physical, chemical, and biological) factors drive MIC.

The models use qualitative, semiquantitative, or quantitative measures to help assess the rate of degradation caused by MIC. Commonly, the rate of degradation is referred to as "MIC risk," implying that the rate of metal degradation is proportional to the risk associated with system failure. This is incorrect as risk is determined from the probability of failure (PoF) and consequence of failure (CoF); the rate of degradation is merely a factor in the estimation of the PoF. CoF is usually determined in conjunction with the asset operator and varies across the asset systems. The recommended practice DNV-RP-G101 (2010) explains how this is done when establishing a risk-based inspection (RBI) analysis. CoF is not considered in any of the MIC models reviewed in this chapter; instead the focus is on the MIC threat. Often the threat of MIC is identified based on physical, chemical, and biological parameters in the system under investigation. However, other important parameters are the design and operating history of the asset.

The models have different approaches to identify the potential for and rate of MIC. Some models base the result on many different inputs, covering physical, chemical, and biological parameters while others only cover biological parameters. Due to scarcity of publicly available data, it is difficult to say which approach gives the most accurate results. In addition, the available information about the specific system under investigation can also differ from one case to another. In other words, decision makers may need to rely on different models when assessing different operating systems. Another important aspect is that, in many cases, the information we have available is incomplete to make a proper decision, and only fragments of data are available. This will also impact the model selection process.

Uncertainty in results must not be underestimated. Therefore, data collection is an absolute necessity to enable evaluation of the model input in order to increase the likelihood that the model output yields a meaningful and reliable outcome. It is also vital to ensure the threat is evaluated over time to gain confidence in the models used and how their output relates to actual asset integrity.

Since the current models only identify the threat of MIC, decision makers need to put their results into context with other degradation mechanisms in the system in order to manage MIC. This can be done by incorporating the result in other asset integrity procedures/analysis where the CoF is considered. Only when this is done can decision makers take action based on the complete view of the asset integrity risk.

This chapter analyzes a number of the key models available and critically discusses them in detail.

14.1 REVIEW OF CURRENT MODELS FOR MIC MANAGEMENT

In this chapter, we present 10 models that can help identify or establish the threat of MIC in a system used for hydrocarbon production or transport. The Checkworks

Predictive Model, Union Electric Callaway MIC Index, and Lutney/Stein MIC Index are described based on an overview presented by Sooknah et al. (2007). The other models are described based on individually published papers.

The earliest model we present here was developed more than 20 years ago (published in 1994). Even so, MIC remains one of the most unpredictable degradation mechanisms to assess and manage in the oil and gas industry today. We have examples in recent time (e.g., Valhall on the Norwegian Continental Shelf 2009) that show that if MIC is inadequately managed it can result in integrity breaches. Incorporating a robust model to assess MIC in a system analysis is an important step toward properly managing MIC (Markoff and Larsen 2010; Mariuou-Dahle and Skovhus 2012). A model can only be considered robust when all aspects contributing to MIC are taken into account. In the case of MIC it is virtually impossible to consider all aspects. The models that are described in this chapter generally rely on looking at only certain aspects. This may be valid in some cases where MIC plays a role if the aspects of the model indeed reflect the most dominant contributors to MIC. However, when factors other than these play a larger role, the outcome of the model may fail to reflect the actual condition of a system or site.

There is no definitive solution to how models can be applied and which models will provide the correct answer in a given situation. Plant operators, asset managers, and corrosion engineers have different availability of data, knowledge of operating conditions, experienced resources, and so on. These are all factors that will influence the model selection process. Another selection criteria may be that a model is chosen based on the available data types that are able to best populate a model.

In Table 14.1, all models and key properties are presented.

14.1.1 Model 1: Checkworks Predictive Model

This model is a susceptibility ranking of corrosion circuits (subsystems) or critical components of a system. The model gives responsibility to the assessor as each factor in the equation below is subjectively established based on system operating condition and properties. Each factor is given a value between 0 and 10, where 0 means no/very low susceptibility and 10 means high/greatest susceptibility.

A given systems susceptibility can be expressed as below:

$$\text{Susceptibility to MIC} = (\text{MWF} \times \text{TF} \times \text{OF} \times \text{BF} \times \text{BD} \times \text{DF})^{1/a} \tag{14.1}$$

MWF is the materials–water factor, which represents the interaction between material of construction and the water. TF is a temperature factor, which accounts for the potential for biofilm establishment and growth at a given temperature. OF (operating factor) accounts for the flow conditions. BF and BD account for the biocide effectiveness and loss of biocide effectiveness, respectively. DF is the discontinuity factor, which accounts for the geometry and design. The constant, a, is not mentioned in the reviewed text. However, as corrosion circuits or components are ranked in relation to each other, as long as it is a constant (and $a > 0$), the value does not matter as the ranking will be the same.

How to establish a quantitative value (between 0 and 10) is not explained in the reviewed text. Most of the factors can initially be determined by reviewing literature,

TABLE 14.1
Summary of the Models Presented in This Chapter

Number	Model	Type of Microorganism	Parameters	Result	Comment
1	Checkworks	Not type specific	Chemical, physical	Qualitative	
2	Union Electric Callaway	SRB, acid-producing bacteria, metal-oxidizing bacteria	Biological, physical	Qualitative	Not suitable for subsea systems or pipelines
3	Lutney/Stein	SRB, slime-forming bacteria, metal-oxidizing bacteria, manganese-oxidizing bacteria, acid-producing bacteria	Biological, physical	Semiquantitative	
4	Pots et al.	SRB	Chemical, physical	Quantitative	
5 and 6	Maxwell, Maxwell and Campbell	SRB	Biological, chemical, physical	Quantitative	Requires usage of MMM
7	Sooknah, Papavinasa and Revie	Not type specific	Chemical, physical	Qualitative	
8	Allison et al.	SRB	Biological, chemical	Qualitative	
9	Skovhus et al., Sørensen et al.	SRP, methanogens	Biological	Quantitative	Requires usage of MMM
10	Taxèn et al.	SRB	Chemical, biological		Not a complete model for estimating the threat of MIC

operator experience, and/or using historical data about growth temperatures, flow conditions, biocide effectiveness, and geometry and design. After a certain time, the assessor can review the assumptions made when establishing the ranking and compare the output of the model to data about actual metal degradation. Such data is commonly collected from time to time by inspection, pigging, or various tests (destructive or nondestructive), depending on the system layout, functions, and availability. The materials–water factor, on the other hand, requires more in-depth knowledge about the water chemistry (e.g., CO_2, H_2S, or O_2 levels) and how the material of construction responds to the environment.

The input parameters should be identifiable in normal operation without a lot of additional resources. This model can be a good preliminary step, for example, a ranking of the PoF of corrosion circuits or components. The ranking can further be used as input in a screening where several degradation mechanisms are assessed and the risk of operating a corrosion circuit or component is established. High-risk areas can be put under further scrutiny and assessed with more advanced models if needed.

For more information about the Checkworks Predictive Model please see EPRI (1994).

14.1.2 Model 2: Union Electric Callaway MIC Index

This model is one of few that are intended to be used over a period of time. Data should be collected periodically and used to identify growth of some specific microorganisms. Combining the growth of the collected cultures with settlement at the sampling site will give a qualitative value, which can be used to rank the system as holding a low, moderate, high, or very high potential for MIC.

Establishing sampling sites and collecting samples makes this model less suitable for assessing MIC in subsea pipelines but more applicable to topside systems. It is not described in the reviewed text whether the samples are from planktonic or sessile microorganisms.

This is one of the models that allows a user to estimate of MIC in existing systems whereas the other models in general are prediction tools.

A qualitative estimate of the likelihood of MIC is calculated by the following equation:

$$\text{Index Value} = 4.5446[(29 - \text{SRB})/4.83 + (6 - \text{Clos}) + (9 - \text{Gall})/1.5 + \text{SF} + \text{VF}] \tag{14.2}$$

SRB, Clos, and Gall are the number of days before sulfate-reducing bacteria (SRB) cultures, *Clostridia* cultures, and *Gallionella* cultures isolated from samples turn positive. It is not clear what types of techniques are used to analyze the cultures. SF is a silt factor, which accounts for the amount of silt at the sampling site. VF is a visual factor, which accounts for the amount of deposits or turbulence formation at the sampling site. Constants 4.83 and 1.5 help balance the equation and weigh factors evenly. Constant 4.5446 is a factor to correct calculation to a worst case of 100. Other constants are not mentioned in the reviewed text. The obtained index value is ranked on a scale from 0 to 100 as follows:

- 0–25 Low potential of MIC
- 26–50 Moderate potential of MIC
- 51–75 High potential of MIC
- 76–100 Very high potential of MIC

In contrast to Checkworks model, the Union Electric Callaway MIC Index relies on biological parameters. Cultures must be collected and monitored over time to obtain input values for the above equation. Even if culturing provides good results on the growth of SRB, *Clostridia*, and *Gallionella* cultures present in the system, one should be very cautious before relying too much on these factors when assessing the potential for and severity of MIC. This is because the cultures are not grown in their original environment (nutrients, biofilm consortia, flow conditions, temperature, etc.), and the outcome of the model (index) is very reliant on the growth of the samples.

For more information about the Union Electric Callaway MIC Index please see Chexal (1997) and Schultz and Hampton (1997).

14.1.3 Model 3: Lutney/Stein MIC Index

In similarity with the previous two models, the Lutney/Stein MIC Index is also a qualitative ranking of a system or site. The model includes biological and physical parameters, but it is unbalanced, putting almost all the emphasis on the physical parameters (see equation below).

This model considers SRB, slime-forming bacteria, metal-oxidizing bacteria, manganese-oxidizing bacteria, and acid-producing bacteria.

$$\text{MIC Index} = (\text{BF}) + (\text{DFF}) + (\text{SF}) \times (\text{MF}) \quad (14.3)$$

All parameters are expressed semiquantitatively:

- BF and DFF 0 (no impact on MIC) to 5 (high impact on MIC)
- SF 0 ($Fv = 0$ ft/s), 1 ($Fv > 10$ ft/s), 3 (4 ft/s $\leq Fv \leq$ 10 ft/s), or 5 ($0 < Fv < 4$ ft/s)
- MF 1 (very resistant) to 9 (very susceptible)

BF is a microbiological factor, which accounts for whether the considered microorganisms are active or not, that is, if the environment facilitates their growth. BF can be determined by use of different techniques; examples are MMM, probes, or pigging debris. Another way to determine microbial growth is to track the number of planktonic cells in the system over time, but be aware microorganisms can be introduced externally (e.g., produced seawater from reservoirs) and growth locations (biofilms) could be located far away from where sampling is possible. It is not stated in the reviewed text if the measure considers sessile or planktonic microorganisms, but if one is to establish the impact of microorganisms in a given environment, sessile sampling provides the best comparative results to planktonic samples (NACE 2012). DFF represents the amount of deposits/biofouling, which can be determined by weight and visual inspection. How these two factors, BF and DFF, are to be combined in order to assess the impact on MIC is not clear from the reviewed text.

SF is a sedimentation factor, which accounts for suspended solids carried at a given flow velocity (Fv). MF is a materials factor, also the most substantial factor, which accounts for the materials' susceptibility to MIC.

It is surprising that stagnant flow in this model gives a low ranking for potential MIC as industry experience shows stagnant flow often creates a favorable environment for MIC. However, several types of microorganisms are motile and can therefore move from one location to another without assistance from the environment. The model does point out that slow-moving systems have the highest potential for MIC ($0 < Fv < 4$ ft/s), and systems reviewed over time very seldom have stagnant flow at all times. Separation tanks, bypasses, drainpipes, similar location can have stagnant flow for shorter periods, the model recommends that assessors should use either an average or worst-case Fv to establish the SF factor in these cases.

The modelers put emphasis on the fact that the microbiological factor (BF) may not cover all microorganisms that can lead to MIC and also state that one should include other potentially harmful microorganisms if data indicates that "MIC is related to other types of microorganisms" (Sooknah et al. 2007). This is an important aspect as MIC is unpredictable, and our knowledge about microorganisms is lacking in terms of a complete and comprehensive understanding regarding their influence on MIC. It is important to survey system input and output and use new data to evaluate how to best apply a model and assess the threat of MIC. Selecting a model and neglecting to adjust it as new data are gathered will, in many cases, lead to flawed decision making.

It is important to be aware that the input data must be evaluated properly. For example, if the model is applied to a system with changing Fv and can have stagnant flow for several weeks at a time. If the assessor fails to see that the stagnant flow scenario is only temporary or uses an average of several periodically measured Fv, the result of the MIC Index could misrepresent the actual potential for MIC in this area as the product SF × DF would render a low output. It is critical to see the difference between continuous stagnant flow and intermittent flow conditions, but at the same time, be aware that some microorganisms are motile. Based on that, it is advised not to rank the threat of MIC in a system/site/corrosion circuit as "no impact" even if Fv = 0 over time.

For more information about the Lutney/Stein MIC Index, see Lutney (1989), Chexal (1997), Lutney and Stein (1999), and EPRI (1999).

14.1.4 Model 4: Pots et al. (2002) Improvements on De Waard-Milliams Corrosion Model

The De Waard-Milliams corrosion model was originally a CO_2 prediction model, but the prediction model from Pots et al. (2002) evolved the corrosion model and enabled assessors to quantify the rate (mm/y) of corrosion caused by microbial activity.

The model and the ranges used are based on where and when SRB grow, and the objective of this model is to calculate a corrosion rate based on a wide range of physical and chemical parameters. The parameters are given a suggested factor ranging from 0 to 5 that reflects their impact on the rate of MIC and provide guidelines on what values to give these factors if certain conditions are met. Further, the parameters are combined in order to semiquantitatively calculate an annual corrosion rate.

Today, several operators and consultants use this model to estimate the rate of corrosion due to MIC. It is easy to adjust the impact of each parameter according to one's own knowledge. In that way, experienced and knowledgeable analysts can apply the model to different systems and fine-tune the model based on historical data, future operating conditions, or both. It should be noted that the model takes not only MIC parameters into account but tries to give a comprehensive holistic model where all parameters contributing to corrosion are being addressed. The model is the only one that looks to place MIC in respect to all the other types of corrosion.

The corrosion rate is derived from Equation 14.4, presented below:

$$\text{Corrosion rate} = \text{Constant} \times \left(\prod_{i=1}^{n} f_i \right)^p \quad (14.4)$$

The constant is set to 2 mm/y, f_i is the value (from 0 to 5) suggested to represent the impact of parameter i, and p is a power law index (0.57). The following parameters are included in the model:

- pH
- Temperature
- Total dissolved solids
- Nutrient content in liquid
- Flow velocity
- Debris
- Pigging frequency
- Prolonged oxygen ingress
- Usage of biocide
- Age of pipeline
- Length of downtime

The model's greatest strength is that, although there are a lot of calculation parameters, they should be easily identifiable by the assessors with input from asset-based operational personnel. The model gives a degradation rate that can be connected to the PoF and thus incorporated in a system analysis (such as risk or reliability).

However, many of the ranges within the model are not justified as they have been established based on operational experience, which has resulted in some of the ranges being rejected by other authors (Maxwell and Campbell 2006). Additionally, the ranges that are attributed to the MIC parameter are only based on SRB characteristics. As we now know, whether directly or indirectly, other types of microorganisms also influence the corrosion processes, and the ranges should ideally be modified to reflect the potential for their activity as well to give the most accurate output.

14.1.5 Models 5 and 6: Maxwell and Campbell (2006) Model for Monitoring the Mitigation of MIC Risk and Maxwell (2006) Model for Predicting MIC in Seawater Injection Pipelines

These two models are described together because they have the same approach to predicting the threat of MIC, and as with the model from Pots et al. (2002), only consider SRB. However, as is described in more detail in the following section, the models include different parameters in the suggested calculation of the corrosion rate (CR) due to MIC.

The methods add a prerequisite to the Pots et al. (2002) model; namely, a biofilm on the metal surface that contains a significant amount of sulfide must be present before MIC can occur. This is in contrast to observations by others where H_2S can react with Fe^0 and form a corrosion-pacifying layer (Nešić et al. 2008). In this model the sulfide is regarded as significant when the concentration is >10 μg sulfide per cm^2. The time until such a biofilm is established can be calculated by the following equation:

$$\text{Time} = \frac{10}{(S \times M_s \times N_s)} \tag{14.5}$$

S is moles sulfide produced per cell per day, suggested by the modelers to be set as a constant (1×10^{-13}[g/mol]/day). M_s is the molecular weight of sulfide, and N_s is the number of cells per cm^2.

Identification of cell numbers per cm^2 is not described in the papers reviewed, but it is understood from other sources that molecular microbiological methods (MMM) are required to establish an understanding about the in situ sessile microbial population and the quantity of cell numbers, that is, N_s (Whitby and Skovhus 2011; Skovhus et al. 2014). The above estimate does not rely on parameters that are specific to any one microbial group but rather the generation of sulfide through their metabolic processes. Thus, the step considers the activity of all sulfate-reducing prokaryotes but not methane-producing microorganisms (methanogens), slime-forming bacteria, or other potentially harmful microorganisms.

According to the model, when a significant sulfide concentration is in reached, MIC will be initiated. The rate of MIC can be calculated by the same equation as presented in the model by Pots et al. (2002). In this calculation, the models include different parameters.

The Maxwell and Campbell (2006) model includes the following parameters:

- Deposits
- Pigging frequency
- Oxygen ingress
- Fluid velocity

The Maxwell (2006) model includes the following parameters:

- Deposits
- Pigging frequency
- Oxygen ingress
- Sulfide
- System age

The reason why methods use different parameters is not described in the publicly available papers that were reviewed for this chapter. However, Maxwell (2006) suggests that system age is not a suitable parameter to include and argues that inclusion of another parameter is required. Maxwell and Campbell (2006) is therefore considered by the authors to be the most mature of the two models as they define the other parameter to be velocity. It is now widely acknowledged that velocity and design (geometry) of a system can impact settlement of microorganisms and thus the potential for MIC.

The prerequisite that a significant sulfide concentration must be in place before MIC is initiated requires that sessile microorganism samples from the asset are available. Maxwell and Campbell (2006) use the term "risk" associated with the model output. From the context of this text it is considered by the authors to be incorrect as mentioned previously. The output of the model is linked to the treat of MIC, used to establish the PoF, which can then be combined with the CoF and thereafter expressed as risk.

14.1.6 Model 7: Sooknah et al. (2007) Pipeline Internal MIC Risk Assessment Model

This model is not a risk assessment model but rather a model to predict the susceptibility of MIC. The model is divided into two steps and follows a two-step approach. First, a preliminary analysis is to take place. In this step, temperature, presence of water, type of nutrients, and antifouling procedures of the system or site are assessed in relation to microbial growth.

If the above parameters show that development of a biofilm is probable, the following parameters are then investigated:

- Flow rate
- Water quality
- Total dissolved solids
- Total settable solids
- Saturation, pH, temperature, alkalinity, calcium
- Cleaning frequency influence

Dividing the model into a two-steps, is a good way to highlight the most critical systems/sites/corrosion circuits and further prioritize further investigation of the areas that are of most concern. It is, however, important that other degradation drivers (e.g., other types of corrosion, oxidation, abrasion, etc.) are also considered because a low threat of MIC does not necessarily mean low PoF and definitely not always low risk of failure.

The data gathered is used to individually establish a qualitative likelihood of MIC based on each parameter.

14.1.7 Model 8: Allison et al. (2008) Strategies for Predicting the Risk of MIC

The three-step model developed by Allison et al. (2008) uses nutrient availability, number of SRBs, and number of general heterotrophic bacteria present to qualitatively establish the potential for MIC in a system. Similarly to many of the before-mentioned models, sulfate-reducing archaea (SRA), methanogens, or other potentially harmful microorganisms are not considered in the model. MIC management is understood to be the primary objective of the model rather than the estimation of the rate of MIC.

Predicting the "risk" of MIC is suggested through a three-step process:

1. Calculate the potential for H_2S generation by SRB activity
2. Identify potential rate of H_2S generation by SRB based on amount of SRB present
3. Identify amount of other bacteria present

Step one is calculated based on water chemistry; steps two and three are identified through microbiological analysis to measure the cells per milliliter, that is, a planktonic rather than sessile count. This is in contrast to most models, which use sessile cells as a numerical measure. There can be a correlation between planktonic cells and the rate of MIC, but general industry opinion is that there is no direct link between the number of planktonic cells and the rate of MIC on metal surfaces, which is a major drawback for this model.

14.1.8 Model 9: Skovhus et al. (2012) and Sørensen et al. (2012) Model for MIC Management

The idea behind the Skovhus et al. (2012) and Sørensen et al. (2012) model is similar to what Maxwell and Campbell (2006) and Maxwell (2006) developed: dividing the calculation into two steps. The first step estimates the processes that occur prior to initiation of MIC, and the second calculates the corrosion rate when MIC has occurred. The model is based on use of MMM to study the in situ microbial consortia and does not include physical or chemical parameters. MMM determination could be incorporated and improve the accuracy of other models, especially where they currently rely on culturing methods.

Solely focusing on biotic components largely differentiates this model from those discussed above, where abiotic components have often been considered the major contributing factors on the corrosion rate. The information gathered in the study is used to calculate stage 1, integrated MIC risk factor (IMRF), and stage 2, potential pit generation rate (PPGR). IMRF represents mechanisms occurring prior to MIC initiation, and PPGR calculates degradation rate after the time limit in IMRF is reached. The method includes sulfate-reducing prokaryotes (SRP) and methanogens.

In what is called a MIC risk assessment tool, the modelers suggest that the rate of iron dissolution caused by microbial activity can be calculated as follows:

$$\text{Rate of iron dissolution} = 4 \times N_{SRB} \times S_{SRB} + 4 \times N_{SRA} \times S_{SRA} + 4 \times N_{MET} \times S_{MET} \tag{14.6}$$

N is the number of cells per cm^2, S is the cell-specific activity of the microbial group (measured in mg/(L·h), and MET is methanogens. In the model, the degradation rate is considered to be proportional to the MIC risk, which is different from the other approaches adopted in this chapter. The model has been successfully implemented in pipeline operations in the Danish sector of the North Sea by Maersk Oil, and they have developed a software module based on the papers/equations by Larsen et al. (2013) and Thomsen et al. (2016).

14.1.9 Model 10: Taxèn et al. (2012) Model for Under-Deposit Corrosion

This model is not directly focused upon MIC but rather corrosion occurring under deposits. Taxèn et al. (2012) state that under-deposit corrosion results from either oxygen corrosion or MIC. The model is a data simulation based on mathematical analysis of the chemical reactions that are expected to take place at the metal surface under a deposit and the result of the reactions. Only MIC considerations in the model are discussed.

Microorganisms influence the environment under deposits as follows: Oxygen is consumed proportional to the following equation:

$$\text{Oxygen consumption} = \frac{cO_2}{1000 \times aSRB} \tag{14.7}$$

The factors included in the equation are not explained, but it is assumed that cO_2 is amount of oxygen, and aSRB is the amount of SRB.

As oxygen is consumed by the microorganisms in a biofilm, its concentration decreases at the pipeline surface. This limits the magnetite formation through reduction in the number of chemical reactions taking place on the pipeline surface. This reduces the protective layer on the pipeline surface, which correspondingly increases the corrosion susceptibility of the pipeline.

In the reviewed paper, Taxèn et al. (2012) does not provide any justification as to why Equation 14.7 is assumed to have any connection to the activity of microorganisms. Microorganism consumption of oxygen is suggested to be proportional to the MIC rate. This view of MIC and the driving parameters for MIC are very specific as the model suggests that the rate of MIC is only influenced by oxygen ingress. However, it does try to examine the role of oxygen in relation to corrosion product formation. So the equation, if properly evaluated and confirmed, could be a significant step to a more comprehensive quantitative model for the MIC threat than we have seen before.

14.2 DISCUSSION OF MODELS FOR MIC MANAGEMENT IN THE OIL AND GAS INDUSTRY

The most important aspect of all the reviewed models is that none of them will categorically give the assessor the risk of failure due to MIC. As defined in the beginning of this chapter, risk is the product of the probability of an event and the consequence of an event. At best the assessor is able to estimate the threat of MIC, which can then be used in other management tools or models to establish the risk of unwanted events, such as failure. For example, low or stagnant flow is, in some cases, associated with high threat of MIC (and possibly high PoF) but will probably not lead to a significant event (low CoF). In this case, the overall risk is often considered low, that is, MIC may not be a significant issue at this part of the system and may therefore not need to be prioritized.

Historically, MIC has not been an area for focused research for the oil and gas industry. With recent significant developments in the area of MIC modeling, this is starting to change. Therefore, authors and companies want to protect business advantages by limiting the published information about their models. The papers reviewed for this chapter are publicly available conference papers that summarize the body of work that may somehow affect the interpretation of the developed models. The outcome of the models reviewed, corrosion rates/potential for MIC, must be seen in a greater context. All degradation mechanisms, such as CO_2 and H_2S corrosion or erosion, effects of chlorine and chlorite addition, should be mapped, and the PoF associated with each significant degradation mechanism should be combined with the CoF at a given corrosion circuit or site. Only then are we able to make decisions based on a holistic view of the system and accurately determine how to best manage it. An example of a management tool that can be used to consider the MIC threat in comparison or in combination with other degradation mechanisms is DNV GLs Recommended Practice DNV-RP-G101 (2010), which describes how risk can be used to plan inspection of offshore topside facilities.

Due to the disagreement among scientists and operators regarding which groups of microorganisms (acid-producing bacteria, SRB, methanogens, etc.) that influence corrosion directly versus those that indirectly influence corrosion, it is recommended to be cautious when using (micro)biology-specific models or at least the ranges that are used in a model based on the specific characteristics of a single or few microorganisms. Not long ago, SRB were the only type of microorganisms that were given any significant notice in MIC management or in the establishment of a MIC threat assessment. Putting too much emphasis on the specific microorganisms and the impact of their metabolic processes can lead to the outcome misrepresenting the actual potential for MIC or the actual corrosion rates. Most of the models base the ranges and values used for evaluating the threat of MIC on environments where one or few microbial groups are known to be active. Caution is recommended when relying on data from one or a few microbial groups. Our knowledge about MIC is not sufficient to allow a focus on the groups of microorganisms that have a proven effect on corrosion (Geissler et al. 2015).

Pots et al. (2002), Maxwell and Campbell (2006), and Maxwell (2006) all argue that the corrosion rate is a best estimate and is to be used as a guide to the potential severity of MIC. All three models use a number of factors to establish the potential and severity of the threat of MIC. This is important because it indicates that the models relying solely on one parameter, such as the models by Taxèn et al. (2012), Skovhus et al. (2012), and Sørensen et al. (2012) could be unfavorably biased and lead to an erroneous estimation of the MIC threat compared to models with a wider spread of input factors.

If further work shows that models relying solely on one input parameter do in fact provide a good representation of the actual MIC rate, then no reason why the models will not be widely adopted. However, at present, these models require evaluation, field testing, and extensive confirmation.

How microorganisms influence corrosion is a much-debated issue. Local loss of material due to corrosion can be a critical result of MIC, and this is the main focus covered in the models reviewed. However, there have been cases where, for example, iron sulfide produced by microorganisms has been transported downstream of a corrosion site and led to plugging of systems and components (Mitchell et al. 2012). Therefore, areas downstream of where the potential MIC threat is identified should also be included within a system analysis for the associated implications.

There are no two systems that are identical with common operating conditions. This provides challenge when applying a model to manage MIC. One model with a given input may give a good representation of the MIC threat associated with operating one type of system but will generate poor results if applied to a different system. Continuous or periodic evaluation of the models will help assessors develop an understanding of whether the applied model is adequate to establish the impact of MIC in a specific system or not.

Most of the models consider "simple" systems, such as transport or water injection pipelines. Commonly, in these types of systems, the environment does not change considerably from one corrosion circuit or site to another, and the system may, in fact, be treated as just one corrosion circuit. MIC management in more complex production systems, such as topside separation systems, can be substantially more demanding. Incorporating one of the reviewed models in a management system for complex oil and gas production systems requires knowledgeable personnel and, above all, the resources and patience to evaluate the models and data over time. In the end, proper management of MIC should yield rewards for the extra effort employed.

Figure 14.1 shows evidence of MIC and pitting attack in carbon steel piping from a closed drain system. This is an example of how MIC can attack a concentrated area and form nonuniform corrosion morphology.

14.3 WHAT'S NEXT?

Development of a mathematical model, perhaps based on MMM or new developments in sensor technology, would be a great step forward in managing MIC. The problems with developing such a model are unfortunately plentiful, including unpredictable nature of microorganisms; limited knowledge; combining a wide range of physical, chemical, and biological parameters; varying operating conditions; and production composition to name a few. To develop a complete mathematical model, if even possible, would likely require a lot of resource. It would, as a first step toward

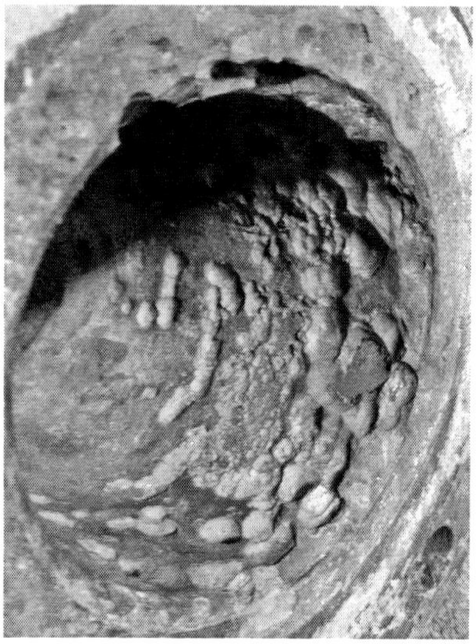

FIGURE 14.1 Evidence of MIC and pitting attack in carbon steel piping from a closed drain system.

a complete model, be interesting to see scientists with the proper skills and knowledge of MIC, electrochemistry, and mathematical modeling of natural processes evaluate the possibly of developing such a model.

Big Data has become a popular concept, and as a result of more comprehensive data management and analysis, many operators are able to collect and process data on a larger scale than ever before. This again, leads to the potential for decision making being based on more comprehensive data, more consistent and automated interpretation of the production properties and environment, and therefore a better understanding of a system. In the case of corrosion, an improved understanding of degradation mechanisms, such as MIC, can be enabled by holistically incorporating the Big Data concept in the operation and management systems.

Using data to create algorithms in a self-learning system that extracts operations and production information and evaluates it to rank the threat of MIC in real time without subjective intervention may be the next exciting step. With sufficient data, perhaps the industry is already on the brink of developing simple mathematical models from experiences and knowledge gained in using basic self-learning algorithms.

In a broader sense, proper assessment of MIC in corrosion management practices is needed to correctly manage MIC. New models should have a direct link to corrosion management, and modelers should keep in mind that the model, which often helps establish the threat of MIC, should fit a comprehensive asset management system. That way, operators can consider all degradation mechanisms and make decisions based on a complete understanding of all threats and the actual risk of operating and maintaining an oil and gas system.

As mentioned in the discussion of some ranking models (e.g., Checkworks Prediction model), the model itself can be used as a preliminary step in a two-step

MIC assessment procedure where a ranking of corrosion circuits is performed first before systems/sites/corrosion circuits with severe threats are put under further scrutiny and assessed more thoroughly. A two-step approach, which allows the MIC threat to be used as one factor in the establishment of the PoF, combined with the CoF and used in a RBI analysis, has been developed by Skovhus et al. (2016). This model is currently used by DNV GL and HESS in Denmark.

REFERENCES

Allison, P W, Sahar, R N R R, Guan, O H, Hain, T S, Vance, I and Thompson, M J. 2008. *The investigation of microbial activity in an offshore production pipeline system and the development of strategies to manage the potential for microbiologically influenced corrosion. Corrosion 2008*—Conference and Exposition: NACE International.

Chexal, V K. 1997. *Proceedings International Corrosion Conference*. Paper IWC-97-84. Eng. Soc. W. PA, Pittsburgh.

DNV-RP-G101. 2010. *Risk Based Inspection of Offshore Topsides Static Mechanical Equipment*. Høvik: DNV GL.

EPRI. 1999. *Checkworks Cooling Water Applications*. CWUG-1999, Electric Power Research Institute, Palo Alto, CA.

EPRI. 1994. *Microbiologically Influenced Corrosion*. TM1001, Electric Power Research Institute, Palo Alto, CA.

Geissler, B, Keller-Schultz, C and Keasler, V. 2015. Don't just blame SRBs and APBs for MIC. *Corrosion*. Paper no. 6083. Conference and Exposition: NACE International.

Larsen, J, Juhler, S, Sørensen, K B and Pedersen, D S. 2013. *The Application of Molecular Microbiological Methods for Early Warning of MIC in Pipelines. Corrosion 2013*. Paper no. 2029. Conference and Exposition: NACE International.

Lutney, R W. 1989. *EPRI MIC Quad Chart TM0089*. Energy Power Research Institute NDE Center, Charlotte, NC.

Lutney, R W and Stein, A. 1999. *Proceedings 13th International Corrosion Conference*. Paper no. 5.6-263. Cape Town, SA. Sept 27.

Mariuou-Dahle, E and Skovhus, T L. 2012. *Produced Water Management Lessons from the Ula & Valhall fields*. 2012 Produced Water Management, Stavanger. Jan 25, 2012.

Markoff, C and Larsen, E. 2010. *Managing MIC at Valhall*. BP Norge, Norway.

Maxwell, S. 2006. Predicting microbiologically influenced corrosion (MIC) in seawater injection systems. *SPE International Oilfield Corrosion Symposium in Aberdeen, UK, 2006*. Paper no. SPE-100519-MS. Society of Petroleum Engineers.

Maxwell, S and Campbell, S. 2006. Monitoring the mitigation of MIC risk in pipelines. *Corrosion 2006*. Conference and Exposition: NACE International.

Mitchell, A F, Liengen, T, Anfindsen, H and Moli S. 2012. Experience of molecular monitoring techniques in upstream oil and gas operations. *Corrosion 2012*—Conference and Exposition: NACE International.

Nešić, S, Wang, S, Fang, H, Sun, W. and Lee, J K-L. 2008. *A New Updated Model for CO_2/H_2S Corrosion in Multiphase Flow. Corrosion 2008*. Conference and Exposition: NACE International.

Pots, B F M, Randy, C J, Rippon, I J, Thomas, M J J S, Kapusta, S D, Girgis, M M and Whitham, T. 2002. *Improvement on De Waard-Milliams Corrosion Prediction and Applications to Corrosion Management. Corrosion 2002*. Paper no. 02235. NACE International.

Schultz, G L and Hampton, B E. 1997. *Proceedings EPRI SWSRI*. Electric Power Research Institute NDE Center, Charlotte NC. June 22–27.

Skovhus, T L, Andersen, E S and Hillier, E. 2016. Management of microbiologically influenced corrosion in risk based inspection analysis (SPE-179930). Society of Petroleum Engineers.

Skovhus, T L, Holmkvist, L, Andersen, K, Pedersen, H and Larsen, J. 2012. MIC Risk Assessment of the Halfdan Oil Export Spool (SPE-155080). *SPE International Conference and Exhibition on Oilfield Corrosion*. Society of Petroleum Engineers.

Skovhus, T L, Caffrey, S M and Hubert, C R J. 2014. *Applications of Molecular Microbiological Methods*. Poole: Caister Academic Press.

Sooknah, R, Papavinasa, S and Revie, R W. 2007. Modelling the occurrence of microbiologically influenced corrosion. *Corrosion 2007*. Paper no. 07514. Conference and Exposition: NACE International.

Sørensen, K B, Thomsen, U S, Juhler, S and Larsen, J. 2012. *Cost Efficient MIC Management System Based on Molecular Microbiological Methods. Corrosion 2012*—Conference and Exposition: NACE International.

Taxèn, C, Comanescu, I and Melchers, R E. 2012. *Framework Model for Underdeposit Corrosion in Water Injection Pipelines*. BIOCOR RSP2: Oil and Gas.

Thomsen, U S, Meng, R L C and Larsen, J. 2016. Monitoring and Risk Assessment for Microbiologically Influenced Corrosion in Offshore Pipelines. *Corrosion*. Paper no. 7194. Conference and Exposition: NACE International.

TM0212-2012. *Detection, Testing and Evaluation of Microbiologically Influenced Corrosion on Internal Surfaces of Pipelines*. NACE International.

Whitby, C and Skovhus, T L. 2011. *Applied Microbiology and Molecular Biology in Oilfield Systems*. New York: Springer Publisher.

Section IV

Field Cases and Laboratory Studies

15 MIC under Conditions of Oxygen or Nitrate Ingress

Jaspreet Mand, Yin Shen, Heike Hoffmann, and Gerrit Voordouw

CONTENTS

ABSTRACT

The majority of microbiologically influenced corrosion (MIC) scenarios occur under anoxic conditions and often implicate sulfate-reducing bacteria (SRB). These bacteria catalyze sulfate respiration and form sulfide as a metabolic end product. While SRB may be one of the most aggressive players in biocorrosion and are by far the most often studied, MIC can involve other organisms as well. The ingress of electron acceptors other than sulfate, such as oxygen or nitrate, into an SRB-dominated environment can lead to changes in the microbial community and in the corrosion threat. For instance, the partial respiration of nitrate by nitrate-reducing bacteria forms nitrite, a corrosive metabolite. The chemical reaction of oxygen or nitrite (a by-product of incomplete nitrate respiration) with SRB-produced sulfide forms polysulfides or elemental sulfur. Sulfur and polysulfides are highly corrosive, and evidence of sulfur-mediated corrosion can be seen when examining the metal surface in the form of severe pitting. Furthermore, changes in microbial community composition may also serve as evidence of sulfur- and polysulfide-mediated corrosion.

DNA pyrotag sequencing was used to determine the microbial community compositions of field samples from a North Sea oil production site and from an Alberta oilfield. In both cases, microbial communities associated with sulfur metabolism were uncovered. An abundance of microorganisms able to oxidize aqueous sulfide using oxygen as an electron acceptor (genera *Arcobacter, Sulfurospirillum, Sulfurimonas*) were discovered. The potential for sulfur formation by these organisms greatly enhances corrosion threat. The fact that other bacteria capable of using elemental sulfur as an electron acceptor (genera *Desulfuromonas, Desulfuromusa*) were also found in these samples serves as evidence that sulfur is often produced at these sites. The presence of these different sulfur cycle microorganisms and the subsequent activity shown under oxygen or nitrate ingress is evidence for their participation in MIC.

15.1 INTRODUCTION

MIC of steel, caused by the presence and activities of microorganisms or by their metabolites, is a serious threat to oil and gas infrastructure. For a steel structure in an aqueous environment, an anodic reaction is as follows:

$$Fe^0 \rightarrow Fe^{2+} + 2e^- \tag{15.1}$$

A counter-cathodic reaction is required, and in the presence of oxygen, the electrons formed in Equation 15.1 are used for the formation of hydroxide ions:

$$\tfrac{1}{2}O_2 + H_2O + 2e^- \rightarrow 2OH^- \tag{15.2}$$

Most often, the environment within pipelines is kept anaerobic to limit oxygen ingress to prevent Equation 15.2 from proceeding. In such cases, the cathodic reaction becomes

$$2H^+ + 2e^- \rightarrow H_2 \tag{15.3}$$

An anoxic environment allows the growth of anaerobic microorganisms, such as SRB. There are several ways in which SRB are associated with MIC. SRB are able to use the hydrogen formed in Equation 15.3 as an electron donor to reduce sulfate and form sulfide (Von Wolzogen Kühr and Van der Vlugt 1934). Some strains of SRB have the ability to use steel as an electron donor directly without the formation of a hydrogen intermediate (Venzlaff et al. 2013; Enning and Garrelfs 2014). The production of sulfide by SRB is a process known as souring; sulfide is a corrosive by-product as well (Gieg, Jack, and Foght 2011). The sulfide can precipitate out of solution as metal sulfides and lead to further anodic depolarization of steel surfaces (Spruit and Wanklyn 1951; Wanklyn and Spruit 1952; King and Miller 1977; Hamilton 1985). Sulfide is also known to promote the entry of hydrogen into a metal surface, which can lead to hydrogen embrittlement of steel.

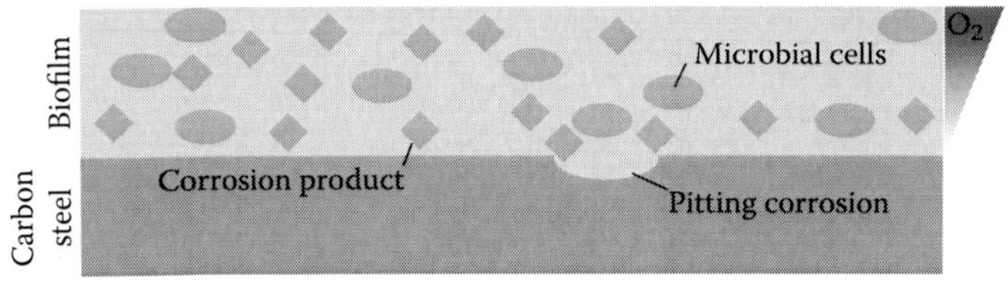

FIGURE 15.1 Schematic of a potential biofilm found on a steel surface, involved in MIC. Concentration gradients of electron acceptors (such as O_2) exist within biofilms, and this allows for numerous types of microorganisms to grow together.

Although SRB are most often connected with MIC, there are many other organisms found in the same environments. Microbial communities found attached to and growing on metal surfaces are generally living as multicellular biofilms. Biofilms consist of microbial cells embedded in an extracellular polymeric substance and are highly diverse in terms of the types of organisms found within (Costerton et al. 1995; Hall-Stoodley, Costerton, and Stoodley 2004; Harrison et al. 2005). Living in a biofilm offers distinct advantages over the planktonic lifestyle. One such advantage is that there is enrichment in carbon and energy sources because nutrients and trace elements are retained and are internally recycled within a biofilm (Beech and Sunner 2004).

Diffusion limitation of nutrients and chemicals within a given biofilm produces concentration gradients within the structure, which contributes to microbial diversity. For example, if there is an ingress of electron acceptors, such as nitrate or oxygen, into a pipeline where a biofilm community persists, the highest concentrations will likely remain near the top of the biofilm. Due to this diffusion limitation, the concentration of nitrate or oxygen will be depleted rather quickly and may not reach the metal surface (Figure 15.1). This leads to a differentiation of anaerobic and aerobic organisms within a single biofilm. In this way, microorganisms other than SRB are able to influence corrosion as well. When biofilms are associated with corrosion, the formation of tubercles overlying pitting in the metal is seen (Ray et al. 2010; Usher, Kaksonen, and MacLeod 2014). MIC almost always leads to pitting corrosion, which is unpredictable and difficult to monitor and mitigate and therefore is the most damaging form of corrosion.

15.2 NITRATE INGRESS

While small concentrations of nitrogen-containing compounds may be indigenous to oil and gas production sites, the addition of nitrate has been implemented in many oilfields worldwide as an alternative to biocides to restrain souring (Hubert et al. 2004; Little, Lee, and Ray 2007). Nitrate is a water-soluble and nontoxic compound used to suppress SRB activity as it stimulates the growth of biocompetitive nitrate-reducing bacteria (NRB). It is effective because NRB outcompete SRB for the use of oil organics, which serve as carbon and energy sources (Gittel et al. 2009). Furthermore, NRB may incompletely reduce nitrate to nitrite, a compound that binds to the dissimilatory sulfite reductase enzyme of SRB with high affinity, prohibiting sulfate reduction (Haveman et al. 2004). Additionally, SRB are inhibited by the increased redox potential caused by oxidation of sulfide with nitrite through a chemical process:

$$4HS^- + 2NO_2^- + 7H^+ \rightarrow NH_4^+ + 4S^0 + \tfrac{1}{2}N_2O + 3\tfrac{1}{2}H_2O \quad (15.4)$$

(Nemati, Jenneman and Voordouw 2001; Myhr et al. 2002; Kjellerup et al. 2005).

15.2.1 Influence of NRB on Steel Corrosion

Since nitrate is often used in field trials, many studies have been done to study its effectiveness in combating souring and its effect on corrosion of carbon steel. Recently, a corrosive NRB strain has been identified that becomes active upon nitrate injection, and in laboratory incubations, iron loss was correlated with anaerobic nitrate reduction (Iino et al. 2015). It was proposed that this NRB (*Prolixibacter* sp. Strain MIC1-1) is able to directly gain electrons from the iron to reduce nitrate (Iino et al. 2015).

In other NRB strains, the incomplete reduction of nitrate results in the formation of nitrite, which is chemically corrosive because it is able to oxidize iron and form ammonium. For example, at the Draugen field, located within the Norwegian Continental Shelf, corrosion was attributed to incomplete nitrate reduction by NRB (Vik et al. 2007). In this instance, nitrate injection was being used to abate microbial souring. However, the start of nitrate application coincided with an increase in nitrite concentrations and a rise in corrosion rates. The subsequent addition of biocide remedied the surge in corrosion rates. Thus, it was concluded that the former corrosion problems were caused by incomplete nitrate reduction to form corrosive nitrite by NRB (Vik et al. 2007). Therefore, an important consideration when using nitrate as an SRB inhibitor is that the dosage should be sufficient to inhibit SRB activity, however not so high as to cause chemical steel corrosion by nitrite, a by-product of nitrate reduction (Kielemoes, De Boever, and Verstraete 2000; Hubert et al. 2005).

15.2.2 Biological Sulfur Formation by soNRB

The addition of nitrate stimulates the sulfide-oxidizing, nitrate-reducing bacteria (soNRB), a group of organisms able to catalyze the oxidation of sulfide (Gevertz et al. 2000; Lambo et al. 2008). When nitrate is injected into a reservoir, there is evidence to show that it is preferentially used as a terminal electron acceptor by soNRB over heterotrophic NRB in many cases. This phenomenon is probably due to the inhibition of heterotrophic NRB by high sulfide concentrations found in some sour reservoirs (Lambo et al. 2008). soNRB are able to biologically oxidize sulfide with nitrate to form elemental sulfur (Equation 15.4) or sulfate, and for some bacteria, this end product is dependent on the nitrate to sulfide (N/S) ratio present (Table 15.1). soNRB preferentially form sulfur when the N/S ratio is low, but when the N/S ratio is high, the likelihood of sulfate formation by these bacteria is greater (Lin, Voordouw, and Coombe 2008). In cases where the N/S ratio is high, sulfur is

TABLE 15.1
Different Sulfide Oxidation Reactions That May Occur Depending on the Nitrate-to-Sulfide (N/S) Ratio

Sulfide Oxidation Reaction	N/S Ratio
$HS^- + 4NO_3^- \rightarrow SO_4^{2-} + 4NO_2^- + H^+$	4
$HS^- + 1.6NO_3^- + 0.6H^+ \rightarrow SO_4^{2-} + 0.8N_2 + 0.8H_2O$	1.6
$HS^- + NO_3^- + H^+ \rightarrow S^0 + NO_2^- + H_2O$	1
$HS^- + 0.4NO_3^- + 1.4H^+ \rightarrow S^0 + 0.2N_2 + 1.2H_2O$	0.4

Source: S. Lin, G. Voordouw, and D. Coombe, Sulfur production associated with souring control by nitrate injection: A potential corrosion risk? *NACE International Corrosion 2008 Conference & Expo*. Paper no. 08662, 2008.

Note: The N/S ratio dictates the formation of sulfur or sulfate.

still formed transiently and correlates with nitrite accumulation, followed by sulfur oxidation to sulfate where nitrite is then consumed (Gadekar, Nemati, and Hill 2006; Lin, Voordouw, and Coombe 2008). The intermediary formation of products nitrite and elemental sulfur is extremely corrosive toward steel (Vik et al. 2007).

The stimulation of soNRB by the addition of nitrates has led to an enhanced corrosion threat in some instances. Under model laboratory conditions, the effect of nitrate addition to an SRB culture and subsequent MIC was assessed. Researchers found that the addition of nitrate and an active soNRB culture to an active SRB culture actually led to higher corrosion rates than those seen for the SRB culture alone (Nemati, Jenneman, and Voordouw 2001). Furthermore, when nitrate and soNRB were added, visible pitting corrosion was witnessed on the carbon steel coupons (Nemati, Jenneman, and Voordouw 2001). The aggressive corrosion was predicted to be caused by the formation of sulfur or polysulfides by the soNRB as shown in Equation 15.4.

The addition of nitrate has also been shown to change the microbial community in oilfield waters. In a field trial, nitrate was added to a test rig, and corrosion rates were compared to those in a parallel rig without nitrate amendment (Drønen et al. 2014). Higher corrosion rates were found with nitrate injection with significant shifts in microbial populations (Drønen et al. 2014). soNRB of the genus *Arcobacter* became prevalent, and nitrite and elemental sulfur were subsequently detected (Drønen et al. 2014). Hence, it was proposed that nitrate injection led to the formation of nitrite through incomplete nitrate reduction. Nitrite reaction with sulfide led to the formation of elemental sulfur, and this sulfur was disproportionated and reduced by other microorganisms. The ingress of nitrate into this system changed the microbial community and led to severe MIC.

15.2.3 Decreased Corrosion Threat due to Nitrate Ingress

Contrary to the studies described above, there are also examples of nitrate ingress leading to lower observed corrosion rates and damage. This result has been demonstrated in a laboratory study where nitrogen-containing compounds were added to an active SRB enrichment from an oilfield to limit SRB-attributed souring. To an SRB culture, one of nitrate, nitrite, or a combination of nitrate and an active culture of soNRB was added, and the effect on souring and the effect on corrosion rates were assessed (Rempel, Evitts, and Nemati 2006). In this study, the addition of nitrite was found to be the most effective for inhibiting sulfide production. Subsequently, the lack of sulfide was attributed to decreased carbon steel corrosion rates as well (Rempel, Evitts, and Nemati 2006).

Even if nitrate injection leads to nitrite formation and subsequent chemical corrosion, nitrate does limit the aggressive pitting corrosion caused by microorganisms (Stott et al. 2008; Stott 2013). In laboratory MIC studies with sulfate-containing medium, sulfide formation was evident, and subsequent localized pitting corrosion was seen. When nitrate was amended into the medium, it appeared to prevent pitting corrosion of steel even though it promoted general corrosion (Dall'Agnol, Cordas, and Moura 2014). General corrosion of steel surfaces is predictable and therefore not as detrimental as pitting corrosion, and in this case, nitrate addition was deemed to lower corrosion threat due to decreased sulfide formation (Dall'Agnol, Cordas, and Moura 2014).

15.3 OXYGEN INGRESS

Although efforts are made to limit oxygen ingress in oil and gas systems, this strong oxidant can still enter pipeline systems. Sources of oxygen entry into a pipeline include ingress through leaky valves or even as a dissolved gas in methanol used to treat hydrates (Boivin and Oliphant 2011).

15.3.1 Influence of Oxygen in SRB-Dominated Environments

Dissolved oxygen concentration is a key parameter that is monitored in order to control corrosion rates (Rosli, Yoon-Seok, and Young 2014). The oxygen concentration in a system can vary greatly, depending on the types of microorganisms that are present. The presence of oxygen-reducing organisms would decrease this critical value whereas the presence of photosynthetic organisms, those that produce oxygen, could lead to an increase in the presence of light. To control chemical corrosion, limiting oxygen ingress is important to prevent (Equation 15.2); however, an anaerobic environment is suitable for SRB growth and subsequent pitting. Similarly, oxygen reduction by aerobic organisms may initially appear beneficial as this could decrease the likelihood of corrosion due to oxygen (Equation 15.2). However, rapid oxygen consumption can actually set up conditions suitable for anaerobic growth associated with MIC later on (Aktas et al. 2013). For example, oxygen consumption by some organisms, such as iron-oxidizing bacteria, has been shown to lower the redox potential in a system so that SRB are able to grow (Videla and Characklis 1992; Beech and Gaylarde 1999). Additionally, within biofilms found on metal surfaces overlying pitting corrosion caused by MIC, the depletion of oxygen can be measured as illustrated in Figure 15.1 (Lee and de Beer 1995; Beech and Gaylarde 1999). In other cases, oxygen ingress will decrease the efficiency of biocides that are being used to control SRB, and thus these bacteria will remain active (Machuca et al. 2014). For example, oxygen decreases the effectiveness of tetrakis hydroxymethyl phosphonium sulfate (THPS), and SRB-mediated pitting corrosion appears to persist as a result (Machuca et al. 2014).

15.3.2 Microbial Sulfide Oxidation to Form Sulfur

When oxygen enters a pipeline system, in addition to the chemical corrosion exhibited in Equation 15.2, it can influence different forms of microbiological corrosion. Oxygen ingress into an anaerobic SRB, sulfide-containing environment, leads to the formation of sulfur through a chemical reaction:

$$2H_2S + O_2 \rightarrow 2S^0 + 2H_2O \tag{15.5}$$

when the concentration of oxygen is low (Soylemezoglu and Harper 1982).

The reaction shown in Equation 15.5 can also be catalyzed by chemolithotrophic microorganisms known as the oxygen-reducing, sulfide-oxidizing bacteria (genera *Thiomicrospira*, *Arcobacter*) (Tang, Baskaran, and Nemati 2009). When concentrations of oxygen are higher, some sulfide-oxidizing bacteria are able to oxidize reduced inorganic sulfur compounds, such as sulfide, completely to sulfate with elemental sulfur serving as an intermediate (Tang, Baskaran, and Nemati 2009):

$$HS^- + 2O_2 \rightarrow SO_4^{2-} + H^+ \tag{15.6}$$

Biological sulfide oxidation can also occur phototrophically. Phototrophic bacteria, such as the green sulfur bacteria or the purple sulfur bacteria, are found in light-exposed, anaerobic environments. These are autotrophic and use carbon dioxide as a terminal electron acceptor to oxidize sulfide to sulfate with the formation of transient sulfur globules:

$$\text{Light energy} + CO_2 + H_2S \rightarrow (CH_2O)_n + S^0 \tag{15.7}$$

While in many species the sulfur globules are stored within the cell, there exist some species (genera *Chlorobium*, *Prosthecochloris*) that excrete sulfur to the outside so it is available to contact metal surfaces, if present (Madigan and Martinko 2006).

15.3.3 Cycling of Anaerobic and Aerobic Conditions

Sulfur corrosion does not necessarily have to occur in systems that contain high concentrations of hydrogen sulfide. Sulfur deposition may occur in systems with small amounts of hydrogen sulfide with oxygen ingress since the sulfur that is formed is very insoluble in aqueous solution (Boivin and Oliphant 2011). Oxygen ingress can lead to the formation of characteristic tubercles on the steel surface that overlie pitting on metal surfaces (Rosli, Yoon-Seok, and Young 2014). In a sulfide-containing, reservoir-simulating bioreactor, pits found on steel coupons within the bioreactor indicate that oxygen ingress may have led to the formation of sulfur from microbiologically produced sulfide (Hubert et al. 2005).

Many studies on the corrosion of carbon steel in the presence of an SRB-dominated biofilm have concluded that it may not be significant until oxygen is introduced into the system (Lee and Characklis 1993). For example, in sulfide-containing systems at ambient temperatures, iron sulfides in the form of makinawite are likely to form. However, when oxygen is introduced into such a system, iron oxides are formed, and the sulfur compounds liberated from makinawite take the form of sulfur or polysulfides (Smith and Craig 2005).

Deoxygenating seawater is used as a successful way to lower the corrosion threat, and furthermore, it is cost effective (Tamburri, Wasson, and Matsuda 2002). However, it appears that any benefit that is gained from deoxygenating water is eliminated if oxygen is reintroduced into the system. Researchers have shown that reintroducing oxygen into an anaerobic or hypoxic system actually increases corrosion rates over those in a continuously hypoxic system (Lee et al. 2006).

Researchers exposed carbon steel to cycles of anoxic and oxic waters and compared the results to those for carbon steel kept in continuously anoxic, continuously oxic, or continuously hypoxic conditions (<0.2 ppm O_2). Less carbon steel corrosion was seen under continuous anaerobic conditions as compared to cycles of oxic and anoxic water (Lee et al. 2005). Even the carbon steel that was continuously exposed to oxygen at all times experienced less corrosion than the steel that was exposed to alternating oxic and anoxic conditions (Lee et al. 2005). This may be attributed to the formation of an iron sulfide corrosion product when the steel is exposed to SRB in an anaerobic environment, but with the ingress of oxygen, the sulfides transfer electrons to the oxygen in solution and accelerate corrosion (Hamilton 2003; Lee et al. 2004). When corrosion of coupons incubated in stagnant seawater were continuously exposed to oxygen, corrosion was seen. However the damage was not as extensive as when anaerobic biofilms formed in seawater were exposed to oxygen (Lee et al. 2004). Similarly, MIC of structures found in seawater environments appears to be influenced by tidal conditions, which involve the fluctuation of oxygenated and deoxygenated seawater (Marty et al. 2014; Marty, van Loosdrecht, and Muyzer 2014).

15.4 CORROSIVITY OF SULFUR AND POLYSULFIDES

Sulfide produced by SRB activity is highly soluble and reactive and can be easily oxidized to other corrosive forms, such as polysulfides or sulfur. Once formed, sulfur deposition may block production tubing, but sulfur itself is not corrosive if there is no water present (Smith and Craig 2005; Sun et al. 2011). However, direct contact between carbon steel and wet elemental sulfur causes both general corrosion and very aggressive pitting corrosion (Sun et al. 2011; Jiang et al. 2012). Elemental chemically formed sulfur is hydrophobic in nature and forms large sulfur crystals in aqueous solution (Johnston and Voordouw 2012). However, biologically formed sulfur globules have a hydrophilic coating allowing their suspension in aqueous solutions (Johnston and Voordouw 2012).

There are many suggested mechanisms for sulfur corrosion of metal (Sun et al. 2011). The first is that sulfuric acid may be formed as result of sulfur hydrolysis:

$$S_8 + 8H_2O \rightarrow 6H_2S + 2H_2SO_4, \qquad (15.8)$$

a mechanism that was originally proposed by Boden and Maldonado-Zagal in 1982.

The second suggested mechanism involves an electrochemical reaction to produce elemental sulfur by reacting iron with polysulfides (MacDonald, Roberts, and Hyne 1978):

$$(x-1)Fe + S_{y-1} \cdot S^{2-} + 2H^+ \rightarrow (x-1)FeS + H_2S + S_{y-x}. \qquad (15.9)$$

The corrosion caused by sulfur may be through an electrochemical reaction and be galvanic in nature due to the deposition of sulfur on to steel surfaces (Fang et al. 2011; Jiang et al. 2012). Last, another electrochemical reaction has also been proposed for wet elemental sulfur corrosion where sulfide is formed from iron sulfide reduction (Schmitt 1991):

$$[FeS_{x+1}] + 2e^- \rightarrow [FeS_x] + S^{2-}. \qquad (15.10)$$

Although the exact mechanism is disputed and may differ depending on the situation, it is agreed that limiting the ingress of oxygen in seawater systems can greatly help to reduce sulfur formation and therefore decrease corrosion threat.

15.5 CASE STUDIES

With the advent of state-of-the-art molecular technologies, potential instances of MIC under conditions of oxygen or nitrate ingress may be further substantiated with results from DNA sequencing studies. Comparative analyses of the 16S rRNA sequences from bacteria and archaea are able to resolve the microbial diversity present in different samples. From this, corrosion threat may be inferred by assigning potential metabolisms to the microbial populations present in oilfield samples.

15.5.1 Albertan Oilfield

The Gryphon oilfield is located in southeastern Alberta, Canada, at 1125–1150 meters below the surface. This reservoir has an in situ temperature of 32°C–34°C. Brackish water containing high bicarbonate concentration is injected for oil production. This field has experienced souring, and high aqueous sulfide concentrations (153–348 ppm) are continually measured. The measured pH levels of water samples from the Gryphon field were above neutral (ranging between 8.2 and 8.7), thereby lowering the concentration of gaseous sulfide. This field was sampled with the goal of determining MIC threat.

The microbial community compositions of six water samples (five production waters and one injection water) were obtained using 454 pyrotag DNA-sequencing

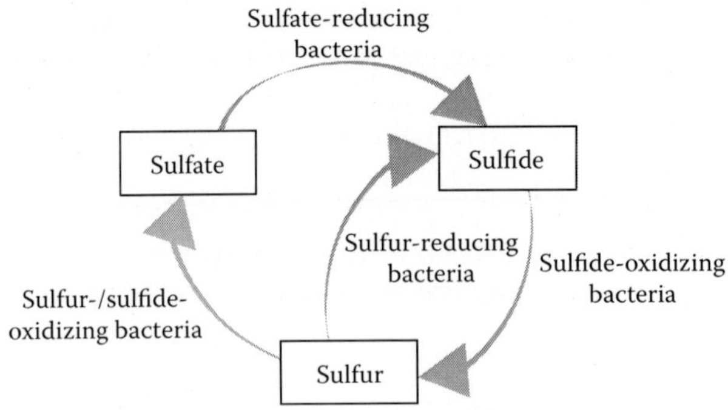

FIGURE 15.2 Potential sulfur cycling that may be occurring at the Gryphon oilfield and at the North Sea oilfield. The formation of elemental sulfur enhances the corrosion threat.

technology. Due to the high concentrations of sulfide, the presence of SRB was predicted. As expected, SRB, such as the genus *Desulfomicrobium* (class *Deltaproteobacteria*), were present in small (3.9% of reads for a PW sample) fractions in these water samples. These bacteria are likely converting aqueous sulfate to sulfide, which appears to persist in this system. But the DNA sequencing also revealed the presence of sulfide-oxidizing bacteria, such as the genus *Sulfurospirillum* (class *Epsilonproteobacteria*). These sulfide oxidizers were present in all samples, comprising between 4.5% and 94% of all pyrosequencing reads. Last, potential sulfur-reducing bacteria (order *Desulfuromonadales*) were also present in all samples. Sulfur-reducing bacteria are similar to SRB, but instead of reducing sulfate, they are able to reduce elemental sulfur back to sulfide.

The presence of sulfide-producing organisms (SRB and sulfur-reducing bacteria) and the presence of sulfide-oxidizing bacteria suggest an active sulfur cycle in the Gryphon oilfield (Figure 15.2). In this cycle, sulfide produced by SRB may be bio-oxidized to sulfur, a potentially corrosive metabolite. The cycling of sulfur species is likely driven by oxygen ingress into the system, and thus, it was concluded that oxygen ingress in aboveground facilities should be limited to decrease the threat associated with sulfur corrosion.

15.5.2 North Sea Oilfield

North Sea oil and gas production systems are rich in sulfur-containing compounds due to the presence of seawater. Seawater is injected to maintain pressure and to stimulate production. SRB are prevalent offshore due to an abundance of sulfate and anaerobic conditions (Cord-Ruwisch, Kleinitz, and Widdel 1987). At one field site that was sampled, produced water was mixed with seawater and subsequently injected into the reservoir to aid in oil and gas production. Therefore, seawater containing approximately 28 mM sulfate was mixed with produced water containing high concentrations of sulfide before it was injected into the reservoir. Chemical analyses of different water samples from the field showed that the produced water contained less sulfate than the injection water (mixture of seawater and produced water), which could indicate microbiological sulfate reduction in situ.

To assess MIC threat, samples from this North Sea site were analyzed using 454 pyrotag DNA-sequencing. SRB were expected to account for potential sulfate reduction, and these were confirmed. The genus *Desulfovibrio* and genus *Desulfobacter* were both identified in the waters. Additionally, sulfide-oxidizing bacteria (genera *Arcobacter*, *Sulfurimonas*, and *Sulfurospirillum*) were also seen. These bacteria can oxidize sulfide produced by SRB back to sulfur or sulfate. Last, the presence of some elemental sulfur-reducing bacteria (order *Desulfuromonadales*) was also detected (Mand et al. 2014). The formation of sulfur and/or polysulfides sets the redox potential to values that are generally too high for anaerobic SRB to thrive and is therefore toxic (Johnston et al. 2009). However, since sulfur-reducing bacteria are prevalent, the sulfur is effectively removed.

The presence of these three groups of sulfur cycle–related microorganisms indicates that the mixing of anaerobic sulfide-containing produced water with oxygen-containing seawater leads to the ideal environment for an active sulfur cycle (Figure 15.2). If these bacteria are thriving, the cycling of sulfate to sulfide may form elemental sulfur as an intermediate, a compound that is very corrosive toward carbon steel.

15.6 CONCLUSIONS

Biological souring of oil and gas operations caused by SRB is intimately linked with MIC in these systems. However MIC is a complex process, and the involvement of different types of microorganisms is dependent on the availability of nutrients. With the ingress of oxygen or nitrate into generally anaerobic environments, oxygen-reducing and nitrate-reducing organisms become active and may even replace the SRB that are generally associated with MIC. These organisms are able to produce nitrite and sulfur, compounds that result in aggressive corrosion of carbon steel surfaces. Therefore the ingress of nitrate and oxygen should be carefully monitored in order to limit corrosion threat.

REFERENCES

Aktas, D. F., J. S. Lee, B. J. Little, K. E. Duncan, B. M. Perez-Ibarra, and J. M. Suflita. 2013. Effects of oxygen on biodegradation of fuels in a corroding environment. *Int. Biodeterior. Biodegrad.* 81: 114–126. doi: 10.1016/j.ibiod.2012.05.006.

Beech, I. B., and C. C. Gaylarde. 1999. Recent advances in the study of biocorrosion: An overview. *Rev. Microbiol.* 30 (3): 177–190. doi: 10.1590/S0001-37141999000300001.

Beech, I. B., and J. Sunner. 2004. Biocorrosion: Towards understanding interactions between biofilms and metals. *Curr. Opin. in Biotechnol.* 15 (3): 181–186. doi: 10.1016/j.copbio.2004.05.001.

Boivin, J., and S. Oliphant. 2011. Sulfur corrosion due to oxygen ingress. *NACE International Corrosion 2011 Conference & Expo*. Paper no. 11120.

Cord-Ruwisch, R., W. Kleinitz, and F. Widdel. 1987. Sulfate-reducing bacteria and their activities in oil production. *J. Pet. Technol.* (January): 97–106.

Costerton, J. W., Z. Lewandowski, D. E. Caldwell, D. R. Korber, and H. M. Lappin-Scott. 1995. Microbial biofilms. *Annu. Rev. Microbiol.* 49: 711–745. doi: 10.1146/annurev.mi.49.100195.003431.

Dall'Agnol, L. T., C. M. Cordas, and J. G. Moura. 2014. Influence of respiratory substrate in carbon steel corrosion by a sulphate reducing prokaryote model organism. *Bioelectrochemistry* 97: 43–51. doi: 10.1016/j.bioelechem.2013.10.006.

Drønen, K., I. Roalkvam, J. Beeder, T. Torsvik, I. H. Steen, A. Skauge, and T. Liengen. 2014. Modeling of heavy nitrate corrosion in anaerobe aquifer injection water biofilm: A case study in a flow rig. *Environ. Sci. and Technol.* 48: 8627–8635. doi: 10.1021/es500839u.

Enning, D., and J. Garrelfs. 2014. Corrosion of iron by sulfate-reducing bacteria: New views of an old problem. *Appl. Environ. Microbiol.* 80 (4): 1226–1236. doi: 10.1128/AEM.02848-13.

Fang, H., B. Brown, D. Young, and S. Nešić. 2011. Investigation of elemental sulfur corrosion mechanisms. *NACE International Corrosion 2011 Conference & Expo*. Paper no. 11398.

Gadekar, S., M. Nemati, and G. A. Hill. 2006. Batch and continuous biooxidation of sulphide by *Thiomicrospira* sp. CVO: Reaction kinetics and stoichiometry. *Water Res.* 40 (12): 2436–46. doi: 10.1016/j.watres.2006.04.007.

Gevertz, D., A. J. Telang, G. Voordouw, and G. E. Jenneman. 2000. Isolation and characterization of strains CVO and FWKO B, two novel nitrate-reducing, sulfide-oxidizing bacteria isolated from oilfield brine. *Appl. Environ. Microbiol.* 66 (6): 2491–2501. doi: 10.1128/AEM.66.6.2491-2501.2000.

Gieg, L. M., T. R. Jack, and J. M. Foght. 2011. Biological souring and mitigation in oil reservoirs. *Appl. Microbiol. Biotechnol.* 92 (2): 263–282. doi: 10.1007/s00253-011-3542-6.

Gittel, A., K. B. Sørensen, T. L. Skovhus, K. Ingvorsen, and A. Schramm. 2009. Prokaryotic community structure and sulfate reducer activity in water from high-temperature oil reservoirs with and without nitrate treatment. *Appl. Environ. Microbiol.* 75 (22): 7086–7096. doi: 10.1128/AEM.01123-09.

Hall-Stoodley, L., J. W. Costerton, and P. Stoodley. 2004. Bacterial biofilms: From the natural environment to infectious diseases. *Nat. Rev. Microbiol.* 2: 95–108. doi: 10.1038/nrmicro821.

Hamilton, W. A. 1985. Sulphate-reducing bacteria and anaerobic corrosion. *Annu. Rev. Microbiol.* 39: 195–217.

Hamilton, W. A. 2003. Microbially influenced corrosion as a model system for the study of metal microbe interactions: A unifying electron transfer hypothesis. *Biofouling* 19 (1): 65–76. doi: 10.1080/0892701021000041078.

Harrison, J. J., R. J. Turner, L. L. R. Marques, and H. Ceri. 2005. Biofilms: A new understanding of these microbial communities is driving a revolution that may transform the science of microbiology. *Am. Sci.* 93 (6): 508–515.

Haveman, S. A., E. A. Greene, C. P. Stilwell, J. K. Voordouw, and G. Voordouw. 2004. Physiological and gene expression analysis of inhibition of *Desulfovibrio vulgaris* Hildenborough by nitrite. *J. Bacteriol.* 186 (23): 7944–7950. doi: 10.1128/JB.186.23.7944.

Hubert, C., M. Nemati, G. Jenneman, and G. Voordouw. 2005. Corrosion risk associated with microbial souring control using nitrate or nitrite. *Appl. Microbiol. Biotechnol.* 68 (2): 272–282. doi: 10.1007/s00253-005-1897-2.

Hubert, C., G. Voordouw, M. Nemati, and G. E. Jenneman. 2004. Is souring and corrosion by sulfate-reducing bacteria in oilfields reduced more efficiently by nitrate or nitrite? *NACE International Corrosion 2004 Conference & Expo*. Paper no. 04762.

Iino, T., K. Ito, S. Wakai, H. Tsurumaru, M. Ohkuma, and S. Harayama. 2015. Iron corrosion induced by nonhydrogenotrophic nitrate-reducing *Prolixibacter* sp. strain MIC1-1. *Appl. Environ. Microbiol.* 81 (5): 1839–1846. doi: 10.1128/AEM.03741-14.

Jiang, X., Y. Zhang, T. Zuo, A. I. Obike, D. Qu, X. Liu, S. Mu, and H. Zhang. 2012. Corrosion of mild steel and inhibition of two field-used inhibitors in the presence of elemental sulfur. *NACE International Corrosion 2012 Conference & Expo*. Paper no. C2012–0001064.

Johnston, S. L., and G. Voordouw. 2012. Sulfate-reducing bacteria lower sulfur-mediated pitting corrosion under conditions of oxygen ingress. *Environ. Sci. Technol.* 46: 9183–9190.

Johnston, S., S. Lin, P. Lee, S. M. Caffrey, J. Wildschut, J. K. Voordouw, S. M. Da Silva, I. A. C. Pereira, and G. Voordouw. 2009. A genomic island of the sulfate-reducing bacterium *Desulfovibrio vulgaris* Hildenborough promotes survival under stress conditions while decreasing the efficiency of anaerobic growth. *Environ. Microbiol.* 11 (4): 981–991. doi: 10.1111/j.1462-2920.2008.01823.15.

Kielemoes, I., P. De Boever, and W. Verstraete. 2000. Influence of denitrification on the corrosion of iron and stainless steel powder. *Environ. Sci. Technol.* 34 (4): 663–671. doi: 10.1021/es9902930.

King, R. A., and J. D. A. Miller. 1977. Corrosion of ferrous metals by bacterially produced iron sulphides and its control by cathodic protection. *Anti-Corrosion Methods Mater.* 24 (8): 9–11. doi: 10.1108/eb007036.

Kjellerup, B. V., R. H. Veeh, P. Sumithraratne, T. R. Thomsen, K. Buckingham-Meyer, B. Frølund, and P. Sturman. 2005. Monitoring of microbial souring in chemically treated, produced-water biofilm systems using molecular techniques. *J. Ind. Microbiol. Biotechnol.* 32 (4): 163–170. doi: 10.1007/s10295-005-0222-5.

Lambo, A. J., K. Noke, S. R. Larter, and G. Voordouw. 2008. Competitive, microbially mediated reduction of nitrate with sulfide and aromatic oil components in a low-temperature, western Canadian oil reservoir. *Environ. Sci. Technol.* 42: 8941–8946. doi: 10.1021/es801832s.

Lee, J. S., R. I. Ray, B. J. Little, and E. J. Lemieux. 2005. Evaluation of deoxygenation as a corrosion control measure for ballast tanks. *Corrosion* 61 (12): 1173–1188. doi: 10.5006/1.3278153.

Lee, J., R. Ray, E. Lemieux, A. Falster, and B. Little. 2004. An evaluation of carbon steel corrosion under stagnant seawater conditions. *Biofouling* 20 (4–5): 237–247. doi: 10.1080/08927010400013274.

Lee, J. S., R. I. Ray, B. J. Little, and E. J. Lemieux. 2006. An evaluation of ballast tank corrosion in hypoxic seawater. *NACE International Corrosion 2006 Conference & Expo.* Paper no. 06300.

Lee, W. C., and D. de Beer. 1995. Oxygen and pH microprofiles above corroding mild-steel covered with a biofilm. *Biofouling* 8 (4): 273–280. doi: 10.1080/08927019509378280.

Lee, W., and W. G. Characklis. 1993. Corrosion of mild steel under anaerobic biofilm. *Corrosion* 49 (3): 186–199. doi: 10.5006/1.3316040.

Lin, S., G. Voordouw, and D. Coombe. 2008. Sulfur production associated with souring control by nitrate injection: A potential corrosion risk? *NACE International Corrosion 2008 Conference & Expo*. Paper no. 08662.

Little, B., J. Lee, and R. Ray. 2007. A review of "green" strategies to prevent or mitigate microbiologically influenced corrosion. *Biofouling* 23 (1–2): 87–97. doi: 10.1080/08927010601151782.

MacDonald, D., B. Roberts, and J. B. Hyne. 1978. The corrosion of carbon steel by wet elemental sulphur. *Corros. Sci.* 18: 411–425. doi: 10.1016/S0010-938X(78)80037-7.

Machuca, L. L., L. Murray, R. Gubner, and S. I. Bailey. 2014. Evaluation of the effects of seawater ingress into 316L lined pipes on corrosion performance. *Mater. Corros.* 65 (1): 8–17. doi: 10.1002/maco.201206897.

Madigan, M. T., and J. M. Martinko. 2006. *Brock Biology of Microorganisms*. 11th edition. Upper Saddle River, NJ: Pearson Prentice Hall.

Mand, J., H. S. Park, T. R. Jack, G. Voordouw, and H. Hoffmann. 2014. Use of molecular methods (pyrosequencing) for evaluating MIC potential in water systems for oil production in the North Sea. *Society of Petroleum Engineers International Oilfield Corrosion Conference and Exhibition*, Aberdeen, Scotland, UK. Paper no. SPE-169638-MS.

Marty, F., H. Gueuné, E. Malard, J. M. Sánchez-Amaya, L. Sjögren, B. A., L. Quillet, M. C. M. van Loosdrecht, and G. Muyzer. 2014. Identification of key factors in accelerated low water corrosion through experimental simulation of tidal conditions: Influence of stimulated indigenous microbiota. *Biofouling* 30 (3): 281–297. doi: 10.1080/08927014.2013.864758.

Marty, F., M. van Loosdrecht, and G. Muyzer. 2014. Molecular characterization of microbial communities associated with accelerated low water corrosion (ALWC) on European harbor structures. In *Applications of Molecular Microbiological Methods*, eds. T. L. Skovhus, S. M. Caffrey, and C. R. J. Hubert. Norfolk, UK: Caister Academic Press.

Myhr, S., B. L. P. Lillebø, E. Sunde, J. Beeder, and T. Torsvik. 2002. Inhibition of microbial H_2S production in an oil reservoir model column by nitrate injection. *Appl. Microbiol. Biotechnol.* 58 (3): 400–408. doi: 10.1007/s00253-001-0881-8.

Nemati, M., G. E. Jenneman, and G. Voordouw. 2001. Impact of nitrate-mediated microbial control of souring in oil reservoirs on the extent of corrosion. *Biotechnol. Prog.* 17 (5): 852–859. doi: 10.1021/bp010084v.

Ray, R. I., J. S. Lee, B. J. Little, and T. L. Gerke. 2010. The anatomy of tubercles: A corrosion study in a fresh water estuary. *Mater. Corros.* 61 (12): 993–999. doi: 10.1002/maco.201005739.

Rempel, C. L., R. W. Evitts, and M. Nemati. 2006. Dynamics of corrosion rates associated with nitrite or nitrate mediated control of souring under biological conditions simulating an oil reservoir. *J. Ind. Microbiol. Biotechnol.* 33 (10): 878–886. doi: 10.1007/s10295-006-0142-z.

Rosli, R. N., C. Yoon-Seok, and D. Young. 2014. Impact of oxygen ingress in CO_2 corrosion of mild steel. *NACE International Corrosion 2014 Conference & Expo.* Paper no. 4299.

Schmitt, G. 1991. Effects of elemental sulfur on corrosion in sour gas systems. *Corrosion* 47 (4): 285–308.

Smith, L., and B. Craig. 2005. Practical corrosion control measures for elemental sulfur containing environments. *NACE International Corrosion 2005 Conference & Expo*. Paper no. 05646.

Soylemezoglu, S., and R. Harper. 1982. Oxygen ingress into geothermal steam and its effect on corrosion of low carbon steel at Broadlands, New Zealand. *Geothermics* 11 (1): 31–42. doi: 10.1016/0375-6505(82)90005-0.

Spruit, C. J. P., and J. N. Wanklyn. 1951. Iron/sulphide ratios in corrosion by sulphate-reducing bacteria. *Nature* 168: 951–952.

Stott, J. F. D., G. Dicken, T. Y. Rizk, C. Gamble, S. Saul, T. Granli, and T. Haegh. 2008. Corrosion inhibition in PWRI systems that use nitrate treatment to control SRB activity and reservoir souring. *NACE International Corrosion 2008 Conference & Expo*. Paper no. 08507.

Stott, J. F. D. 2013. Implementation of nitrate treatment for reservoir souring control: Complexities and pitfalls. *Society of Petroleum Engineers International Conference & Workshop on Oilfield Corrosion, Aberdeen, Scotland, UK.* Paper no. 155155. doi: 10.2118/155155-MS.

Sun, H., H. Fang, J. Davis, and R. Hudgins. 2011. Elemental sulfur corrosion and inhibition in the presence of sulfur solvent. *NACE International Corrosion 2011 Conference & Expo.* Paper no. 11125.

Tamburri, M. N., K. Wasson, and M. Matsuda. 2002. Ballast water deoxygenation can prevent aquatic introductions while reducing ship corrosion. *Biol. Conserv.* 103 (3): 331–341. doi: 10.1016/S0006-3207(01)00144-6.

Tang, K., V. Baskaran, and M. Nemati. 2009. Bacteria of the sulphur cycle: An overview of microbiology, biokinetics and their role in petroleum and mining industries. *Biochem. Eng. J.* 44 (1): 73–94. doi: 10.1016/j.bej.2008.12.011.

Usher, K. M., A. H. Kaksonen, and I. D. MacLeod. 2014. Marine rust tubercles harbor iron corroding archaea and sulphate reducing bacteria. *Corros. Sci.* 83: 189–197. doi: 10.1016/j.corsci.2014.02.014.

Venzlaff, H., D. Enning, J. Srinivasan, K. J. J. Mayrhofer, A. W. Hassel, F. Widdel, and M. Stratmann. 2013. Accelerated cathodic reaction in microbial corrosion of iron due to direct electron uptake by sulfate-reducing bacteria. *Corros. Sci.* 66: 88–96. doi: 10.1016/j.corsci.2012.09.006.

Videla, H. A., and W. G. Characklis. 1992. Biofouling and microbially influenced corrosion. *Int. Biodeterior. Biodegrad.* 29: 195–212. doi: 10.1016/0964-8305(92)90044-O.

Vik, E. A., A. O. Janbu, F. Garshol, L. B. Henninge, S. Engebretsen, C. Kuijvenhoven, D. Oliphant, and W. P. Hendriks. 2007. Nitrate-based souring mitigation of produced water—Side effects and challenges from the Draugen produced water reinjection pilot. *Society of Petroleum Engineers International Symposium on Oilfield Chemistry, Houston, T15.* Paper no. 106178.

Von Wolzogen Kühr, C. A. H., and L. S. Van der Vlugt. 1934. De grafiteering van gietijzer als electrobiochemich process inanaerobe gronden. *Water* 18: 147–165.

Wanklyn, J. N., and C. J. P. Spruit. 1952. Influence of sulphate-reducing bacteria on the corrosion potential of iron. *Nature* 169: 928–929.

16 Integrated Methodology to Characterize Microbial Populations and Functions across Small Spatial Scales in an Oil Production Facility

Christopher R. Marks, Joshua T. Cooper, Vincent Bonifay, Blake W. Stamps, Huynh M. Le, Brian H. Harriman, Annette De Capite, Kathryn R. Brown, Deniz F. Aktas, Jan Sunner, Bradley S. Stevenson, Kathleen E. Duncan, Michael J. McInerney, Yves Gunaltun, Pierre Souquet, and Joseph M. Suflita

CONTENTS

ABSTRACT

Petroleum reservoir microorganisms directly influence the economics of energy recovery operations through deleterious activities such as souring and biocorrosion. We investigated potential differences in chemistry and microbial populations between stagnant areas and bulk process fluids from three processing modules (low-pressure separator, desalter, and water separator) within an oil-processing facility. Samples were collected before and after the flushing of a sample valve at each site. Microbial community surveys were conducted through 16S rRNA gene sequencing and paired with analyses of fluid chemical composition through chromatographic methods and mass spectrum–based metabolite profiling. The production fluids were high saline brines (1.5–4.0 M [Cl^-]) with a pH range of 4.5–6.2, varying both within and between sampling sites. Sulfate (0.1–8.0 mM) and acetate (0.5–9.3 mM) were present in all samples, but nitrate was below detection levels. Marked chemical differences were observed between bulk fluids and stagnant dead-leg fluids at each sampling point. Stagnant dead-leg samples exhibited chemical stratification with a chloride gradient of ~3 M in the desalter module as well as sulfate and acetate gradients in the water separator module. Molecular surveys revealed that the microbial communities were predominantly composed of three taxonomic groups: *Halanaerobiales*, *Campylobacterales*, and *Desulfovibrionales*. An increase in species diversity was found in the stagnant fluids from each dead leg relative to the respective bulk fluid. MPN determinations of heterotrophic fermenting, sulfate-reducing, and thiosulfate-reducing organisms were all greater than 1×10^4 cells mL^{-1}. Targeted mass spectral analysis of the production fluids revealed the presence of catechols, phenols, and dihydrodiols, indicative of aerobic hydrocarbon biodegradation. Metabolites associated with anaerobic hydrocarbon biodegradation were not detected. Untargeted metabolomic screening revealed over 1000 identified compounds that mapped primarily to known lipid, carbohydrate, and amino acid metabolic pathways. Through the pairing of molecular microbial surveys and advanced metabolite profile analyses, these results suggest a systemic colonization of the facility by anaerobic halophiles, commonly associated with petroleum reservoirs. These organisms were likely cross-fed by organic electron donors produced during the transformation of hydrocarbons by aerobic microorganisms. Furthermore, these results highlight the importance of interrogating the small volumes of stagnated fluids within dead legs for an accurate assessment of the chemical and biological processes occurring within these problematic sites.

16.1 INTRODUCTION

Microorganisms colonize diverse habitats throughout the natural and engineered environments associated with the oil and gas industry. The metabolic processes of these organisms can have substantive economic and environmental impacts on the production and processing of petroleum through reservoir souring, product biofouling, and infrastructure corrosion. The latter process, also termed microbially influenced corrosion (MIC), can result in the hazardous release of hydrocarbons and brines to the surrounding environment. The association of corrosive sulfide formation and viable sulfate-reducing bacteria (SRB) in oil production fluids was established almost a century ago through the pioneering work of Bastin et al. (1926). Increased appreciation for the metabolic activities of the oilfield microflora led to the promulgation of formal standards for monitoring populations through the cultivation-based serial dilution technique (API 1965). Methods for the selective enumeration of microorganisms exhibiting specific metabolic traits (e.g., sulfate-reduction, acid production, etc.) have been expanded upon and remain an industry monitoring standard (NACE International 2014). Despite their widespread use, microbial population monitoring through growth in serial dilutions of selective cultivation media tends to underrepresent the size of a physiological group within a sample for two critical reasons: (i) only organisms capable of growth with the nutrients provided are assayed, and (ii) commercially available test formulations often do not adequately represent the sample under investigation. Enumeration media for acid-producing bacteria (APB), thiosulfate-reducing bacteria (TRB), and sulfate-reducing bacteria are typically selected on the basis of sample salinity and potentially other factors routinely determined by bulk fluid analyses. However, our study demonstrates that the chemical and biological characteristics of process fluids can vary dramatically over small spatial scales both within and between process modules. Thus, results obtained from cultivation-based monitoring of microbial populations in bulk fluids may not be representative of the communities in chronic problem areas, such as pipeline "dead legs." These are piping segments that may be continuously or intermittently exposed to bulk process fluids but generally receive limited or only intermittent flow. These areas are known to be sites of numerous incidents each year with localized corrosion as one of the primary causes of dead-leg integrity failures (Sloley 2011; Murata, Benaquisto, and Storey 2015).

In an effort to overcome such inherent limitations, cultivation-independent approaches for microbial monitoring, including quantitative polymerase chain reaction (qPCR) assays targeting diagnostic genes, microarray analyses, and 16S rRNA gene library sequencing, have become more widely accepted as part of an overall asset integrity assurance program (Maxwell, Hoffman, and Divine 2007; Eckert and Skovhus 2011). These methods utilize genomic DNA from field samples for a snapshot of the genetic potential and membership of microbial communities within the asset. Among the major advantages of these techniques are their universal applicability across different sample matrices, eliminating the need for multiple media formulations, and a generally shorter analysis time relative to growth-based assays (Maxwell, Hoffman, and Divine 2007). As with any assay, molecular methods also have their particular interpretational limits. The sample collection is from the total community, including both viable and actively metabolizing cells as well as inactive members. In addition, interrogation for specific genes is also limited by the specificity of the probe or primer utilized. Such probes are continuously improved as the discipline evolves; they are often initially designed based on reference molecular sequence data available in curated databases and may not possess the coverage or specificity required to make accurate determinations of the desired in situ microbial populations.

In addition to genomic data, fundamental information on the predominant pathways associated with aerobic and anaerobic hydrocarbon metabolism is consistently emerging. Methods for the survey and detection of metabolites are being developed and tested in many petroliferous environments. Metabolites can often be conclusively identified using mass spectral methods (e.g., GC-MS, HPLC-MS/MS, HPLC-qTOF) that can target "signature metabolites" indicative of specific bioconversions or global surveys of all small organic molecules within a sample matrix (Beller 2000; Gieg and Suflita 2005; Bonifay et al. 2013; Bian et al. 2015). As with the molecular methods, results are interpreted relative to curated databases (e.g., Kyoto Encyclopedia of Genes and Genomes; for an additional listing, see the Metabolomic Society) that link information on genes and transformation products within metabolic pathways. Such information is rapidly growing as the techniques are applied in an increasing number of studies and environments (Kanehisa et al. 2012). Recently, mass spectral–based metabolite profiling was used in conjunction with other techniques to investigate microbial functioning and corrosion processes in a North Slope pipeline system by interrogating resident communities that were planktonic (Duncan et al. 2009), differentially dislodged during pigging operations (Stevenson et al. 2011), and pipeline section cut-outs (e.g., "cookie" samples) (Lenhart et al. 2014). Collectively, the resulting information implicated anaerobic hydrocarbon biodegradation in supporting corrosive biofilms driving MIC within a high-temperature oil production pipeline. In consistent fashion, oilfield-associated microbial assemblages were found to metabolize crude oil under the conditions prevalent in the same reservoirs (Gieg et al. 2010). The integration of field and laboratory samples through the strategic combination of traditional, molecular, and metabolite analyses can, in our opinion, lead to substantive insights into the predominant mechanisms influencing important ecological processes in the oil and gas sector.

Our investigation sought to evaluate the potential for the involvement of microorganisms in the corrosion of carbon steel infrastructure within an oil-processing facility through the integration of geochemical analyses, molecular surveys, and metabolite profiling of production brines. The study site is a facility handling production from a sour oilfield experiencing aggressive corrosion and pipe failures at dead legs in separator oily water waste lines (Gunaltun and Kumar 2014). The asset receives high-salinity fluids (2.8–4.2 M Cl^-; 100–150 g L^{-1} Cl^-) as well as H_2S and CO_2. Asset managers estimated corrosion rates to be ~2–40 mpy (0.05–1 mmpy) based on fluid composition and operational parameters though rates as high as 315 mpy (8 mmpy) were documented in this system (personal communication). Despite the relatively high salinity of the produced fluids, sulfate-reducing and thiosulfate-reducing microorganisms were historically and repeatedly detected in the bulk separator fluid waste lines through cultivation-based enumerations. Through the combination of chemical and microbiological analyses, the role of partial or complete stagnation in microbial community composition and function were assessed, and a potential model for the involvement of microorganisms in the failures of infrastructure integrity is proposed. The information emanating from the data collection and analyses exemplified here may ultimately lead to the generation of models for the a priori evaluation of mitigation strategies to control deleterious microorganisms.

16.2 EXPERIMENTAL PROCEDURES

16.2.1 Site Description and Sampling

Fluid samples were obtained from near stagnant areas of dead leg sampling points at the 6 o'clock position of 16 inch (40.64 cm) oily water disposal lines from three processing modules within the facility (Figure 16.1). These included an oil–water separator, a low-pressure separator, and a desalter (Figure 16.1). Average brine composition throughout the facility was reported by the asset manager as (mg L^{-1}): 144718 Cl^-, 493 SO_4^{2-}, 159 HCO_3^-, 0 CO_3^{2-}, 86428 Na^+, 5577 K^+, 2270 Ca^{2+}, 2087 Mg^{2+}, 2 Fe^{2+}, 563 Sr^{2+}, and 15 Ba^{2+}. The bulk fluid and dead leg temperatures were reported as 50°C and 30°C, respectively. System pressure ranges between 1 and 3 bar_g with H_2S and CO_2 concentrations of 4% and 3% mole-basis, respectively. Near-stagnant fluids were obtained by collecting three successive aliquots of 30 mL samples into separate screw cap vials from each oily water line. The bulk fluid sample (1 L) was obtained from the same sampling points after flushing several liters of production water through the sampling valve. All samples were stored at 4°C during shipping and until use.

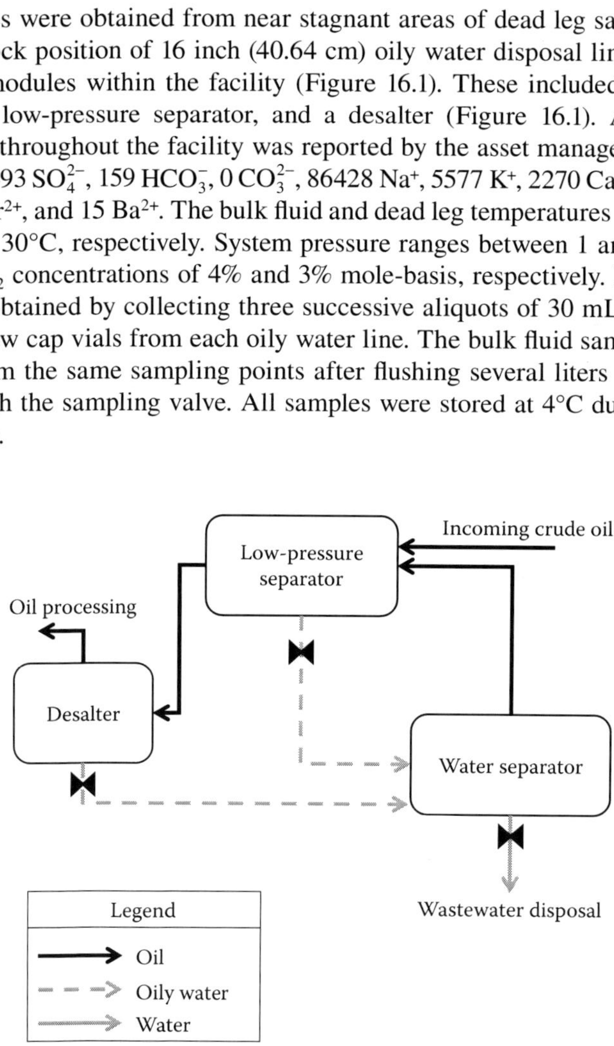

FIGURE 16.1 Schematic diagram depicting simplified connectivity of selected processing modules in an oilfield production facility. Sampling points are designated by black triangles.

16.2.2 Chemical Analyses

Sample fluids were analyzed for various chemical characteristics, including pH, alkalinity and common anion (chloride, nitrate, sulfate, and acetate) concentration. The pH of bulk fluid samples was determined with a pH meter (Accumet® Basic AB15; Fisher Scientific, Pittsburgh, PA) at 18°C. The limited volume of the dead-leg sample necessitated the addition of a small aliquot (~20 μL) to a pH test strip at 18°C. Alkalinity was determined by titration for the bulk fluid samples using a Hach Alkalinity Test Kit (Hach Company, Loveland, CO) as described by the manufacturer. Inorganic anions (chloride, nitrate, and sulfate) were quantified by suppressed ion chromatography on a Dionex ICS-3000 system (Dionex, Sunnyvale, CA) equipped with Dionex AS4A-SC guard (4 mm × 50 mm) and analytical (4 mm × 250 mm) columns. Separations were performed isocratically at mobile phase of carbonate buffer (1.7 mM HCO_3^-; 1.8 mM CO_3^{2-}; 2.0 mL min^{-1}) and a suppression current of 35 mA. Chloride was determined on a 1:10,000 dilution of sample fluids. Samples for nitrate and sulfate analysis were pretreated with Dionex OnGuardII Ag and Na cartridges to remove interfering halides as instructed by the manufacturer. Acetate was quantified on a high-pressure liquid chromatograph system (Beckman System Gold; Solvent Delivery Module 126, Detector Module 166, and NEC PC-8300 controller) equipped with an organic acid column (Prevail; 5 μm; 4.6 mm × 250 mm; Grace Davidson Discovery Science, Deerfield, IL) and operated isocratically with a potassium phosphate buffer mobile phase (25 mM; pH 2.5; 1.0 mL min^{-1}) and detection by UV absorbance at 214 nm.

16.2.3 DNA Extraction, Amplification, and Analyses

Bulk fluid samples were filtered (250 mL; 0.45 micron pore size polyethersulfone filter) in the field. The filters were immediately treated with 1 mL DNAzol (Molecular Research Center, Inc., Cincinnati, OH) to aid cell lysis and prevent nucleic acid degradation, shipped to the laboratory, and frozen at −80°C until required. Prior to DNA extraction, the filters were briefly thawed, incubated with proteinase K (Qiagen, Venlo, Limburg) for 15 minutes, and rinsed by vortexing successively with nuclease-free water, RNA lysis buffer, and RNA dilution buffer from the Maxwell®16 (Promega, Madison, WI) Tissue LEV Total RNA Purification Kit (AS1220). The rinses were pooled and transferred to AS1220 cartridge for DNA extraction. Modifications to the manufacturer's standard protocol were used to extract DNA as previously described (Oldham et al. 2012). Amicon 30K Ultra-15 centrifugal filter units (EMD Millipore, Billerica, MA) were used to concentrate 10 mL of each dead-leg sample and to decrease the sample salinity. The eluted DNA was cleaned to remove potential PCR inhibitors using the MO BIO PowerDNA Cleanup Kit (MO BIO Laboratories, Inc., Carlsbad, CA).

The population densities of total bacteria/archaea and selected physiological groups in the bulk fluid samples were estimated using qPCR analysis. Bacterial and archaeal 16S rRNA genes were targeted using the primer pairs Bac27F/338R (Stevenson et al. 2011) and Arc333F/958R (Reysenbach and Pace 1995), respectively. The genetic potential for sulfate reduction and dissimilatory sulfide production were assessed by the quantification of the adenoside-5'-phosphosulfate reductase gene (*apsA*) by RH1apsF/RH2apsR and the dissimilatory sulfite reductase gene (*dsrA*) by RH1-dsr-F/RH3-dsr-R, respectively (Ben-Dov, Brenner, and Kushmaro 2007). The StepOnePlus™ Real-Time PCR System was used for thermal cycling and StepOne Software v2.1 (Life Technologies Carlsbad, CA) for data acquisition and analysis. Thermal cycling conditions were as described in the references for the primers. A 1:10 dilution series of a control DNA plasmid containing a reference gene insert for each assay was used to generate a 5- to 7-point standard curve. Standards were assayed in duplicate and the samples in triplicate. Libraries of amplified 16S rRNA genes were prepared as described by Liang et al. (2016). Samples were sequenced on the Illumina MiSeq using V2 PE250 chemistry (Illumina, Inc., San Diego, CA).

16.2.4 Bioinformatic and Statistical Methods

Sequences were joined using the computer program PEAR 0.9.51 (Zhang et al. 2014) prior to being processed using QIIME 1.9.1 (Caporaso et al. 2010) and UPARSE v7.0.1090 (Edgar 2013). Taxonomy was assigned using mothur 1.36.0 (Schloss et al. 2009) and the SILVA r119 SSU RNA gene database (Pruesse et al. 2007) formatted for use in QIIME. Operational taxonomic units (OTUs), defined at 97% similarity, that represented greater than 5% of any library were further aligned and classified within SINA (Pruesse, Peplies, and Glöckner 2012) using the SILVA r123 database and ARB 6.0-rc3 (Ludwig et al. 2004). Alpha diversity measures included Shannon's diversity index, Shannon's evenness, and the number of observed OTUs. Beta diversity was calculated as a weighted Unifrac distance matrix (Lozupone and Knight 2005).

Correlation analyses were calculated as Spearman's ranked correlation coefficients using the vegan package in R (Oksanen et al. 2016). Redundancy analysis ordinations (RDA) were computed for both community compositional and untargeted metabolite profile data through the vegan R package within the GUSTA ME web-based application (Buttigieg and Ramette 2014; Oksanen et al. 2016). Analyses of community compositions were performed on relative OTU abundance for bulk fluid and dead leg #2 samples from each module. Comparisons of untargeted metabolite profiles were conducted on calculated relative peak areas for each resolved feature in the bulk fluid and all dead-leg samples from each module. RDA ordinations were based on the Bray–Curtis distance metric and were constrained by Z-score transformed values of explanatory variables measured for each respective sample (pH, salinity, [sulfate] and [acetate]) with Type I scaling.

16.2.5 Metabolite Extraction and Analyses

The organic constituents in sample fluids were investigated by mass spectrometry using both targeted and untargeted metabolite profiling. The targeted assessment of "signature metabolites" diagnostic of selected microbial metabolic processes was conducted on the bulk fluid samples. A subsample of each brine (1 L) was acidified in the field (pH ≤2; 6N HCl), extracted with three aliquots of ethyl acetate (200 mL), and the organic fractions were separated and retained. The pooled organic fractions were dried over anhydrous sodium sulfate and concentrated to a volume of 100 μL under flowing N_2. Concentrated organic extracts were reacted with *N,O*-bis(trimethylsilyl) trifluoracetamide to yield trimethylsilyl-derivatized compounds. Metabolites were separated and analyzed by GC-MS, and putative identifications were made by comparisons to authentic derivatized standards and the NIST library (Elshahed et al. 2001; Gieg and Suflita 2002; Gieg and Suflita 2005; Duncan et al. 2009).

A more global survey of organic components of both bulk fluid and dead-leg brines was conducted through an untargeted mass spectral analysis. An aliquot from each sample (5 mL) was acidified (pH ≤2), extracted with an equal volume of ethyl acetate, and the solvent was evaporated to dryness under N_2, reconstituted with isopropanol (100 μL), and analyzed by HPLC (Agilent 1290) interfaced to a high-resolution MS (Agilent G6538A quadrupole time-of-flight [QToF] MS). A 5-μL injection was made onto a zic-HILIC analytical column for the separation when in positive ion detection mode and onto a reverse phase C_{18}-column when analyzing in negative ion mode. Analysis using this combination of columns has, in our hands, proven effective at increasing the array of compounds detected at usable and reliable abundances.

Mass spectrometry data was turned into feature information by processing through the IDEOM v.19 workflow (Creek et al. 2012), using XCMS centWave (Tautenhahn, Böttcher, and Neumann 2008) for peak detection and mzMatch.R (Scheltema et al. 2011) for alignment of samples. Detected features were matched against the Kyoto Encyclopedia of Genes and Genomes (KEGG) metabolite database (Kanehisa et al. 2012). Statistical analyses applied to global putative metabolite profiles were performed by a metabolomics package in R (De Livera and Bowne 2015). Putatively identified metabolites were mapped to known metabolic pathways using Pathos (Leader et al. 2011).

16.3 RESULTS

16.3.1 Chemical Characteristics of Sample Fluids

The physicochemical properties of bulk and dead-leg fluids from production waters of three oil-processing modules (Figure 16.1) were interrogated by chromatographic and titration methods. All samples were found to be high-salinity brines with varying pH and total chloride concentrations between bulk and dead-leg samples within and between modules (Table 16.1). Bulk fluid pH and acetate concentrations were very similar between processing modules with an average of 6.03 ± 0.20 and 1.04 ± 0.03 mM, respectively. The pH values agree with those predicted by asset managers of 5.91–6.36 based on bulk fluid chemical modeling estimations (personal communication). The salinity and sulfate concentrations of the low-pressure separator (4.1 M Cl^-; 4.81 mM SO_4^{2-}) and water separator (4.5 M Cl^-; 4.50 mM SO_4^{2-}) bulk fluids were also very similar though they were dramatically higher than that observed in the desalter fluids (1.5 M Cl^-; 1.03 mM SO_4^{2-}). Nitrate was below detection in all samples retrieved from the oil production platform.

In contrast with the general similarities in the circulating bulk fluids, the relatively stagnant samples in the dead legs exhibited apparent gradients in their characteristics. The desalter dead-leg sample was highly stratified with respect to salinity with a

TABLE 16.1
Characterization of the Physicochemical Properties of Bulk and Dead-Leg Production Fluid Samples

	pH	Salinity [Cl^-] (M)	[SO_4^{2-}] (mM)	[Acetate] (mM)
		Desalter		
Bulk	6.2	1.5	1.03	1.01
Dead leg #1	5.2	4.0	3.17	3.29
Dead leg #2	6.0	3.4	2.10	6.63
Dead leg #3	6.0	1.7	1.18	6.87
		Low-Pressure Separator		
Bulk	5.8	4.1	4.81	1.02
Dead leg #1	4.5	3.1	0.20	7.58
Dead leg #2	4.5	3.1	0.11	9.27
Dead leg #3	4.5	2.9	0.16	9.03
		Water Separator		
Bulk	6.2	4.5	4.50	1.08
Dead leg #1	6.1	4.0	7.98	0.59
Dead leg #2	6.1	4.0	4.15	1.23
Dead leg #3[a]	ND	ND	ND	ND

[a] Values not determined (ND) for water separator dead leg #3 due to high oil fraction.

gradient ranging from 4.0 M (closest to the bulk fluids) to 1.7 M Cl^- near the sampling port. Corresponding gradients in sulfate (3.17 to 1.18 mM), acetate (3.29 to 6.87 mM), and pH (~5.0 to 6.0) were also evident in the desalter dead-leg fluids (Table 16.1). The stagnant fluids in the low-pressure separator dead legs are starkly different in physicochemical properties than the bulk fluids though no gradation is apparent. The pH of the dead-leg fluid is lower than the bulk brine (pH = 4.5 versus 5.7) coincident with the increase in acetate concentration (~7- to 9-fold) greater than that measured in the bulk production fluid from that module. Finally, the water separator samples did not show a gradation in pH or salinity within the dead leg and are representative of the bulk fluids. No chemical data could be determined for water separator dead-leg sample #3 due to the high oil content. Within this dead leg, there was also an inverse correlation between sulfate loss and acetate production in samples #1 and #2. Interestingly, sample #2 was very similar in chemical properties to the bulk fluid and may represent mixing of the bulk process fluids during the sampling event.

Relationships between brine properties were investigated by calculating Spearman correlation coefficients for each parameter across all samples (Figure 16.2). Several parameters exhibited strong positive or negative correlations, including salinity and sulfate (0.78), pH and acetate (–0.79), and between acetate and sulfate (–0.81). A weaker negative correlation was observed between flow and acetate (–0.58). The relationships between flow, acetate, and sulfate suggests that the partial stagnation in the dead legs may be coincident with acetogenesis potentially linked to sulfate utilization.

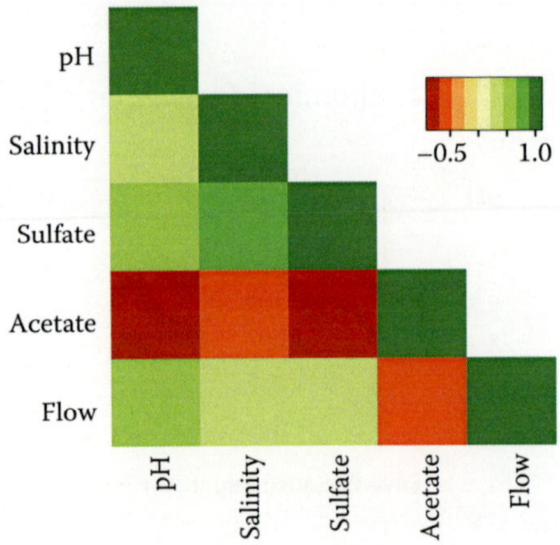

FIGURE 16.2 Brine chemical characteristic correlation analysis showing Spearman ranked correlation coefficients between measured fluid properties. Increasingly positive correlations are shown in green, negative shown in red, and no observed correlation is indicated in yellow.

16.3.2 Microbial Community Characterization

The microbial community composition was investigated by quantifying genetic markers by qPCR. The trends in population structure and membership were assessed via high-throughput sequencing of 16S rRNA genes. Amplicons were successfully produced from all bulk fluid samples and several of the dead-leg samples despite the relatively small sample volume (10 mL) available for analysis. The total microbial population in each of the bulk fluid samples was quantified based on qPCR assays targeting the bacterial and archaeal 16S rRNA genes. Microbial population density ranged from 10^5 to 10^8 copies 16S rRNA gene mL^{-1} process brine with the largest populations found in the desalter bulk fluid (Table 16.2). No archaeal 16S rRNA gene or *mcrA* gene amplicons and no mitochondrial 16S rRNA genes (fungal or other) were obtained from any of the samples assayed, suggesting that microbial populations were composed exclusively of bacteria. The genetic potential for sulfate reduction was quantified by *aprA* abundance and found to be present in all bulk fluid samples ranging from 10^1 to 10^4 copies *aprA* gene mL^{-1} brine with the greatest abundance in the low-pressure separator. The potential for dissimilatory sulfide production was quantified by *dsrA* genes that were found to be present in all bulk fluid samples ranging from 10^4 to 10^6 copies *dsrA* gene mL^{-1} brine with the greatest abundance also in the low-pressure separator. The gene abundance for *dsrA* was several orders of magnitude greater than *aprA* within each of the bulk fluids, suggesting that sulfoxy-anion utilization other than sulfate may potentially support more biological sulfide production.

The diversity and composition of each of the microbial communities in the facility were interrogated through the analysis of 16S rRNA gene amplicon libraries for the bulk fluid and dead-leg sample #2 for each processing module. Metrics of community diversity for each library are shown in Table 16.2. In each module, the dead-leg community had greater species richness (number of OTUs observed at the rarified

TABLE 16.2
Summary of Community Diversity Metrics and Population Sizes for Bulk Fluid and Dead Leg #2 Brines from Each Module

	Desalter		Low-Pressure Separator		Water Separator	
	Bulk	**Dead Leg #2**	**Bulk**	**Dead Leg #2**	**Bulk**	**Dead Leg #2**
Diversity Metrics						
$OTU_{0.03}$[a]	73	91	83	134	87	98
Shannon evenness	0.34	0.54	0.52	0.80	0.52	0.68
Shannon diversity	2.11	3.53	3.33	5.62	3.33	4.51
qPCR Analyses[b]	**Desalter Bulk**		**Low-Pressure Sep. Bulk**		**Water Sep. Bulk**	
Bacterial 16S	3.70×10^8 (2.31×10^8)		5.98×10^5 (5.22×10^4)		1.17×10^5 (7.78×10^4)	
Archaeal 16S	ND		ND		ND	
aprA	1.27×10^4 (1.15×10^4)		1.79×10^3 (3.26×10^2)		8.86×10^1 (4.07×10^1)	
dsrA	1.20×10^6 (1.41×10^5)		4.43×10^4 (5.61×10^3)		1.65×10^4 (7.14×10^3)	
mcrA	ND		ND		ND	

[a] Number of OTUs observed within each sample. Libraries were subsampled without replacement to 1400 sequences.

[b] Quantitative PCR values reported as copies of target gene mL^{-1} brine. Target genes are Bacterial 16S rRNA gene (Bacterial 16S); Archaeal 16S rRNA gene (Archaeal 16S); Adenosine-5'-phosphosulfate reductase subunit A (*aprA*); Dissimilatory sulfite reductase subunit A (*dsrA*); Methylcoenzyme-M reductase (*mcrA*). Values reported are averages of three technical replicate measures with standard deviation in parentheses. ND, not detected.

sequencing depth), evenness, and total diversity relative to the corresponding bulk fluid microbial assemblages. The low-pressure separator dead leg harbored the most diverse microbial population with respect to both number of taxa detected and the evenness of their abundance distribution despite having the lowest measured pH value of 4.5.

Analysis of entire microbial communities based on the taxa observed in each sample as well as their relative abundance in each library allowed for the comparison of compositional differences. Pairwise dissimilarities calculated between each of the bulk fluid and dead leg #2 sample are shown in Figure 16.3. The order-level taxonomic composition of each microbial population revealed that members of the *Halanaerobiales* accounted for ~50% of all reads obtained in this study and, along with the *Campylobacterales* and *Desulfovibrionales*, were present at substantive levels (>1.5% relative abundance) in all samples (Figure 16.3). The low-pressure separator bulk fluid community was predominantly composed of *Halanaerobiales* (73.4%), *Desulfovibrionales* (9.4%), and *Campylobacterales* (3.6%) while the more evenly distributed and species rich dead leg was populated by the *Coriobacteriales* (10.8%), *Halanaerobiales* (9.7%), *Flavobacteriales* (7.6%), *Campylobacterales* (5.2%), and *Desulfovibrionales* (1.8%). Similarly, *Halanaerobiales* (66.6%), *Campylobacterales* (6.6%), and *Desulfovibrionales* (4.0%) were the most abundant taxa in the water separator bulk fluid assemblage. As in the low-pressure separator but unlike the respective bulk fluids, the water separator dead-leg community is highly enriched in members of the *Gammaproteobacteria* with unclassified *Gammaproteobacteria* (28.0%), *Bacillales* (12.0%), *Halanaerobiales* (9.1%), and *Pasteurellales* (7.9%) among the most abundant orders. Despite the strong similarities in bulk fluid community structure and membership between the low-pressure and water separator communities (95.1% pairwise similarity), partial stagnation did not enrich for similar microbial populations within the dead legs of these modules. In contrast, the desalter bulk fluid community harbored relatively few taxa with the *Campylobacterales* (71.5%) as the dominant taxonomic group along with the *Halanaerobiales* (16.1%) and the *Desulfovibrionales* (1.6%). This was distinctly different than all other analyzed assemblages with pairwise dissimilarities exceeding 19.7% in all cases. For comparison, the dead-leg community at the desalter is highly similar to the bulk fluids of the low-pressure and water separators (pairwise dissimilarities of 2.9% and 4.6%, respectively) and was again primarily populated by *Halanaerobiales* (67.6%), *Desulfovibrionales* (11.1%), and *Campylobacterales* (5.6%).

Further phylogenetic analysis was performed on select ubiquitous OTUs present in each sample to provide increased taxonomic resolution. Two closely related OTUs (0.03% pairwise distance) classified within the *Halanaerobiales* (OTU 2 and OTU 296) accounted for the majority of sequences affiliated with this order observed in this study. *Halanaerobium congolense* is the closest described type strain to OTU 2 (Figure 16.4a), and relative abundances ranged from 5.1% in the water separator dead leg to 53.8% in the water separator bulk fluid. In contrast, OTU 296 was most abundant in the desalter dead leg (33.2%) and low-pressure separator bulk fluid (32.7%) with the lowest abundance found in the low-pressure separator dead leg (0.28%) and water separator dead leg (0.31%). This OTU is equivalently related to *H. saccharolyticum* and *H. lacusrosei* (Figure 16.4a). The *Campylobacterales* are represented by OTU 1, affiliated with *Sulfurospirillum arcachonense* (Figure 16.4b), which accounts for 3.4% to 71.4% relative abundance in the low-pressure separator bulk

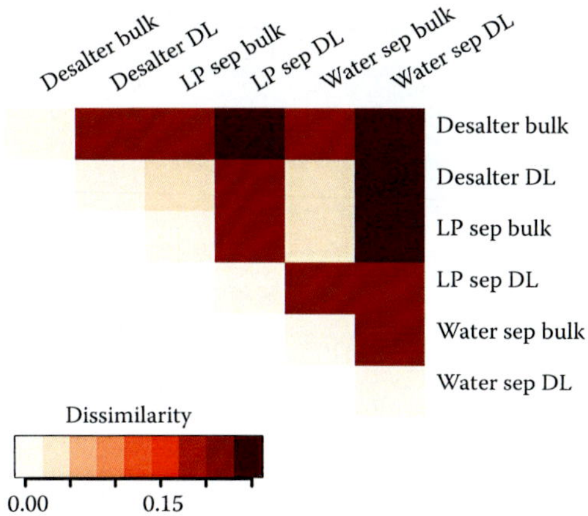

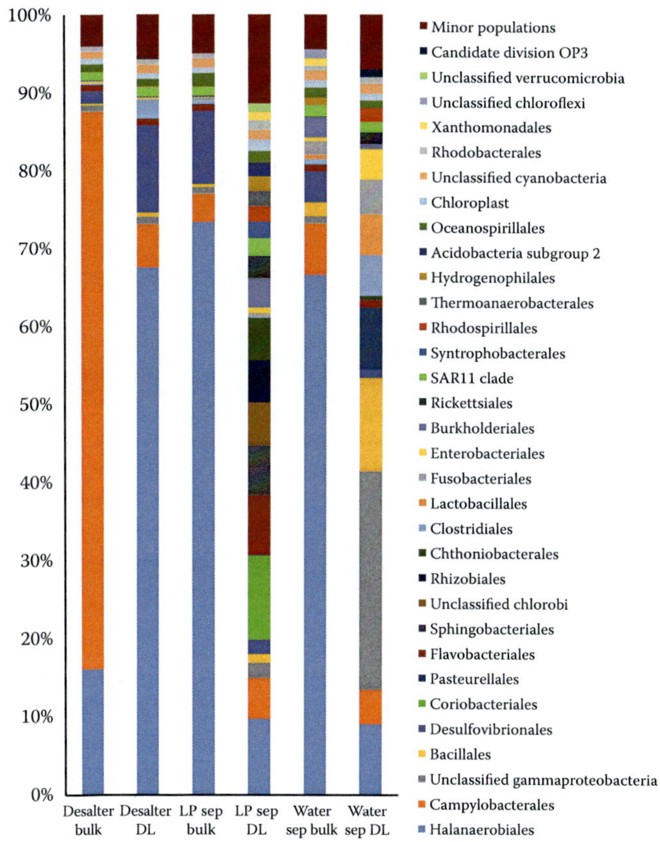

FIGURE 16.3 Top: Pairwise dissimilarity matrix heat map depicting Unifrac pairwise distances between microbial communities in each sample. Greater pairwise dissimilarity is shown by increasing red shading. Bottom: Relative abundance of order-level phylogenetic composition of brine communities.

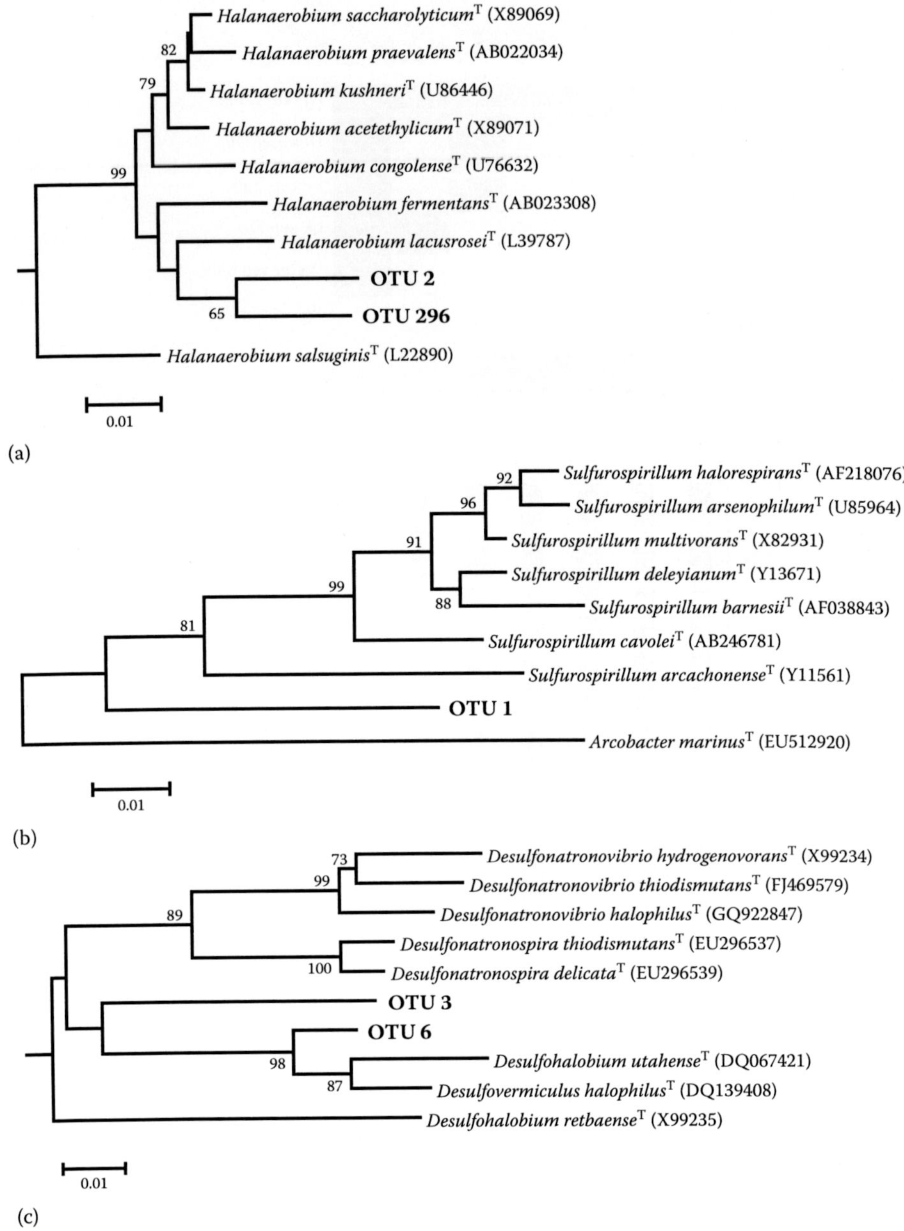

FIGURE 16.4 Phylogenetic analyses of selected abundant OTUs from the (a) *Halanaerobium* spp., (b) *Sulfurospirillum* spp., and (c) *Desulfohalobiaceae* spp. Neighbor-joining dendrograms were constructed from alignments of the OTU representative sequences with full-length type strains using the Tajima-Nei model with pairwise deletion in MEGA6 (Tamura et al. 2013). Genbank accession numbers are designated in parentheses and sequences obtained in this study are designated in bold. Bootstrap values (n = 500 replicates) greater than 65 are shown.

fluid and desalter bulk fluid, respectively. Finally, OTUs 3 and 6 are the most abundant representatives of the *Desulfovibrionales* and are most closely related to members of the *Desulfohalobiaceae*, *Desulfonatronovibrio thiodismutans*, and *Desulfohalobium utahense*, respectively (Figure 16.4c). Both OTUs 3 and 6 follow similar enrichment patterns between samples having the highest abundance in the desalter dead-leg and low-pressure separator bulk fluids and the lowest in the water separator dead leg brines.

Redundancy analysis was performed to examine the relationship between the geochemical parameters measured for each sample and the community composition, based on OTU relative abundance (Figure 16.5a). The explanatory variables

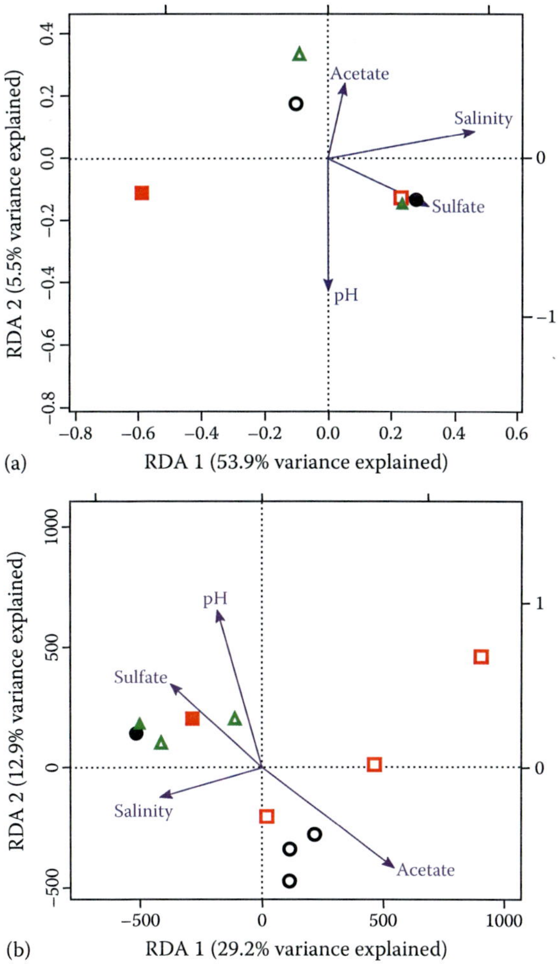

FIGURE 16.5 Redundancy analyses (RDA) of (a) community composition and (b) untargeted metabolite profiles. Ordinations are constrained by environmental parameters (Table 16.1) for each sample, shown in blue vectors. Bulk fluid and dead-leg samples are indicated by closed and open markers, respectively. Desalter, low-pressure separator, and water separator samples are shown in red squares, black circles, and green triangles, respectively.

pH, salinity, acetate, and sulfate concentrations accounted for 61.66% of the total community data set variance, and components RDA 1 and RDA 2 together explained 59.3%. Salinity is strongly correlated to RDA 1 and pH to RDA 2 with biplot scores of 0.91 and −0.82, respectively. The ordination shows that highly similar brine chemistries have enriched very similar microbial populations within the desalter dead leg #2 and the low-pressure and water separator bulk fluids despite the difference in fluid flow.

16.3.3 Metabolite Analyses

Metabolite profiling was performed on each of the brine samples by HPLC-QToF mass spectrometry to survey the low molecular weight organic compounds (presumably putative metabolites) present in each sample while targeted analyses for compounds indicative of selected microbial catabolic processes was by GC-MS. The untargeted survey resolved 10,115 chemical features and tentatively identified 1,112 compounds based on the comparison of exact masses to a reference database. The brines contain a suite of primary aliphatic acids, ranging from C_{10} to greater than C_{22}, and numerous organonitrogen compounds (e.g., amino acids, amines, and amides). Features annotated as $C_{12}H_{17}N$ and $C_{14}H_{21}N$ were detected in most of the sample fluids but were prominent components of the water separator bulk and dead-leg fluids. These features were not able to be identified as specific compounds but may be secondary or tertiary amines. Two compounds associated with carbohydrate metabolism (tartaric acid and L-arabinonate) were found in high abundance in the dead-leg samples but were not detected in any of the bulk fluid samples nor water separator dead leg #2.

Redundancy analysis was conducted on metabolite profile feature relative peak areas for each sample and constrained by the corresponding geochemical parameters: pH, salinity, acetate, and sulfate concentrations (Figure 16.5b). The explanatory variables represented for 61.74% of the total community data set variance and components RDA 1 and RDA 2 accounted for 42.1%. Acetate concentration showed the strongest correlation with RDA 1 and pH with RDA 2 with biplot scores of 0.79 and 0.95, respectively. RDA indicated essentially identical metabolite profiles were observed for the low-pressure separator and water separator bulk fluids, and the water separator dead leg #2 was similar to these bulk brines. Low-pressure separator dead-leg samples contain similar metabolite composition and have similar geochemical characteristics despite a gradient in acetate concentrations. The metabolite profiles for the desalter dead-leg brines are highly variable and are strongly dissimilar between each other (in agreement with strong chemical gradation) and most other samples interrogated. Overall, ordination analysis revealed that despite some geochemical differences, the bulk fluids within these processing modules had a largely similar metabolite profile in contrast to their microbial assemblages. Conversely, dead-leg fluid metabolite compositions are highly variable within and between processing units.

Targeted metabolite profiling was used to screen for hydrocarbon biotransformation products in the bulk fluid samples. All bulk fluid samples contained alkylated monoaromatic hydrocarbons, aromatic alcohols (e.g., cresols, phenols) and a suite of alkanoic acids. Alkyl-, benzyl-, and naphthylmethyl-succinic acids indicative of anaerobic hydrocarbon activation by fumarate addition were not detected in any sample. Alkylated catechols and other benzenedihydrodiols were identified in the desalter bulk fluid, indicating aerobic biodegradation of monoaromatic hydrocarbons at or near this site.

16.4 DISCUSSION

This case study represents an integration of traditional and modern investigative tools to assess the contribution of microbial activity to carbon steel corrosion in dead-leg areas within several processing modules in a sour oilfield facility. Analysis of brine chemistry, molecular surveys of microbial community composition and functional potential in conjunction with metabolite-profiling efforts revealed compositional differences, both chemical and biological, within and between the flowing bulk fluids and partially stagnated dead-leg areas in each of the oil-processing modules. The interconnectedness of the various processing modules (Figure 16.1) was verified by the chemical properties of the bulk fluids from each module (Table 16.1)—that is, the bulk oily processing waters were generally very similar in their chemical characteristics with the expected exception of decreased salinity and sulfate content in the desalter. Despite the similarity in bulk fluid properties, the partially stagnated brines in the dead legs had significantly different and distinct chemical characteristics (Table 16.1; Figure 16.3). The inverse correlation between sulfate loss, acetate formation, and pH decrease in the low-pressure separator and desalter dead legs suggested that biological sulfate reduction may have occurred over small spatial scales.

Molecular surveys identified microbial assemblages inhabiting the brine in both the bulk and dead-leg fluids in all modules. Despite the similar chemical characteristics in the low-pressure separator and the water separator, partial stagnation did not promote the enrichment of a common microbial community in the respective dead legs on those modules. The dead-leg communities were found to be more diverse than the respective bulk fluid in each module (Table 16.2). This increase in species richness and evenness may be promoted by stagnation, removing shear forces at the pipe wall as a selective pressure, and/or the decrease in temperature of 20°C in the dead legs relative to the bulk process fluids. Throughout the oilfield processing facility, dead legs primarily served as monthly sampling points for system monitoring, but sampling efforts on individual modules need not be coordinated (personal communication). Given the potential differences in fluid residence time between these dead legs, it is possible that these samples represent different points along a shared successional trend toward a common microbial community. Though this trend is speculative at this time, it constitutes an area deserving of further research and may yield valuable information regarding the rise of deleterious communities within dead legs.

Regardless of structure and diversity differences between oilfield module communities, members of a few select taxonomic orders, *Halanaerobiales*, *Campylobacterales*, and *Desulfovibrionales*, were ubiquitous in all samples. Members of the *Halanaerobiales* accounted for a substantive fraction and often the largest portion of each community interrogated except the desalter bulk fluids. These anaerobic halophiles are commonly detected and isolated from petroleum-impacted environments, including oil reservoirs and production infrastructure (Bhupathiraju et al. 1994; Ravot et al. 1997; Bhupathiraju et al. 1999; Neria-González et al. 2006; Sette et al. 2007; Gales et al. 2011; Liang et al., 2016). Members of the *Halanaerobiales* are well known halophiles that thrive in highly saline habitats and typically produce acetate as a major fermentation end product of complex organic substrates, including carbohydrates and yeast extract. To date, no species has been shown to utilize sulfate as an electron acceptor though many have the capacity to reduce thiosulfate to sulfide during heterotrophic growth (Whitman et al. 2015). Despite their common association with petroleum-laden environments, they are not known to metabolize hydrocarbons, and the ecological role of these organisms in these habitats remains unknown. A previous study of corrosive biofilms on steel coupons in a Mexican oilfield pipeline revealed the presence of taxa related to the genus *Halanaerobium* in conjunction with the more predominant *Enterobacteriaceae* at a brine salinity of 60–80 g L^{-1} NaCl (Neria-González et al. 2006). More recently, Liang et al. found highly saline (77–200 g L^{-1} NaCl) production fluids from hydraulically fractured natural gas wells harbored a microbial community dominated by *Halanaerobiales* (Liang et al. 2016). Subsequent corrosion studies conducted on the isolate *Halanaerobium* sp. DL-01 implicated these taxa in the corrosion of carbon steel infrastructure through the production of acetate and sulfide from the metabolic degradation of guar gum coupled to the reduction of thiosulfate within the formation. Murali Mohan et al. (2013) examined the changes in chemistry and microbial populations in production fluids with time after well completion in the Marcellus Shale formation. The salinity of produced brines increased from ~0.2 to 2.6 M NaCl (10 to 150 g L^{-1} NaCl) and microbial community composition changes mirrored the salinity dynamics with the *Halanaerobium* spp. initially being relatively minor populations but increased in relative abundance to represent >99% of observed taxa at ~150 g L^{-1} NaCl (Murali Mohan et al. 2013).

In contrast, the bulk fluids from the desalter module harbored the only community predominantly populated by *Epsilonproteobacteria* rather than *Halanaerobiales.* This broad taxonomic group is frequently observed in production fluids from petroleum reservoirs (Voordouw et al. 1996; Grabowski et al. 2005; Hubert et al. 2012) and members of the genera *Sulfurospirllum* (Kodama, Ha, and Watanabe 2007), *Sulfurimonas* (Gevertz et al. 2000), *Acrobacter* (Gevertz et al. 2000), and *Sulfuricurvum* (Kodama 2004) have been cultivated from these habitats. Many of these organisms are capable of the oxidation of reduced sulfur species coupled with the reduction of nitrate. Such metabolic capabilities spur interest in the use of nitrate for reservoir souring control purposes. Sulfate-reducing prokaryotes are known to be inhibited by the metabolic activities of nitrate-reducing, sulfur-oxidizing bacteria of the *Epsilonproteobacteria* (Hubert and Voordouw 2007; Telang, Jenneman, and Voordouw 2011; Voordouw et al. 2013). *Sulfurimonas denitrificans* CVO, formerly *Thiomicrospira denitrificans* (Takai 2006), is known to oxidize sulfide via the sulfur-oxidation (Sox) pathway to elemental sulfur or sulfate under both denitrifying and microaerophilic conditions (Gevertz et al. 2000; Sievert et al. 2008). Though *S. denitrificans* CVO, *Arcobacter* sp. FWKO B, *Sulfurospirillum cavolei*, and *Sulfuricurvum kujiense* were all isolated from fluids associated with subsurface petroleum habitats, growth of each organism is inhibited at >1% salt. Clearly, our results attest that other members of the *Epsilonproteobacteria* are able to tolerate more highly saline conditions (~1.5 M Cl^-) in the desalter production fluids. In this regard, *Epsilonproteobacteria* increased in relative abundance from <2% to 17% when produced brine salinity reached ~10% NaCl (~1.8 M Cl^-) (Murali Mohan et al. 2013). Since nitrate was below detection in all production samples, this likely suggests a local oxygen ingress point, the presence of a microaerophilic population of halotolerant sulfide-oxidizing bacteria, and potentially an area of active sulfur cycling within the facility. There is a pump immediately upstream of the desalter to increase the low-pressure separator outflow. A weak or faulty seal in this unit could conceivably account for localized available oxygen in the desalter bulk fluids.

16.4.1 Metabolites and Putative Metabolic Activities

Untargeted metabolite profiling showed the brines to contain a plethora of small organic compounds of numerous chemical classes. The observation of acidic fermentation intermediates such as tartaric acid and L-arabinonate exclusively in the dead-leg samples and at high concentrations is of particular note. Some species of gram-negative soil taxa have been shown to produce these carboxylic acids to solubilize phosphate minerals (Yi, Huang, and Ge 2007). These organic acids were also shown to be highly effective agents for bioleaching iron, nickel, and cobalt from mineral ores under acidic conditions (McKenzie, Denys, and Buchanan 1987).

Targeted metabolite analysis revealed the presence of alkanoic and aromatic acids, BTEX hydrocarbons, ethylene glycols, and hydroxylated aromatic compounds (e.g., cresols and phenol) within the bulk fluids from each of the modules. Interestingly, catechols and alkylated aromatic dihydrodiols were identified solely in the desalter bulk fluid. These molecules are produced by the incorporation of molecular oxygen into the aromatic ring of BTEX hydrocarbons by mono- and dioxygenase enzymes and represent signature metabolites indicative of the aerobic oxidation of aromatic hydrocarbons.

Taken together with the identification of taxa phylogenetically associated with known sulfur-oxidizing microaerophiles at this site in contrast to the strictly anaerobic taxa found within all other brines from each module, this suggests that oxygen ingress is occurring within the desalter module. These metabolites and potentially other oxygen-bearing organic molecules (e.g., aromatic acids, cresols, phenols, and di/tri-ethylene glycols) that were detected in each of the processing modules may serve as electron donors to the resident microflora, driving localized production of acetate and sulfide throughout the facility. Similarly, Liang et al. proposed a scenario where *Halanaerobium* spp. produced acetate and sulfide during metabolism of guar gum in a hydraulically fractured natural gas reservoir, ultimately resulting in the pitting corrosion of transport pipelines downstream of the wellhead and far removed from the catalytic organism(s) (Liang et al. 2016).

Comparison of the community and metabolite profile ordinations revealed interesting trends between microbial population membership and metabolic activity (Figure 16.5). The water separator surge drum receives oily water from both the desalter and the low-pressure separator, and some fluids are specifically recirculated between the water and low-pressure separators (Figure 16.1). Thus, it is not surprising that the bulk fluids of the low-pressure and water separators share very similar chemical characteristics (Table 16.1),and have enriched for essentially identical populations of microorganisms (Figure 16.3a). Further, the metabolite profiles of these biologically equivalent bulk brines indicate identical metabolic processes occurring within these spatially segregated communities, and a shared ecology exists between certain processing modules (Figure 16.5). The decreased salinity and potential oxygen ingress at the desalter would account for the dramatically different taxa and metabolite profiles observed in the bulk fluid of this module compared to the low-pressure and water separators despite the brine circulation pattern. Interestingly, significantly different metabolite profiles were recovered from the desalter dead-leg samples despite possessing an identical microflora to the low-pressure and water separator bulk fluids (Figure 16.5). Thus, community membership alone should not be regarded as a surrogate for metabolic activity in MIC monitoring regimes.

16.4.2 Proposed Mitigation Strategies and Hypothesized Impact on Communities

Through the combination of geochemical, molecular, and metabolomics investigations, a model for microbial involvement in corrosion activities within this asset can be proposed. Asset managers had previously predicted corrosion rates if no mitigation actions were taken to be 0.5 to 1 mmpy, but these models cannot account for microbial activity or the biological formation of iron sulfides within the system and frequently underrepresent observed pitting rates in the field (Bonis and MacDonald 2015). As fluids stagnate within the dead legs, *Halanaerobiaceae* spp. and *Desulfohalobiaceae* spp. may oxidize organic substrates circulating in the bulk fluids to acetate coupled to sulfidogenesis from thiosulfate and sulfate. Bonis and MacDonald (2015) examined cases of recurrent corrosion within fields producing H_2S and CO_2 to elucidate factors determining the risk to infrastructure degradation. Their analysis indicated that impedances in the kinetics of FeS precipitation, iron sulfide deposition, and galvanic effects induced by local chemical differences promoted aggressive localized corrosion within dead legs with little to no impact to the surrounding infrastructure (Bonis and MacDonald 2015). Based on the proposed activities of the indigenous microflora, metabolic production of sulfide would drive the formation of inorganic sulfide particles and a decrease in local fluid pH associated with the excretion of acidic metabolic end products. The incomplete oxidation of organic electron donors results in the production of organic acids (e.g., acetic acid, L-arabinonic acid, tartaric acid, etc.), or their corresponding conjugate bases depending upon environmental pH within the stagnated zones. The consequences of these metabolic activities could, in concert with local chemical changes within the dead legs, act to impede the precipitation kinetics of FeS at the metal–fluid interface. As mentioned previously, dead legs experience a cooling of approximately 20°C relative to the bulk fluids. In addition, the process brines contain high concentrations of divalent cations (e.g., Ca^{2+} and Mg^{2+}). Cooling and precipitation with alternative cations tend to impede the formation of protective FeS layers (Crolet and Bonis 2010; Bonis and MacDonald 2015). These factors and locally decreased pH values, arising from both biotic and abiotic reactions, would favor the formation of less protective scales (i.e., mackinawite) rather than pyrite. Furthermore, the composite nature of mixed inorganic anions, variable cation mineral molecular volumes, limited diffusion of inhibitors from bulk fluids, and the physical presence of microbial cells may likely result in the formation of a highly porous scale layer, thus allowing for continued under-deposit dissolution of the pipe surface. This suggests that microorganisms indirectly exacerbate corrosion within the dead legs of this asset by creating differential chemical environments, preventing the formation of a protective FeS layer and probably promoting galvanic coupling with the surrounding FeS. Ultimately, the combination of biotic and abiotic factors promotes aggressive localized anodic corrosion beyond predicted rates.

Surveys of microbial communities and/or metabolites can be correlated with field metadata (e.g., brine chemistry, operational parameters, documented corrosion rates, etc.) through recursive portioning and/or constrained ordination analyses, such as canonical correspondence analysis (CCA), principal components analysis (PCA), and redundancy analysis (RDA). These statistical analyses can offer initial insights into relationships between community composition and environmental variables and are reviewed elsewhere (Legendre and Legendre 2012). The observed correlations can then serve as the basis for hypothesis generation and in conjunction with multiple linear regressions and generalized linear models targeting response variables (specific taxa, community structures, etc.) of interest. These methods provide models for the extrapolation of the selected variable in response to environmental parameter changes a priori. In the future, this approach could allow asset managers to evaluate the potential impacts of operational changes upon resident microflora and potentially be used to further refine corrosion risk models to incorporate activities driven by microbial processes.

The combined application of molecular methods for microbial community surveys, geochemical analyses, and mass spectral–metabolite profiling provides deep insights into the microbial processes occurring within an environment. This case study utilized these paired data types to investigate biological processes occurring within an engineered environment that ultimately led to the proposal of a mechanism for microbiologically influenced corrosion within an oil-processing facility experiencing aggressive localized corrosion of carbon steel. Continued research involving expansive surveys of brines throughout field assets is necessary to further develop and test hypotheses linking microbial populations to deleterious activities. This study was conducted exclusively on planktonic fluid communities. Biofilms are also widely recognized as catalysts for biocorrosion, and this same investigative approach can be used to explore the ecology and physiology of such assemblages and their impact on infrastructure integrity. Several specific questions remain, such as whether dead legs select for common microbial communities through a predictable succession pattern, and can specific taxa or chemicals serve as indicators of corrosive conditions in situ. Ultimately, an increased understanding of the microbial ecology of these engineered habitats will offer valuable data for asset protection planning and production assurance.

REFERENCES

American Petroleum Institute. 1965. "API Recommended Practice for the Biological Analysis of Subsurface Injection Waters." In 2nd ed. New York: American Petroleum Institute.

Bastin, Edson S., Frank E. Greer, C. A. Merritt, and Gail Moulton. 1926. "The Presence of Sulphate-Reducing Bacteria in Oil Field Waters." *Science (New York, N.Y.)* 63 (1618): 21–24. doi: 10.1126/science.63.1618.21.

Beller, Harry R. 2000. "Metabolic Indicators for Detecting in Situ Anaerobic Alkylbenzene Degradation." *Biodegradation* 11 (2–3): 125–139. doi: 10.1023/A:1011109800916.

Ben-Dov, Eitan, Asher Brenner, and Ariel Kushmaro. 2007. "Quantification of Sulfate-Reducing Bacteria in Industrial Wastewater, by Real-Time Polymerase Chain Reaction (PCR) Using *dsrA* and *apsA* Genes." *Microbial Ecology* 54 (3): 439–451. doi: 10.1007/s00248-007-9233-2.

Bhupathiraju, V. K. K., Aharon Oren, P. K. Sharma, Ralph S. Tanner, Carl R. Woese, and Michael J. McInerney. 1994. "*Haloanaerobium salsugo* sp. nov., a Moderately Halophilic, Anaerobic Bacterium from a Subterranean Brine." *International Journal of Systematic Bacteriology* 44 (3): 565–572. doi: 10.1099/00207713-44-3-565.

Bhupathiraju, V. K. K., Michael J. McInerney, Carl R. Woese, and Ralph S. Tanner. 1999. "*Haloanaerobium kushneri* sp. nov., an Obligately Halophilic, Anaerobic Bacterium from an Oil Brine." *International Journal of Systematic Bacteriolology* 49 (3): 953–960.

Bian, Xin-Yu, Serge M. Mbadinga, Yi-Fan Liu, Shi-Zhong Yang, Jin-Feng Liu, Ru-Qiang Ye, Ji-Dong Gu, and Bo-Zhong Mu. 2015. "Insights into the Anaerobic Biodegradation Pathway of N-Alkanes in Oil Reservoirs by Detection of Signature Metabolites." *Scientific Reports* 5 (January). Nature Publishing Group: 9801. doi: 10.1038/srep09801.

Bonifay, Vincent, Jan A. Sunner, Whitney Smith, and Iwona B. Beech. 2013. "Metabolomics Study of Corroding Microbial Systems." In *ASMS Conference on Mass Spectrometry and Allied Topics*. June 9–13, Minneapolis, MN.

Bonis, Michel, and Reg MacDonald. 2015. "$H_2S + CO_2$ Corrosion: Additional Learnings from Field Experience," *Corrosion 2015*. Houston, TX: NACE International.

Buttigieg, Pier L., and Alban Ramette. 2014. "A Guide to Statistical Analysis in Microbial Ecology: A Community-Focused, Living Review of Multivariate Data Analyses." *FEMS Microbiology Ecology* 90 (3): 543–550. doi: 10.1111/1574-6941.12437.

Caporaso, J. Gregory, Justin Kuczynski, Jesse Stombaugh, Kyle Bittinger, Frederic D. Bushman, Elizabeth K. Costello, Noah Fierer et al. 2010. "QIIME Allows Analysis of High-Throughput Community Sequencing Data." *Nature Methods* 7 (5): 335–336. doi: 10.1038/nmeth.f.303.

Creek, Darren J., Andris Jankevics, Karl E. V. Burgess, Rainer Breitling, and Michael P. Barrett. 2012. "IDEOM: An Excel Interface for Analysis of LC-MS-Based Metabolomics Data." *Bioinformatics (Oxford, England)* 28 (7): 1048–1049. doi: 10.1093/bioinformatics/bts069.

Crolet, J.-L., and Michel R. Bonis. 2010. "Algorithm of the Protectiveness of Corrosion Layers 1—Protectiveness Mechanisms and CO_2 Corrosion Prediction." *Corrosion 2010*. Houston, TX: NACE International.

De Livera, Alysha M., and Jarius B. Bowne. 2015. Analysis of Metabolomics Data. R package version 0.1.4.

Duncan, Kathleen E., Lisa M. Gieg, Victoria A. Parisi, Ralph S. Tanner, Susannah G. Tringe, J. Bristow, and Joseph M. Suflita. 2009. "Biocorrosive Thermophilic Microbial Communities in Alaskan North Slope Oil Facilities." *Environmental Science & Technology* 43 (20): 7977–7984. doi: 10.1021/es9013932.

Eckert, Richard, and Torben L. Skovhus. 2011. "Methods to Investigate MIC in the Oil and Gas Industry." *Materials Performance* 50 (8): 2–6.

Edgar, Robert C. 2013. "UPARSE: Highly Accurate OTU Sequences from Microbial Amplicon Reads." *Nature Methods* 10 (10): 996–998. doi: 10.1038/nmeth.2604.

Elshahed, Mostafa S., Lisa M. Gieg, Michael J. McInerney, and Joseph M. Suflita. 2001. "Signature Metabolites Attesting to the in Situ Attenuation of Alkylbenzenes in Anaerobic Environments." *Environmental Science & Technology* 35 (4): 682–689. doi: 10.1021/Es001571u.

Gales, G., N. Chehider, C. Joulian, F. Battaglia-Brunet, J. L. Cayol, A. Postec, J. Borgomano et al. 2011. "Characterization of *Halanaerocella petrolearia* gen. nov., sp. nov., a New Anaerobic Moderately Halophilic Fermentative Bacterium Isolated from a Deep Subsurface Hypersaline Oil Reservoir." *Extremophiles* 15 (5): 565–571. doi: 10.1007/s00792-011-0387-y.

Gevertz, D., A. J. Telang, Gerrit Voordouw, and Gary E. Jenneman. 2000. "Isolation and Characterization of Strains CVO and FWKO B, Two Novel Nitrate-Reducing, Sulfide-Oxidizing Bacteria Isolated from Oil Field Brine." *Applied and Environmental Microbiology* 66 (6): 2491–2501.

Gieg, Lisa M., and Joseph M. Suflita. 2002. "Detection of Anaerobic Metabolites of Saturated and Aromatic Hydrocarbons in Petroleum-Contaminated Aquifers." *Environmental Science & Technology* 36 (17): 3755–3762. doi: 10.1021/Es0205333.

Gieg, Lisa M., Irene A. Davidova, Kathleen E. Duncan, and Joseph M. Suflita. 2010. "Methanogenesis, Sulfate Reduction and Crude Oil Biodegradation in Hot Alaskan Oilfields." *Environmental Microbiology* 12 (11): 3074–3086. doi: 10.1111/j.1462-2920.2010.02282.x.

Gieg, Lisa M., and Joseph M. Suflita. 2005. *Petroleum Microbiology*. Edited by Bernard Ollivier and Michel Magot. Washington, DC: American Society of Microbiology Press. 337–356. doi: 10.1128/9781555817589.

Grabowski, Agnès, Olivier Nercessian, Françoise Fayolle, Denis Blanchet, and Christian Jeanthon. 2005. "Microbial Diversity in Production Waters of a Low-Temperature Biodegraded Oil Reservoir." *FEMS Microbiology Ecology* 54 (3): 427–443. doi: 10.1016/j.femsec.2005.05.007.

Gunaltun, Yves, and Kamesh Kumar. 2014. "Changes in Water Chemistry in Dead Ends Potentially Influenced by Bacterial Activity." *Corrosion 2014*. Houston, TX: NACE International.

Hubert, Casey R. J., Thomas B. P. Oldenburg, Milovan Fustic, Neil D. Gray, Stephen R. Larter, Kevin Penn, Arlene K. Rowan et al. 2012. "Massive Dominance of *Epsilonproteobacteria* in Formation Waters from a Canadian Oil Sands Reservoir Containing Severely Biodegraded Oil." *Environmental Microbiology* 14 (2): 387–404. doi: 10.1111/j.1462-2920.2011.02521.x.

Hubert, Casey, and Gerrit Voordouw. 2007. "Oil Field Souring Control by Nitrate-Reducing *Sulfurospirillum* Spp. That Outcompete Sulfate-Reducing Bacteria for Organic Electron Donors." *Applied and Environmental Microbiology* 73 (8): 2644–2652. doi: 10.1128/AEM.02332-06.

Kanehisa, Minoru, Susumu Goto, Yoko Sato, Miho Furumichi, and Mao Tanabe. 2012. "KEGG for Integration and Interpretation of Large-Scale Molecular Data Sets." *Nucleic Acids Research* 40 (D1): D109–114. doi: 10.1093/nar/gkr988.

Kodama, Yumiko. 2004. "*Sulfuricurvum kujiense* gen. nov., sp. nov., a Facultatively Anaerobic, Chemolithoautotrophic, Sulfur-Oxidizing Bacterium Isolated from an Underground Crude-Oil Storage Cavity." *International Journal of Systematic and Evolutionary Microbiology* 54 (6): 2297–2300. doi: 10.1099/ijs.0.63243-0.

Kodama, Yumiko, Le Thu Ha, and Kazuya Watanabe. 2007. "*Sulfurospirillum cavolei* sp. nov., a Facultatively Anaerobic Sulfur-Reducing Bacterium Isolated from an Underground Crude Oil Storage Cavity." *International Journal of Systematic and Evolutionary Microbiology* 57 (Pt 4): 827–831. doi: 10.1099/ijs.0.64823-0.

Leader, David P., Karl Burgess, Darren Creek, and Michael P. Barrett. 2011. "Pathos: A Web Facility That Uses Metabolic Maps to Display Experimental Changes in Metabolites Identified by Mass Spectrometry." *Rapid Communications in Mass Spectrometry: RCM* 25 (22): 3422–3426. doi: 10.1002/rcm.5245.

Legendre, Pierre, and Louis Legendre. 2012. *Numerical Ecology. Developments in Environmental Modelling*. Vol. 24. Developments in Environmental Modelling. Amsterdam: Elsevier. doi: 10.1016/B978-0-444-53868-0.50009-5.

Lenhart, Tiffany R., Kathleen E. Duncan, Iwona B. Beech, Jan A. Sunner, Whitney Smith, Vincent Bonifay, Bernadette Biri, and Joseph M. Suflita. 2014. "Identification and Characterization of Microbial Biofilm Communities Associated with Corroded Oil Pipeline Surfaces." *Biofouling* 30 (7): 823–835. doi: 10.1080/08927014.2014.931379.

Liang, Renxing, Irene A. Davidova, Christopher R. Marks, Blake W. Stamps, Brian H. Harriman, Bradley S. Stevenson, Kathleen E. Duncan, and Joseph M. Suflita. 2016. "Metabolic Capability of a Predominant *Halanaerobium* sp. in Hydraulically Fractured Gas Wells and its Implication in Pipeline Corrosion." *Frontiers in Microbiology* 7: 988.

Lozupone, Catherine, and Rob Knight. 2005. "UniFrac: A New Phylogenetic Method for Comparing Microbial Communities." *Applied and Environmental Microbiology* 71 (12): 8228–8235. doi: 10.1128/AEM.71.12.8228-8235.2005.

Ludwig, Wolfgang, Oliver Strunk, Ralf Westram, Lothar Richter, Harald Meier, Yadhukumar, Arno Buchner et al. 2004. "ARB: A Software Environment for Sequence Data." *Nucleic Acids Research* 32 (4): 1363–1371.

Maxwell, Stephen, Heike Hoffman, and Carol Divine. 2007. "Application of Molecular Microbiology Techniques as Tools for Monitoring Oilfield Bacteria." *Corrosion 2007*. Houston, TX: NACE International.

McKenzie, Denis I., Lydia Denys, and Alan Buchanan. 1987. "The Solubilization of Nickel, Cobalt and Iron from Laterites by Means of Organic Chelating Acids at Low pH." *International Journal of Mineral Processing* 21 (3–4): 275–292. doi: 10.1016/0301-7516(87)90059-7.

Murali Mohan, Arvind, Angela Hartsock, Kyle J. Bibby, Richard W. Hammack, Radisav D. Vidic, and Kelvin B. Gregory. 2013. "Microbial Community Changes in Hydraulic Fracturing Fluids and Produced Water from Shale Gas Extraction." *Environmental Science & Technology* 47 (22): 13141–13150. doi: 10.1021/es402928b.

Murata, René, Janet Benaquisto, and Cecily Storey. 2015. "A Methodology for Identifying and Addressing Dead-Legs and Corrosion Issues in a Process Hazard Analysis (PHA)." *Journal of Loss Prevention in the Process Industries* 35 (May): 387–392. doi: 10.1016/j.jlp.2015.02.004.

NACE International, TM0194-2014. 2014. "Field Monitoring of Bacterial Growth in Oil and Gas Systems." Houston, TX: NACE International.

Neria-González, Isabel, En Tao Wang, Florina Ramírez, Juan M. Romero, and César Hernández-Rodríguez. 2006. "Characterization of Bacterial Community Associated to Biofilms of Corroded Oil Pipelines from the Southeast of Mexico." *Anaerobe* 12 (3): 122–133. doi: 10.1016/j.anaerobe.2006.02.001.

Oksanen, Jari, F., Guillaume Blanchet, Roeland Kindt, Pierre Legendre, Peter R. Minchin, R. B. O'Hara, Gavin L. Simpson, Peter Solymos, M. Henry H. Stevens, and Helene Wagner. 2016. "Vegan: Community Ecology Package." http://cran.r-project.org/package=vegan.

Oldham, Athenia L., Heather S. Drilling, Blake W. Stamps, Bradley S. Stevenson, and Kathleen E. Duncan. 2012. "Automated DNA Extraction Platforms Offer Solutions to Challenges of Assessing Microbial Biofouling in Oil Production Facilities." *AMB Express* 2 (1): 60. doi: 10.1186/2191-0855-2-60.

Pruesse, Elmar, Christian Quast, Katrin Knittel, Bernard Fuchs, and Wolfgang Ludwig. 2007. "SILVA: A Comprehensive Online Resource for Quality Checked and Aligned Ribosomal RNA Sequence Data Compatible with ARB." *Nucleic Acids Research* 35 (21): 7188–7196.

Pruesse, Elmar, Jörg Peplies, and Frank O. Glöckner. 2012. "SINA: Accurate High-Throughput Multiple Sequence Alignment of Ribosomal RNA Genes." *Bioinformatics* 28 (14): 1823–1829. doi: 10.1093/bioinformatics/bts252.

Ravot, G., Michel Magot, B. Ollivier, B. K. C. Patel, E. Ageron, P. A. D. Grimont, P. Thomas, and J. L. Garcia. 1997. "*Haloanaerobium congolense* sp. nov., an Anaerobic, Moderately Halophilic, Thiosulfate- and Sulfur-Reducing Bacterium from an African Oil Field." *FEMS Microbiology Letters* 147 (1): 81–88. doi: 10.1016/S0378-1097(96)00508-3.

Reysenbach, Anna-Louise, and Norman R. Pace. 1995. "Reliable Amplification of Hyperthermophilic Archaeal 16S rRNA Genes by the Polymerase Chain Reaction." In *Archaea: A Laboratory Manual—Thermophiles*. Edited by F. T. Robb and A. R. Place. Cold Spring Harbor, NY: Cold Spring Harbor Laboratory Press.

Scheltema, Richard A., Andris Jankevics, Ritsert C. Jansen, Morris A. Swertz, and Rainer Breitling. 2011. "PeakML/mzMatch: A File Format, Java Library, R Library, and Tool-Chain for Mass Spectrometry Data Analysis." *Analytical Chemistry* 83 (7): 2786–2793. doi: 10.1021/ac2000994.

Schloss, Patrick D., Sarah L. Westcott, Thomas Ryabin, Justine R. Hall, Martin Hartmann, Emily B. Hollister, Ryan A. Lesniewski et al. 2009. "Introducing Mothur: Open-Source, Platform-Independent, Community-Supported Software for Describing and Comparing Microbial Communities." *Applied and Environmental Microbiology* 75 (23): 7537–7541. doi: 10.1128/aem.01541-09.

Sette, Lara D., Karen C. M. Simioni, Suzan P. Vasconcellos, Lucia J. Dussan, Eugênio V. S. Neto, and Valéria M. Oliveira. 2007. "Analysis of the Composition of Bacterial Communities in Oil Reservoirs from a Southern Offshore Brazilian Basin." *Antonie van Leeuwenhoek, International Journal of General and Molecular Microbiology* 91: 253–266. doi: 10.1007/s10482-006-9115-5.

Sievert, Stefan M., Kathleen M. Scott, Martin G. Klotz, Patrick S. G. Chain, Loren J. Hauser, James Hemp, Michael Hügler et al. 2008. "Genome of the Epsilonproteobacterial Chemolithoautotroph *Sulfurimonas denitrificans*." *Applied and Environmental Microbiology* 74 (4): 1145–1156. doi: 10.1128/AEM.01844-07.

Sloley, Andrew. 2011. "Watch Out for Dead-Legs." *Chemical Processing*, September. http://www.chemicalprocessing.com/articles/2011/watch-out-for-dead-legs/?show=all.

Stevenson, Bradley S., Heather S. Drilling, Paul A. Lawson, Kathleen E. Duncan, Victoria A. Parisi, and Joseph M. Suflita. 2011. "Microbial Communities in Bulk Fluids and Biofilms of an Oil Facility Have Similar Composition but Different Structure." *Environmental Microbiology* 13 (4): 1078–1090. doi: 10.1111/j.1462-2920.2010.02413.x.

Takai, Ken. 2006. "*Sulfurimonas paralvinellae* sp. nov., a Novel Mesophilic, Hydrogen- and Sulfur-Oxidizing Chemolithoautotroph within the Epsilonproteobacteria Isolated from a Deep-Sea Hydrothermal Vent Polychaete Nest, Reclassification of *Thiomicrospira denitrificans* as *Sulfurimonas denitrificans* comb. nov. and emended description of the genus *Sulfurimonas*." *International Journal of Systematic and Evolutionary Microbiology* 56 (8): 1725–1733. doi: 10.1099/ijs.0.64255-0.

Tamura, Koichiro, Glen Stecher, Daniel Peterson, Alan Filipski, and Sudhir Kumar. 2013. "MEGA6: Molecular Evolutionary Genetics Analysis Version 6.0." *Molecular Biology and Evolution* 30 (12): 2725–2729. doi: 10.1093/molbev/mst197.

Tautenhahn, Ralf, Christoph Böttcher, and Steffen Neumann. 2008. "Highly Sensitive Feature Detection for High Resolution LC/MS." *BMC Bioinformatics* 9 (1): 504. doi: 10.1186/1471-2105-9-504.

Telang, Anita J., Gary E. Jenneman, and Gerrit Voordouw. 2011. "Sulfur Cycling in Mixed Cultures of Sulfide-Oxidizing and Sulfate- or Sulfur-Reducing Oil Field Bacteria." *Canadian Journal of Microbiology* 45 (11): 905–913. doi: 10.1139/cjm-45-11-905.

Voordouw, Gerrit, S. M. Armstrong, M. F. Reimer, B. Fouts, Anita J. Telang, Y. Shen, and D. Gevertz. 1996. "Characterization of 16S rRNA Genes from Oil Field Microbial Communities Indicates the Presence of a Variety of Sulfate-Reducing, Fermentative, and Sulfide-Oxidizing Bacteria." *Applied and Environmental Microbiology* 62 (5): 1623–1629.

Voordouw, Gerrit, Brenton Buziak, Shiping Lin, Alexander Grigoriyan, Krista M. Kaster, Gary E. Jenneman, and Joseph J. Arensdorf. 2013. "Use of Nitrate or Nitrite for the Management of the Sulfur Cycle in Oil and Gas Fields." In *International Symposium on Oilfield Chemistry*. Society of Petroleum Engineers. doi: 10.2118/106288-MS.

Whitman, William B., Fred Rainey, Peter Kämpfer, Martha Trujillo, Jonsik Chun, Paul DeVos, Brian Hedlund, and Svetlana Dedysh, eds. 2015. *Bergey's Manual of Systematics of Archaea and Bacteria*. Chichester, UK: John Wiley & Sons, Ltd.

Yi, Yanmei, Weiyi Huang, and Ying Ge. 2007. "Exopolysaccharide: A Novel Important Factor in the Microbial Dissolution of Tricalcium Phosphate." *World Journal of Microbiology and Biotechnology* 24 (7): 1059–1065. doi: 10.1007/s11274-007-9575-4.

Zhang, Jiajie, Kassian Kobert, Tomáš Flouri, and Alexandros Stamatakis. 2014. "PEAR: A Fast and Accurate Illumina Paired-End reAd mergeR." *Bioinformatics* 30 (5): 614–620. doi: 10.1093/bioinformatics/btt593.

17 Determining the Source of H_2S on an Offshore Oil Production Platform

John J. Kilbane II

CONTENTS

ABSTRACT

Periodic monitoring by an oil company of hydrogen sulfide (H_2S) concentrations in gas samples collected at various locations on an offshore oil production platform indicated that H_2S was becoming a pervasive problem that was getting worse in wells that had previously not contained detectable concentrations of H_2S. This case study demonstrated that only by obtaining comprehensive data from multiple samples that included H_2S quantification, sulfur isotopic ratios in gas and water samples, microbiological data from growth and genetic testing, and water chemistry testing to quantify organic acids was it possible to conclude that there were two sources of H_2S in samples from this offshore platform. The oil company now knows that thermogenic H_2S is migrating into some oil production zones in this reservoir and that the topside water handling system is vulnerable to microbial contamination, and can plan accordingly.

17.1 INTRODUCTION

The presence of hydrogen sulfide during oil or gas production creates safety and corrosion issues, increases the cost of oil or gas production, and can decrease the value of the oil or gas obtained (Tabari et al. 2011). Therefore, it is particularly concerning if H_2S is detected at locations where it was previously absent. If H_2S is to be avoided or minimized, the source of the H_2S must be determined. An offshore oil production facility successfully produced low sulfur oil for several years with no detectable H_2S in the production fluids; however, over time, H_2S was observed, and the concentration was seen to be variable and increasing. The oil company wanted to determine the source of the H_2S and take appropriate action.

H_2S can be formed by thermochemical or biochemical processes (Thode 1991; Struchtmeyer et al. 2011). At high temperatures and pressures, thermochemical processes can transform inorganic sulfur present in minerals and/or water or organic sulfur present in petroleum into H_2S. Thermochemical reduction of sulfate to H_2S (TSR) is the reaction of sulfate minerals (primarily anhydrite) and hydrocarbons (beginning at temperatures of 120°C–140°C) to form H_2S and calcium carbonate. Similarly, thermal decomposition of sulfides in kerogen and/or oil can convert organically bound sulfur into H_2S. The formation of thermogenic H_2S occurs over geological time frames in the course of the conversion of organic sediment material into petroleum and natural gas. It is not generally considered that the thermogenic production of H_2S is an ongoing process that continues during oil and gas production from a geological deposit. The biochemical conversion of sulfate to H_2S is performed by microorganisms and could be an ongoing process in an oil deposit and usually is limited to reservoirs with a temperature <80°C (Habicht and Canfield 1997; Detmers et al. 2001; Gittel et al. 2009).

Reservoir souring is a term used to describe an increasing change in the concentration of H_2S in the oil/gas from a reservoir and is most commonly caused by microbial sulfate reduction (Struchtmeyer et al. 2011; Tabari et al. 2011). Reservoir souring is often associated with water injection into the subsurface to maintain reservoir pressure and aid in the secondary production of oil (Tabari et al. 2011). Some oil deposits were successfully produced for years prior to water injection without observing any H_2S in the produced fluids, but after water injection, H_2S was observed due to the proliferation of sulfate-reducing bacteria (SRB) and/or sulfate-reducing archaea collectively referred to as sulfate reducing prokaryotes (SRP) in the formation. Similar concerns have been raised with the injection of hydraulic fracturing fluids into the subsurface, subsequently resulting in the appearance of H_2S in the produced fluids due to microbial conversion of sulfate (Struchtmeyer et al. 2011).

If the source of H_2S is from microorganisms, then efforts to kill or influence the metabolism of microorganisms (such as the injection of nitrate; Schwermer et al. 2008; DeGusseme et al. 2009; Gittel et al. 2009) will be needed to control the situation. If the source of the H_2S is thermochemical, then the only means of dealing with it are to employ chemicals to scavenge the H_2S or to limit oil/gas production to only those locations that do not contain thermochemical H_2S. One of the best ways of determining if the source of H_2S is thermochemical or biochemical is to determine the isotopic ratio of $\delta\ ^{32}S$ to $\delta\ ^{34}S$ in the sulfur (Thode 1991).

There are several isotopes of sulfur with $\delta\ ^{32}S$ and $\delta\ ^{34}S$ being the most abundant and $\delta\ ^{32}S$ being the primary sulfur isotope. Chemical processes generally do not discriminate between different isotopes whereas biochemical processes often will preferentially metabolize one isotope versus another, resulting in what is termed isotopic fractionation (Habicht et al. 1997). Sulfur isotopes $\delta\ ^{34}S$ and $\delta\ ^{32}S$ are quantified using mass spectrometry, and the data are generally reported as the change in the concentration of $\delta\ ^{34}S$ per thousand (‰) as compared with an internationally recognized standard (Krouse and Coplen 1997). In microbiological sulfate reduction, those sulfate molecules that contain $\delta\ ^{32}S$ are metabolized preferentially versus sulfate molecules containing the $\delta\ ^{34}S$ isotope.

This results in the formation of hydrogen sulfide molecules that have lower amounts of $\delta\ ^{34}S$ than was present in the sulfate. The amount of reduction of the $\delta\ ^{34}S$ content of hydrogen sulfide gas as compared with the sulfate source that is caused by microbiological cultures ranges from no detectable change to –40‰ (Habicht et al. 1997; Detmers et al. 2001; Stam et al. 2010). However, for environmental samples typically found in oil industry samples that are mixed microbial cultures that contain several species of SRP as well as other bacteria, the amount of isotopic fractionation (reduced amount of $\delta\ ^{34}S$ in the H_2S gas) is about –20‰ (Balliger et al. 2001). The sulfur isotope abundance in various materials in nature can vary considerably so that merely obtaining data for the sulfur isotopic ratio in a given sample will generally not allow conclusions to be made regarding the origin of a given material as having resulted from chemical or from biochemical reactions. However, if the sulfur isotopic ratios of both the starting material and the end product of a reaction are known, then the presence or absence of changes in the isotopic ratio can indicate if the reaction was chemical or biochemical. In this study, H_2S gas from the sour offshore production well was collected using an isotrap for subsequent analysis by gas chromatography mass spectrometry to determine the ratio of sulfur isotopes as compared with values for sulfur isotopes in sulfate from produced water.

Other types of data besides sulfur isotopic data can also be useful in determining if the source of H_2S is thermochemical or biochemical, such as the likely thermal exposure history of the oil, the sulfur content of the oil, the temperature of the oil reservoir, microbiological data, and water chemistry data. Geological analyses of rock cores can help to determine the age of the rock and the likely environment of the oil deposit. Chemical analysis of the oil to determine the hydrocarbon composition and the isotopic analyses of carbon and hydrogen in hydrocarbons, particularly methane, can be useful in determining if microbial activity has occurred in the reservoir. Microbial growth tests (NACE Standard TMO194) can be performed to detect and quantify microorganisms, such as SRP, present in produced water, and microbial metabolism in the subsurface may result in organic acids, such as acetic, propionic, and butyric acids, that can be detected by chemical analysis of the produced water (Larsen 2014; Fichter et al. 2015).

In recent times, the study of microbiology has benefitted from the use of genetic techniques that overcome some of the limitations of traditional microbial growth tests and have led to the conclusion that many microbial species present in environmental samples can't be grown in traditional microbial growth tests (Haveman and Pedersen 2002). The myriad species of microorganisms in nature can occupy a wide range of environments such that different microbes use different nutrients (biochemical pathways) for growth. Therefore, the nutrients present in traditional microbial growth media may not be suitable for all types of microbial species. Moreover, even if the correct combination of nutrients are present, the growth conditions such as temperature, salinity, and redox potential also need to be matched to the needs of each type of microbial community (Fichter et al. 2015). For all these reasons, genetic testing, such as the quantitative polymerase chain reaction (qPCR) or DNA sequencing, are increasingly being used to detect and quantify microorganisms in oilfield samples (Zhu and Kilbane 2005; Larsen 2014; Fichter et al. 2015).

An offshore oil production facility observed that H_2S was becoming a pervasive problem that was getting worse in wells that had previously not contained detectable concentrations of H_2S. This case study describes the results of investigations of that problem and discusses lessons learned.

17.2 RESULTS

The initial investigation of the source of the H_2S that was performed prior to the current project included geological data; chemical analyses of the oil; isotopic analyses of sulfur, carbon, and hydrogen; water analysis; reservoir temperature; and microbial growth testing. The geological analyses indicated that the oil was derived from an early Cretaceous lacustrine Type 1 kerogen source rock deposited in a hypersaline environment (Cubitt et al. 2004). The oil appears to be non-biodegraded as it contains high concentrations of low molecular weight hydrocarbons/alkanes and has an inferred thermal stability of "early oil." It contains less than 20 ppm diamondoids consistent with not being thermally cracked. The isotopic analyses of carbon and hydrogen indicated that methane present in production fluids was formed along with the oil and not due to thermal degradation of the oil or to microbial production of methane (Cubitt et al. 2004). It was concluded that H_2S is unlikely to be derived from kerogen or oil thermal decomposition because of the low organic sulfur content of the oil and the low temperatures inferred to have been experienced by the oil in the subsurface. Similarly, thermochemical sulfate reduction was considered to be an unlikely source of the H_2S because of the relatively low temperatures experienced by this oil during its formation and the current reservoir temperature of 65°C.

By the process of elimination, this makes microbial sulfate reduction a likely source of the H_2S, but no hydraulic fracturing or water flooding occurred in this oil well, and other data seemed inconsistent with a microbial source for the H_2S. The reservoir temperature is low enough to allow microbial growth, and there is abundant sulfate present in produced water. However, the oil does not appear to be biodegraded, and isotopic data indicate that the methane does not derive from microbial activity, and microbial growth tests performed with produced water samples failed to detect viable SRP or bacteria of any kind (genetic testing was not performed in the initial investigation). Isotopic analyses of sulfur were performed on anhydrite

TABLE 17.1
Sulfur Isotopic Ratio Data

Sample Description	H_2S, $\delta\,^{34}S$‰	Sulfate, $\delta\,^{34}S$‰	H_2S versus Sulfate, $\delta\,^{34}S$‰ Change
First sampling: Separator gas, anhydrite mineral	28.2	15.1	+13.1
Second sampling: Gas and sulfate from water system	8.7	27.8	–19.1
Third sampling: Separator gas, sulfate from water	27.7	27.9	–0.2
Third sampling: Gas and sulfate from water system	8.3	27.9	–19.6

mineral from rock cores (the presumed source of sulfate in the reservoir) and on the sulfur in H_2S gas yielding values of 15.1‰ and 28.2‰, respectively as shown in Table 17.1. If microbiological/biochemical processes converted the sulfate into H_2S, the concentration of $\delta\ ^{34}S$ should have decreased, yielding a lower sulfur isotopic ratio value (Detmers et al. 2001), but when the values of the anhydrite and the H_2S are compared, the sulfur isotopic ratio value increases by 13.1‰. This result makes it difficult to conclude what the source of H_2S is, but this result is more consistent with a thermochemical source for the H_2S.

Consequently, the oil company had no conclusive information about the source of the H_2S and authorized additional investigation. The sulfur isotopic analyses were repeated with new samples. For this second sampling event, sulfate was precipitated from production water and used to determine the sulfur isotopic composition of sulfate instead of obtaining data from anhydrite mineral in rock samples. The value obtained was 27.8‰, which is significantly higher than the value of 15.1‰ obtained for anhydrite. A new sample of H_2S gas was collected, and sulfur isotopic analysis resulted in a value of 8.7‰ as shown in Table 17.1. This value is significantly lower than the value obtained from the produced water and is consistent with microbiological production of H_2S (Detmers et al. 2001; Stam et al. 2010). However, microbial growth tests again failed to detect SRP (no genetic tests were performed at this time). Moreover, there was a great discrepancy between sulfur isotopic values obtained from the first and the second H_2S samples, which were 28.2‰ and 8.7‰, respectively. The oil company could form no clear conclusions from these data and were no closer to identifying the source of the H_2S.

At this point, Intertek became involved to provide independent test data in a third set of samples to confirm the source of the H_2S and better characterize the situation. Samples of pressurized fluid were obtained directly from the production well and were subsequently used to obtain oil, gas, and water samples for chemical and microbiological testing. Nonpressurized water samples were also obtained from the separator and from the produced water handling system on the oil production platform. Gas samples were obtained from two locations: the primary separator and from the produced water handling system. Discussions with the personnel who had performed/observed prior sampling indicated that the concentration of H_2S obtained from the separator was somewhat variable and often lower than the H_2S concentration obtained from a nearby location associated with the produced water handling system. The first H_2S sample collected was derived from the separator while the second H_2S sample was derived from the produced water handling system. Accordingly, Intertek obtained gas samples from both locations, and sulfate derived from the produced water was used as the source of the sulfate in the third round of sampling, and the results are shown in Table 17.1.

The data in Table 17.1 for the third sampling event show the sulfur isotopic data for the concentration of $\delta\ ^{34}S$ in H_2S obtained from the primary separator and in sulfate from produced water shows no significant depletion of $\delta\ ^{34}S$, which is consistent with thermogenic rather than biogenic processes (Thode 1991). The data for the gas sample obtained from the produced water system in the third set of samples shows substantial depletion of $\delta\ ^{34}S$ of about −20‰, which is consistent with biogenic rather than thermogenic processes (Balliger et al. 2001).

These data are consistent with the interpretation that H_2S collected from separator samples are from thermogenic processes, but once produced water is brought topside, microbial sulfate reduction generates increased concentrations of H_2S in the water handling system. This interpretation is further supported by the results of microbiological and water chemistry analyses. Microbial growth tests and quantitative polymerase chain reaction (qPCR) failed to detect microbes in the produced water obtained directly from the wellhead or in the primary separator while microbes, including SRP, were detected in produced water samples obtained from the topside water handling system as shown in Table 17.2. Similarly, no organic acids were detected in produced water samples obtained directly from the wellhead or in the primary separator, but low concentrations of acetic acid (76 ppm) were detected in water samples obtained from the produced water handling system. If microorganisms are growing in the subsurface metabolizing hydrocarbons, then organic acids, chiefly acetic acid, would accumulate. Thus, the water chemistry data indicates a lack of microbial activity in the subsurface. Additionally, isotopic data for carbon and hydrogen in methane, ethane, propane, and carbon dioxide (data not shown) show a thermogenic origin for all these gases from every well sampled. This is consistent with a thermogenic origin for the H_2S in these samples.

Genetic testing was performed using the quantitative polymerase chain reaction (qPCR) (Zhu and Kilbane 2005; Klindworth et al. 2013) to quantify the total concentration of microbes as well as the concentration of some types of microorganisms, such as SRP, acid-producing bacteria (APB), and archaea (such as thermophiles as methanogens). The qPCR reactions work by targeting specific DNA sequences that are unique to groups of bacteria or are unique to conserved regions of genes that encode for enzymes in specific biochemical pathways, such as sulfate reduction or the formation of organic acids. These genetic tests do not require any growth of microorganisms and instead rely on the analysis of deoxyribonucleic acid (DNA) present in the sample, and qPCR assays yield data similar to microbial growth tests. As shown in Table 17.2, no microorganisms were detected by genetic testing of water from the primary separator, but microorganisms, including SRP, were detected in samples from the downstream produced water handling system. Moreover, only microbial growth tests incubated at 30°C yielded positive results while samples incubated at 65°C did not. Since the reservoir temperature is 65°C but the temperature is significantly cooler in the produced water handling system, these results are consistent with the conclusion that microorganisms are not growing in the reservoir, but they are growing in the produced water brought to the surface.

17.3 DISCUSSION

The results reported here demonstrate the importance of obtaining data from more than one type of test to aid in understanding a situation and being able to form conclusions with confidence. Furthermore, these results highlight the importance of attention to detail in collecting samples and interpreting data.

The first samples analyzed to determine sulfur isotopic ratios utilized rock core mineral matter (anhydrite) to obtain a value that was thought to be representative of the sulfate in the reservoir. Subsequent testing of sulfate precipitated from the

TABLE 17.2
Microbiological and Chemical Analyses of Water Samples

	Growth Tests			qPCR Results				Chemical
Sample	**GHB**	**APB**	**SRB**	**Bacteria**	**Archaea**	**APB**	**SRP**	**Acetic Acid**
Separator	$<0.3 \times 10^0$	$<0.3 \times 10^0$	$<0.3 \times 10^0$	$<10^2$	$<10^2$	$<10^2$	$<10^2$	<1.0 ppm
Downstream	4.0×10^5	2.5×10^0	1.4×10^0	8.3×10^5	$<10^2$	1.5×10^2	1.1×10^2	76 ppm

produced water demonstrated that the actual sulfur isotopic value for sulfate in the petroleum reservoir was not accurately represented by the data obtained from the rock core and differed by 12.7‰. Because of this discrepancy, the interpretation of data from the first set of samples was hampered by the puzzling finding that the δ ^{34}S‰ concentration in H_2S gas appeared to be increased, and this result was not consistent with either a thermogenic or a biogenic origin for the H_2S. Similarly, when the sulfur isotopic data from the second set of samples was obtained, it was not appreciated that the location on the platform where the H_2S sample was collected could have a profound influence on the results obtained and the interpretation of the data. Collecting a sufficient mass of H_2S using an isotrap to enable the analysis of sulfur isotopes can take 24 hours or more if the concentration of H_2S in the gas stream is low, but the required mass can be collected in minutes from a gas stream containing a high concentration of H_2S. When the second set of samples was collected, a platform worker that had observed higher H_2S readings in the produced water system downstream from the separator thought that he was facilitating the completion of the sample collection task by obtaining the gas sample from the downstream location rather than from the separator as had been done previously. Since the scientist(s) charged with interpreting the data did not know that the second sample was obtained from a different location, the results were thought to be in complete disagreement with the results from the first set of samples. No meaningful conclusions could be drawn, and the accuracy of the analyses was questioned. Only after data from the third set of samples was obtained was it possible to sort out the confusion and determine that there were two sources of H_2S. H_2S of thermogenic origin was present in the formation and could be detected in samples from the wellhead and the primary separator while H_2S of biogenic origin was present in the downstream produced water handling system due to the growth of microorganisms.

These conclusions could be made with confidence because data from several types of analyses lead to similar conclusions. The oil analyses showed that the oil did not appear to be biodegraded and that methane associated with the oil was not derived from microbial methane production. This and the fact that no microorganisms or organic acids were detected in the water from the separator were consistent with the conclusion that H_2S from the petroleum reservoir was of thermogenic origin. Since the temperature of the reservoir is well below that required for the thermochemical reduction of sulfate and H_2S was not originally observed from this well during the first several years of production, it is probable that the H_2S is not from the production zone of this oil well. Instead, it is possible that the observed H_2S is from another deeper/distant geological deposit that is in communication with the hydrocarbon deposit under production by this well. As oil is produced, the pressure of the formation decreases and apparently allowed H_2S-containing gas/fluid to migrate into the production zone of this well, and that would account for the initial absence of H_2S followed by the gradually increasing and variable concentration of H_2S observed over time. Future testing can be performed to confirm this conclusion because if gas/fluids from a different geological deposit are migrating into the production zone then it is likely that the percentage of original versus new oil/gas/water will change over time as production continues, and detailed chemical analyses should be able to detect changes in the composition of oil/gas/water from this well over time.

The presence of microorganisms growing in the downstream produced water handling system was documented by detecting microorganisms in microbial growth tests and genetic tests by the presence of acetic acid (a by-product of microbial metabolism) and the presence of decreased concentrations of the $\delta\ ^{34}S$ isotope in H_2S from this downstream location. The concentrations of microbes in the downstream water sample were low, particularly for SRP, but only water samples and no biofilm samples were available. Microorganisms prefer to grow as biofilms attached to surfaces, and it has commonly been observed that the concentration and composition of microbes present in a water sample differs from a biofilm sample collected from the same location (Larsen 2014; Xu and Gu 2015; Brown et al. 2016; Chen et al. 2016). When studying microbiologically influenced corrosion, it is preferable to obtain biofilm samples because microbial concentrations, particularly SRP, can be significantly higher in biofilms. However, in this study, biofilm samples were not available, but it is likely that higher concentrations of SRP than are reported in Table 17.2 are present as biofilms in the topside water handling system and are responsible for the observed H_2S. Currently, no biocide is added to the produced water system, and there is no corrosion-monitoring plan in place. Now that the oil company knows that its produced water system is at risk for microbiologically influenced corrosion (and the production of H_2S creating a safety hazard) it was recommended that biocide addition should be practiced and that a corrosion-monitoring program should be implemented.

17.4 CONCLUSIONS

- Data from different types of analyses aid in understanding a situation and in forming conclusions with confidence.
- Microbial growth tests and genetic tests failed to detect sulfate-reducing bacteria or other microbes in produced water obtained from the primary separator.
- No organic acids (by-products of microbial metabolism) were present in produced water from the separator.
- Sulfur isotopic data indicates that the source of the H_2S produced directly from the well is thermogenic.
- Biogenic H_2S is produced locally in the produced water handling system.
- Carbon and hydrogen isotopic analyses of hydrocarbons indicate a thermogenic origin for the oil and gas.

REFERENCES

Balliger, C., M. H. Schroth, S. M. Bernasconi, J. Kleikemper, J. Zeyer. (2001) "Sulfur isotope fractionation during microbial sulfate reduction by toluene-degrading bacteria" *Geochemica et Cosmochemica Acta* 65(19):3289–3298.

Brown, D., S. Johnston, R. Turner, P. Teevens. (2016) "Planktonic cell density is independent of biofilm initiation and cell density" *Corrosion 2016*, Houston, TX: NACE International.

Chen, R., Y. Luo, J. X. Chen. (2016) "Evolution of the microbial community of the biofilm in a methane-based membrane biofilm reactor reducing multiple electron acceptors" *Environmental Science & Pollution Research* 23(10):9540–9548.

Cubitt, J. M., W. A. England, S. R. Larter. (2004) "Understanding petroleum reservoirs: Towards an integrated reservoir engineering and geochemical approach" Geological Society Special Publication 237.

DeGusseme, B., P. DeSchryver, M. DeCooman, K. Verbekan, P. Boeckx, W. Verstraete, N. Boon. (2009) "Nitrate-reducing, sulfide-oxidizing bacteria as microbial oxidants for rapid biological sulfide removal" *FEMS Microbial Ecology* 67:151–161.

Detmers, J., V. Bruchert, K. S. Habicht, J. Kuever. (2001) "Diversity of sulfur isotope fractionations by sulfate-reducing prokaryotes" *Applied & Environmental Microbiology* 67(2):888–894.

Fichter, J., E. Summer, C. Janes, G. Hamblin. (2015) "Direct comparison of microbial population analysis using ATP quantification, culture based enumeration in bug bottles, epiflourescence microscopy, and shotgun population analysis by sequencing 16S amplicons" NACE 2015 paper C2015-5555.

Gittel, A., K. B. Sorensen, T. L. Skovhus, K. Ingvorsen, A. Schramm. (2009) "Prokaryotic community structure and sulfate reducer activity in water from high-temperature oil reservoirs with and without nitrate treatment" *Applied & Environmental Microbiology* 75:7086–7096.

Habicht, K. S., D. E. Canfield. (1997) "Sulfur isotope fractionation during bacterial sulfate reduction in organic-rich sediments" *Geochemica et Cosmochemica Acta* 61:5351–5361.

Haveman, S. A., K. Pedersen. (2002) "Distribution of culturable microorganisms in Fennoscandian Shield groundwater" *FEMS Microbial Ecology* 39:129–137.

Klindworth, A., E. Pruesse, T. Schweer, J. Peplies, C. Quast, M. Horn, F. O. Glockner. (2013) "Evaluation of general 16S ribosomal RNA gene PCR primers for classical and next-generation sequencing-based diversity studies" *Nucleic Acids Research* 41(1):1–11. doi: 10.1093/nar/gks808.

Krouse, H. R., T. B. Coplen. (1997) "Reporting of Relative Sulfur Isotope-Ratio Data" *Pure and Applied Chemistry* 69(2):293–295.

Larsen, K. R. (2014) "A closer look at microbiologically influenced corrosion" *Materials Performance* 53(1):32–40.

Schwermer, C. V., G. Lavik, R. M. M. Abed, B. Dunsmore, T. G. Ferdelman, P. Stoodley, A. Gieseke, D. deBeer. (2008) "Impact of nitrate on the structure and function of bacterial biofilm communities in pipelines used for injection of seawater into oil fields" *Applied & Environmental Microbiology* 74:2841–2851.

Stam, M. C., P. R. D. Mason, C. Pallud, P. Van Cappellen. (2010) "Sulfate reducing activity and sulfur isotope fractionation by natural microbial communities in sediments of a hypersaline soda lake (Mono Lake California)" *Chemical Geology* 278:23–30.

Struchtmeyer, C. G., J. P. Davis, M. S. Elshahed. (2011) "Influence of the drilling formulation process on the bacterial communities in thermogenic natural gas wells of the Barnett Shale" *Applied & Environmental Microbiology* 77(14):4744–4753.

Tabari, K., M. Tabari, O. Tabari. (2011) "Investigation of reservoirs souring reasons after water flooding" *Australian Journal of Basic & Applied Sciences* 5(12):952–954.

Thode, H. G. (1991) "Sulfur Isotopes in Nature and the Environment: An Overview" In: Krouse, H. R. and V. A. Grinenko, *Stable Isotopes in the Assessment of Natural and Anthropogenic Sulphur in the Environment*, SCOPE, New York: John Wiley & Sons Ltd. pp. 1–26.

Xu, D., T. Gu. (2015) "The war against problematic biofilms in the oil and gas industry" *Microbial & Biochemical Technology* 7:e124. doi: 10.4172/1948-5948.1000e124.

Zhu, X. Y., J. J. Kilbane. (2005) "Faster and more accurate data collection for microbiologically influenced corrosion" SPE International Symposium on Oilfield Chemistry, Paper no. SPE-93089-PP.

18 Microbiologically Induced Corrosion Associated with the Wet Storage of Subsea Pipelines (Wet Parking)

Laura L. Machuca

CONTENTS

ABSTRACT

Before and after commissioning of subsea equipment and pipelines, it is typically the case that operational challenges and delays can lead to prolonged periods of equipment shutdown. To ensure pipelines remain within design specifications, pipelines are filled with treated fluids until commissioning takes place or until operations are restored. These lay-up periods are referred to as wet parking and are common to any subsea operation. The implications of this practice lie in the associated risk of chemical degradation when the wet lay-up period is prolonged beyond the anticipated date of start-up. In addition, if wet parking or pipeline suspension is expected to take place after several years of production, the risk of corrosion increases due to contamination of the pipe internal surface with hydrocarbons, deposits, and biofilms. Therefore, the selection of an appropriate chemical treatment for preservation of subsea pipelines and subsequent wet lay-up poses a significant dilemma for offshore engineers. This chapter discusses several challenges related to the wet parking of subsea pipelines, in particular, the risk of microbiologically influenced corrosion in pipelines filled with treated seawater and monoethylene glycol.

18.1 INTRODUCTION

Offshore exploration and production of oil and gas have increased steadily during the past years. Subsea production has advanced from shallow water to water depths of up to 3,000 meters and represents one of the world's largest energy supplies. The subsea production system involves unique challenges and demands on engineering. It must operate using the most reliable and cost-effective practices to ensure safe and pollution-free operations in an offshore environment. Therefore, it requires specialized designs, equipment, and vessels as well as particular production assurance and asset integrity procedures. In particular, subsea pipelines are increasingly being required to operate at high pressures and temperatures (HP/HT). Challenges associated with HP/HT operations include corrosion and flow assurance issues, which can compromise the viability of any development.

Internal corrosion is the main factor affecting wall thinning during operation. Corrosion is a complex mechanism that depends on fluid composition, water content, flow regime, and operating conditions. In hydrocarbon pipeline systems, internal corrosion damage may be caused by CO_2 corrosion, H_2S stress corrosion cracking (SSCC), and microbiologically induced corrosion (MIC). Material grade selection and the application of production chemicals are essential to control corrosion and flow assurance problems. Production chemicals are broadly applied and include corrosion inhibitors, biocides, hydrate and paraffin inhibitors, demulsifiers, solvents, and glycols. Material selection is critical in the design of a production system and depends on the composition of the produced fluids. Although carbon steel remains the main construction material for oil and gas production facilities, the application of corrosion-resistant alloys (CRAs) for subsea applications have increased drastically.

The complexity of subsea developments can lead to circumstances where the expected performance of materials can be compromised. Before and after commissioning of subsea equipment and pipelines, it is typically the case that operational challenges and delays can lead to prolonged periods of equipment shutdown. Under these conditions, the most practical procedure to ensure pipelines remain within design specifications is to leave the pipelines filled with treated fluids until commissioning takes place or until operations are restored. These lay-up periods are referred to as wet parking and are common to any subsea operation. The implications of this practice lie in the associated risk of chemical degradation when the wet lay-up period is prolonged beyond the anticipated date of start-up. In addition, if wet parking or pipeline suspension is expected to take place after several years of production, the risk of corrosion increases due to contamination of the pipe internal surface with hydrocarbons, deposits, and biofilms. Should corrosion occur in equipment exposed to these fluids during wet parking, corrosion may, in extreme cases, render the equipment unfit for the task it was designed for.

This chapter discusses several challenges related to the wet parking of subsea pipelines, in particular the risk of microbiologically influenced corrosion in pipelines filled with treated seawater.

18.2 WET PARKING AND THE ASSOCIATED RISK OF MIC

18.2.1 Hydrostatic Testing

The most common practical alignment to wet parking is prolonged retention of hydrostatic test fluids. Before any production can start, subsea pipelines and pressure equipment must be hydrotested to verify that equipment does not leak or have manufacturing flaws. This procedure involves filling the system with fluid and pressurizing it above its design maximum allowable operating pressure (MAOP), commonly to 125% MAOP for no more than 8 hours. Seawater is the fluid of choice in the hydrotesting of subsea pipelines because of its low cost and the large volumes required to fill large pipelines. This practice contaminates the interior surface of the lines with microorganisms, salts, and particulate matter. In addition, it is often the case that hydrostatic test water is left in the system for many months and even years before the system is actually commissioned, increasing the likelihood of corrosion. Examples include general corrosion, crevice corrosion, pitting corrosion, differential aeration corrosion, and MIC, depending upon exposure conditions. Corrosion caused by any one or a combination of these mechanisms may reduce pipeline and equipment service life and, in extreme cases, make it unfit for its purpose. The severity of the problem appears to depend upon the chemistry and microbiology of the water that is used, the length of time that the water remains in the line, and the temperature of the system.

Microbiological corrosion of pipelines and tanks during lay-up periods after hydrotesting has been widely documented. In a previous study, two cases of failures were reported to have resulted when hydrotest water was not treated (Darwin et al. 2010). In one case, failure took place before commissioning within a few months of wet parking. For the second case, failure occurred after 3 years of production. In both cases, microorganisms were detected at the 6 o'clock position at areas of localized corrosion. For this reason, a general industry rule of thumb for hydrotesting is that chemical treatments should be applied if the wet parking periods is sought to last beyond 30 days.

18.2.2 Tie-Ins of Subsea Equipment

Ocean water can get into wet-parked pipelines during the installation and commissioning of subsea facilities. Hydrotested subsea pipelines laid up on the ocean floor and sealed up with treated seawater need to have the end caps removed for tie-ins with subsea equipment. This operation will require the connection point on the flowline to be opened, thus exposing the flowline directly to natural seawater. Typically, this procedure is limited to a few hours of exposure. However, unexpected setups can lead to prolonged exposures that result in higher seawater volumes mixing with treated fluids inside pipelines. Unforeseen events delaying tie-in operations may include vessel position loss and vessel breakdown, tie-in tooling failure, or damage to subsea hardware and unfavorable weather conditions. In the case of breakdown of dive systems or poor weather/cyclones, it is possible to experience durations of greater than 1 week exposure. Recently, a laboratory study was conducted to investigate the likelihood of localized corrosion during tie-ins of subsea equipment (Machuca et al. 2014b). Results showed that mixtures of seawater and treated fluids led to localized corrosion of 316L stainless steel mainly due to an increase in oxygen content and microbial activity. The presence of oxygen not only induced pitting corrosion directly, but it also decreased biocide efficiency by chemical degradation. Results from immersion tests indicated that under high concentrations of dissolved oxygen, the efficiency of the THPS was considerably reduced. This was confirmed by the large bacterial colonization of the steel surface regardless of the concentration of biocide applied. Under these conditions, pitting was initiated after as little as 7 days of exposure. Results also showed that if oxygen is restricted to levels below 20 ppb in the system, 316L is protected against pitting at the recommended dosages of chemical treatment for up to 14 days exposure even in the presence of bacteria. At concentrations of DO below 60 ppb, which were obtained by mixtures of 80% treated seawater and 20% raw seawater, 316L exhibited protection against pitting corrosion for up to 28 days. A concentration of 2 ppm of DO was sufficient to induce pitting corrosion on 316L from the first 7 days of exposure to treated seawater. The injection of a sealing gel into the ends of the pipelines before the end cap is removed has been proposed as a method to minimize the ingress of raw seawater into pipelines during tie-ins. It would be assumed that a small quantity of seawater will bypass the gel seal and form a reservoir of raw seawater in the end of the pipeline. It is also the case that after tie-in procedures, subsea pipelines are left in this state for 2–5 years.

18.2.3 Pipeline Suspension and Shut-Ins after Production

In certain instances, parent facilities are required to be shut down for maintenance or refurbishment after several years of production. During this time, pipelines can be left suspended for several months or years while the refurbishment takes place. At the start of this operation, it should be assumed that there is a normal background level of production bacteria in the pipeline along with small amounts of solid deposits, for example, sand. Microorganisms in oil and gas production systems are usually the result of contamination after seawater flooding of the reservoirs or during drilling processes. However, many microorganisms can be indigenous to these ecosystems (Magot et al. 2000). Due to the in situ presence of substantial amounts of electron donors and electron acceptors for bacteria growth, diverse populations of microorganisms with a wide range of physiological and metabolic activities have been found in these systems. The activity of microorganisms has been shown to have detrimental effects on oil and gas production activities. For instance, biological souring and MIC in oilfield systems are commonly associated with the activity of thermophilic sulfate, sulphur, and thiosulfate reducers (TRB) as well as methanogenic archaea and fermenting microorganisms (Liang et al. 2014).

To prevent microbial corrosion issues, subsea pipelines can be flooded with oxygen-free treated fluids before suspension. Ideally, the flooding will also include mechanical pigging to remove deposits that could have accumulated during operations. Small pools of condensate and caps of gas in the top of the line can be expected to remain in the line. During suspension, pipelines will remain depressurized to ambient pressure and temperature. This change in temperature can, however, have a negative impact on microbial activity.

18.3 MIC OF SUBSEA CONSTRUCTION MATERIALS IN SEAWATER

To understand the risk of MIC during wet lay-up periods in subsea pipelines, it is important to review how microorganisms interact with steel surfaces in seawater. Seawater is one of the most severe natural corrosive environments. The aggressiveness of seawater lies in its chloride content, which is known to deteriorate the steel surface. In the majority of ocean waters, the total proportion of dissolved salts in seawater is about 35 parts per thousand and may vary slightly between different sea locations. The corrosive effects of seawater on metals have always been considered to be greater than the corrosion caused by salt solutions having the same composition. This difference is usually attributed to the presence of biological activity in natural environments and some other organic and inorganic molecules that form metallic complexes, which can also catalyze corrosion reactions.

Microorganisms are known to influence metallic corrosion processes, facilitating, initiating, or accelerating them, particularly when the microorganisms are in close contact with the metal surfaces (Beech and Sunner 2004). In natural seawater, metallic surfaces are easily covered by biofilms within a few hours of exposure (Gaylarde and Videla 1995). Both microbial and electrochemical processes simultaneously influence the metal–solution interface (Lewandowski et al. 1997). The role of the microorganisms is either to assist in the establishment of an electrolytic cell (indirect) or to stimulate the anodic or cathodic reactions (direct) (Gu et al. 2011). Both the corrosion mechanism and microorganisms are affected by the nature of the substratum material and environmental conditions with oxygen being one of the most critical variables affecting the structure and composition of biofilms as well as the MIC mechanisms (Machuca et al. 2011a, 2013). Also, mechanisms that rule MIC in freshwater environments seem to be quite different from those observed in seawater and other natural water bodies. Most studies report MIC in seawater as mode of localized corrosion.

Regardless of the particular reactions and processes governing MIC in different environments, there is general agreement that microbial adhesion and the consequent biofilm formation is the determining step in MIC (Beech and Coutinho 2003). Biofilms are sessile communities of microbial cells and extracellular products associated with a substratum (Lens et al. 2003). Virtually, any industrial system where water contacts a surface is susceptible to biofilm formation. Biofilm seem to be a preferential mode of life for microorganisms as it offers a more sustainable environment for their survival, growth, and reproduction. In addition, biofilm development allows mutualistic and synergistic interactions between microorganisms and affords protection from external harsh conditions (Purevdorj and Stoodley 2003).

18.3.1 Carbon Steel

Most intensive MIC research has been conducted on carbon steel as it has been the dominant construction material for subsea pipelines mainly because of its techno-economic feasibility. Carbon steel exposed to temperate natural seawater will rapidly corrode and be covered by layers of corrosion products and microorganisms (Figure 18.1). Previously, the effect of dissolved oxygen on the bacterial community structure developed on carbon steel in seawater was evaluated (Machuca et al. 2011a). The study showed that continuous supply of oxygen favored microbial diversity and uniform corrosion processes in the short term. The water was maintained for up to 4 weeks without replenishment.

It was shown that corrosion was not influenced by the microbial community established on the steel under aerobic conditions. In the absence of oxygen, rust layers were not formed, and a different biofilm community was identified on the steel. However, localized corrosion was observed under oxygen-free conditions. Using denaturing gradient gel electrophoresis (DGGE) and DNA sequencing, nitrate-reducing bacteria (NRB) were identified as the predominant populations in the anaerobic biofilms.

In another study, corrosion of carbon steel was evaluated in stagnant seawater for over a period of 1 year (Lee et al. 2004). Corrosion was shown to be more aggressive under strictly anaerobic conditions compared to aerobic conditions as measured by weight loss and polarization resistance. The authors concluded that aggressive corrosion of steel under anaerobic conditions was triggered by SRB activity.

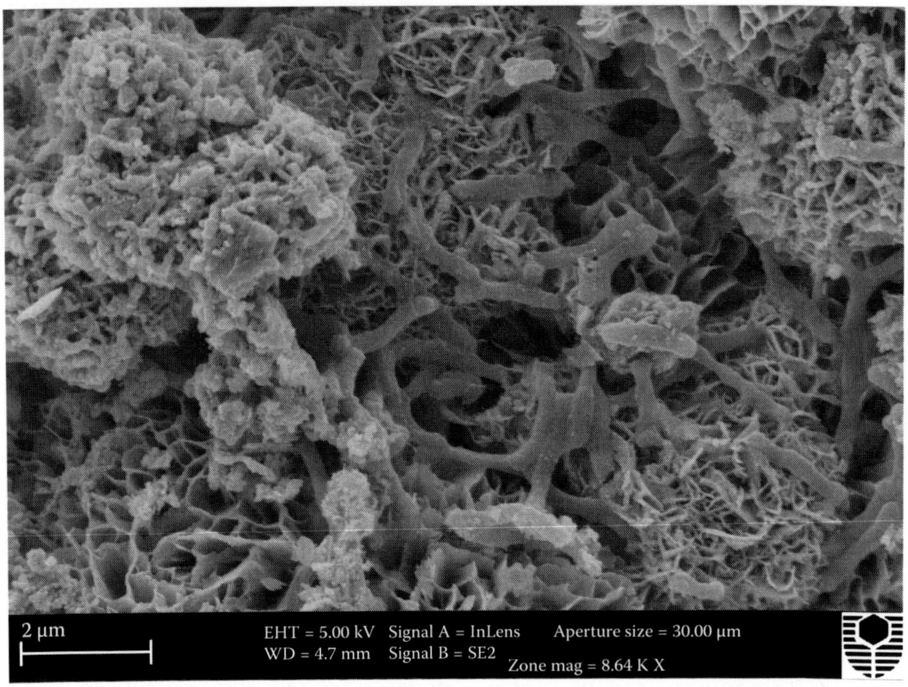

FIGURE 18.1 Microorganisms embedded in corrosion products formed on carbon steel after a few days of exposure to natural seawater.

Conversely, it has been proposed that maximal corrosion activity should be expected where oxygen is available as terminal electron acceptor and oxic/anoxic heterogeneities are created within biofilms (Hamilton 2003). It was postulated that under oxygen-free deposits, SRB can thrive and sulfide production by SRB contributes to accelerate corrosion even if there is oxygen available in the bulk solution.

Other variables, such as flow regime, pH, and ionic strength, have been found to influence MIC of carbon steel in seawater. The effect of flow conditions on MIC has been investigated previously (Zhao et al. 2006). The study reported an increase in cell growth and MIC in solutions under low flow conditions compared to stagnant conditions. The authors attributed this effect to both mass transfer and nutrient distribution limitations under stagnant conditions. Cell growth and MIC were not observed under high flow velocity conditions.

It has been observed that the growth of SRB can be highly influenced by the availability of ferrous ion in the solution (Marchal et al. 2001). In the absence of Fe^{+2}, the incorporation of a steel coupon into the growth medium promoted SRB growth once the pH was lowered to acidic values. The release of Fe^{+2} ions by the corrosion process in acidic medium favored SRB growth. The authors proposed that the excretion of acidic species from bacterial metabolism, for example, acetic acid, can promote the release of Fe^{+2} ions from steel necessary to sustain both corrosion and microbial activity.

The long-term corrosion of carbon steel in seawater has been studied previously (Melchers 2007). The long-term corrosion of carbon steel in seawater seems to have been associated with the activity of SRB, particularly toward the later stages of exposure where corrosion build-up creates suitable conditions for the growth of anaerobes. The authors emphasized that SRB activity is conditioned by the availability of nutrients, particularly Fe^{+2} ions, organic carbon, and nitrogen.

Fe^{2+} oxidation driven by iron-oxidizing bacteria leads to the formation of Fe^{3+} precipitates close to the microbial cells in a process called biomineralization. Iron mineralization can occur in the presence or absence of oxygen, depending on the type of microorganisms involved. Iron oxidation in neutral pH anoxic habitats is carried out by iron-oxidizing nitrate-reducing bacteria (Miot et al. 2009). Due to their low solubility at neutral pH, the Fe (III) by-products of this reaction precipitate rapidly in the direct vicinity of the cells. Biomineralization on corroding steel has been associated with the formation of tubercles and rusticles in seawater environments (Little et al. 2014). Although this process is thought to accelerate corrosion, the presence of tubercles on carbon steel has not always correlated with the onset of localized corrosion (Ray et al. 2011).

18.3.2 Corrosion-Resistant Alloys (CRAs)

In today's oil and gas production, there is a major focus on the application and development of CRAs for tubular and line pipe materials, particularly for infield and downhole piping of subsea structures. This focus has come with the advent of more aggressive service conditions, improved performance of CRAs, and better cost recovery and reliability over traditional carbon steel inhibition systems. It is known that the excellent general corrosion resistance of these alloys is due to the formation of a stable, predominantly chromium oxide layer on the surface that renders them passive (Machuca et al. 2012). However, these alloys are still susceptible to localized corrosion in chloride environments, that is, pitting, crevice corrosion, and MIC. Accordingly, research on the interactions of microorganisms with CRAs has increased dramatically over the past years, particularly on low-grade stainless steels (Acuña et al. 2006; Machuca et al. 2011b).

Several mechanisms have been proposed to describe how microorganisms can affect the integrity of protective oxide films on corrosion-resistant alloys. Proposed mechanisms include microbial production of acidic species, formation of differential concentration cells on the surface, metal-binding properties of the microbial extracellular polymeric substances, and metal deposition, among others. In addition, the main mechanisms of CRA corrosion in seawater, pitting and crevice corrosion, can be initiated and accelerated by biofilms (Zhang and Dexter 1995; Machuca et al. 2013). One of the most interesting mechanisms for MIC of CRAs in seawater is ennoblement of the corrosion potential (E_{corr}), a shift of the corrosion potential to more positive values (Motoda et al. 1990). Since ennoblement does not occur when an alloy is exposed to sterile seawater, the phenomenon has been attributed to microorganisms. Ennoblement is often associated with an increase in cathodic reaction rate that results from either the catalysis of the preexisting cathodic reaction, such as the oxygen reduction in aerated waters, or the reduction of an alternative oxidizer produced by the biofilm. Either thermodynamic or kinetic effects can cause an ennoblement of the corrosion potential (Dickinson et al. 1996; Dexter and Chandrasekaran 2000; Washizu et al. 2004). The significance of ennoblement lies in its influence on the susceptibility to corrosion of anodic materials in galvanic couples and the initiation and propagation of localized corrosion (Mollica 1992). More recently, it has been shown that ennoblement can be the result of direct electron transfer between biofilm and steel in the absence of soluble electron donors or acceptors (Mehanna et al. 2009).

Most studies where ennoblement has been reported used open systems where oxygen and nutrients were plentiful and biofilm activity was not limited. One study reported ennoblement of steels in the presence of sulfate-reducing bacteria (SRB) in an oxygen-free solution (Sahrani et al. 2008). The authors attributed the observed phenomenon to a decrease in pH under the biofilm. However, this was not experimentally demonstrated.

Machuca et al. studied ennoblement of different CRAs in natural seawater under different experimental conditions. Under stagnant closed conditions, biofilms did not ennoble the corrosion potential of CRAs in seawater regardless of the presence of oxygen (Machuca et al. 2013). Conversely, ennoblement of CRAs was observed within few days of exposure to seawater under conditions where seawater was continuously replenished and nutrients and oxygen were not exhausted from the medium (Machuca et al. 2014a). These results suggest that continuous supply of nutrients, rather than oxygen presence, is the prerequisite for ennoblement to take place.

The application of molecular microbiology methods to study biofilm communities on surfaces has proved to be a powerful approach to study microbe–steel interactions. Studies of microbial diversity on CRAs have shown that biofilm microorganisms are highly specific in their preference for substratum surface and oxygen conditions. In the presence of oxygen, microbial diversity is expected to increase due to the presence of both aerobic and anaerobic species.

In terms of substratum surface, biofilms formed on nickel-based alloys have been shown to be structurally different to those formed on stainless steels in seawater (Machuca et al. 2014a). Biofilms on nickel alloys have been shown to be particularly copious, containing significant amounts of EPS (Figure 18.2).

The influence of substratum surface on bacterial adhesion was also studied by Lopes et al. (2005). The adhesion ability of an SRB bacterium onto metallic nickel was higher compared to that on 304 stainless steel and polymethyl methacrylate (PMMA), a nonmetallic material used as experimental control. The SRB growth rate was also shown to be stimulated by the increase in nickel concentrations. The mechanisms of adhesion were not investigated in detail, but the authors suggested that chemotaxis toward a nickel-releasing surface may play a role in microbial adhesion.

Although ennoblement has been observed on different grades of CRAs, no correlation has been found between the extent of ennoblement and localized corrosion susceptibility in these materials. This observation is associated with the known potential temperature dependence of CRAs to initiate localized corrosion (Machuca et al. 2012). Both a critical potential and a critical temperature are required to stabilize pit growth on the alloys' surface with highly resistant alloys exhibiting higher critical potential and temperatures for pit initiation. This indicates that for CRAs such as superduplex and super austenitic steels exposed to temperature seawater MIC does not represent a risk since high temperatures required for passivity breakdown will be detrimental to seawater microorganisms.

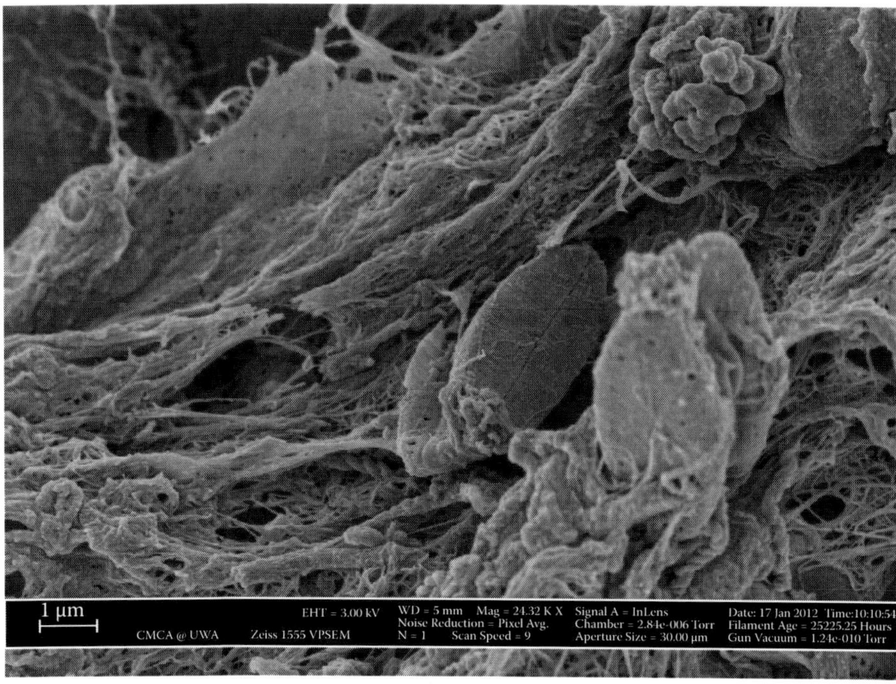

FIGURE 18.2 Biofilm on a nickel-based alloy exposed to natural seawater. Biofilms on nickel alloys have been shown to contain copious amounts of EPS.

18.4 SEAWATER TREATMENTS USED FOR THE WET PRESERVATION OF SUBSEA PIPELINES

In order to protect against corrosion, seawater used for wet parking needs to be treated. Seawater treatments include filtration and right dosages of chemical treatments. Filtration reduces the amount of sediment and nutrients entering the pipeline, decreasing the risk of MIC and under-deposit corrosion (UDC). Chemical treatments may include an oxygen scavenger, a biocide, and a corrosion inhibitor. Biocides have different modes of action against microorganisms, but ultimately they cause changes in the proper functioning of cells and consequently death of the microorganisms at appropriate dosages. If biocides are not applied to hydrotest waters, populations of organisms will thrive in the stagnant water inside the lines, and MIC can occur.

Chemical additives should be compatible with their exposure environment, added to the fluid in a concentration that compensates for foreseeable mixing with any untreated fluids, such as seawater ingress, and be compatible with other additives or added to the fluid in a procedure that prevent their immediate incompatibility, for example, addition of an oxygen scavenger prior to the biocide (Moore et al. 2010). If there is a possibility that temperature of the hydrotest fluid will fall to 0°C or lower during testing or storage, glycol or alternative antifreeze should be added to the hydrotest water. If glycol is used, it should be at a solution concentration of 60% or greater to inhibit microbial corrosion processes.

18.4.1 Biocides

Common biocides used in the treatment of wet parking fluids include glutaraldehyde, quaternary ammonium compounds (quats), and tetrakis hydroxymethyl phosphonium chloride (THPS). These biocides should effectively reach biofilms and limit its activity on surfaces rather than merely disrupt planktonic cells. This is particularly due to the fact that microbial colonization and the consequent biofilm formation has proved to be practically inevitable on a long-term basis, especially in large-scale industrial operations. If preservation fluids remain in contact with equipment or pipelines past their planned time of exposure, the preservation plan should be reconsidered to determine requirement for further action.

There is no single biocide that is the optimum choice for all conditions encountered. Choice of biocide will be a function of the factors including the following:

- Environmental toxicity and ease of disposal: THPS and glutaraldehyde are considered the least harmful to the environment. Treatment of either biocide is equally easily accomplished in the event that neutralization is required.
- Biocidal effectiveness: Glutaraldehyde has been shown to provide better protection than quaternary amines and THPS during wet parking.
- Persistency and half-life of biocides (time before half of the biocide has degraded): This will vary depending on the pH and temperature of the seawater treated.

- Presence of oxygen: In a closed system where a significant amount of oxygen is present due to voids, quaternary amines seem to be the most suitable option to provide protection.

Where it is not possible to ensure sufficient oxygen scavenger mixing time before biocide addition, biocide should be added to the preservation fluid in solid form. Alternatively, an excess concentration of glutaraldehyde, THPS, or quaternary amine should be added.

18.4.2 Oxygen Scavengers

Oxygen scavengers are used to prevent oxygen-related corrosion of carbon steel and crevice corrosion of CRAs. They are only suitable for systems where a limited amount of oxygen is present. The use of an oxygen scavenger, however, is not normally required for carbon steel due to rapid consumption of oxygen in a closed system, resulting in a minimal amount of oxygen-related corrosion. Oxygen scavengers should be used with carbon steel in circumstances where any corrosion is unacceptable due to concerns regarding wall loss or the impact of formation of corrosion products and scales.

Common oxygen scavengers include ammonium and sodium bisulfite. Particular attention should be given to the effect of the oxygen scavenger on microbial corrosion. It has been reported that the sulfite-/bisulfite-based oxygen scavenger can act as a nutrient for bacteria, in particular sulfate-reducing bacteria. Park et al. investigated the effect of sodium bisulfite (SBS) on the microbial community in a brackish water pipeline (Park et al. 2011). Upstream of the SBS injection, corrosion product primarily consisted of $FeCO_3$ with some iron carbonate hydroxide and sand. Conversely, downstream of the SBS injection, corrosion product consisted of 55%–65% $FeCO_3$, 20%–30% sulfur, and 10%–20% FeS. Sulfate, the product from the oxygen removal reaction by SBS, can be reduced by SRB to form sulfide, which precipitates as FeS in the presence of ferrous irons.

18.4.3 Corrosion Inhibitors

Corrosion inhibitors are added to carbon steel systems that are exposed to a significant amount of oxygen and/or acidic conditions. Acidic conditions may result from the biocide dosage used or exposure of glycol to oxygen (acidic degradation products). Water-soluble corrosion inhibitors containing carboxylic acid, and ester-based compounds were shown to enhance the proliferation of the bacteria *Serratia marcescens* ACE 2 and *Bacillus cereus* ACE 4 (Muthukumar et al. 2006). It was shown that bacteria reduced the efficiency of the corrosion inhibitor. Likewise, *Desulfovibrio vulgaris* was reported to assimilate imidazoline, a common corrosion inhibitor, as a source of nutrients and induce localized corrosion of carbon steel (González-Rodríguez et al. 2008).

18.4.4 Monoethylene Glycol (MEG)

In gas pipelines, hydrate formation is one of the main flow assurance concerns that can cause production interruption. MEG is one of the preferred hydrate inhibitors because it has low solubility in the oil phase and is not as volatile as methanol. Therefore, loss of MEG into gas and condensate phase is limited. The use of MEG as a wet parking fluid has recently been proposed due to its known biocidal properties. However, risks associated with its application are not well understood and documented.

Ethylene glycol has been widely used in different industry applications and is known to be highly recalcitrant to biodegradation in high concentrations. Several studies have been conducted to evaluate the ability of microorganisms to biodegrade MEG solutions in low concentrations. However, only a few studies have been reported that specifically address the biocidal effect of MEG.

A field observation on the biocidal effect of MEG mainly used as a heating fluid in an annulus pipe (API 5L X65) to prevent wax formation in a 27-km pipeline was reported (Esaklul et al. 2003). The heating fluid was made up of 30 wt.% MEG + 70 wt.% DI water, and its temperature was 82°C at the inlet and 35°C at the outlet. MEG solution was not treated with biocide. Results showed that MIC was not observed for the duration of the trial (1 year of operation). The authors concluded that glycol in excess of 20 wt.% has a biocidal effect. The solution in this application was constantly flowing.

A laboratory test program was undertaken to determine the biocidal and/or bacteriostatic effects of various concentrations of common alcohols and kinetic hydrate inhibitors used in gas pipelines (Report 2011). Studies were conducted using conventional culture test methods for SRB and mixed cultures of mesophilic bacteria originated from petroleum production water systems (30°C–35°C). Results were based on 28 days exposure to test chemicals. For planktonic microbes (free-swimming), results showed that minimum bacteriostatic concentrations for the cultures were about 12%–14% MEG. In the case of biofilms (sessile microbes), the same bacteriostatic concentrations were observed. The study showed that minimum bactericidal concentrations corresponded to 50% MEG. The authors concluded that wet gas pipelines routinely operating at >15% v/v glycol should be ranked low risk for MIC. Any wet gas line operating at 12%–15% glycol should be ranked moderate risk for MIC. The authors also pointed out the possibility of bacteria adaptation to the toxic habitat and their tolerance to higher MEG concentrations.

In a previous study by Lee et al. (2003), two consortia of microorganisms were used to evaluate bacterial growth in glycol/artificial seawater (ASW) mixtures. Results showed that bacteria were able to use low concentrations of glycol (10%) as nutrient source. Microbial degradation of glycols decreased when increasing glycol concentration from 20% to 50%. The authors concluded that bacteria can grow in glycol concentrations of 10% or less; however, higher concentrations will have bacteriostatic effects.

Microorganisms have also been shown to rapidly adapt to degrade MEG when preferred carbon sources are exhausted (Fincher and Payne 1962). Mrklas et al. (2004) reported that significant levels of MEG were aerobically biodegraded in bioreactors. It was noticed that biodegradation rates were limited by the exhaustion of nutrients, particularly phosphates. Biodegradation rates increased by one order of magnitude after phosphate addition.

Kataoka et al. (2001) studied the capabilities of several yeasts and acetic acid bacteria to oxidize MEG. Higher yields were observed at 10% MEG concentration compared to 20% MEG. Aeration was required for degradation, which indicated that oxygen is required for the oxidation of ethylene glycol.

Evans and David (1974) evaluated biodegradation of glycols in river waters. Concentrations of glycol evaluated were 0, 2, and 10 mg/L. Experiments showed that glycols degraded according to the bacterial state and temperature of the river water. MEG degraded completely at 20°C within 3 days, and bacterial degradation rates decreased at lower temperatures. The variation in the rate of degradation at low temperatures indicates that MEG breakdown is dependent on the microbial numbers and types in individual rivers. At winter river temperatures (<8°C), MEG was degraded partially or completely within 7 days.

Ray et al. (2000) studied crevice and pitting corrosion in the presence of bacteria in 35% glycol with different sources of water for up to 6 months. Results showed that some bacteria grow in the MEG/water mixture (source of water include tap, distilled, and RO). Crevice corrosion was not found in 316L equipped with artificial crevices exposed to MEG solutions inoculated with bacteria. The authors suggested that these results were due to the low conductivity of the solution, which contained up to 0.08% chlorides. It was concluded that bacteria can grow under 35% MEG and that MEG was actually degraded biologically. There was no reference as to whether the bacteria in MEG/water can lead to MIC.

Starosvetsky et al. (2007) reported that at 36°C and under aerobic conditions, proliferation of heterotropic bacteria (HAB) in 0%–10% MEG in tap water increased rapidly in the first 20 days. In 20% and 50% MEG, bacteria number (CFU/mL) decreased in the first 20 days and then gradually increased. The authors suggested that bacteria adapted to the toxic media, which explains their regrowth on more concentrated glycol mixtures. It was indicated that HAB utilized MEG solutions for their metabolism and cell production.

18.5 CURRENT CHALLENGES AND LIMITATIONS

Currently, there are a number of procedures and treatments available from various chemical vendors to prevent corrosion and, in particular, MIC during wet storage of subsea pipelines. However, current concerns for subsea operators are related to the efficiency of treatment chemicals for long lay-up periods (years). This is because it is difficult to accurately predict the effectiveness of the chemical treatments in the long term. Likewise, there is very limited experience and research data on the long-term preservation of subsea pipelines. It is predicted that chemical treatments will naturally degrade when exposed to system conditions. For THPS biocide degradation, a 20% loss every 30 days of exposure has been indicated. Willmon (2010) reported an initial degradation of 25%–55% of the applied biocide concentration over the first 14 days of deployment. This initial degradation was shown to be dependent on the source of water. The bioloading or initial bacterial numbers in the water, temperature, and oxygen content of the seawater were shown to have a direct impact on the initial degradation of the biocide. Mechanisms of biocide degradation include chemical incompatibilities with other chemical additives, such as oxygen scavengers and corrosion inhibitors; light exposure; biodegradation; and oxidation. An increase in seabed temperatures can also increase the degradation kinetics of biocides and stimulate microbial metabolic activity. At doses lower than biostatic, microorganisms can become adapted to chemical biocides and eventually can degrade them (Laopaiboon et al. 2006).

One of the main reasons why the risk of MIC during extended wet parking periods remains uncertain is due to the complexity and lack of understanding of the microbiology in these contained environments. Most of the research intended to address key questions related to MIC during wet parking have been limited to short-term exposures. Under the stagnant conditions developed in contained environments, diffusion and uniform distribution of the biocide inside the lines may not be fully achieved. If the appropriate concentration of biocides does not evenly reach all microorganisms in the seawater and throughout the pipeline, certain microorganisms are likely to survive in the lines even if their metabolic activity becomes restricted. It is worthwhile to keep in mind that microorganisms and, in particular, bacteria within biofilms can develop effective and versatile defense mechanisms against environmental stresses, such as chemical treatments and nutrient scarcity. Microbial cells embedded and physically sheltered in a biofilm matrix are known to be highly resistant to biocides. Metabolic activity in these systems will also depend on the abundances, functional capabilities and physiological states of the microorganisms present in the seawater.

A fascinating and important phenomenon in the field of microbiology is known as dormancy. Dormancy is a physiological state found in certain microorganisms in response to environmental stress, such as nutrient deprivation, where cells differentiate into a slow- or nongrowing condition. A dormant cell is characterized by impaired culturability or its inability to grow on growth media normally conducive to proliferation. Resuscitation from dormancy has been observed when substrate availability is restored. However, an adaptation process, also called the resuscitation phase, seems to be required for the bacteria to fully recover and reactivate in the environment. Dormant cells appear to require specific stimuli or triggers before they become active and culturable. The significance of this is particularly obvious in the assessment of microbiological content in treated samples using cultivation methods. If a big population of microorganisms is in a dormant state, the microbial content of a sample could be grossly underestimated. The culturability of dormant microorganisms have been described as being as little as 10^{-5} (Kell and Young 2000). Resuscitation-promoting factors (Rbf) have been found to play a key role in reactivation of dormant human pathogens (Davandra, Kana, and Mizrahi 2010). The availability of nutrients as well as compounds released by contiguous growing bacteria have also been described as key signals that stimulate resuscitation (Kell et al. 1998). However, little is known about the microbial pathways involved in transitions from dormancy to active states and vice versa in field microorganisms and the environmental cues triggering these transitions.

18.6 CONCLUDING REMARKS

Microbiologically influenced corrosion (MIC) of oil and gas production equipment and pipelines remain an unsolved problem that incurs significant costs and causes serious environmental damage worldwide. In particular, the extended wet preservation of subsea pipelines poses a significant dilemma for offshore engineers. The complexity of recent developments in oil and gas subsea facilities is making it more likely that subsea infrastructure and pipelines are kept under wet storage for long periods of time (months and even years). Over this holding time, also known as wet parking, the activity of residual bacteria and newly introduced bacteria can increase as the effectiveness of the preservation chemicals decays. Today, there is limited experience and experimental data available on the long-term performance of preservation fluids, and as a result, misconceptions, biases, and knowledge gaps continue to persist in the understanding and control of MIC in wet-parked assets. The success of bacteria is given by their enormous diversity and ability to adapt to and thrive in varying and often hostile environments. The activity of microorganisms is controlled by many external factors, including nutrients, physical and chemical variables, and the presence of toxic substances; hence, microbes must be able to adapt to the conditions in their immediate surroundings. In this regard, studies on alternative metabolic pathways and stress response mechanisms in bacteria under unfavorable conditions for bacterial growth are required. Attention should be given to the study of microbial dormancy, a metabolically inactive state that allows bacterial long-term survival in unfavorable circumstances, such as nutrient deprivation, desiccation, and the presence of antimicrobial substances. The presence of nutrients has been considered a key signal that stimulates exit from dormancy. Our limited knowledge of the conditions in which bacteria exist in the environment precludes the identification of key factors affecting microbial growth within contained environments, such as those found in wet-stored subsea pipelines. This should encourage further research on this topic.

Future research in the field should also focus on the application of alternative treatment methods to control microbial activity. Environmental concerns and current legislation associated with disposal of treatment fluids continue to encourage the replacement of toxic biocides with more environmentally friendly antimicrobial approaches that are compatible with system operation and less detrimental to the environment.

REFERENCES

Acuña, N., Ortega-Morales, B. O. and Valadez-Gonzalez, A. 2006. Biofilm colonization dynamics and its influence on the corrosion resistance of austenitic UNS S31603 stainless steel exposed to Gulf of Mexico seawater. *Marine Biotechnology*, 8, 62–70.

Beech, I. B. and Coutinho, C. M. L. M. 2003. Biofilms on corroding materials. *In:* P. Lens, A. P. Moran, T. Mahony, P. Stoody and V. O'Flaherty, (eds.) *Biofilms in Medicine Industry and Environmental Biotechnology: Characteristics, Analysis and Control.* London: IWA Publishing, pp. 115–131.

Beech, I. B. and Sunner, J. 2004. Biocorrosion: Towards understanding interactions between biofilms and metals. *Current Opinion in Biotechnology*, 15, 181–186.

Darwin, A., Annadorai, K. and Heidersbach, K. 2010. Prevention of corrosion in carbon steel pipelines containing hydrotest water—An overview. *Corrosion 2010, Paper N. 10401.* Houston, TX: NACE International.

Davandra Kana, B. and Mizrahi, V. 2010. Resuscitation-promoting factors as lytic enzymes for bacterial growth and signaling. *FEMS Immunology and Medical Microbiology*, 58, 39–50.

Dexter, S. C. and Chandrasekaran, P. 2000. Direct measurement of pH within marine biofilms on passive metals. *Biofouling*, 15, 313–325.

Dickinson, W. H., Caccavo Jr, F. and Lewandowski, Z. 1996. The ennoblement of stainless steel by manganic oxide biofouling. *Corrosion Science*, 38, 1407–1422.

Esaklul, K., Fung, G., Harrison, G. and Perego, R. 2003. Active heating for flow assurance control in deepwater flowlines system, in: Offshore Technology Conference 2003, OTC 15188 May 5–8, Houston, TX, pp. 1–8.

Evans, W. H. and David, E. J. 1974. Biodegradation of mono-, di- and triethylene glycols in river waters under controlled laboratory conditions. *Water Research*, 8, 97–100.

Fincher, E. L. and Payne, W. J. 1962. Bacterial utilization of ether glycols. *Applied Microbiology*, 10(6), 542–547.

Gaylarde, C. C. and Videla, H. A. 1995. *Bioextraction and Biodeterioration of Metals*, Cambridge, UK: Cambridge University Press.

González-Rodríguez, C. A., Rodríguez-Gómez, F. J. and Genescá-Llongueras, J. 2008. The influence of *Desulfovibrio vulgaris* on the efficiency of imidazoline as a corrosion inhibitor on low-carbon steel in seawater. *Electrochimica Acta*, 54, 86–90.

Gu, J. D., Ford, T. E. and Mitchell, R. 2011. Microbiological corrosion of metallic materials. *In:* Revie, R. W. (ed.) *Uhlig's Corrosion Handbook*. Third Edition, Hoboken, NJ: John Wiley & Sons, Inc, pp. 549–558.

Hamilton, W. A. 2003. Microbially influenced corrosion as a model system for the study of metal microbe interactions: A unifying electron transfer hypothesis. *Biofouling*, 19, 65–76.

Kataoka, M., Sasaki, M., Hidalgo, A. G. D., Nakano, M. and Shimizu, S. 2001. Glycolic acid production using ethylene glycol-oxidizing microorganisms. *Bioscience, Biotechnology, and Biochemistry*, 65, 2265–2270.

Kell, D. B., Kaprelyants, A. S., Weichart, D. H., Harwood, C. R. and Barer, M. R. 1998. Viability and activity in readily culturable bacteria: A review and discussion of the practical issues. *Antonie van Leeuwenhoek*, 73, 169–187.

Kell, D. B. and Young, M. 2000. Bacterial dormancy and culturability: The role of autocrine growth factors. *Current Opinion in Microbiology*, 3, 238–243.

Laopaiboon, L., Phukoetphim, N. and Laopaiboon, P. 2006. Effect of glutaraldehyde biocide on laboratory-scale rotating biological contactors and biocide efficacy. *Electronic Journal of Biotechnology*, 9, 358–369.

Lee, J., Ray, R., Lemieux, E., Falster, A. and Little, B. J. 2004. An evaluation of carbon steel corrosion under stagnant seawater conditions. *Biofouling*, 20, 237–247.

Lee, J. S., Ray, R. I., Lowe, K. L., Jones-Meehan, J. and Little, B. J. 2003. An evaluation of microbial growth and corrosion of 316L SS in glycol/seawater mixtures. *Biofouling*, 19, 151–160.

Lens, P., Moran, A. P., Mahony, T., Stoody, P. and O'flaherty, V. 2003. *Biofilms in Medicine Industry and Environmental Biotechnology: Characteristics, Analysis and Control*. London: IWA Publishing.

Lewandowski, Z., Dickinson, W. and Lee, W. 1997. Electrochemical interactions of biofilms with metal surfaces. *Water Science and Technology*, 36, 295–302.

Liang, R., Grizzle, R. S., Duncan, K. E., Mcinerney, M. J. and Suflita, J. M. 2014. Roles of thermophilic thiosulfate-reducing bacteria and methanogenic archaea in the biocorrosion of oil pipelines. *Frontiers in Microbiology*, 5 (89), 1–12.

Little, B. J., Gerke, T. L. and Lee, J. S. 2014. Mini-review: The morphology, mineralogy and microbiology of accumulated iron corrosion products. *Biofouling*, 30, 941–948.

Lopes, F. A., Morin, P., Oliveira, R. and Melo, L. F. 2005. The influence of nickel on the adhesion ability of *Desulfovibrio desulfuricans*. *Colloids and Surfaces B: Biointerfaces*, 46, 127–133.

Machuca, L. L., Bailey, S. I. and Gubner, R. 2012. Systematic study of the corrosion properties of selected high-resistance alloys in natural seawater. *Corrosion Science*, 64, 8–16.

Machuca, L. L., Bailey, S. I., Gubner, R., Watkin, E., Ginige, M. P. and Kaksonen, A. 2011a. Bacterial community structure in natural marine biofilms and the corrosion of carbon steel in seawater. *18th International Corrosion Congress, Paper 371*. Perth, Australia.

Machuca, L. L., Bailey, S. I., Gubner, R., Watkin, E. and Kaksonen, A. 2011b. Microbiologically influenced corrosion of high-resistance alloys in seawater. *Corrosion 11, Paper N. 11230*. Houston, TX: NACE International.

Machuca, L. L., Bailey, S. I., Gubner, R., Watkin, E., Kaksonen, A. and Heidersbach, K. 2013. Effect of oxygen and biofilms on crevice corrosion of UNS S31803 and UNS N08825 in natural seawater. *Corrosion Science*, 67, 242–255.

Machuca, L. L., Jeffrey, R., Bailey, S. I., Gubner, R., Watkin, E., Ginige, M. P., Kaksonen, A. H. and Heidersbach, K. 2014a. Filtration-UV irradiation as an option for mitigating the risk of microbiologically influenced corrosion of subsea construction alloys in seawater. *Corrosion Science*, 79, 89–99.

Machuca, L. L., Murray, L., Gubner, R. and Bailey, S. 2014b. Evaluation of the effect of seawater ingress into 316L lined pipes on corrosion performance. *Materials and Corrosion*, 65, 8–17.

Magot, M., Ollivier, B. and Patel, B. K. C. 2000. Microbiology of petroleum reservoirs. *Antonie van Leeuwenhoek*, 77, 103–116.

Marchal, R., Chaussepied, B. and Warzywoda, M. 2001. Effects of ferrous ion availability on growth of a corroding sulfate-reducing bacterium. *International Biodeterioration & Biodegradation*, 47, 125–131.

Mehanna, M., Basseguy, R., Délia, M. L. and Bergel, A. 2009. Effect of Geobacter sulfurreducens on the microbial corrosion of mild steel, ferritic and austenitic stainless steels. *Corrosion Science*, 51, 2596–2604.

Melchers, R. E. 2007. The effects of water pollution on the immersion corrosion of mild and low alloy steels. *Corrosion Science*, 49, 3149–3167.

Miot, J., Benzerara, K., Morin, G., Kappler, A., Bernard, S., Obst, M., Ferard, C. et al. 2009. Iron biomineralization by anaerobic neutrophilic iron-oxidizing bacteria. *Geochimica et Cosmochimica Acta*, 73, 696–711.

Mollica, A. 1992. Biofilm and corrosion on active-passive alloys in seawater. *International Biodeterioration & Biodegradation*, 29, 213–229.

Moore, J., Keasler, V. and Bennett, B. 2010. Compatibility of tetrakis(hydroxymethyl) phosphonium sulfate (THPS) and ammonium bisulfite (ABS). *Corrosion 2010, Paper N. 10407*. Houston, TX: NACE International.

Motoda, S., Suzuki, Y., Shinohars, T. and Tsujikawa, S. 1990. The effect of marine fouling on the ennoblement of electrode potential for stainless steels. *Corrosion Science*, 31, 515–520.

Mrklas, O., Chu, A., Lunn, S. and Bentley, L. R. 2004. Biodegradation of monoethanolamine, ethylene glycol and triethylene glycol in laboratory bioreactors. *Water, Air, and Soil Pollution*, 159, 249–263.

Muthukumar, N., Maruthamuthu, S. and Palaniswamy, N. 2006. Water-soluble inhibitor on microbiologically influenced corrosion in diesel pipeline. *Colloids and Surfaces B: Biointerfaces*, 53, 260–270.

Park, I., Chatterjee, X., Dong, S. H., Wang, C. W. S., Caffrey, S. M., Jack, T. R., Boivin, J., and Voordouw, G. 2011. Effect of sodium bisulfite injection on the microbial community composition in a brackish-water-transporting pipeline. *Applied and Environmental Microbiology*, 77, 6908–6917.

Purevdorj, B. L. and Stoodley, P. 2003. The role of signaling in biofilm development. *In:* P. Lens, A. P. Moran, T. Mahony, P. Stoody & V. O'flaherty (eds.), *Biofilms in Medicine Industry and Environmental Biotechnology: Characteristics, Analysis and Control*. London: IWA Publishing, pp. 63–78.

Ray, R. I., Little, B. J. and Meehan, J. J. 2000. An evaluation of microbiologicallly influenced corrosion in potential fluids for between-hull spaces. *Corrosion 2010, Paper N. 02453*. Houston, TX: NACE International.

Ray, R. I., Lee, J. S. and Little, B. J. 2011. The anatomy of tubercles on steel. *NACE Corrosion Conference & Expo*. NACE International.

Report, R. 2011. The effect of alcohols and kinetic hydrate inhibitors in gas pipelines on the risk of microbial-influenced corrosion (MIC). Energy Institute.

Sahrani, F. K., Aziz, M., Ibrahim, Z. and Yahya, A. 2008. Open circuit potential study of stainless steel in environment containing marine sulphate-reducing bacteria. *Sains Malaysiana*, 37, 359–364.

Starosvetsky, J., Starosvetsky, D. and Armon, R. 2007. Identification of microbiologically influenced corrosion (MIC) in industrial equipment failures. *Engineering Failure Analysis*, 14, 1500–1511.

Washizu, N., Katada, Y. and Kodama, T. 2004. Role of H2O2 in microbially influenced ennoblement of open circuit potentials for type 316L stainless steel in seawater. *Corrosion Science*, 46, 1291–1300.

Willmon, J. 2010. THPS degradation in the long-term preservation of sub-sea flow-lines and risers. *Corrosion 2010, Paper N. 10402*. NACE International.

Zhang, H. J. and Dexter, S. C. 1995. Effect of biofilm on crevice corrosion of stainless steels in coastal seawater. *Corrosion*, 51, 56–66.

Zhao, W. J., Gu, K. and Nesic, S. 2006. Effects of mass transfer and flow conditions on SRB corrosion of mild steel. *Corrosion 06, Paper N. 06666*. San Diego, CA: NACE International.

19 How Production Chemicals Can Influence Microbial Susceptibility to Biocides and Impact Mitigation Strategies

Timothy J. Tidwell, Victor Keasler, and Renato De Paula

CONTENTS

ABSTRACT

Microorganisms can create multiple challenges for oilfield production, such as microbiologically influenced corrosion (MIC), reservoir souring, and biofouling. Biocides are often applied to production systems to remove biofilm and prevent H_2S production, but the misapplication of biocides can lead to significant issues, including resistance and/or lack of susceptibility to biocide treatment. In addition to biocides, many other chemicals are often injected into produced fluids to mitigate the risks of corrosion, scale, paraffin, asphaltenes, plugging, and phase-separation issues. These chemicals can be easily overlooked by microbiologists when diagnosing and prescribing a biocide to treat a system. In this study, we present a situation that gave a unique opportunity to study the effects of prolonged treatment with water-quality improving chemistries such as solid-controlling chemistries and the negative effects they can have on a microbial mitigation strategy. Our research showed that the continuous, low-dosage application of chemicals to treat suspended solids was causing microbes to become dormant and thus more resistant to biocide treatments.

19.1 INTRODUCTION

Hydrocarbon production systems often require chemical treatments to mitigate issues related to flow assurance, asset integrity, and phase separation. A large body of work exists detailing the practice of production chemistry in oil and gas production systems. A few examples of treatment chemicals include additives for drilling and completion, corrosion inhibitors, scale dispersants, solvents and crystal modifiers for treatment of paraffin, asphaltenes and hydrates, emulsion breakers, foamers, and viscosity modifiers (Kelland 2015; Fink 2011; Stewart 2011). Biocides are also commonly used in oil and gas production systems to treat microbial contamination that can result in (1) degradation of hydrocarbon products, (2) mechanical failures and loss of containment from microbially influenced corrosion (MIC), (3) plugging of filters and exchangers with biomass, and (4) production of hydrogen sulfide, which can lead to health concerns for personnel and increased treatment costs (Boopathy and Daniels 1991; De Paula and Keasler 2012).

The interaction of production chemicals can sometimes result in the deactivation of one or more of the chemical products, resulting in an ineffective treatment or increased chemical treatment cost. Additionally, chemical treatments can also cause unintended consequences, such as creating a stable oil/water emulsion or causing foaming upon application. Specifically for biocidal products, deactivation of a biocide by another unrelated chemical treatment, such as bisulfite-based oxygen scavengers, can result in poor biocidal efficacy, resulting in a loss of microbial control. Production chemists are all too aware of issues around chemical compatibility but often do not consider the potential unintended consequences that these chemicals can create with microbes that can survive and adapt to the harsh microenvironments of a production system. Many production chemicals have a high bioavailability and can cause changes in microbial physiology or may be metabolized for energy. For example, methanol, commonly used as a solvent, and methylamines frequently present in H_2S scavengers are readily degraded by methylotrophic archaea (Costa and Leigh 2014). Moreover, chemicals used for solids dispersion or corrosion inhibition, such as surfactants and metal chelators, may even have a low level of biocidal activity, which can lead to the inducement of microbial dormancy, pressuring microbes to migrate into protective biofilms, or selecting for microbial strains that are more resistant to biocidal treatments altogether.

Often, due to operational constraints, chemical compatibility, or economic concerns, the application of biocides in production systems is carried out at sublethal dosages (Campbell 2011; Tidwell 2015). Previously, studies on the effects of low-dosage biocidal compounds have not been possible due to limitations in microbiological field techniques. As a result, very little research exists on the effects of these low-dosage chemical injections and their impact on microbial populations in the field. Historically, culturing techniques were used to determine microbial activity in production systems and to qualify biocides, but the ability of culture-dependent techniques to reveal more nuanced microbial survival strategies is very limited. Culture methods can only detect 0.1% to 1% of the total microbial population in environmental samples, and many archaeal species are unlikely to be detected as most are virtually unculturable in simple growing conditions (Keasler 2013; Tidwell 2015). As a result, culturing techniques are highly suspect for false negatives due to unculturability, microbial dormancy, and cells existing in a viable but nonculturable (VBNC) state (Oliver 2005, 2010; Keasler 2012). Due to the limitations of culturing methods, culture-independent methods have advanced as the preferred methods for the enumeration of microbes in oilfield systems. Concomitant ATP and AMP quantification is often used to determine the viability and health of a microbial population (Chatterjee 2012; Keasler 2013). The quantification of AMP can help determine the energy state of a population by comparing the ratio of ATP to AMP as the AMP level is a sensitive indicator of the energy status of a cell as seen in Figures 19.1 and 19.2. Using the ratio of ATP to AMP, we can get an index of the ratio called the AMP index (AMP*i*). Using these assays in tandem with culturing methods can better determine if a population is truly losing viability, entering a dormant state, or is unculturable.

Flow cytometry for microbial analysis in the oilfield is a fairly recent development (Tidwell 2015). In brief, flow cytometry allows real-time enumeration of total, viable, and damaged bacterial cells through the use of fluorescent viability stains. For most viability stains, when bacteria cells are damaged (e.g., during biocide treatment), increasing amounts of an added fluorescent stain will accumulate inside the cells. The resulting dimly or brightly fluorescent cells can be individually counted

$$AMPi = \frac{[cATP + cAMP] - [cATP]}{[cATP]}$$

FIGURE 19.1 Relationship between cellular stress and the ATP/AMP ratio.

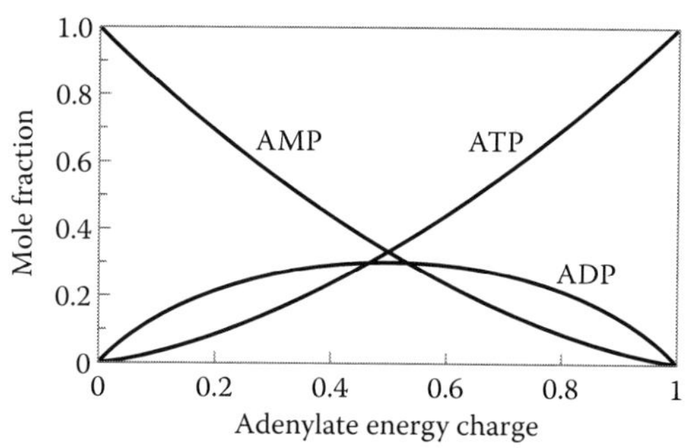

FIGURE 19.2 Adenylate energy charge based on mole fractions of ATP, ADP, and AMP.

when the cells pass through a laser detection system. Fluorescent stains allow for many measures of viability, such as membrane permeability, esterase activity, dehydrogenase activity, and intracellular polarization. Many of these stains, such as those used to measure cellular polarization, allow for indirect chemical measurement of the (*trans*)-membrane potential that can subtly differentiate between a range of physiologic states beyond alive and dead (growth/no growth). Given the complexity of oilfield microbial populations, flow cytometry is an ideal culture-independent method for understanding the effects of sublethal biocide treatments and dormancy on a cell-by-cell basis (Davey and Kell 1996; Freire 2015; Tidwell 2015).

19.2 MICROBIAL BIOCIDE RESISTANCE, DORMANCY, AND PERSISTER CELLS

Most organisms live in unpredictable environments where they typically experience conditions that are less than ideal for growth and reproduction. A very common response to environmental stress is for an organism to enter a reversible state of reduced metabolic activity or dormancy (Lewis 2007; Lennon 2011; Orman 2013). By doing so, these organisms can lower their energy expenditures considerably and evade unfavorable conditions, such as exposure to biocides, that might otherwise wipe out an entire population. A summary of these states is listed in Figure 19.3. Organisms must quickly decide to invest resources into generating resting structures and the machinery that is needed for transitioning into and out of a dormant state (Lewis 2007). These dormant organisms must be able to detect and respond to signals associated with the return of favorable conditions, or they may miss narrow opportunities for growth and reproduction before returning to a dormant state. Despite these trade-offs, microbial species from all domains of life have evolved an ability to use dormancy during periods of environmental stress. This strategy is also known as the storage effect, an ecological hypothesis stating that environmental fluctuations drive

Resistance: Microbial community adaptation to toxic chemicals, such as biocides and surface actives, may occur by selectively enriching a population for a resistant strain or gene that can be swapped with the community.

Dormancy: Is a bet-hedging strategy used by microorganisms. It refers to an organism's ability to enter a reversible state of low metabolic activity when faced with unfavorable environmental conditions. Dormant microorganisms generate a seed bank, which comprises individuals that are capable of being resuscitated following environmental change.

Persister cells: Are a different dormant variant of microbial cells that are highly tolerant (if not totally resistant) to antibiotics and biocides. The function of these specialized dormant cells is to ensure the survival of the population in the presence of lethal factors.

Viable but nonculturable (VBNC): Refers to bacteria and archaea that are in a state of very low metabolic activity and do not divide, but are alive and have the ability to become culturable once resuscitated.

FIGURE 19.3 Definitions of different survival phenomena.

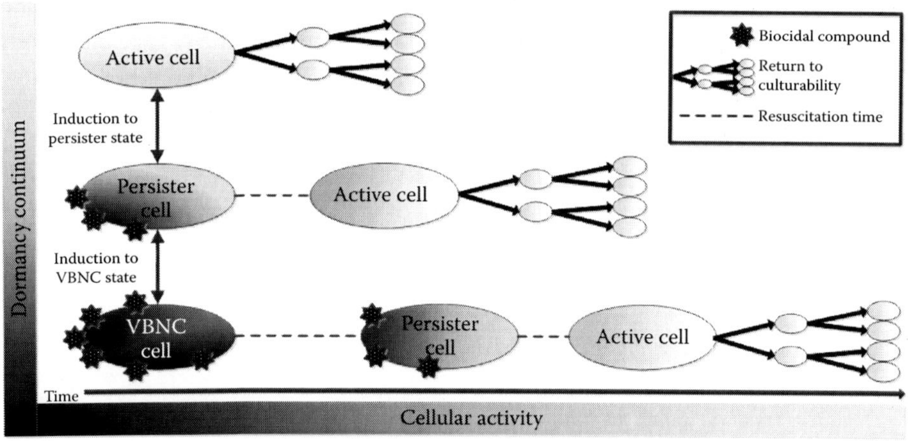

FIGURE 19.4 Microbial dormancy continuum hypothesis. (Adapted from Ayrapetyan, M., Williams, T. C., Oliver, J. D. Bridging the gap between viable but non-culturable and antibiotic persistent bacteria. *Cell* 23(1): 7–13. January 2015.)

temporal variations in population growth that produce long-lived individual organisms, thus promoting multispecies coexistence (Lewis 2007; Lennon 2011; Orman 2013).

Persister cells are a different dormant variant of microbial cells that are highly tolerant (if not totally resistant) to antibiotics and biocides. These specialized dormant cells seen in Figure 19.4 take responsibility to ensure the survival of the population in the presence of lethal factors. Microbial persisters are not a new phenomenon. Persisters were first described in 1944 during the initial studies on penicillin (Lewis 2007; Kaplan 2011). Species-specific phenotypic switches have been found in model organisms, such as *E. coli*, that drastically alter the metabolism of the cell to induce a hibernating-like behavior. These phenotypic switches are not limited to model organisms, and there are likely countless more metabolic survival strategies yet to be discovered. Stress responses, such as exposure to low-dosage biocides, high temperature, or oxygen, may act as general activators of persister formation (Balaban 2004).

Resistance to biocides can also result from traditional adaptation mechanisms. Microbial community adaptation to toxic chemicals, such as biocides and surface actives, can occur by selectively enriching a population for toxin-resistant strains or genes encoding degradation pathways or antitoxins that can be swapped with the community via horizontal gene transfer (Oh 2013).

19.3 UNINTENDED CONSEQUENCES FROM PRODUCTION CHEMICALS

The Eagle Ford region of South Texas is a sedimentary rock formation consisting of organic-rich, pyritic, fossiliferous marine shale. The development of hydraulic fracturing allowed the recent development of the field for recovery of tight oil, condensate, and gas. The formation temperatures for the Eagle Ford Shale are notoriously high, ranging from 190°F–325°F. It was initially thought that these very high temperatures would inhibit microbial growth, and many wells were completed with poor quality water with little to no consideration of a microbial control program; however, despite the very high bottom-hole temperatures, microbial issues have plagued the region.

In this study, we examined an oil-producing field near Kennedy, TX, that presented with a severe MIC problem. After numerous corrosion failures throughout the

FIGURE 19.5 MIC damage to the production system. (Photos courtesy of Matt Butler.)

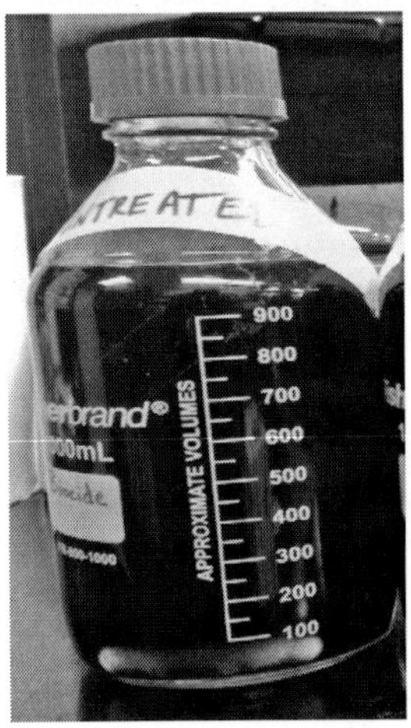

FIGURE 19.6 Produced water sample with very high levels of suspended iron sulfide.

system (Figure 19.4), particularly after oil/water separation where the temperature had dropped to 30°F–40°F (Figures 19.5 and 19.6).

Initially, field-based biocide efficacy testing trials were initiated using standard oilfield microbial culture methods on produced water samples to identify effective biocide treatments. However, when a qualified biocide was implemented, it failed to adequately control microbial growth in the system. It was then determined that a systematic approach was needed to study the issue.

19.4 INITIAL KILL STUDY

Initially, the field engineers identified that the current biocide program was not providing good control of the MIC process. Utilizing microbial growth studies and ATP quantification, a new biocide was identified with significantly better efficacy than the incumbent biocide (Figure 19.7). However, upon implementation of the new biocide in the field (Biocide A), the treatment still showed limited success.

In an attempt to determine the reason for the discrepancy between small-scale testing and large-scale field chemical application of the new biocide, a more thorough laboratory study was initiated using ATP/AMP quantification and flow cytometry.

Generally, a high AMP*i* after biocide treatment is typically indicative of dormancy, stress coping, or another persistence mechanism in microbes. Although ATP/AMP quantification can suggest the physiological state of the entire population, it cannot determine the effects of the biocide on specialized niches of the consortia. Flow cytometry allowed a more detailed look at how the biocide was affecting individual cells.

The dosage response of the incumbent biocide showed that with increasing biocide concentration, the percentage of viable microbes never fell below 7.2%. However, treatment of the same system with Biocide A at 500 ppm allowed 12.6% to survive. Doubling the concentration to 1000 ppm caused a decrease in viable microbes from 12.6% to 1.2%. Doubling the concentration even more to 2000 ppm resulted in only 0.4% surviving, which is a much better biocidal efficacy that was observed for the incumbent biocide.

Despite a somewhat large increase in AMP*i*, indicating the tendency of the microbial population to shift toward dormant or stressed physiology following

FIGURE 19.7 Kill study set up in 30 mL sealed bottles.

treatment with Biocide A, the flow cytometry results demonstrated an effective microbial kill (99.6%) in correlation with field testing results but contrary to actual production data results, which continued to show the biocide was not effective.

Upon a deeper dive to understand the chemical program from the top down, research efforts were then focused on other chemical treatments in place in the system that had been added in conjunction with the biocidal application to determine if these chemicals could possibly be deactivating Biocide A or if something else was occurring. The main additive of concern was a surface active product aimed to control solid deposition (SC), which was injected into the water phase at low doses in the production system to control plugging from suspended iron sulfide solids (Figure 19.6).

SC was introduced into the experimental design described above and analyzed using flow cytometry as seen in Figure 19.8. The data showed that SC alone reduced microbial viability by 50% but increased the number of slightly depolarized microbes, which would indicate a stressed, but viable population of microbes.

In order to simulate the field conditions in the lab, the solids control chemistry was added at 100 ppm as a pretreatment prior to the addition of a biocide. The data in Figure 19.9 demonstrates that when the same study was run with Biocide A, pretreatment with SC seemed to antagonize biocidal efficacy. ATP quantification showed a similar 3-log reduction as before, but AMP quantification indicated that the addition of SC induced a large increase in the AMP*i*, indicating a more stressed/dormant state. A similar result was seen with flow cytometry. Histograms from the study seen in Figure 19.10 show the persistence rate of Biocide A at 2000 ppm was initially 0.4%, but the addition of the SC caused a tenfold increase in the number of viable/persistent microbes. This result showed that the biocide would not be effective unless other changes to the chemical program were made.

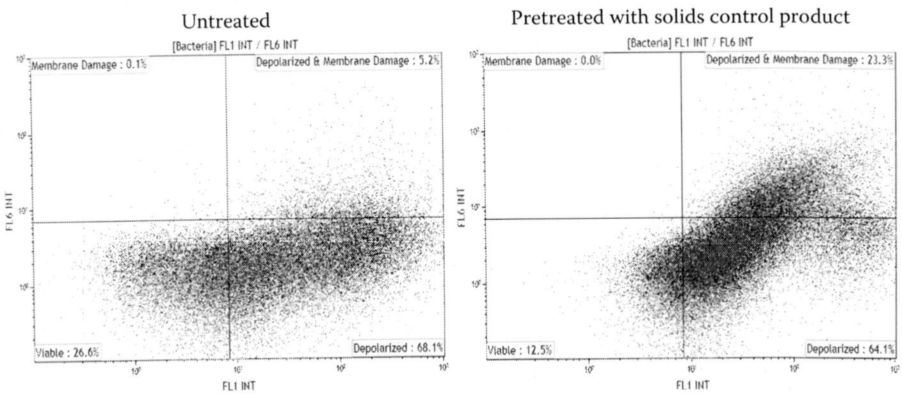

FIGURE 19.8 Flow cytometry histogram showing the response of microbes after being exposed to the solids control chemistry.

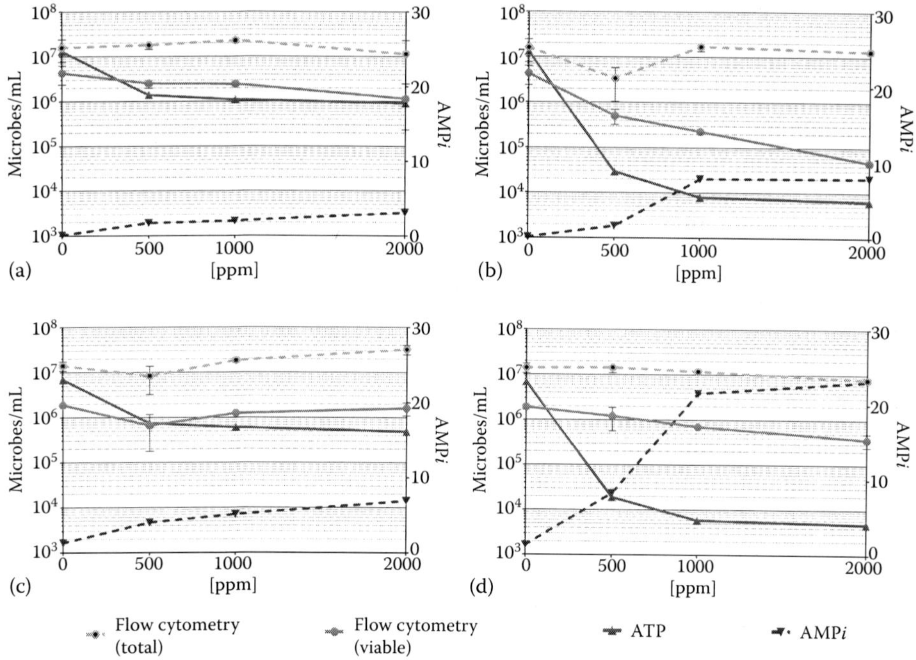

FIGURE 19.9 Initial kill study results. (a) Incumbent biocide, (b) Biocide A, (c) Incumbent biocide + SC, and (d) Biocide A + SC.

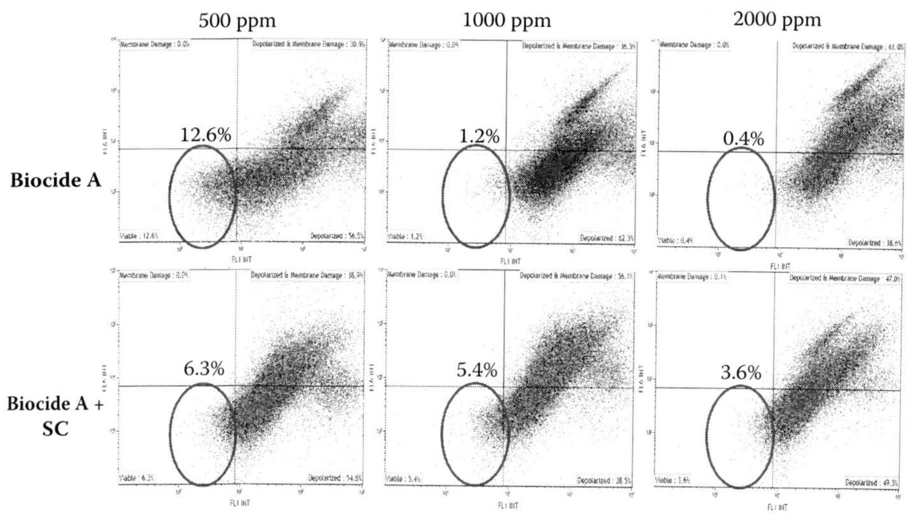

FIGURE 19.10 Initial kill study results showing the flow cytometry histograms after biocide treatment.

19.5 IDENTIFICATION OF ACCEPTABLE SOLIDS CONTROL CHEMISTRIES

Microbial influenced corrosion (MIC) often results in the formation of hydrogen sulfide and iron sulfide (FeS). Therefore, FeS can be an indicator of sulfate reduction by sulfate-reducing prokaryotes (SRP) (Gramp 2010). The ability to control FeS in a fouled system must be combined simultaneously with a requirement to maintain microbial control.

To determine if a solids control chemistry could work to mitigate the issues of the suspended iron sulfide without negatively impacting the efficacy of the biocide, a study looking at four fundamentally different solids control chemistries was

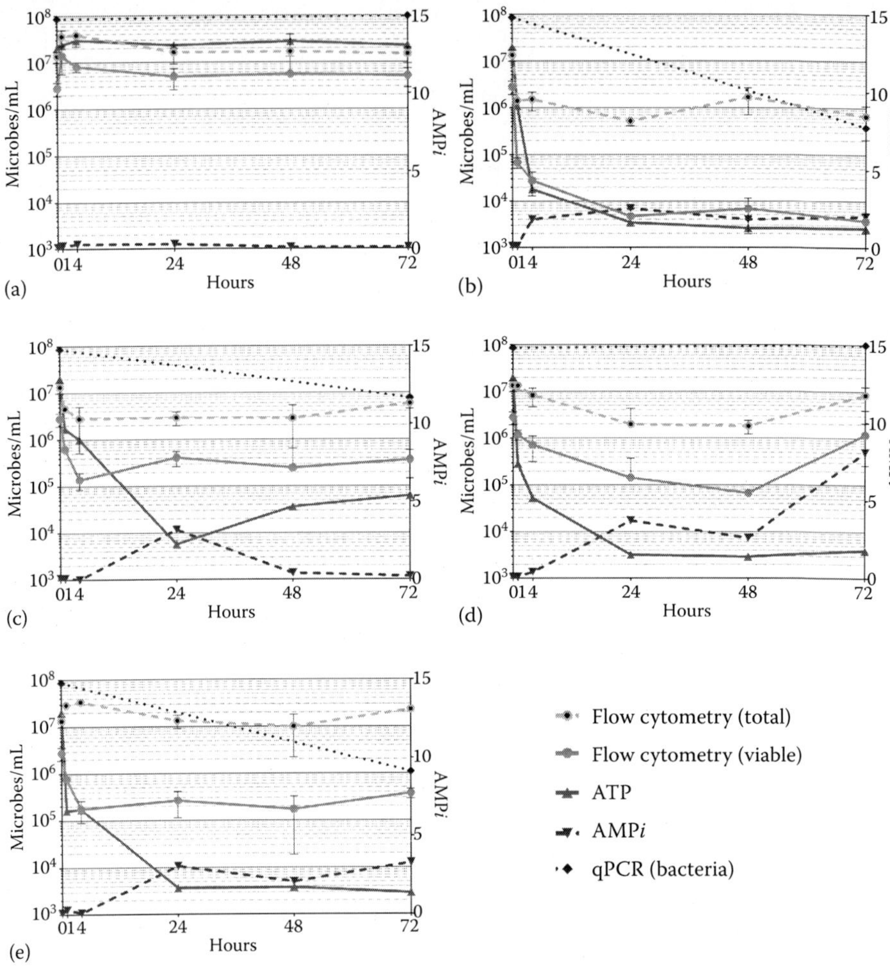

FIGURE 19.11 Results from 72-hour kill study with various solids control products. (a) Untreated, (b) Successful kill, (c) Persistance mechanism 1, (d) Persistance mechanism 2, and (e) Persistance mechanism 3.

performed. For the study, 500 ppm of Biocide A was added after 250 ppm pretreatments of four different candidate solids control products over 72 hours.

Figure 19.11a through e shows the quantitative results from flow cytometry, ATP/AMP testing, and qPCR over the 72-hour exposure to biocides and various SC products. In these tests, samples from triplicate cultures were taken at 0, 1, 4, 24, 48, and 72 hours and analyzed by ATP/AMP quantification and flow cytometry. qPCR was performed at the beginning and end of the study. Figure 19.12a shows the flow cytometry histograms from the duration of the kill study. This data shows a large, stable population of microbes in various metabolic states that were apparently not affected by the experimental setup. One product, seen in Figure 19.11b, demonstrated the ability to treat the water and avoid the dormancy effect. Microbial kill was very effective as measured with ATP and flow cytometry. A rapid 4-log reduction in microbial viability was seen after 24 hours, and the AMP*i* results did not indicate a high level of dormancy induced by the biocide/SC combination. Figure 19.12 shows that the persistence rate fell to only 0.9% after just 4 hours and continued to decline to less than 0.2%

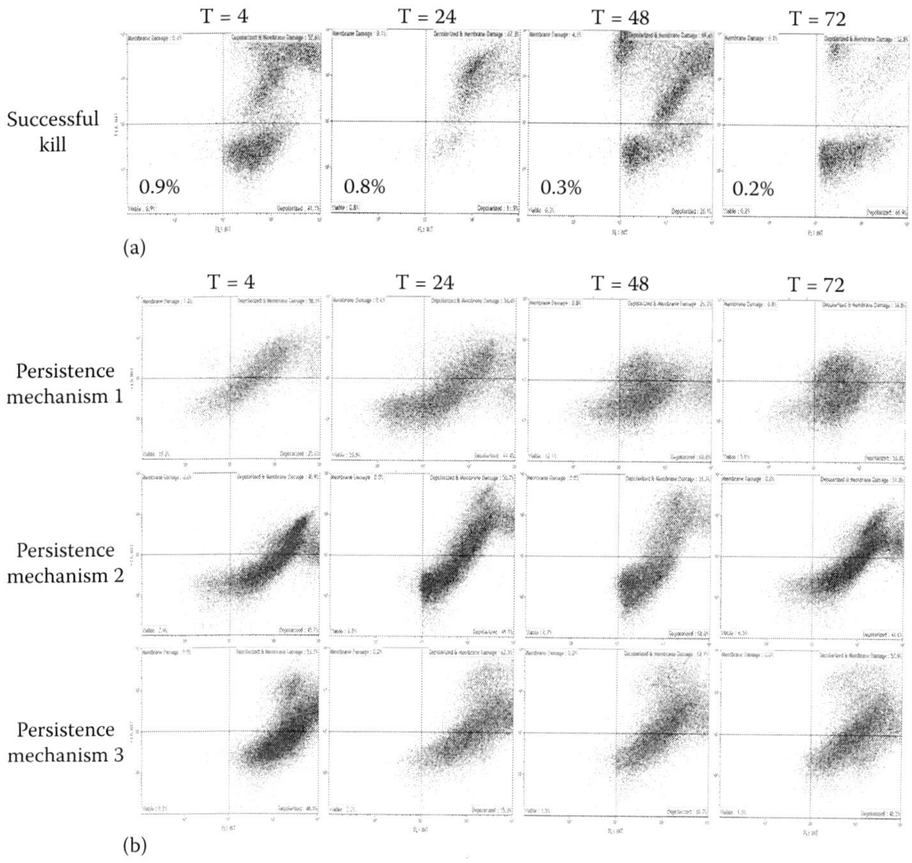

FIGURE 19.12 (a) Flow cytometry histograms showing a successful kill over 72 hours and (b) flow cytometry histograms showing persistence over 72 hours.

after 72 hours. Quantitative PCR (qPCR) also indicated a decrease in total DNA after 72 hours. While these experiments showed it would be possible to treat both problems, the other three SC products produced results that are consistent with a phenotypically different population that would be much more difficult to control (Figure 19.12b).

19.6 COMPARISON OF MICROBIOLOGICAL PERSISTENCE MECHANISMS

It is important to understand the survival mechanisms behind this chemical triggering and how to interpret them with the methods available. A detailed accounting of the data and other testing completed was published in 2015 (Tidwell and Ryther 2015).

Each of the four products produced a distinctive data pattern that we labeled as either a *successful kill* or a *persistence mechanism*. While the precise chemical and biological mechanisms for this observed behavior were not studied here, it is important to oil and gas production chemists to understand how a microbial consortium may behave in a system and understand how to look for these patterns so as to not be surprised by poor biocidal results when treating systems with multiple chemicals. In brief, the failed treatments did produce noticeable 2- to 3-log decreases in ATP. However, flow cytometry was able to parse out more of the story and showed that large decreases with ATP were usually accompanied by dramatic increases in the AMP*i* that often better correlated with viability seen with flow cytometry.

19.7 DISCUSSION

Exposure of microbes to low-dosage chemistries with high bioavailability can cause negative consequences, such as microbial dormancy. Microbial dormancy and persistency is achieved through a diverse set of genetic and cellular mechanisms and can lead to systems that are very difficult to gain control over. While these mechanisms have been extensively studied in human pathogens, the consequences of sublethal biocide treatment in industrial systems are less understood (Lewis 2007; Lennon 2011; Orman 2013). These survival mechanisms may also include those linked with increased microbial biofilm formation, which has been shown to be induced by low-dosage antibiotic exposure (Kaplan 2011). Although this mechanism has not been studied for systems such as oilfield production systems, it seems plausible that exposure to continuous low-dosage biocidal or biostatic chemistries could cause a higher MIC risk by stressing microbes and pushing them to the biofilm phenotype where they are much more difficult to treat and are more likely to cause severe localized corrosion. Although, these studies show that under biocidal treatment conditions, the dormant microbes were not very metabolically active, the ability of microbes to quickly recover and become active was illustrated.

19.8 CONCLUSIONS

Ultimately, these studies proved that systemic changes in the entire chemical program were needed in addition to changes in the biocide to control microbes plaguing the system. The operator of this facility was able to implement the changes

FIGURE 19.13 The effects of MIC on a massive scale in the Eagle Ford Shale in Texas. (Photos courtesy of Matt Butler.)

to the chemical program, but significant challenges remain as years of microbial onslaught have damaged the system extensively (Figure 19.13a through c). While we were not able to definitely prove that our changes in our approach to solids control will result in less MIC, field monitoring has shown that the amended program has provided better control and subsequently decreased the need for solids control as the amount of suspended iron sulfide has dramatically declined since the change.

REFERENCES

Ayrapetyan, M., Williams, T. C., Oliver, J. D. Bridging the Gap between Viable but Non-Culturable and Antibiotic Persistent Bacteria. *Cell*. 23(1): 7–13. January 2015.

Balaban, N. Q. et al. Bacterial Persistence as a Phenotypic Switch. *Science*. 305: 1622–1625. 2004.

Boopathy, R., Daniels, L. Effect of pH on Anaerobic Mild Steel Corrosion by Methanogenic Bacteria. *Applied and Environmental Microbiology*. 57(7): 2104–2108. 1991.

Campbell, S., Duggleby, A., Johnson, A. Conventional Application of Biocides May Lead to Bacterial Cell Injury Rather Than Bacterial Kill Within a Biofilm. NACE. 11234:1–15. 2011.

Chatterjee, I. et al. A New Development for the Microbial Monitoring Toolkit: Real-Time Adenosine Monophosphate (AMP) Technology for Oilfield Applications. Corcon International. 2012.

Costa, K. C., Leigh, J. A. Metabolic Versatility in Methanogens. *Curr Opin Biotechnol.* 2014 Oct; 29: 70–75. doi: 10.1016/j.copbio.2014.02.012. Epub Mar 21, 2014.

Davey, H. M., Kell, D. B. Flow Cytometry and Cell Sorting of Heterogeneous Microbial Populations: The Importance of Single-Cell Analyses. *Microbiological Reviews.* 60: 641–696. 1996.

De Paula, R. M. et al. Optimization of a Microbial Control Program in an Aging Gulf of Mexico Asset to Minimize the Risk of Corrosion. NACE International. C2012-0001195. 2012.

Duncan, K. E. et al. The Effect of Corrosion Inhibitors on Microbial Communities Associated With Corrosion in a Model Flow Cell System. *Applied Microbial Biotechnology.* 98: 907–918. 2014.

Fink, J. *Petroleum Engineer's Guide to Oil Field Chemicals and Fluids.* Gulf Professional Publishing, May 13, 2011.

Freire, J. M. et al. Monitoring of Antibacterial Permeabilization in Real Time Using Time-Resolved Flow Cytometry. *Biochimica et Biophysica Acta.* 1848: 554–560. 2015.

Gramp, J. P. et al. Formation of Fe-Sulfites in Cultures of Sulfate-Reducing-Bacteria. *Journal of Hazardous Materials.* 175: 1062–1067. 2010.

Kaplan, J. B. Antibiotic-Induced Biofilm Formation. *International Journal of Artificial Organs.* 34(9): 737–751. 2011.

Keasler, V. et al. Expanding the Microbial Monitoring Tookit: Evaluation of Traditional and Molecular Monitoring Methods. *International Biodeterioration & Biodegradation.* 81: 51–56. 2013.

Keasler, V. V. et al. Multi-Faceted Approach for Optimizing a Microbial Control Program. SPE International. SPE-154645. 2012.

Kelland, M. A. *Production Chemicals for the Oil and Gas Industry,* Second Edition. CRC Press, Jan 9, 2015.

Kip, N., A van Veen, J. The Dual Role of Microbes in Corrosion. *International Society for Microbial Ecology.* 1–10. 2014.

Lennon, J. T., Jones, S. E. Microbial Seed Banks: The Ecological and Evolutionary Implications of Dormancy. *Nature Reviews Microbiology.* 9: 119–130. 2011.

Lewis, K. Persister Cells, Dormancy and Infectious Disease. *Nature Reviews Microbiology.* 5: 48–56. 2007.

Oh, S. et al. Microbial Community Adaptation of Quaternary Ammonium Biocides as Revealed By Metagenomics. *Environmental Microbiology.* 15(10): 2850–2864. 2013.

Oliver, J. D. Recent Finding on the Viable but Nonculturable State in Pathogenic Bacteria. *FEMS Microbiol Rev.* 34: 415–425. 2010.

Oliver, J. D. The Viable but Nonculturable State in Bacteria. *The Journal of Microbiology.* 43: 93–100. Feb 2005.

Orman, M. A., Brynildsen, M. P. Dormancy Is Not Necessary or Sufficient for Bacterial Peristence. *Antimicrobial Agents and Chemotherapy.* 57(7): 3230–3239. 2013.

Stewart, M., Arnold, K. *Produced Water Treatment Field Manual.* Houston, TX: Gulf Professional Publishing. ISBN: 9781856179843. 2011.

Tidwell, T. J., De Paula, R., Smadi, M. Y., Keasler, V. V. Flow Cytometry as a Tool for Oilfield Biocide Efficacy Testing and Monitoring. *International Biodeterioration & Biodegradation.* 98: 26–34. 2015.

Tidwell, T. J., Ryther, R., Keasler, V. Misapplication of Solids Control Chemistries May Stress Microbes Toward Dormancy. SPE Produced Water Handling & Management Symposium 2015. SPE-17406MS.

20 Two Case Studies of Corrosion from an Injection Water Pipeline in the North Sea

Corrosion Control due to Operation Management and High Corrosion Potential due to Nitrate Mitigation

Irene Roalkvam and Karine Drønen

CONTENTS

ABSTRACT

Oil rigs and installations have a high risk of corrosion due to temperature and pressure conditions, a natural saline environment, and high CO_2 concentration. In addition, high concentrations of sulfate and organic carbon stimulate anaerobic microorganisms that may cause microbially induced corrosion. However, the type of corrosion and severity of corrosion development vary at different oilfields.

Here we present two case studies from a corroding injection water pipeline on an oil rig in the North Sea, and also the adjustments made in operational conditions that provided lower corrosion rates, lower production maintenance costs, and higher security. At the study site, Utsira formation water from the subsea aquifer was used for oil reservoir injection, due to its anaerobic qualities and low sulfate content. However, corrosion-monitoring studies revealed high corrosion rates and aggressive pitting corrosion. The problem was addressed with optimization of the mixing ratio between production water and aquifer water and changing of the carbon source in added chemicals, thereby forming a stable noncorrosive microbial community in the pipeline. Hence, the pitting corrosion in the injection pipeline decreased to low rates.

Furthermore, the effect of NO_3^- mitigation was studied by using advanced in situ flow rig systems in combination with extensive monitoring of chemical and microbial parameters in the pipeline. Experiments showed that NO_3^- mitigation caused aggressive pitting corrosion, and qPCR data confirmed changes in the microbial community such as increased abundance of nitrate reducers and sulfur reducers. We hypothesize that microbially produced nitrite and sulfur-related processes play a main role in this corrosion situation. Hence, NO_3^- can have a severe effect on corrosion development and should not be universally used for mitigation in similar systems without a thorough assessment.

20.1 INTRODUCTION

Since the first successful oil drilling projects in the North Sea during the 1960s, several oilfields have been discovered, surveyed, and exploited by drilling rigs and vessels. In general, the oil rigs have a high risk of chemical corrosion due to the temperature and pressure conditions as well as the high CO_2 concentration. In addition, offshore installations with seawater injection systems are typically vulnerable to microbially induced corrosion (MIC) due to the continuous flow of tepid, sulfate-rich water in the steel pipes. Such conditions are ideal for the growth of sulfate-reducing bacteria (SRB), which are the most studied MIC-causing microbes (Little and Lee 2007; Lenhart et al. 2014). Although the MIC process is not fully understood, microorganisms such as sulfate reducers, iron oxidizers, sulfur oxidizers, iron reducers, nitrate reducers, and acid producers are shown to play a role. Researchers also acknowledge that different mechanisms drive MIC under different environmental conditions (Kakooei et al. 2012; Enning and Garrelfs 2014; Kip and van Veen 2015).

In our studies, samples were retrieved from the oil rig LS in the North Sea. The formation is composed of middle Jurassic sandstone that is part of the Brent group, and the reservoirs are situated at 2300–2700 m deep (Johannesen and Nøttvedt 2008; Norwegian Petroleum Directorate 2015). Drilling was initiated from oil rig LS in 2000, connecting the four wells from the geological structure with seafloor installations. In addition, seafloor installations started production from a second geological structure located further north in 2004. The main production from this field is oil, but gas is also obtained (Norwegian Petroleum Directorate 2015). The main exploiting strategy is water alternating gas (WAG) injection, which provides pressure support during the secondary phase of oil recovery. Aquifer water from the Utsira formation has been used as the injection water instead of seawater typically used on other oil rigs. The Utsira formation is found in the thick Cenozoic postrift succession, approximately 550–1500 m below sea level (Eidvin et al. 1999). The salinity of the Utsira aquifer water is 28.5‰–44.2‰ (with an average of 34.2‰). The temperature measured at the wellheads is 26°C, and the pH varies from 6.3 to 7.4 (with an average of 6.8) as measured after gas equalization with the atmosphere (Drønen et al. 2015). The pristine water is anaerobic and fully saturated with CH_4 and CO_2 and has a low sulfate concentration (<5 mg/L) (Beeder et al. 2007). The corrosion potential of the water was, therefore, expected to be low. Instead, heavy corrosion was experienced. Here we present two case studies of corrosion in the injection water pipeline from the oil rig LS where the Utsira aquifer water is included. In the first case study, the pitting corrosion rate in the injection water pipeline decreased from heavy to low due to operation management of the injection water composition, that is, the mixing ratio between produced water and Utsira aquifer water. In the second case study, heavy corrosion developed in a flow rig with nitrate mitigation while the low corrosion situation remained in the parallel rig without nitrate addition. We hypothesize that nitrite produced by nitrate-reducing bacteria led to metabolic inhibition of other key players in the microbial community, thereby causing a starvation situation that allowed a MIC process of direct electron uptake and disproportionation of sulfur. Our model suggests that this development was caused by a cascade of effects where the nitrate reducing *Arcobacter* seemed to play a key role. We therefore isolated a novel *Arcobacter* strain from the flow rig experiment, which we then characterized and genome sequenced in order to understand its metabolic capacity and a potential role in MIC.

20.2 CASE STUDY 1: CORROSION CONTROL DUE TO OPERATION MANAGEMENT IN INJECTION WATER LINES

20.2.1 Injection Water Pipelines and Corrosion History

Bioprobes and weight loss coupons were installed in the pipelines and semiannually monitored with respect to general and pitting corrosion as well as corrosion product formation. In the period between March 2001 and April 2004, the injection water comprised ~100% Utsira aquifer water in injection water line Alpha. This resulted in a general corrosion of 0.05 mm/year (Figure 20.1c), which is considered a moderate corrosion rate according to NACE standard RP0775 (2005). However, an aggressive pitting corrosion of up to 0.5 mm/year was measured (Figure 20.1c), which is considered a severe corrosion rate. The corrosion products formed in this period were

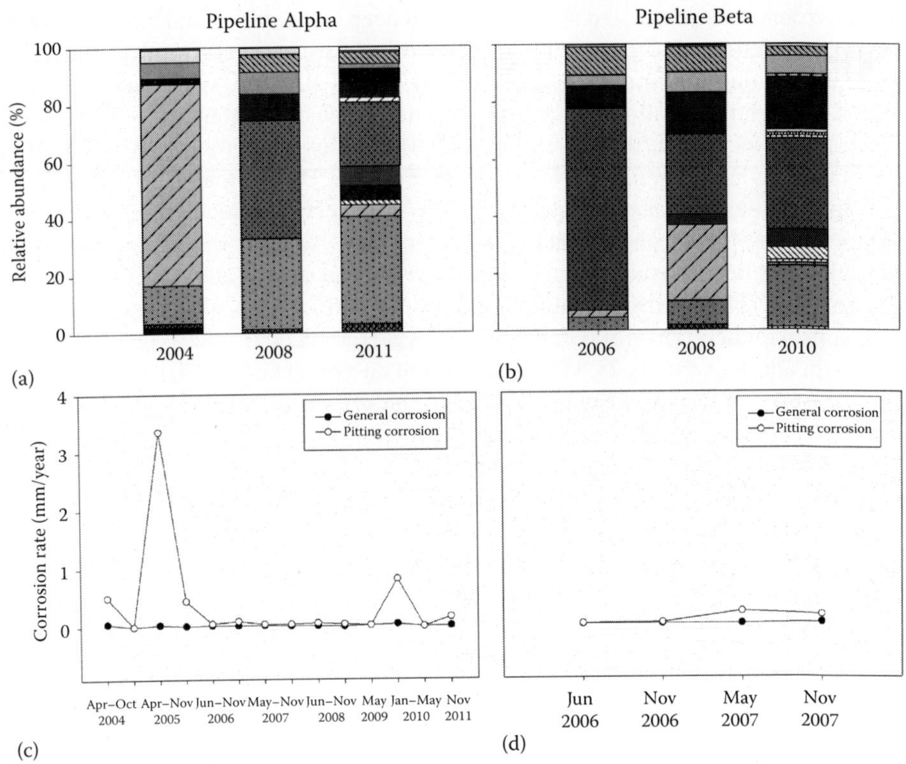

FIGURE 20.1 Taxonomic classification of microbial communities and corrosion rates in lines Alpha and Beta. The taxonomic classification is based on 16S rRNA gene amplicon libraries and results are shown as relative abundance. Community compositions are shown for the years 2004, 2008, and 2011 for line Alpha (a) and for the years 2006, 2008, and 2010 for line Beta (b). *Gammaproteobacteria* ▧; *Deltaproteobacteria* ▢; *Epsilonproteobacteria* ▦; *Zetaproteobacteria* ▧; *Clostridia* ■; *Caldilineae* ▢; *Halobacteria* ▨; *Spiochaetes* ▦; *Synergistia* ▦; *Fusobacteria* ▥; *Thermotogae* ■; *Thermodesulfobacteria* ■; *Thermoplasmata* ▧; *Thermococci* ▢; *Methanomicrobia* ▦; *Methanobacteria* ▨; *Korarchaeota* ▥; *Methanococci* ▩; *Marine Group Ia* ■; <1% at class level ▢; Unclassified ▢. General corrosion and pitting corrosion were calculated from weight loss coupons, analyzed semiannually, from line Alpha (c) and line Beta (d). Only change periods with changing corrosion rates are shown in the graphs.

mainly siderite ($FeCO_3$) but also amorphous iron sulfides. During 2005, the volume of Utsira aquifer water in the injection water decreased steadily to approximately 70% (88% on average), the remainder being produced water (0.1 mg/L SO_4^{2-}, 31.0‰ salinity; Liengen 2016, pers.com). This resulted in an extreme pitting corrosion rate of 3.35 mm/year. To decrease the corrosion rate to acceptable levels, the amount of Utsira aquifer water in the injection water was steadily reduced during the following year while monitoring the corrosion rate semiannually until the Utsira aquifer water concentration stabilized on an average percentage of 59% in 2006. This mixing ratio of Utsira aquifer water to produced water (59:41) seemed to constitute the best injection water quality and has been used since with reported pitting corrosion rates between 0.05 and 0.09 mm/year (Figure 20.1c). After 2006, the general corrosion rate also decreased to 0.01–0.02 mm/year (Figure 20.1c), which is considered a low corrosion condition. A wider range of corrosion products formed in this period, including siderite ($FeCO_3$), magnetite (Fe_3O_4), and iron hydroxyl carbonate ($Fe_6(OH)_{12}(CO_3)$).

The parallel injection water line Beta opened in 2004 when ~100% Utsira aquifer water was used as the sole source for injection water. Despite the changing quality of the injection water during 2004–2006, severe corrosion was never experienced in this pipeline. Thus, general and pitting corrosion were constantly low with a corrosion rate of 0.01–0.03 mm/year and 0.02–0.03 mm/year, respectively. The only exception was a peak in 2007 when the pitting corrosion (0.22 mm/year) reached high rates (Figure 20.1d). Few corrosion products formed until 2007; mainly siderite and magnetite were detected.

20.2.2 High-Throughput Sequencing

In order to study changes in the microbial communities over time, 16S rRNA gene amplicon libraries were constructed from a series of biofilms sampled on bioprobes in the injection water pipeline. DNA was extracted from the bioprobes from the Alpha line (sampled in 2004, 2008, and 2011) and the Beta line (sampled in 2006, 2008, and 2010). In addition, DNA was extracted from the injection water (59:41 mixing ratio of Utsira aquifer water and production water) (sampled in 2008) and two different pipeline inspection gauge (PIG) samples from the Alpha line (sampled in 2010 and 2011). The universal primers Un787F (Roesch et al. 2007) and a modified version of Un1392R (Lane et al. 1985) were used in a two-step PCR that amplified the 16S rRNA gene region of interest and marked each subsample with an individual tag (see Roalkvam et al. 2012, for details). The amplicons were pooled and sequenced using the Roche/454 GS-FLX system at the Norwegian Sequencing Centre. The software AmpliconNoise (Quince et al. 2011) and lowest common ancestor (LCA) algorithm in MEGAN (Huson et al. 2007) were used as bioinformatics tools for the filtering, chimera removal, and taxonomical classification. The amplicon libraries comprised 5,620 reads on average for each sample after filtering. The results are presented as relative abundance and include both archaea and bacteria; a consequence of the universal primers used to target both archaeal and bacterial 16S rRNA genes. However, the varying copy numbers of 16S rRNA genes in different taxa were not taken into consideration. Over- or underestimation of certain taxa due to optimal annealing of primers for some groups but not all groups can unfortunately not be prohibited. Furthermore, the microbial community structures are only shown at class level as the fraction of unclassified reads increases at lower taxonomical levels. This is a result of using the LCA algorithm taxonomic classification as it is based on the upper 10% of the blast scores, thus making it difficult to classify unknown environmental taxa at species level. However, a deeper taxonomical classification was achieved for a small selection of sequences that were individually aligned to the ref_seq database by using Blastn.

20.2.3 The Microbial Community in Injection Water Pipelines Alpha and Beta

The first subsample from the Alpha line was taken in 2004, during a period when ~100% Utsira aquifer water was used in the injection water and the installation suffered pitting corrosion problems. The microbial community in this sample was dominated by Archaea (88.7%), where the main contributors were methanogens within the classes *Methanobacteria* (70.4%), *Methanomicrobia* (13.4%), and *Methanococci* (2.1%) (Figure 20.1a). These taxa are known to produce methane from H_2 and CO_2 or small organic molecules, such as alcohols or formate (Whitman et al. 2014). Methanogens are shown to play an important role in MIC due to their hydrogenotrophic metabolisms. Early hypotheses concerning the MIC process included cathodic depolarization of the steel, followed by removal of the hydrogen layer from the metal surface by the methanogens (Daniels et al. 1987; Zhang et al. 2003; Enning and Garrelfs 2014). More recent studies have also shown that some species among the methanogens can retrieve electrons directly from the metal using an unknown mechanism (Dinh et al. 2004; Mori et al. 2010; Uchiyama et al. 2010). Such a mechanism would be more sustainable and effective compared to the diffusion-dependent depolarization mechanism. In addition, the amplicon library also comprised bacterial reads with *Deltaproteobacteria* (5.6%) and *Epsilonproteobacteria* (2.1%) being most prominent (Figure 20.1a). The representatives of the first taxon are found within *Desulfobacteraceae* and *Desulfobulbaceae*, which are known to a have corrosive metabolisms due to sulfide production and utilization of hydrogen (Muyzer and Stams 2008; Rabus et al. 2013). However, other uncultured representatives within *Deltaproteobacteria* may also have sulfate-reducing capabilities. Therefore, we hypothesize that taxa within *Methanobacteria* and *Deltaproteobacteria* present in the injection water system are responsible for the pitting corrosion observed in the Alpha line in 2004.

Due to the introduced changes in injection water composition after 2004, a different microbial community structure was observed in remaining Alpha line samples (2008 and 2011) and all the Beta line samples (2006, 2008, and 2010). The archaea population in both lines decreased and stabilized with a relative abundance of between 7.8% and 48%. The bacterial communities in both pipelines were dominated by fermenting bacteria (Figure 20.1a and b), such as *Anaerobaculum* and *Thermovirga* within *Synergistia*, *Caminicella* within *Clostridia*, and *Pelobacter* within *Deltaproteobacteria* and *Thermotogae* (Schink and Stieb 1983; Davey et al. 1993; Alain et al. 2002; Menes and Muxí 2002; Dahle and Birkeland 2006; Narasingarao and Häggblom 2007; Frock et al. 2010; Maune and Tanner 2012). Of these, the *Thermovirga* clade was by far the most abundant (<49%). The transition to a bacteria-dominated community could be related to the increasing concentration of organic compounds in the injection water originating from the Utsira aquifer or operational processes at oil rig LS. The pristine aquifer water is naturally low in organic compounds; however, the Utsira formation has been used as a deposit site for slope, cuttings, organic waste, and CO_2 from the oil industry since 1996 (MIT Energy Initiative 2015; Petroleum Safety Authority Petroleum 2015), and leaks have been detected (Petroleum Safety Authority Norway 2015). The cultivated representative, *Thermovirga lienii*, ferments proteinaceous substrates and amino acids but not sugars, fatty acids, or alcohols (Dahle and Birkeland 2006). Thus, fermentation of a protein–amino acid source constantly supplied through the injection water was probably the dominating process in the biofilms. We suggest three possible sources for such a supply: (1) In 2010, it was observed that particulate material (>0.2 μm) sampled at the wellhead of the Utsira aquifer water could support an anaerobe bacterial culture that produced large amounts of nitrate and nitrite. The nitrogen-rich material may originate from the deposited cuttings and slope waste that has been injected into the aquifer in the oil rig LS area. Cutting material in particular can have a high concentration of organic amines, which are commonly used to avoid bacterial growth and souring (Global Drilling Fluids and Chemicals Ltd. 2015; MI Swaco 2015). (2) Cell debris in the injection water, caused by bacterial death in response to the dramatic changes when waters of two very different qualities are mixed, can also represent a protein source. (3) Since 2006, an antiscaling solution containing polyaspartate (20–30 ppm final concentration) (Birketveit 2015, pers. com.) has been added to the injection water. Although growth on aspartate was not tested during the characterization of *T. lienii*, the genomic sequencing information, as presented in the Kegg pathway database (Kanehisa and Goto 2000), indicates that the metabolic trait of aspartate fermentation is present in the species. Fermentation of aspartate would result in formation of fumarate, which could be utilized by members of the *Clostridia* clade (Brisbarre et al. 2003). Furthermore, *Anaerobaculum* is known to ferment a number of sugars and amino acids; however, it did not grow on aspartate (Maune and Tanner 2012). A recent paper reported a wide range of amino acid utilization, including aspartate, when a *Synergistia* strain was cocultured with a methane-producing archaea (Qiu et al. 2014). Hence, *Anaerobaculum* might be involved in fermentation of aspartate in a biofilm with a H_2-scavenging archaeal partner given a homologous mechanism to the one described by Qui and colleges in the *Anaerobaculum* strain. Moreover, fermentation of polysaccharides or sugars of bacterial origin within the biofilm may also serve as a food source for this taxon, and once again, this activity could possibly be enhanced by a H_2-scavenging archaeal partner.

Bacterial taxa with organotrophic or lithotrophic metabolisms were also present in the pipelines, such as *Pseudoalteromonas* within *Gammaproteobacteria*, the sulfate-reducing bacteria within *Deltaproteobacteria*, and the methanogens within *Methanomicrobia* and *Methanobacteria* (Figure 20.1a and b) (Ivanova et al. 1998; Thauer et al. 2008; Rabus et al. 2013). Of these, the *Deltaproteobacteria* are probably responsible for the relatively low corrosion rate (<0.09 mm/year) observed in the Alpha line between 2006 and 2008 (Figure 20.1c).

The microorganisms could feed on the hydrocarbon residues from production water, emulsion inhibitors, corrosion inhibitors, glycol-based drilling fluids and other chemicals that are added to the injection water over time during operation. For example, glycols were initially added to a final concentration of 0.1–0.3 mM in the form of monoethylene glycol (MEG) and ethylene glycol monobutyl ether (EGME) until 2006 when the glycols were replaced by polyaspartate as a scaling inhibitor. Low concentrations of glycol still remain in the system and may originate, for example, from the cuttings and deposits in the Utsira aquifer.

The results suggest that the methanogens and sulfate-reducing bacteria causing MIC in the Alpha line in 2004 were less abundant after 2006, when production water was included in the injection water. Furthermore, these groups might also have changed metabolism from hydrogen consumption from the steel to utilization of hydrogen or small organic molecules produced by the fermenting bacteria. The fermenting bacteria are probably supported by the available hydrocarbon residues or various chemicals in the injection water. Thus, operation management resulted in the formation of a stabile microbial community with low MIC potential in the injection pipeline.

20.2.4 Comparing Microorganisms in Pipeline Biofilm and Water

In order to evaluate how representative the microbial community that established on the bioprobes and pipe wall were compared to the natural community in the injection water pipelines, 16S rRNA gene sequences from analogous subsamples were compared. The samples included bioprobes from the Alpha and Beta lines as well as injection water, all sampled in 2008. PIG samples, that is, pipe wall material from 2010 and 2011, were also included. In short, 16S rRNA gene sequences from all samples were aligned using the MAFFT (multiple alignment using fast Fourier transform at EMBL-EBI) software (Katoh and Standley 2013) with default settings, and representative operational taxonomical units (OTUs) from the samples were taxonomically classified by searching highly similar sequences in the NCBI Reference RNA sequences (refseq_rna) collection using Blastn. Dominating and coinciding OTU sequences were compared between the samples.

The number of OTUs in injection water was 134, which is significantly higher than the 84 and 85 OTUs detected in the Alpha and Beta lines, respectively, indicating a higher diversity in the water compared to the biofilm on the bioprobes. In the Alpha line, 87 OTUs were also present in the wall debris (PIG sample) from 2010 but had diminished to 60 OTUs the year after. The <13 dominant OTUs from each sample were extracted from the data set and studied more thoroughly, revealing that the dominant sequences in the bioprobes and PIG samples were not abundant in the injection water (Figure 20.2). Also, the dominant OTUs in the injection water were not abundant in the other samples except OTU1, which was also abundant in the Beta line (Figure 20.2). Thus, detachment of the PIG material into the injection water was not taking place at measurable levels, and the pipe wall surface therefore seemed to be strongly selective. The injection water was dominated by Archaea where the most abundant OTUs had highest similarity with *Methanothermobacterium* sp./*Methanobacterium* sp. Also, OTUs affiliated with other methanogenic archaea, such as *Methanolobus bombayensis*, *Methanogenium* sp., *Methanococcus thermolithotrophicus*, *Methanoplanus limnicola*, *Methanomicrococcus shengliensis*, and *Methanocalculus pumilus*, were prominent (Figure 20.2). Thus, methane production from H_2 and CO_2 was most likely the dominating process in the water.

In contrast to the injection water, the biofilms from bioprobes and PIGs were dominated by three coinciding sequences with highest similarities to the fermenting *Thermovirga lienii* and *Anaerobaculum hydrogeniformans* within *Synergistia* (Figure 20.2). *Anaerobaculum* and *Thermovirga* have previously been found in pigging material and in biofilms on a metal coupon (Stevenson et al. 2011; Liang et al.

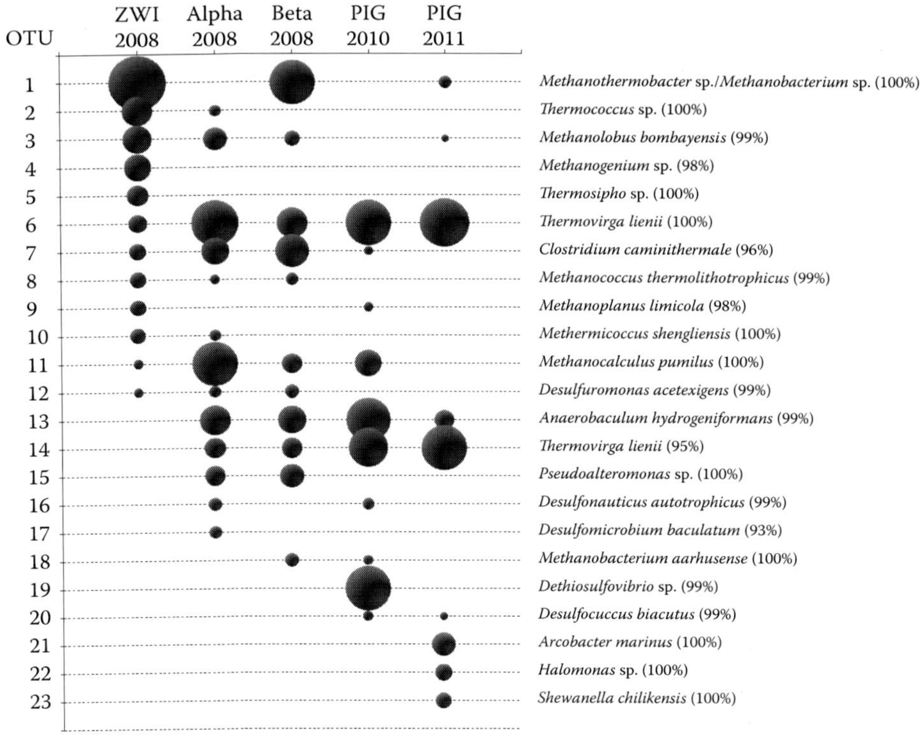

FIGURE 20.2 Comparisons of 23 dominating operational taxonomic units (OTUs) from injection water, bioprobes from the Alpha and Beta pipelines, and two PIG material samples. The 16S rRNA genes of the samples were aligned, and <13 most abundant OTUs in each sample were compared. The OTUs were identified through a Blastn search in the ref_seq database in NCBI, which was based on the sequence similarity and score of the top three hits from the database. The areal of the bubbles represent the relative abundance of the OTUs in each sample.

2014), indicating that members of *Synergistia* could be responsible for the biofilm formation although biofilm formation has not been shown for cultivated representatives of this class. Cells vary in their ability to form biofilms, and the metal surface itself may represent a selective foundation. Furthermore, substrate limitations in the water may stimulate biofilm formation where growth of microbes on the pipe wall is supported by the passage of large amounts of water over time (Drønen et al. 2014). This shows that the microbial community established on the pipe wall was well represented on the bioprobes. The MIC potential of a biofilm on the steel surface is much higher than for microbes in the flowing water, which makes the bioprobes a valuable tool for the MIC monitoring of pipes. The results also revealed a high similarity of dominating OTUs in all the biofilms sampled in 2008–2011. This verifies our previously proposed hypothesis of biofilm stability over the period since 2006.

The microbial communities in biofilm and water were also compared with respect to optimal temperature for growth of the dominating taxa. Besides the moderate thermophilic *Thermovirga* and *Anaerobaculum* clades (Dahle and Birkeland 2006; Maune and Tanner 2012), the PIG samples had high abundance of mesophilic taxa, such as

Dethiosulfovibrio, *Methanocalculus*, *Arcobacter*, *Shewanella*, and *Halomonas* (Mori et al. 2000; Surkov et al. 2001) (Figure 20.2). In general, the thermophilic fermenting bacteria seemed to exist in syntrophy with a mesophilic H_2 scavenging partner. The *Dethisulfovibrio* population was detected in high abundances in the Alpha line following a peak in the pitting corrosion measurements in 2010 (Figure 20.1c). Interestingly, the type strain of *Dethiosulfovibrio*, *D. peptidovorans*, was obtained from a corroding offshore oil well and is known to produce sulfide from elemental sulfur or thiosulfate when growing on amino acids (Magot et al. 1997). The injection water was dominated by thermophilic and hyperthermophilic taxa within *Methanothermobacterium* and *Thermococci*, which have an optimal growth temperature of >65°C (Blotevogel and Fischer 1985; Grote et al. 1999; Wasserfallen et al. 2000; Gorlas et al. 2013). This is higher than the in situ temperature of the injection water (45°C–55°C), indicating that some of the microorganisms in this water originate from the warmer production water. The injection water also comprised high abundances of mesophilic clades, such as *Methanolobus* and *Methanogenium* (Kadam et al. 1994; Chong et al. 2002) (Figure 20.2). However, future studies are needed to reveal if the differences between the dominating water and biofilm microbes are caused by temperature differences, that is, cooling effects of the steel pipe, or if other factors, such as capability of biofilm formation or available syntrophic partners in fermentation, play a more important role.

20.2.5 Concluding Remarks

Heavy corrosion occurred in the injection water Alpha line at oil rig LS when Utsira aquifer water was the sole water source. A 16S rRNA gene amplicon library from the biofilm on the pipe wall in the Alpha line in 2004 revealed a microbial community dominated by *Methanobacteria* and with some *Deltaproteobacteria*, both with MIC potential. Studies of dominating OTUs at a deeper taxonomical level showed a higher similarity between biofilm samples from the Alpha and Beta pipelines or PIGs compared to the injection water itself. Changes in operation management protocols, such as adding production water to the injection water and altering the source of organic compounds, led to a stabilized system with low corrosion rates after 2006. The microbial community was dominated by bacteria with fermentative and heterotrophic metabolisms, in addition to methanogenic *Methanomicrobia*, after 2006.

This case study indicates that if the concentration of organic compounds is kept high enough to enhance growth of fermenting taxa within Bacteria, these taxa could outcompete the growth of the microorganisms with highest MIC potential or feed them with fermentation products, such as H_2 and small organic acids. These substrates should be more favorable than steel.

20.3 CASE STUDY 2: DEMONSTRATING A HIGH CORROSION POTENTIAL IN INJECTION WATER AS TRIGGERED BY NITRATE

20.3.1 Flow Rig Setup and Corrosion Rates

Addition of nitrate is a known strategy to stimulate the nitrate-reducing bacteria to outcompete the SRBs for resources and thereby avoid reservoir souring and corrosion due to lower H_2S concentration (Telang et al. 1997; Myhr et al. 2002; Thorstenson et al. 2002; Larsen et al. 2004). However, nitrate can behave unpredictably in the presence of H_2S and sulfur compounds (Stott et al. 2008). In order to test the effect of nitrate mitigation on our system, two parallel flow rigs were connected to the injection water pipeline (99% Utsira aquifer water, 1% produced water) on oil rig LS in 2003. One rig was supplemented with 5 mM nitrate while the other remained unchanged. Both flow rigs were equipped with weight loss coupons, an electrical resistance (ER) probe, and a linear polarization resistance (LPR) probe for corrosion rate measurements as well as a vessel with biostuds for biomass harvesting for microbial analyses (see Beeder et al. 2007, for details). The experiment lasted for 3 months.

The continuous corrosion rate measurements, using the linear polarization resistance (LPR) probes, revealed very different corrosion development in the two flow rigs. The nonamended rig had a stabile corrosion rate of <0.1 mm/year, which is categorized as moderate (Beeder et al. 2007). The nitrate-amended rig had an initial phase of ~50 days with relative stabile corrosion rates <0.5 mm/year. This was followed by 15 days of aggressive corrosion where the average rate was 3.7 mm/year, but peaks of >200 mm/year also occurred in this period (Beeder et al. 2007). The corrosion rate eventually decreased to <0.5 mm/year in the terminal phase of the experiment. The weight loss coupons confirmed the different corrosion rates in the two rigs with an average rate of 1.4 mm/year in the nitrate-amended rig and only 0.06 mm/year in the nonamended rig. Furthermore, no corrosion products formed in the nonamended rig whereas, 1- to 2-mm thick layers of greigite (Fe_3S_4) and siderite ($FeCO_3$) formed on the coupons in the nitrate-amended rig. Abiotic formation of greigite is possible through the reaction between iron sulfide (FeS) on the steel surface and elemental sulfur, but at neutral pH and in the short time scale of the experiment, the chemical formation would be too modest to explain the greigite formation. An alternative and more efficient mechanism of greigite formation is oxidation of iron sulfide by microbially produced nitrite, similar to the process described by Lin et al. (2009). The sulfide needed for the initial iron sulfide formation could originate from sulfate/sulfur reduction or by disproportionation of sulfur. Sulfate reduction was rejected as the source of sulfide due to the low sulfate concentration and the low sulfate reduction rate measurements (Drønen et al. 2014). However, the presence of sulfur in the injection water has been confirmed by chemical analysis (data Statoil from 2000 to 2002, not published), direct microscopy, and bioassays (Drønen et al. 2014). The latter analysis indicated 1.4 mM sulfur in the Utsira aquifer water in 2010 (Drønen et al. 2014). It was also shown that sulfide and sulfate occurred, indicating sulfur disproportionation ($4S^0 + 4H_2O \rightarrow SO_4^{2-} + 3HS^- + 5H^+$) by the Utsira aquifer water community. Thus, the most likely source of sulfide for the iron sulfide formation during the heavy corrosion was linked to sulfur disproportionation. In addition, high levels of sulfide production from disproportionation probably stimulated polysulfide formation, which made the sulfur more biologically accessible and promoted the formation of greigite by nitrite.

20.3.2 Microbial Community

The main members of the microbial community were identified and quantified using genera-specific primers to target their 16S rRNA genes in quantitative PCR

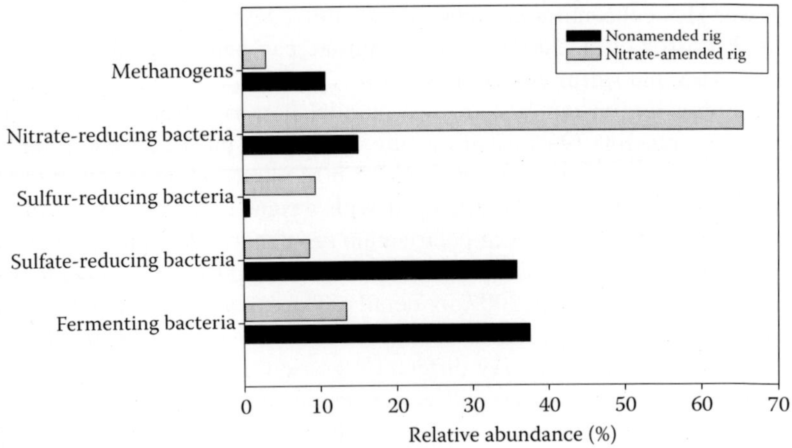

FIGURE 20.3 Bacterial and archaeal taxa quantified with qPCR were divided into functional groups based on known metabolisms of closely related species. The relative abundance of methanogens, sulfate-reducing bacteria, and fermenting bacteria decreased after adding nitrate to the system while nitrate-reducing bacteria and sulfur-reducing bacteria increased.

(qPCR). The 11 primers covered 74% and 87% of the community members in the nitrate-amended rig and nonamended rig, respectively (see Drønen et al. 2014, for details). The highest cell numbers were detected in the nonamended rig, where there were 1.8×10^7 bacterial cells/cm^2 and 2.1×10^6 archaeal cells/cm^2. The nonamended rig was dominated by bacterial taxa associated with the functions of fermentation (*Pelobacter* and *Acetobacter*) and sulfate reduction (*Desulfovibrio* and *Desulfobacter*) (Figure 20.3). In addition, a high abundance of taxa with a possible role in nitrate reduction (*Arcobacter* and *Sulfurospirillum*) and methanol/methylamine degradation (*Methanolobus*) was observed. The quantity of bacteria performing sulfur reduction (*Desulfurimonas*) was relatively low.

In the nitrate-amended rig, bacterial cell numbers were more than threefold lower (5.2×10^6 cells/cm^2), and the archaeal cell numbers were more than 17-fold lower (1.2×10^4 cells/cm^2). When comparing the microbial groups from the two rigs, there were considerably fewer fermenting bacteria, sulfate-reducing bacteria, and methanogens in the nitrate-amended rig (Figure 20.3). The total abundance of the nitrate-reducing bacteria and sulfur-reducing bacteria remained the same. The relative abundance of nitrate-reducing bacteria was then 65.5% as compared to 15.1% in the nonamended rig, and these taxa represented the main metabolic processes resulting in nitrite formation from added nitrate. In contrast, the abundance of sulfur-reducing bacteria increased from 0.8% to 9.4% as a result of the nitrate addition (Figure 20.3).

20.3.3 The Model

Based on the observed corrosion rate development, formation of corrosion products, and knowledge of the dominating microbial taxa in the nitrate-amended rig, a model was formulated to explain heavy corrosion that developed in the system

(Drønen et al. 2014). According to our model, the heavy corrosion rate developed through four steps: (1) The added nitrate encouraged the *Arcobacter* and *Sulfurospirillum* populations to shift metabolism from sulfur reduction to nitrate reduction. This led to increased concentrations of nitrite that metabolically inhibited the fermentation process by restraining the H_2 scavengers in the community, such as the sulfate-reducing *Desulfovibrio* clade. This limited the fermentation process and reduced the availability of small organic compounds for other taxa in the biofilm if not inhibited directly by nitrite. The sulfur-reducing *Desulfuromonas* clade that probably utilized acetate during nonamended conditions also lost its organic source. Instead, these bacteria initiated direct electron uptake from the metal for energy support, a capability confirmed using bioassay experiments (Drønen et al. 2014). (2) The corrosion process was fueled by sulfur disproportionation, where the end product sulfide reacted with anodically available Fe^{2+} to form iron sulfide. In addition, the sulfide probably accelerated the dissolution of sulfur and enhanced polysulfide formation. Possible species involved in sulfur disproportionation were not revealed using our methods. (3) The polysulfide and greigite formation was reinforced by nitrite produced by the nitrate-reducing *Arcobacter* and *Sulfurospirillum* strains. (4) The corrosion proceeded through the electron-conducting greigite layer; however, the buildup of a nonconducting siderite outer layer probably resulted in a decrease in the heavy corrosion rate in the late phase of the experiment.

20.3.4 The Nitrate-Reducing Arcobacter Strain

The genus *Arcobacter* was the most numerous in the nitrate-amended rig, comprising 2.1×10^6 cells/cm^2 or 53.3% of the total community. According to our model, this nitrate-reducing bacterium seemed to play a crucial role in the development of the aggressive corrosion rate. In order to identify properties that could support its role in MIC, we have isolated, characterized, and sequenced the genome of the *Arcobacter* strain from the experiment with the nitrate-amended rig.

The new strain of *Arcobacter anaerophilus*, designated strain IR-1, was isolated from enrichment cultures with Utsira aquifer water, added injection water, iron foil, and elemental sulfur (Roalkvam et al. 2015). Single colonies of strain IR-1 were transferred from plates with Marine Broth 2216 added 5 mM nitrate to liquid culture of the same medium. Thereafter, a growth medium based on anaerobic mineral medium for nitrate reducers (NRB medium) described by Myhr and Torsvik (2000) was used. The cells were gram-negative, curved rods (1.5–2 μm long) and motile by a single polar flagellum. Their temperature range was 15°C–40°C (T_{opt} 37°C), the salinity range was 0.5%–6% NaCl (S_{opt} 2% NaCl), and the optimal pH was 7.2–7.5 (Roalkvam et al. 2015). Maximum growth on acetate and nitrate at optimal growth conditions was calculated to include cell doubling every 1.26 hours. Strain IR-1 had a versatile metabolism and could respire anaerobically with nitrate, elemental sulfur, or ferric citrate and aerobically with 3%–10% oxygen. Inorganic compounds, such as hydrogen and hydrogen sulfide, could be utilized when CO_2 was a carbon source while utilization of elemental sulfur and thiosulfate required small amounts of acetate (1 mM). In addition, the strain grew on acetate, lactate, pyruvate, peptone, tryptone, and yeast extract. Substrates, such as yeast extract, peptone, and pyruvate, were also fermented without an external electron acceptor. Sugars were not utilized.

20.3.5 Implications for the Role of Strain IR-1 in MIC

Although the characterization of the isolate revealed its metabolic potential, genomic sequencing was used in order to understand the processes ongoing inside the cells and also to study the possible roles of *A. anaerophilus* IR-1 in MIC. The reads obtained by sequencing technology from Pacific Biosciences were assembled into seven contigs, comprising 3.257 Mbp of genomic information in total (Roalkvam et al. 2015). The Prokka annotation tool was used to identify 3596 protein coding genes. *A. anaerophilus* IR-1 had the largest genome compared to the other sequenced genomes of *Arcobacter* species, such as the free living *A. nitrofigilis* (Pati et al. 2010) and the pathogen *A. butzleri* (Miller et al. 2007).

The strain could not utilize Fe^0 in culture experiments. In addition, we did not find genes for nanowires or a mechanism for direct electron uptake, such as the one described for taxa within *Desulfuromonas*, *Desulfobacterium*, *Desulfovibrio*, and *Desulfopila* (Dinh et al. 2004; Venzlaff et al. 2013; Drønen et al. 2014). Thus, it is unlikely that strain IR-1 plays a primary role in electrical MIC. Furthermore, genes for disproportionation were not found in the genome. Instead, *A. anaerophilus* IR-1 could be involved in chemical MIC due to metabolic activity, mainly by the production of nitrite, hydrogen sulfide, and sulfuric acid or detoxification by hydrogen sulfide removal. One of the most important steps in the MIC process is the colonization of the steel surface, and biofilm formation has been shown in both pathogenic and free-living genera within *Arcobacter* (Macalady et al. 2008; Ferreira et al. 2013; Pereira-Medrano et al. 2013). During the cultivation experiments of *A. anaerophilus* IR-1, aggregation of cells and biofilm formation was stimulated when sulfur species (elemental sulfur, polysulfide, hydrogen sulfide) were present in the media. This is in accordance with the properties of Utsira aquifer water and injection water where such compounds have been found. In addition, genes encoding pathways for polymerase synthesis could probably provide the extracellular polymeric substances (EPS) needed for biofilm formation. Regarding the metabolic activity by *Arcobacter* in a biofilm, nitrite was considered the most corrosive metabolite produced. Most nitrate-reducing *Epsilonproteobacteria* will produce either N_2 or ammonium during growth (Nakagawa et al. 2007; Kern and Simon 2009; Sikorski et al. 2010; Grote et al. 2012); however, the genomic analysis revealed that *A. anaerophilus* IR-1 has an incomplete denitrification pathway. Only the genes for nitrate reductase (Nap) and nitrous oxide reductase (Nos) were identified, meaning that the reduction of nitrate would form nitrite as the end product in strain IR-1. Sodium nitrite is commonly used as a corrosion inhibitor where the nitrite can form a protective oxidation layer on the steel surface by turning iron hydroxide ($Fe(OH)_2$) into more stable iron oxide compounds (Fe_3O_4 or Fe_2O_3) (Hayyan et al. 2012; Dariva and Galio 2014). The actual concentration of nitrite in the system is critical as lower nitrite concentrations can create an uneven oxidation layer with the formation of anodic and cathodic zones that can cause pitting corrosion. Due to the incomplete denitrification pathway in *A. anaerophilus* IR-1, it is likely that the nitrite concentration would increase in a biofilm dominated by this taxon. This characteristic fits well with the model presented earlier where extreme MIC was observed only in the nitrate-amended rig dominated by *Arcobacter*. Furthermore, the *A. anaerophilus* IR-1 could produce hydrogen sulfide if sulfur was provided as an electron acceptor via the expression of the sulfide:quinone oxidoreductase (Sqr) gene. In a MIC situation, the hydrogen sulfide can react with iron (Fe^{2+}) released at the anodic site, a mechanism that can drive the MIC process effectively (Enning and Garrelfs 2014). The protons released in this reaction can form hydrogen gas or diffuse into the metal to cause sulfide stress cracking (SSC), a form of hydrogen embrittlement that greatly reduces the quality of the steel. A third possible mechanism is the production of sulfuric acid by strain IR-1 in the biofilm due to oxidation of elemental sulfur, thiosulfate, and possibly sulfite, using the SOX complex. The sulfuric acid could result in an acid attack on the steel where hydrogen gas and corrosion products comprising iron sulfate are generated (Panossian et al. 2012). The hydrogen could be utilized by the *Arcobacter* strain or other microorganisms in the biofilm. Finally, the IR-1 strain could perform detoxification of sulfide from disproportionation during oxidation of sulfide coupled to nitrate reduction. This could cause sulfur or polysulfide formation with concomitant greigite transformation.

20.3.6 Concluding Remarks

Different corrosion regimes developed in the two flow rigs at oil rig LS with an extreme corrosion rate in the nitrate-amended rig and a low corrosion rate in the nonamended rig. These results demonstrate that nitrate mitigation is not an optimal solution to corrosion problems in all systems. Based on electrochemical corrosion monitoring and microbial community analysis using molecular methods (such as qPCR), a model was developed hypothesizing that the increasing nitrite concentration from the nitrate-reducing bacteria in the system initiated a cascade of events, including changes in the relative abundance of different taxa in the microbial community and also driving them to use other metabolic processes. A novel nitrate-reducing *Arcobacter* strain IR-1 was isolated from the nitrate-amended rig. Through physiological characterization and genomics of *A. anaerophilus* IR-1, we gained new knowledge about the metabolic properties of environmental *Arcobacter* strains. We hypothesize that *A. anaerophilus* IR-1 could play a role in MIC by nitrite production due to an incomplete denitrification pathway. The *A. anaerophilus* IR-1 may also produce corrosive metabolites, such as hydrogen sulfide or sulfuric acid, or produce sulfur or polysulfide from the oxidation of hydrogen sulfide that can support formation of conductive greigite.

REFERENCES

Alain, K., Pignet, P., Zbinden, M., Quillevere, M., Duchiron, F., Donval, J.-P., Lesongeur, F. et al. 2002. "*Caminicella sporogenes* gen. nov., sp. nov., a novel thermophilic spore-forming bacterium isolated from an East-Pacific Rise hydrothermal vent." *Int. J. Syst. Evol. Microbiol.* 52 (5): 1621–28.

Beeder, J., Andersen, T. R., Liengen, T., Drønen, K., and Torsvik, T. 2007. "Corrosion as a side effect during nitrate treatment of produced water and aquifer water injection." *Corrosion* 2007, 1–15. Tennessee, USA: Corrosion Conference Expo. Paper No. 07512.

Birketveit, Ø. 2015. "M-I SWACO Production Technologies, A Schlumberger Company." Personal communication.

Blotevogel, K. H., and Fischer, U. 1985. "Isolation and characterization of a new thermophilic and autotrophic methane producing bacterium *Methanobacterium thermoaggregans* sp. nov." *Arch. Microbiol.* 142 (3): 218–22.

Brisbarre, N., Fardeau, M. L., Cueff, V., Cayol, J. L., Barbier, G., Cilia, V., Ravot, G., Thomas, P., Garcia, J. L., and Ollivier, B. 2003. "*Clostridium caminithermale* sp. nov., a slightly halophilic and moderately thermophilic bacterium isolated from an Atlantic deep-sea hydrothermal chimney." *Int. J. Syst. Evol. Microbiol.* 53: 1043–49.

Chong, S. C., Liu, Y., Cummins, M., Valentine, D. L., and Boone, D. R. 2002. "*Methanogenium marinum* sp. nov., a H_2-using methanogen from Skan Bay, Alaska, and kinetics of H_2 utilization." *Antonie Van Leeuwenhoek* 81: 263–70.

Dahle, H., and Birkeland, N. K. 2006. "*Thermovirga lienii* gen. nov., sp. nov., a novel moderately thermophilic, anaerobic, amino-acid-degrading bacterium isolated from a North Sea oil well." *Int. J. Syst. Evol. Microbiol.* 56: 1539–45.

Daniels, L., Belay, N., Rajagopal, B. S., and Weimer, P. J. 1987. "Bacterial methanogenesis and growth from CO_2 with elemental iron as the sole source of electrons." *Science* 237 (4814): 509–11.

Dariva, C. G., and Galio, A. F. 2014. "Corrosion inhibitors -principles, mechanisms and applications." In *Developments in Corrosion Protection*, edited by M. Aliofkhazraei, 365–79. Rijeka, Croatia: InTech.

Davey, M. E., Wood, W. A., Key, R., Nakamura, K., and Stahl, D. A. 1993. "Isolation of three species of *Geotoga* and *Petrotoga*: Two new genera, representing a new lineage in the bacterial line of descent distantly related to the *Thermotogales*." *Syst. Appl. Microbiol.* 16 (2): 191–200.

Dinh, H. T., Kuever, J., Muszmann, M., Hassel, A. W., Stratmann, M., and Widdel, F. 2004. "Iron corrosion by novel anaerobic microorganisms." *Nature* 427 (6977): 829–32.

Drønen, K., Roalkvam, I., Beeder, J., Torsvik, T., Steen, I. H., Skauge, A., and Liengen, T. 2014. "Modeling of heavy nitrate corrosion in anaerobe aquifer injection water biofilm: A case study in a flow rig." *Environ. Sci. Technol.* 48 (15): 8627–35.

Drønen, K. A., Liengen, T., Bødtker, G., Heldal, M., Sandaa, R. A., Reigstad, L., Lanzen, A., Øvreaas, L., and Roalkvam, I. 2017. In prep.

Eidvin, T., Riis, F., and Rundberg, Y. 1999. "Upper cainozoic stratigraphy in the central North Sea (Ekofisk and Sleipner fields)." *Norsk Geol. Tidsskr* 79 (2): 97–127.

Enning, D., and Garrelfs, J. 2014. "Corrosion of iron by sulfate-reducing bacteria: New views of an old problem." *Appl. Environ. Microbiol.* 80 (4): 1226–36.

Ferreira, S., Fraqueza, M. J., Queiroz, J. A., Domingues, F. C., and Oleastro, M. 2013. "Genetic diversity, antibiotic resistance and biofilm-forming ability of *Arcobacter butzleri* isolated from poultry and environment from a Portuguese slaughterhouse." *Int. J. Food Microbiol.* 162 (1): 82–8.

Frock, A. D., Notey, J. S., and Kelly, R. M. 2010. "The genus *Thermotoga*: Recent developments." *Environ. Technol.* 31 (10): 1169–81.

Global Drilling Fluids and Chemicals Ltd. 2015. "Amine Biocide Liquid." Accessed October 28. http://www.oil-drilling-fluids.com/biocides/amine-biocide.

Gorlas, A., Alain, K., Bienvenu, N., and Geslin, C. 2013. "*Thermococcusprieurii* sp. nov., a hyperthermophilic archaeon isolated from a deep-sea hydrothermal vent." *Int. J. Syst. Evol. Microbiol.* 63 (8): 2920–6.

Grote, J., Schott, T., Bruckner, C. G., Glöckner, F. O., Jost, G., Teeling, H., Labrenz, M., and Jürgens, K. 2012. "Genome and physiology of a model *Epsilonproteobacterium* responsible for sulfide detoxification in marine oxygen depletion zones." *Proc. Natl. Acad. Sci. U.S.A.* 109 (2): 506–10.

Grote, R., Li, L., Tamaoka, J., Kato, C., Horikoshi, K., and Antranikian, G. 1999. "*Thermococcus siculi* sp. nov., a novel hyperthermophilic archaeon isolated from a deep-sea hydrothermal vent at the Mid-Okinawa Trough." *Extremophiles* 3 (1): 55–62.

Hayyan, M., Sameh, S. A., Hayyan, A., and Alnashef, I. M. 2012. "Utilizing of sodium nitrite as inhibitor for protection of carbon steel in salt solution." *Int. J. Electrochem. Sci.* 7 (8): 6941–50.

Huson, D. H., Auch, A. F., Qi, J., and Schuster, S. C. 2007. "MEGAN analysis of metagenomic data." *Genome Res.* 17 (3): 377–86.

Ivanova, E. P., Kiprianova, E. A., Mikhailov, V. V., Levanova, G. F., Garagulya, A. D., Gorshkova, N. M., Vysotskii, M. V., Nicolau, D. V., Yumoto, N., Taguchi, T., and Yoshikawa, S. 1998. "Phenotypic diversity of *Pseudoalteromonas citrea* from different marine habitats and emendation of the description." *Int. J. Syst. Bacteriol.* 48: 247–56.

Johannesen, E. P., and Nøttvedt, A. 2008. "Norway encircled by coastal plains and deltas." In *The making of a Land—Geology of Norway*, edited by I. B. Ramberg, I. Bryhni, A. Nøttvedt and K. Rangnes, 356–83. Trondheim: Norwegian Geological Association.

Kadam, P. C., Ranade, D. R., Mandelco, L., and Boone, D. R. 1994. "Isolation and characterization of *Methanolobus bombayensis* sp. nov., a methylotrophic methanogen that requires high concentrations of divalent cations." *Int. J. Syst. Evol. Microbiol.* 44 (4): 603–7.

Kakooei, S., Ismail, M. C., and Ariwahjoedi, B. 2012. "Mechanisms of microbiologically influenced corrosion: A review." *World Appl. Sci. J.* 17 (4): 524–31.

Kanehisa, M., and Goto, S. 2000. "KEGG: Kyoto encyclopedia of genes and genomes." *Nucleic Acids Res.* 28 (1): 27–30.

Katoh, K., and Standley, D. M. 2013. "MAFFT multiple sequence alignment software version 7: Improvements in performance and usability." *Mol. Biol. Evol.* 30 (4): 772–80.

Kern, M., and Simon, J. 2009. "Electron transport chains and bioenergetics of respiratory nitrogen metabolism in *Wolinella succinogenes* and other *Epsilonproteobacteria*." *Biochim. Biophys. Acta* 1787 (6): 646–56.

Kip, N., and van Veen, J. A. 2015. "The dual role of microbes in corrosion." *ISME J.* 9 (3): 542–51.

Lane, D. J., Pace, B., Olsen, G. J., Stahl, D. A., Sogin, M. L., and Pace, N. R. 1985. "Rapid determination of 16S ribosomal RNA sequences for phylogenetic analyses." *Proc. Natl. Acad. Sci. U.S.A.* 82 (20): 6955–9.

Larsen, J., Rod, M. H., and Zwolle, S. 2004. "Prevention of reservoir souring in the Halfdan Field by nitrate injection." Corrosion 2004, Louisiana.

Lenhart, T. R., Duncan, K. E., Beech, I. B., Sunner, J. A., Smith, W., Bonifay, V., Biri, B., and Suflita, J. M. 2014. "Identification and characterization of microbial biofilm communities associated with corroded oil pipeline surfaces." *Biofouling* 30 (7): 823–35.

Liang, R., Grizzle, R. S., Duncan, K. E., McInerney, M. J., and Suflita, J. M. 2014. "Roles of thermophilic thiosulfate-reducing bacteria and methanogenic archaea in the biocorrosion of oil pipelines." *Front. Microbiol.* 5 (89): pp 89.

Liengen, T. 2016. "Statoil." Personal communication.

Lin, S., Krause, F., and Voordouw, G. 2009. "Transformation of iron sulfide to greigite by nitrite produced by oil field bacteria." *Appl. Microbiol. Biotechnol.* 2 (83): 369–76.

Little, B. J., and Lee, J. S. 2007. *Microbiologically influenced corrosion*. Edited by R. W. Revie, *Wiley Series in Corrosion*: pp. 28–33. New Jersey: John Wiley & Sons Inc.

Macalady, J. L., Dattagupta, S., Schaperdoth, I., Jones, D. S., Druschel, G. K., and Eastman, D. 2008. "Niche differentiation among sulfur-oxidizing bacterial populations in cave waters." *ISME J.* 2 (6): 590–601.

Magot, M., Ravot, G., Campaignolle, X., Ollivier, B., Patel, B. K., Fardeau, M. L., Thomas, P., Crolet, J. L., and Garcia, J. L. 1997. "*Dethiosulfovibrio peptidovorans* gen. nov., sp. nov., a new anaerobic, slightly halophilic, thiosulfate-reducing bacterium from corroding offshore oil wells." *Int. J. Syst. Bacteriol.* 47 (3): 818–24.

Maune, M. W., and Tanner, R. S. 2012. "Description of *Anaerobaculum hydrogeniformans* sp. nov., an anaerobe that produces hydrogen from glucose, and emended description of the genus *Anaerobaculum*." *Int. J. Syst. Evol. Microbiol.* 62 (4): 832–8.

Menes, R. J., and Muxí, L. 2002. "*Anaerobaculum mobile* sp. nov., a novel anaerobic, moderately thermophilic, peptide-fermenting bacterium that uses crotonate as an electron acceptor, and emended description of the genus *Anaerobaculum*." *Int. J. Syst. Evol. Microbiol.* 52 (1): 157–64.

MI Swaco, a Schlumberger Company. 2015. "Biocides." Accessed October 28. http://www.slb.com/services/miswaco/services/drilling_fluid/df_products/biocides.aspx.

Miller, W. G., Parker, C. T., Rubenfield, M., Mendz, G. L., Wosten, M., Ussery, D. W., Stolz, J. F., Binnewies, T. T., Hallin, P. F., Wang, G. L., Malek, J. A., Rogosin, A., Stanker, L. H., and Mandrell, R. E. 2007. "The complete genome sequence and analysis of the *Epsilonproteobacterium Arcobacter butzleri*." *PLoS ONE* 2 (12): e1358.

MIT Energy Initiative. 2015. "Sleipner fact sheet: Carbon dioxide capture and storage project." Accessed October 28. https://sequestration.mit.edu/tools/projects/sleipner.html.

Mori, K., Tsurumaru, H., and Harayama, S. 2010. "Iron corrosion activity of anaerobic hydrogen-consuming microorganisms isolated from oil facilities." *J. Biosci. Bioeng.* 110 (4): 426–30.

Mori, K., Yamamoto, H., Kamagata, Y., Hatsu, M., and Takamizawa, K. 2000. "*Methanocalculus pumilus* sp. nov., a heavy-metal-tolerant methanogen isolated from a waste-disposal site." *Int. J. Syst. Evol. Microbiol.* 50 (5): 1723–9.

Muyzer, G., and Stams, A. J. 2008. "The ecology and biotechnology of sulphate-reducing bacteria." *Nature Rev. Microbiol.* 6 (6): 441–54.

Myhr, S., Lillebo, B. L., Sunde, E., Beeder, J., and Torsvik, T. 2002. "Inhibition of microbial H_2S production in an oil reservoir model column by nitrate injection." *Appl. Microbiol. Biotechnol.* 58 (3): 400–8.

Myhr, S., and Torsvik, T. 2000. "*Denitrovibrio acetiphilus*, a novel genus and species of dissimilatory nitrate-reducing bacterium isolated from an oil reservoir model column." *Int. J. Syst. Evol. Microbiol.* 50 (4): 1611–9.

NACE International. 2005. Preparation, installation, analysis and interpretation of corrosion coupons in oilfield operations. In *NACE Standard RP0775.*

Nakagawa, S., Takaki, Y., Shimamura, S., Reysenbach, A.-L., Takai, K., and Horikoshi, K. 2007. "Deep-sea vent epsilonproteobacterial genomes provide insights into emergence of pathogens." *Proc. Natl. Acad. Sci. U.S.A.* 104 (29): 12146–50.

Narasingarao, P., and Häggblom, M. M. 2007. "*Pelobacter seleniigenes* sp. nov., a selenate-respiring bacterium." *Int. J. Syst. Evol. Microbiol.* 57 (9): 1937–42.

Norwegian Petroleum Directorate. 2015. "Factpages about norwegian oil fields." Accessed July 6. http://factpages.npd.no/FactPages/.

Panossian, Z., Almeida, N. L. d., Sousa, R. M. F. d., Pimenta, G. d. S., and Marques, L. B. S. 2012. "Corrosion of carbon steel pipes and tanks by concentrated sulfuric acid: A review." *Corros. Sci.* 58: 1–11.

Pati, A., Gronow, S., Lapidus, A., Copeland, A., Del Rio, T. G., Nolan, M., Lucas, S. et al. 2010. "Complete genome sequence of *Arcobacter nitrofigilis* type strain (CI^T)." *Stand. Genomic Sci.* 2 (3): 300–8.

Pereira-Medrano, A. G., Knighton, M., Fowler, G. J. S., Ler, Z. Y., Pham, T. K., Ow, S. Y., Free, A., Ward, B., and Wright, P. C. 2013. "Quantitative proteomic analysis of the exoelectrogenic bacterium *Arcobacter butzleri* ED-1 reveals increased abundance of a flagellin protein under anaerobic growth on an insoluble electrode." *J. Proteomics* 78: 197–210.

Petroleum Safety Authorety Norway. 2015. "Injection wells - status on the NCS." Accessed October 28. http://www.psa.no/well-integrity/injection-wells-status-on-the-ncs-article 6941-900.html.

Qiu, Y.-L., Hanada, S., Kamagata, Y., Guo, R.-B., and Sekiguchi, Y. 2014. "*Lactivibrio alcoholicus* gen. nov., sp. nov., an anaerobic, mesophilic, lactate-, alcohol-, carbohydrate- and amino-acid-degrading bacterium in the phylum *Synergistetes*." *Int. J. Syst. Evol. Microbiol.* 64 (6): 2137–45.

Quince, C., Lanzen, A., Davenport, R. J., and Turnbaugh, P. J. 2011. "Removing noise from pyrosequenced amplicons." *BMC Bioinformatics* 12 (38): 1–18.

Rabus, R., Hansen, T. A., and Widdel, F. 2013. "Dissimilatory sulfate- and sulfur-reducing prokaryotes." In *The Prokaryotes*, edited by E. I. Rosenberg, E. F. DeLong, S. Lory, E. Stackebrandt and F. Thompson, 309–404. Springer Berlin Heidelberg.

Roalkvam, I., Dahle, H., Chen, Y., Jorgensen, S. L., Haflidason, H., and Steen, I. H. 2012. "Fine-scale community structure analysis of ANME in Nyegga sediments with high and low methane flux." *Front. Microbiol.* 3: 216.

Roalkvam, I., Drønen, K., Stokke, R., Daae, F. L., Dahle, H., and Steen, I. H. 2015. "Physiological and genomic characterization of *Arcobacter anaerophilus* IR-1 reveals new metabolic features in *Epsilonproteobacteria*." *Front. Microbiol.* 6: 987.

Roesch, L. F., Fulthorpe, R. R., Riva, A., Casella, G., Hadwin, A. K., Kent, A. D., Daroub, S. H., Camargo, F. A., Farmerie, W. G., and Triplett, E. W. 2007. "Pyrosequencing enumerates and contrasts soil microbial diversity." *ISME J.* 1 (4): 283–90.

Schink, B., and Stieb, M. 1983. "Fermentative degradation of polyethylene glycol by a strictly anaerobic, gram-negative, nonsporeforming bacterium, *Pelobacter venetianus* sp. nov." *Appl. Environ. Microbiol.* 45 (6): 1905–13.

Sikorski, J., Munk, C., Lapidus, A., Ngatchou Djao, O. D., Lucas, S., Glavina Del Rio, T., Nolan, M. et al. 2010. "Complete genome sequence of *Sulfurimonas autotrophica* type strain (OK10^T)." *Stand. Genomic Sci.* 3 (2): 194–202.

Stott, J. F. D., Dicken, G., Rizk, T. Y., Gamble, C., Saul, S., Granli, T., and Haegh, T. 2008. "Corrosion inhibition in PWRI systems that use nitrate to control SRB activity and reservoir souring". Corrosion 2008, New Orleans.

Stevenson, B. S., Drilling, H. S., Lawson, P. A., Duncan, K. E., Parisi, V. A., and Suflita, J. M. 2011. "Microbial communities in bulk fluids and biofilms of an oil facility have similar composition but different structure." *Environ. Microbiol.* 13 (4): 1078–90.

Surkov, A. V., Dubinina, G. A., Lysenko, A. M., Glöckner, F. O., and Kuever, J. 2001. "*Dethiosulfovibrio russensis* sp. nov., *Dethosulfovibrio marinus* sp. nov. and *Dethosulfovibrio acidaminovorans* sp. nov., novel anaerobic, thiosulfate- and sulfur-reducing bacteria isolated from 'Thiodendron' sulfur mats in different saline environments." *Int. J. Syst. Evol. Microbiol.* 51 (2): 327–37.

Telang, A. J., Ebert, S., Foght, J. M., Westlake, D. W. S., Jenneman, G. E., Gevertz, D., and Voordouw, G. 1997. "Effect of nitrate injection on the microbial community in an oil field as monitored by reverse sample genome probing." *Appl. Environ. Microbiol.* 63 (5): 1785–93.

Thauer, R. K., Kaster, A. K., Seedorf, H., Buckel, W., and Hedderich, R. 2008. "Methanogenic archaea: Ecologically relevant differences in energy conservation." *Nat. Rev. Microbiol.* 6 (8): 579–91.

Thorstenson, T., Sunde, E., Bodtker, G., Lillebo, B. L., Torsvik, T., and Beeder, J. 2002. "Biocide replacement by nitrate in sea water injection systems." Corrosion 2002, Colorado.

Uchiyama, T., Ito, K., Mori, K., Tsurumaru, H., and Harayama, S. 2010. "Iron-corroding methanogen isolated from a crude-oil storage tank." *Appl. Environ. Microbiol.* 76 (6): 1783–8.

Venzlaff, H., Enning, D., Srinivasan, J., Mayrhofer, K. J. J., Hassel, A. W., Widdel, F., and Stratmann, M. 2013. "Accelerated cathodic reaction in microbial corrosion of iron due to direct electron uptake by sulfate-reducing bacteria." *Corros. Sci.* 66: 88–96.

Wasserfallen, A., Nölling, J., Pfister, P., Reeve, J., and Conway de Macario, E. 2000. "Phylogenetic analysis of 18 thermophilic *Methanobacterium* isolates supports the proposals to create a new genus, *Methanothermobacter* gen. nov., and to reclassify several isolates in three species, *Methanothermobacter thermautotrophicus* comb. nov., *Methanothermobacter wolfeii* comb. nov., and *Methanothermobacter marburgensis* sp. nov." *Int. J. Syst. Evol. Microbiol.* 50 (1): 43–53.

Whitman, W., Bowen, T., and Boone, D. 2014. "The Methanogenic Bacteria." In *The Prokaryotes*, edited by E. Rosenberg, E. F. DeLong, S. Lory, E. Stackebrandt, and F. Thompson, 123–63. Springer Berlin Heidelberg.

Zhang, T., Fang, H. H., and Ko, B. C. 2003. "Methanogen population in a marine biofilm corrosive to mild steel." *Appl. Microbiol. Biotechnol.* 63 (1): 101–6.

21 Effects of Environmental Conditions on External Microbiologically Influenced Corrosion of Underground Pipelines

A Laboratory and Field Study Using Electrical Resistance Bioprobes

Stefan Jansen, Mirjam van Burgel, Berthil Slim, and Jan Gerritse

CONTENTS

ABSTRACT

Corrosion is one of the leading causes of failures in on- and offshore pipelines. Microbiologically influenced corrosion (MIC) has been identified as one of the major origins of corrosion and can result in locally high metal loss rates. Despite advances in the understanding of MIC, it remains difficult to accurately detect MIC and to predict where and when it will occur. External MIC is strongly affected by local environmental conditions. Relevant factors include soil and groundwater conditions as well as protective measures, like coatings and cathodic protection (CP). In this chapter, case studies are presented of the application of ER bioprobes to assess the effects of environmental conditions on external MIC.

An ER bioprobe was developed based on an electrical resistance (ER) probe. On the metal coupon of an ER probe, a corroding biofilm of sulfate-reducing microorganisms is grown under laboratory conditions. The bioprobe is subsequently inserted in groundwater or soil. Corrosion rates are instantaneously measured with the ER bioprobe, making it possible to simulate and study the development of external MIC of buried pipelines under realistic field conditions.

The effects of soil and groundwater composition on MIC rates were determined both in controlled laboratory conditions as well as in the field. Availability of organic matter, sulfate, and nitrate were shown to be key parameters for MIC. Fluctuating conditions, like the sequential exposure to groundwater with and without oxygen, resulted in a large increase of MIC rates. Groundwater flow had a strong impact on the MIC rates, showing the importance of mass transfer of nutrients and oxygen as a rate-determining factor. The ER bioprobes were also used to study the effectiveness of various levels of CP on MIC.

This research provides practical tools to assess the risk of MIC in specific types of groundwater. Also, the detailed information obtained provides mechanistic understanding enabling the prediction of MIC risk under varying conditions.

21.1 INTRODUCTION

In the oil and gas industry, corrosion is one of the leading causes of failures in on- and offshore pipelines, threatening pipeline integrity both internally and externally. Besides financial consequences, pipeline failures or leakages can have a major impact on safety and the environment (Kruger 2011). Of the different origins of external corrosion, microbiologically influenced corrosion has been identified as one of the major causes. MIC can result in locally high metal loss rates (Flemming 1994). Despite advances in the understanding of MIC, it remains difficult to accurately detect MIC and to predict where and when it will occur (Enning and Garrelfs 2014). In combination with high metal loss rates, these characteristics present challenges to implement effective corrosion management of underground pipeline systems for which external MIC is a serious threat.

The activity of microorganisms in biofilms at the metal surface can enhance corrosion by an alteration of the local conditions. External MIC is strongly affected by environmental conditions like soil and groundwater composition (e.g., Jack et al. 1996). In Dutch soil, many subsurface pipelines are exposed to conditions favorable for anaerobic sulfate-reducing microorganisms, which play an important role in microbiological corrosion processes (Von Wolzogen 1934; Enning and Garrelfs 2014). Despite a lot of research, many questions remain about the microorganisms causing MIC and the circumstances under which it occurs in the subsurface environment (Enning and Garrelfs 2014). A better understanding of MIC could help the industry to increase safety and reduce the high costs for protection, inspection, and reparation of underground metal constructions and pipelines.

A bioprobe was developed based on an electrical resistance probe. On the metal coupon of an ER probe a corroding biofilm with sulfate-reducing microorganisms is grown under laboratory conditions in a relatively fast and reproducible way. This bioprobe is exposed to groundwater or soil, making it possible to simulate and study the development of external MIC of buried pipelines under realistic field conditions. The field results can be compared with those from experiments performed under stable and controlled lab conditions to identify key factors that increase the risk of MIC and to quantify their influence. Besides that, the effectiveness of mitigating measures, like cathodic protection, can be determined.

In this chapter, case studies of the application of bioprobes are presented, performed both in the laboratory as well as in the field, to assess the effects of environmental conditions on MIC. First, the used bioprobes and experimental setups are introduced. Then the effects are described of different groundwater parameters on MIC, like oxygen, sulfate and nitrate concentrations and temperature.

21.2 EXPERIMENTAL APPROACH

21.2.1 The MIC ER Probe

ER probes with sulfate-reducing biofilms were used as "bioprobe tools" to study the effects of environmental conditions and cathodic protection on external MIC. Under typical field conditions, the development of a MIC biofilm on metal surfaces usually takes place over a long period of time, normally on the order of years. With the MIC ER probes that were developed, the formation of a biofilm on an ER probe is accelerated, enabling the study of effects of groundwater conditions and countermeasures on corrosion rates on a relatively short time scale.

ER probes were obtained from MetriCorr (type 100171; www.metricorr.com). The ER probes consist of a 500-μm-thick coupon of mild steel with a surface area of 1 cm^2, which can be exposed to a chosen environment. The probe contains a reference steel coupon, which is not exposed to compensate for temperature effects that can distort the corrosion rate calculations. By measuring electrical resistance, the thickness of the steel, and therefore the corrosion rate, can be determined over time. Electrical parameters, such as DC and AC currents and potentials, were measured with respect to an Ag/AgCl reference electrode (SE11, Sensortechnik Meinsberg, Waldheim, Germany).

The probe was modified by stimulating growth of a biofilm on the steel surface with sulfate-reducing microorganisms responsible for MIC. For this purpose, the probe was exposed to anoxic groundwater from a chosen field location. In this way, the biofilm formed consisted of microorganisms from that specific location. The growth of sulfate-reducing microorganisms was stimulated by the dosage of lactate and sulfate at an ambient temperature of about 20°C. The growth of a MIC biofilm at the metal surface was further stimulated by enhancing transport of nutrients (such as lactate and sulfate) into the biofilm by creating a continuous flow of groundwater along the steel surface. This flow enhancement was realized by pumping groundwater along the metal surface or by placing one or more probes in a stirred reactor vessel. Figure 21.1 schematically shows three different possible designs of an ER bioprobe: (a) unmodified, (b) in a flow-through setting enveloped by a tube, (c) with flow via a tube alongside the probe.

For most of the experiments described in this chapter, the source of groundwater used was Location A, a location in the north of the Netherlands where MIC was reported to have occurred on the external side of underground pipelines. Table 21.1 gives the chemical composition of the groundwater from this location used for most experiments. When groundwater from another location was used, this is indicated in the text.

Since the variables influencing MIC in the field can be numerous, field experiments were combined with experiments under stable and controlled lab conditions. The qualitative effects of groundwater conditions and countermeasures were

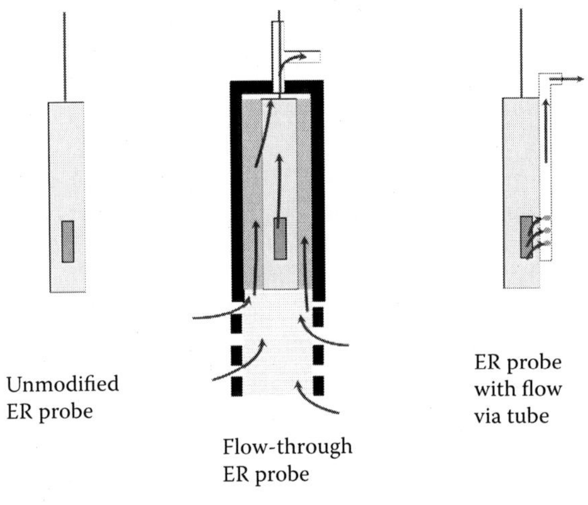

FIGURE 21.1 Schematic depiction of various MIC ER probe designs.

TABLE 21.1
Sulfate, Nitrate, Electrical Conductivity, and pH of Groundwater in Groundwater-Monitoring Wells at Location A Where Five Identical MIC ER Probes Were Placed (Groundwater Sampled December 8, 2015)

Groundwater Monitoring Well	Sulfate (mg/l)	Nitrate (mg/l)	Electrical Conductivity (μS/cm) at 20°C	pH at 19.5°C
1	110	29	625	5.5
2	110	50	632	5.4
3	19	<0.5	275	6.4
4	130	100	786	5.4
5	13	<0.5	241	6.4

thus studied in the field and in more detail in laboratory setups under controlled conditions.

21.2.2 The Laboratory Setup

Figure 21.2 gives a schematic overview of the laboratory setups used. The ER probe was placed in a closed anoxic reactor unit. Site-specific groundwater was pumped through the reactor system at a fixed rate. To maintain anaerobic and pH-buffered conditions similar to that in the field, the influent vessel was kept under an N_2/CO_2 atmosphere (80%/20% v/v). The data of the ER probe were recorded one to four

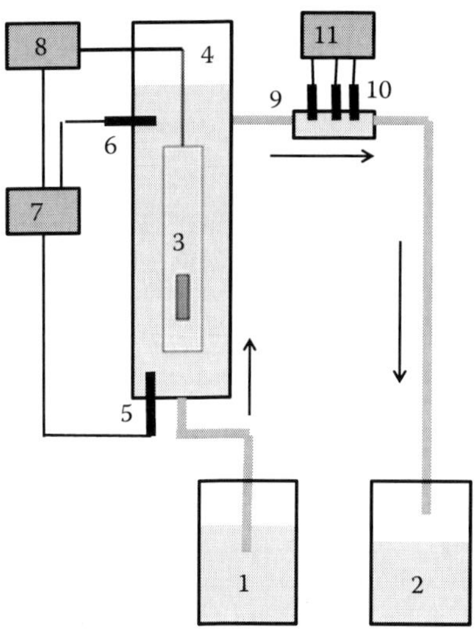

1. Influent vessel
2. Effluent vessel
3. ER probe
4. Anaerobic reactor unit
5. Counter electrode
6. Reference electrode (Ag/AgCl)
7. Potentiostat
8. ER probe logger
9. Flow-through unit
10. Electrodes for pH, oxygen concentration, and redox potential
11. Logger for pH, oxygen concentration and redox potential

Arrows denote flow direction.
Units 1 and 2 are filled with location-specific groundwater, and unit 1 is continuously kept under N_2/CO_2 atmosphere.

FIGURE 21.2 Laboratory setup.

times per hour with a data logger (ICL-02i, www.metricorr.com). In the effluent stream, a flow-through unit was installed, featuring sensors with pH, redox potential, and oxygen sensors. A steel pin counter electrode and an Ag/AgCl reference electrode were placed in the system and were connected to a potentiostat to enable manipulation of the electric potential at the ER probe, such as application of CP.

21.2.3 The Field Setup

Field experiments were performed at Location A in the north of the Netherlands. This location was chosen because it is located in an area where external MIC has been discovered on several spots on underground pipelines. The soil parameters of this area are largely determined by the presence of peat. The top layer of a few decimeters consists of humus-rich soil and, largely excavated, peat. Underneath is yellow sand.

On the field site, 25 groundwater-monitoring wells were installed. These consisted of PVC tubes with an external diameter of Ø75 mm and had depths ranging from 1.6 to 3.2 meters below surface level. The deepest end of the PVC tubes was provided with a well screen consisting of vertical slits (20 cm × 0.5 mm) to allow the groundwater to flow from the relevant depth into the well. A filter wick was covering the slotted part of the tube. Thus, by inserting the ER (bio)probes in the monitoring wells, it was possible to carry out measurements in the groundwater at a controlled depth and location.

The effect of cathodic protection on (microbiological) corrosion on the ER probes was studied with the setup depicted in Figure 21.3. With a rectifier (Delta ES030-5), a cathodic protection potential was applied between the probe and its environment. A steel counter electrode, serving as anode, was located in a nearby ditch. The applied CP potential was measured with an Ag/AgCl reference electrode, which was located in the monitoring well adjacent to the probe and was connected to the Metricorr logger to measure the thickness of the probe, electric potential, and current density (DC and AC).

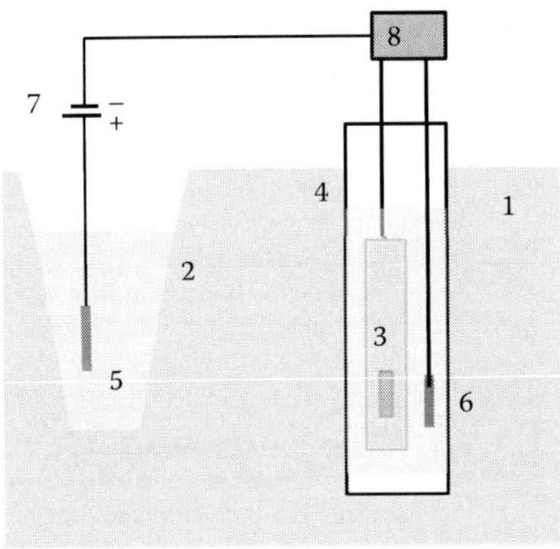

1. Soil
2. Ditch
3. ER probe
4. Groundwater testing well
5. Counter electrode
6. Reference electrode (Ag/AgCl)
7. Potentiostat
8. ER probe logger

FIGURE 21.3 Field setup.

21.2.4 Parameters Influencing MIC Investigated in This Research

In the experiments described in this chapter, the effects on MIC of the following parameters was investigated:

- Groundwater composition (pH, electrical conductivity [EC], oxygen concentration, redox potential, lactate concentration, sulfate concentration, molybdate [inhibitor for sulfate-reducing microorganisms], and BRES [inhibitor for methanogenic microorganisms])
- Groundwater flow
- Temperature
- Pulses of oxygen
- Groundwater from different locations
- Cathodic protection

In this chapter, a selection of the results is presented.

21.3 RESULTS

21.3.1 Laboratory Experiments

21.3.1.1 Effects of Groundwater Composition

The effect of groundwater composition was studied in the laboratory by exposing the probe to groundwater from Location A and adding various compounds.

The experiments from which the results are reported here were carried out using an enveloped ER probe (Figure 21.1b).

A typical example of the results obtained is given in Figure 21.4. Coupon thickness was measured continuously over time, and corrosion rate was calculated from that.

The results can be summarized as follows (the numbers refer to the successive parts of Figure 21.4):

1. Without adding nutrients, no significant corrosion took place within 5 days (corrosion rate decreasing to 30 μm/year).
2. The addition of sulfate alone did not lead to an increase of the corrosion rate within 20 days.
3. The addition of lactate led to a slight increase of the corrosion rate from 30 to about 70 μm/year.
4. Discontinuation of lactate dosage led to a decrease of the corrosion rate.
5. After addition of lactate, the corrosion rate rose again, and this increase accelerated until 460 μm/year.
6. Addition of molybdate (a sulfate-reducing microorganisms inhibitor) led to a 50% decrease in the corrosion rate.
7. Discontinuation of the molybdate supply resulted in recovery of the corrosion rate.

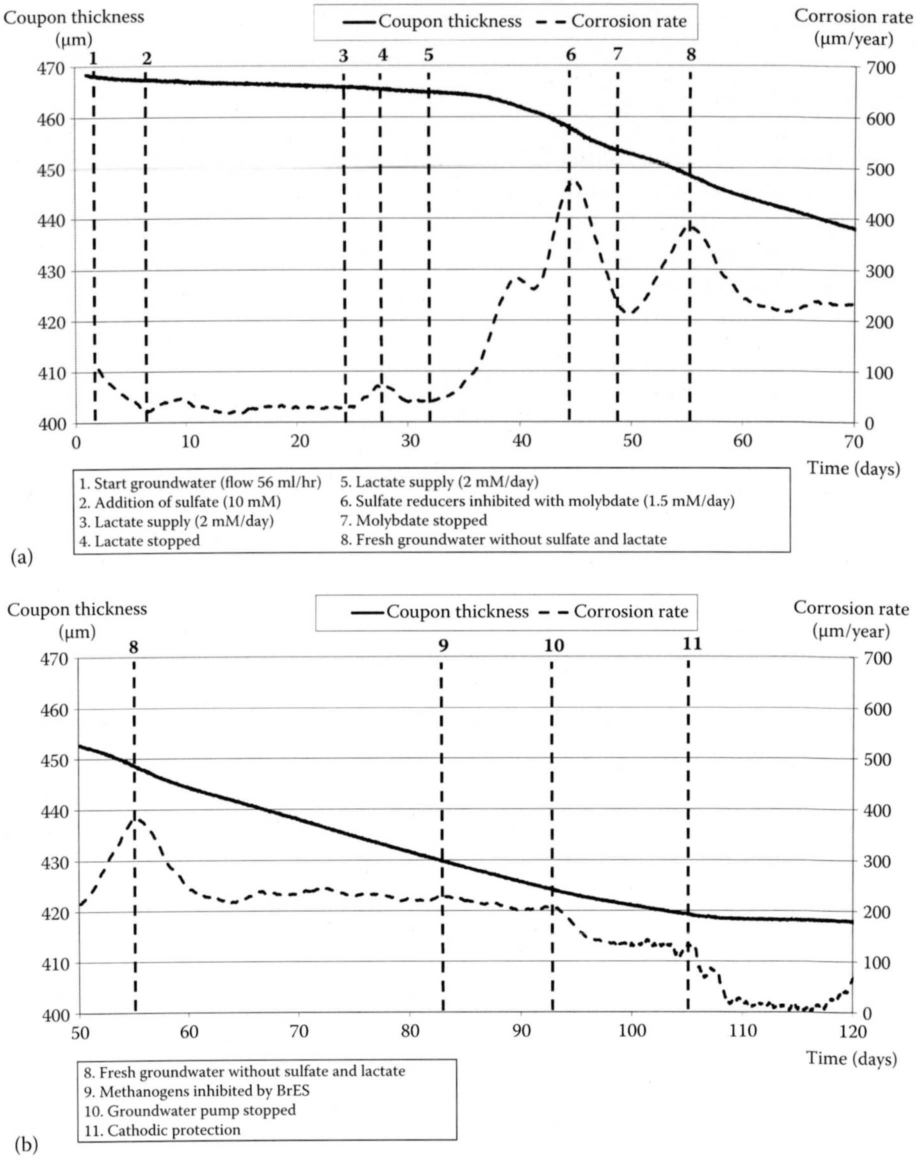

FIGURE 21.4 Example of coupon thickness and corrosion rate of ER probe exposed to a succession of various conditions: (a) day 0–70; (b) day 50–120.

8. The corrosion rate decreased when fresh groundwater without sulfate and lactate was supplied but remained above the level before the increase in the corrosion rate after point 5.
9. Addition of bromoethanesulfonate (BRES) (inhibitor of methanogenic microorganisms) did not lead to significant change of the corrosion rate.

10. Stopping the flow of groundwater led to a reduction of the corrosion rate from 200 to 130 μm/year.
11. Applying a cathodic protection potential corresponding to –500 mV with respect to the free corrosion potential, the corrosion rate could be reduced to <10 μm/year.

The following conclusions about the effects of relevant environmental factors can be drawn:

1. Sulfate and lactate

   Sulfate and lactate are nutrients for sulfate-reducing microorganisms and are thus expected to have a crucial influence on the rate of MIC caused by growth of these organisms. The results of this experiment confirm this influence since MIC was initiated after supply of sulfate plus lactate.
2. Development of an active biofilm

   It took some time (ca. 10 days) after point 5 (after lactate was supplied again) for the corrosion rate to increase to a maximum value. This is consistent with the hypothesis that the corrosion at the surface is caused by a biofilm that needs time to develop.
3. Molybdate (inhibitor of sulfate-reducing microorganisms)

   The decrease in the corrosion rate after dosing of molybdate confirms the active role of sulfate-reducing microorganisms in MIC with this groundwater.
4. BRES (inhibitor of methanogenic microorganisms)

   Corrosion rate after dosing of BRES did not decrease, which indicates that, for this groundwater, methanogenesis did not play a significant role in the MIC process.
5. Groundwater flow

   The stimulation of MIC by groundwater flow was confirmed by these experiments.

21.3.1.2 Effects of Temperature

The effects of temperature were studied by exposing an active MIC sensor (Design 2, Figure 21.1b) to different temperatures (Figure 21.5).

For this experiment, first a MIC ER probe was started at 16°C. Subsequently, this setup was placed in a refrigerator and was successively exposed to temperatures of 4°C, 9°C, and 2°C, respectively. The pipelines on which this research was based are typically exposed to temperatures that range around 10°C (The Netherlands, ca. 1.5 m depth, e.g., Jacobs et al. 2011). For this reason, in these experiments, temperatures around this value were chosen (from 4°C to 16°C).

The comparison between the corrosion rates at different temperatures was complicated by the fact that the corrosion rate increased in time independently from temperature, suggesting biological growth (see the range from day 130 to 150). The effect of temperature can most clearly be seen from changes in corrosion rate upon changes of temperature (Table 21.2). Each time when the temperature was lowered,

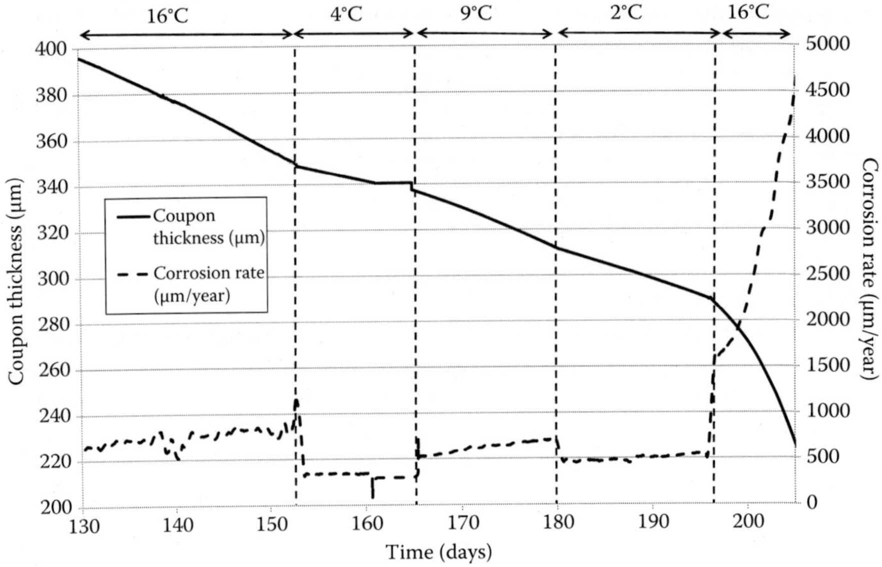

FIGURE 21.5 Effect of temperature on coupon thickness and corrosion rate.

TABLE 21.2
Effects of Temperature Changes on Corrosion Rates

Temperature Change	Change in Corrosion Rate (Before → After)	Change in Corrosion Rate (%)
16°C → 4°C	850 μm/year → 340 μm/year	40%
4°C → 9°C	300 μm/year → 530 μm/year	177%
9°C → 2°C	700 μm/year → 480 μm/year	69%
2°C → 16°C	550 μm/year → 1600 μm/year	291%

a decrease in corrosion rate occurred, and increasing the temperature led to an increase in the corrosion rate. In a realistic field situation, these results probably imply that a rapid change in temperature will induce a change in corrosion rate while a gradual change in temperature will have less effect because the biofilm seems to be able to adapt to these changes.

21.3.1.3 Effects of Pulses of Oxygen

To investigate the effect of oxygen, an airflow (100 ml/hour) was supplied for periods of 1 to 2 hours (Figure 21.6). Aeration of the groundwater (dotted black line 1) led to a strong increase of the corrosion rate. When oxygen supply was stopped, the corrosion rate remained higher than before. For the case shown here, corrosion rate was

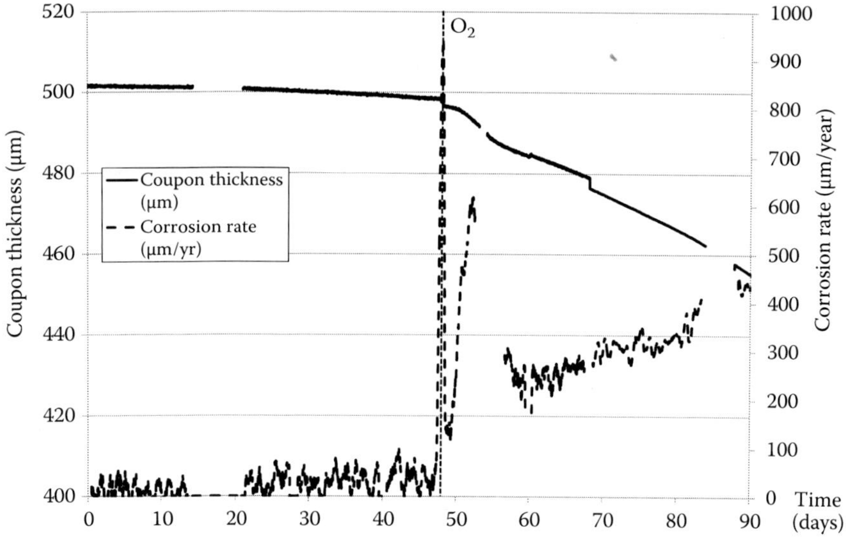

FIGURE 21.6 Effect of pulses of oxygen (indicated with dotted line and O_2).

increased from approximately 20 μm/year before aeration to approximately 300 μm/year after aeration.

We hypothesize the following mechanism to explain the observed pattern. In the initial system, the ER probe is exposed to a sulfate-reducing environment. The formation of sulfide and ferrous iron might lead to a layer of iron sulfide (FeS) and siderite ($FeCO_3$) covering the metal surface and acting as a protective layer hampering further corrosion. Oxygen that enters the solution will oxidize this layer, thus removing the protective layer and creating a reactive metal surface, prone to corrosion. This could thus remove the protection by layers of corrosion products and promote the establishment and activity of a MIC biofilm. Another potential cause for the increased corrosion rate is the additional cathodic reaction, stimulated by the oxidizing potential of oxygen, driving anodic dissolution and overall corrosion.

This finding can have important consequences for locations with changing groundwater levels. At these locations, the steel surface might be exposed to alternating conditions, including the sequential presence and absence of oxygen. These results indicate that this can have an accelerating effect on MIC rates.

21.3.1.4 MIC in Groundwater from Three Different Locations

The potential for MIC was assessed with groundwater from three different locations in the Netherlands where MIC on pipelines was found to have occurred in various degrees (ranging from severe to no MIC, Table 21.3). For these three locations, groundwater composition was determined, and MIC was measured in the laboratory using ER probes.

TABLE 21.3
Groundwater Characteristics and MIC Observed at Subsurface Pipelines for Three Locations in the Northeast of the Netherlands

Location	Nitrate (mg/L)	Dissolved Organic Carbon (mg/L)	Sulfate (mg/L)	pH (–)	Redox Potential (mV)	Electrical Conductivity (μS/cm)	Oxygen Content (mg/L)	MIC Degree
B	<0.5	51	<0.5	6.4	–60	701	0.2	Severe
C	140	8	85	7.33	94	999	0.1	Low
D	<0.5	8.6	8.2	6.5	–22	242	1.6	Severe

Groundwater was collected from all locations on October 21, 2013. The pH, redox potential, electrical conductivity, and oxygen content were measured on site. After collection, the groundwater samples were stored cool and analyzed within a week for sulfate, nitrate, and dissolved organic carbon (Table 21.3).

As a reference, groundwater from Location A was used.

For the four types of groundwater, the risk of corrosion was determined by means of the previously described laboratory setup. The ER probe with flow via a tube probe was used (Figure 21.1c).

The approach was identical for each type of groundwater:

1. Start with groundwater only.
2. Flush with groundwater containing additional sulfate and lactate, encouraging sulfate reduction.
3. If MIC did not develop to a significant extent, additional measures to stimulate MIC (pulses of oxygen, additional nutrients).
4. Flush with groundwater + sulfate (lactate).
5. Flush with only groundwater (no added sulfate or lactate).
6. Shutdown of the corrosion with cathodic protection (applied potential: –500 mV with respect to the free corrosion potential).

In all cases, a flow rate was applied of 25 ± 10 ml/hour. Groundwater influent was kept under a N_2/CO_2 atmosphere. The pH, oxygen, and redox potential of the effluent were measured continuously.

21.3.1.4.1 Location B (Figure 21.7)

During phase 1 (groundwater only) the color of the groundwater in the reactor remained the same as the influent (bright yellow), and there was no visible sulfate reduction or corrosion. When lactate and sulfate were added to the groundwater, there was no change of medium color or corrosion rate at first. From day 30 onward, the surface of the sensor turned black, and from day 43, the solution was black as well. Thereafter, the formation of black precipitate was constantly increasing, indicating sulfate reduction and formation of FeS. The increase in sulfate reduction coincided with a decrease in the redox potential from –200 mV to –350 mV. The corrosion rate increased sharply to a maximum value of approximately 1500 μm/year. In the absence of lactate, the corrosion rate remained high, which is consistent with the hypothesis that sulfate reducers cause MIC to utilize steel as the electron donor, either directly or via hydrogen as an intermediate. When both lactate and sulfate were no longer administered, the corrosion rate dropped from about 1500 to 1000 μm/year. Imposing a CP potential of –500 mV with respect to the free corrosion potential completely stopped the corrosion.

The electrical potential between sensor surface and the solution showed a remarkable pattern. At the moment that the corrosion by MIC increased to a large extent, the potential measured on the exposed steel surface of the ER probe changed from about –0.7 V to about –0.55 V. We suspect that this change was caused by an oxidation reaction by MIC at the sensor surface.

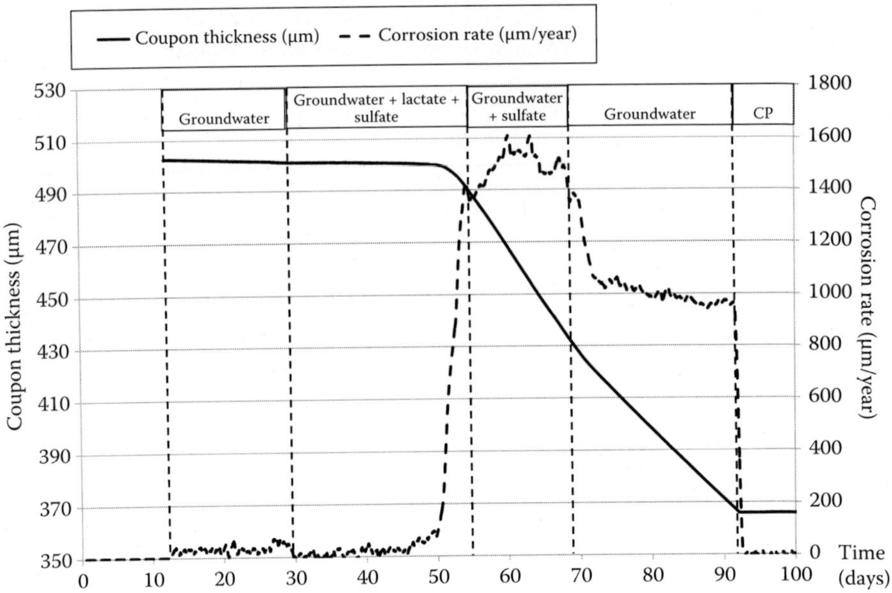

FIGURE 21.7 Coupon thickness (left axis) and corrosion rate (right axis) of MIC ER probe for groundwater Location B.

21.3.1.4.2 Location C

The experiments using groundwater from Location C showed a completely different result compared to the experiments using groundwater from Location B. During the whole experiment, there was no sulfate reduction observed. The liquid remained clear yellow, and the redox potential did not drop below a value of –20 mV (in comparison, the redox potential at Location B reached –350 mV). This is probably due to the high amount of nitrate remaining in this groundwater. For sulfate reduction to occur, all nitrate must be removed first since nitrate is known to inhibit sulfate reduction (e.g., Jenneman et al. 1986; Davidova et al. 2001).

It was tried to achieve MIC by supply of sufficient lactate to remove nitrate. Not even this led to reduction of all nitrate. In this groundwater, there was apparently not sufficient biological activity for this conversion. Inhibition of nitrate reduction by oxygen can be ruled out as a cause: No oxygen was detected during the whole period. The steel coupon of the ER probe remained visually unaffected throughout the experiment. Finally, no changes in electrical potential between the metal surface and solution were observed, contrary to the potential change observed at the steel coupon of the ER probe exposed to groundwater from Location B.

21.3.1.4.3 Location D

The experiments using groundwater from Location D showed sulfate reduction and MIC but not as strong as in groundwater from Location B. From day 30 onward, FeS was visible in the reactor. However, there was hardly any corrosion at the sensor surface. To increase MIC, on day 60, a pulse of oxygen was added. From the moment this was done, the MIC rate increased from 50 to 120 μm/year and kept increasing until the end of the experiment. The changes in conditions that were imposed during this period only appeared to have a small or temporary effect on the increase of the corrosion rate. Therefore, in this case, the increase of the corrosion rate over time seemed to indicate a gradual growth of an active MIC biofilm. By the end of the experiment (90 days), the corrosion rate had reached a value of 275 μm/year and appeared still to be increasing. The change in the potential at the sensor surface appeared to reflect this (from –0.75 V to –0.63 V) although the change was not as quick and extensive as in similar experiments using the groundwater from Location B.

21.3.1.4.4 Exposing a Biofilm from (Corrosive) Groundwater A to (Noncorrosive) Groundwater C

The setup with groundwater from Location A (reference location Northeast Netherlands) with an ER probe already covered with a biofilm from this location was used during the year prior to these experiments for different experiments. At the beginning of the experiments discussed in this paragraph, there was an active MIC biofilm on the sensor surface, which was visible to the naked eye. During the 3 months prior to these experiments, the corrosion of this probe was successfully stopped by means of an applied CP potential of approximately –500 mV with respect to the free corrosion potential. During that time, the setup remained at room temperature.

After restarting this setup, MIC activity returned quickly although the initial corrosion rate was modest (about 100 μm/year). From day 45 onward, active sulfate reduction occurred, and the corrosion rate increased to approximately 600 μm/year. This situation remained reasonably stable when the sensor surface was exposed to groundwater.

The next step in this experiment differed from those in the other experiments. To observe the response of an active MIC biofilm to groundwater having a high nitrate concentration, this setup was exposed to groundwater from Location C from day 60 onward. This did not lead to a decline as might be expected in case the nitrate would cause inhibition of sulfate reduction and of MIC. Instead, it led to an increase in the corrosion rate of up to about 1300 μm/year. The corrosion rate continued to rise until the corrosion was effectively stopped by applying a CP potential of –500 mV with respect to the free corrosion potential.

With the flow of nitrate-rich water, the appearance of the setup changed. At first, the liquid was black gray and cloudy with black and brown precipitates near and on the metal surface. After exposure to groundwater from Location C, the solution became clear, and orange-brown precipitates appeared, indicating the oxidation of the iron sulfides formed previously.

The potential at the sensor surface showed changes upon changes of groundwater, corresponding to different mechanisms of corrosion. After development of active sulfate reduction, the potential changed from –0.7 V to –0.3 V. When nitrate-rich groundwater was administered, a sharp change was observed from about –0.3 V to –0.6 V.

21.3.1.4.5 Comparison of Corrosion Measurements Using Groundwater from Various Locations

The corrosion experiments using groundwater from four locations showed clear differences in MIC rates. The groundwater from Location B showed strong sulfate reduction and MIC. The groundwater from Location C showed no sulfate reduction and no MIC. The groundwater from Location D showed moderate sulfate reduction and MIC. The corrosion rate increased slowly during the experiment. The corrosion rate using groundwater from Location A was in between the rates of Location B and Location D. When an active MIC biofilm from Location A was exposed to water with high nitrate concentrations (from Location C), the corrosion rate increased sharply. The question is whether this is MIC or not; corrosion in this case could also have been caused by chemical oxidation processes caused by nitrate.

The differences observed in MIC for the different types of groundwater corresponded to the expectation (Location C: no MIC; Locations B and D: MIC). There also seemed to be a logical relationship between the measured groundwater characteristics and MIC risk: This risk is highest for the location with a low redox potential and a low nitrate concentration.

For the effect of the nitrate concentration, the type of exposure seemed to be important. In case of continuous exposure to nitrate, MIC by sulfate reduction cannot develop (see the results for Location C), but in situations in which sulfate-reducing conditions dominate a large part of the time and temporarily higher concentrations of nitrate occur, the corrosion rate can be strongly increased (see the results for exposing biofilm from Location A to groundwater from Location C).

21.3.2 Field Tests

21.3.2.1 Transferring Probes from the Laboratory to the Field

To test the applicability of the ER bioprobes to monitor MIC in the field, the following experiment was carried out. Five probes (Design 3, Figure 21.1c) were exposed to groundwater with sulfate and lactate in one anoxic reactor vessel to grow an active MIC biofilm (Figure 21.1c). Coupon thickness and corrosion rate were measured over time. Once a significant MIC rate had developed, the probes were transferred to field location A. The probes were transported using anoxic bags (Merck, anaerocult) stored in a cool box. Within 3 days, the probes were inserted into five separate groundwater-monitoring wells. The groundwater composition of these wells was sampled at one time point (Table 21.1).

The probes showed comparable behavior in the laboratory (Figure 21.8). To increase the MIC rate, a pulse of oxygen was imposed. At the end of September, just before transferring to the field, the corrosion rates were 220 to 420 μm/year. After transferring into the field, the corrosion patterns for all probes first showed a sudden change, probably due to adaptation to the new conditions. Soon after this, all probes were recovered and showed comparable corrosion behavior. Absolute differences in corrosion rate between the probes, however, became more distinct after some time in the field. The corrosion rate for some probes increased over time whereas for others it decreased. The spread in the corrosion rates increased,

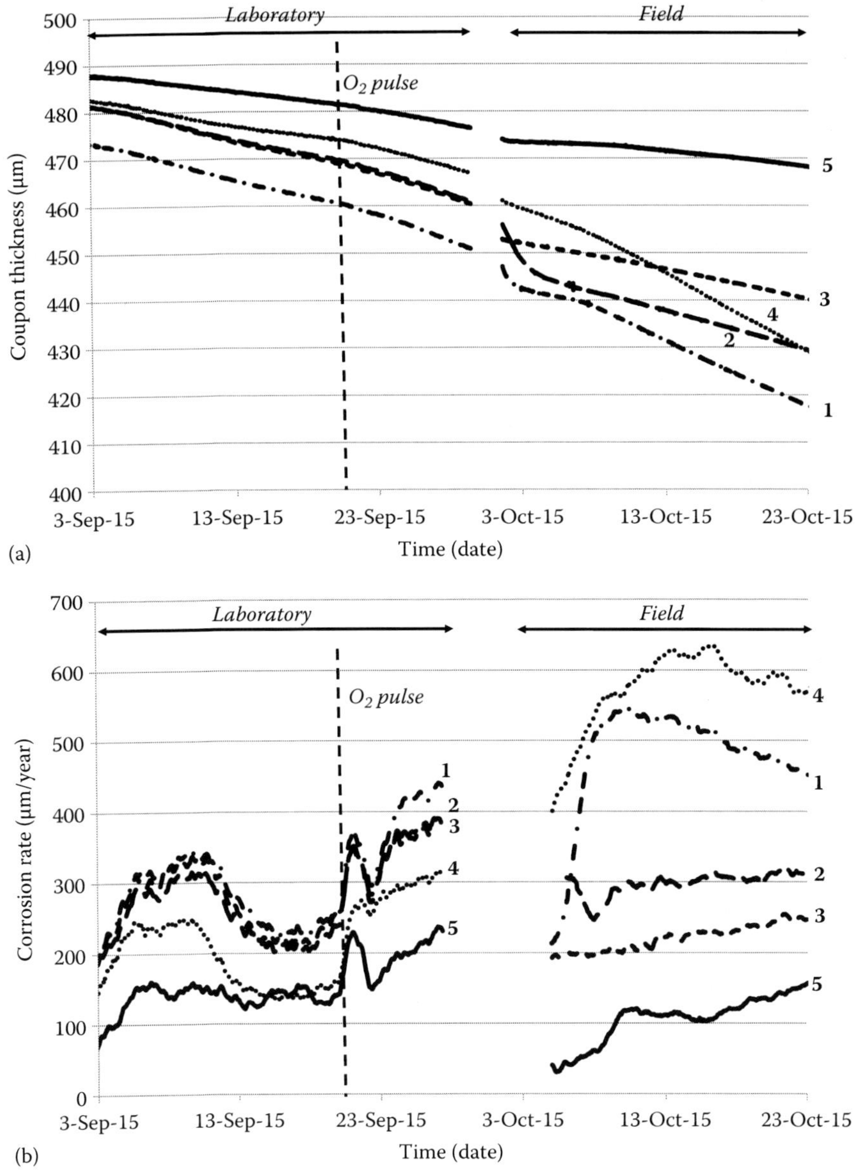

FIGURE 21.8 Effect of transfer of various ER bioprobes from the laboratory to the field coupon thickness (a) and corrosion rate (b) as a function of time. The numbers 1 to 5 refer to five different ER bioprobes, cultivated in one reactor vessel in the laboratory and transferred to five different groundwater-monitoring wells in the field (details: see Table 21.1).

extending to a range of values from 150 to 650 μm/year at the end of the experiment. The explanation for the differences in corrosion behavior may lie in the differences in groundwater composition to which the probes were exposed. Table 21.1 shows that parameters that are expected to affect the MIC rates, such as nitrate concentration, pH, and sulfate concentration, varied considerably between the wells even if the wells are located at short distances from each other (10 to 100 meters). It is difficult to determine which factor is decisive for the differences observed. These data, however, do make clear that (1) Differences in environmental conditions in the field will lead to differences in MIC rates and (2) these differences vary at a locally small scale due to substantial variations of the groundwater conditions at this scale.

21.3.2.2 Effects of Mass Transfer: (Changing) Groundwater Flow and Groundwater versus Sand Conditions

In the laboratory, the effect of mass transfer was observed in several ways. During one experiment, the MIC rates of an ER probe placed in a saturated sand layer were compared to those of an ER probe placed in a liquid groundwater medium (data not shown). In another experiment, flow conditions were manipulated by switching the groundwater pumping on and off (Figure 21.4). Both experiments clearly showed the importance of mass transfer: Under conditions of stimulated mass transfer, significantly higher MIC rates were found. This is in accordance with expectations: For MIC, transport of nutrients and corrosion products is necessary, and by affecting this, mass transfer can be rate limiting.

To assess the importance of mass transfer as a rate-limiting factor in the field, several experiments were carried out.

First, groundwater movement along the surface of the ER bioprobes (Design 3, Figure 21.1c) was stimulated by using groundwater pumps connected to the tubes alongside the probe. Figure 21.9 demonstrates the effect of groundwater pumping. When the groundwater pump was switched on, the MIC rate increased from approximately 70 μm/year to approximately 250 μm/year. This result is comparable to that seen in the laboratory when flow conditions were manipulated by switching the groundwater pumping on and off (Figure 21.4). Both experiments clearly show the importance of mass transfer: Under conditions of stimulated mass transfer, significantly higher MIC rates were found. This is in accordance with expectations: For MIC, transport of nutrients and corrosion products is necessary, and by affecting this, mass transfer can be rate limiting.

Figure 21.10 demonstrates the effect of the permeability of the environment to which a MIC ER probe is exposed. During this experiment, two MIC ER probes were first exposed to a liquid groundwater environment in a groundwater-monitoring well. To investigate the effect of exposure to a saturated sand layer, the plastic monitoring wells around the probes were removed, and the resulting gaps were filled with enclosing soil, leading to the probes being surrounded with predominantly sand. Figure 21.10 shows that after changing the environment of the MIC ER probes from liquid to sand, the MIC rates dropped substantially (see Table 21.4).

This experiment also shows that under field conditions, mass transfer limitation can form a rate-determining factor for MIC.

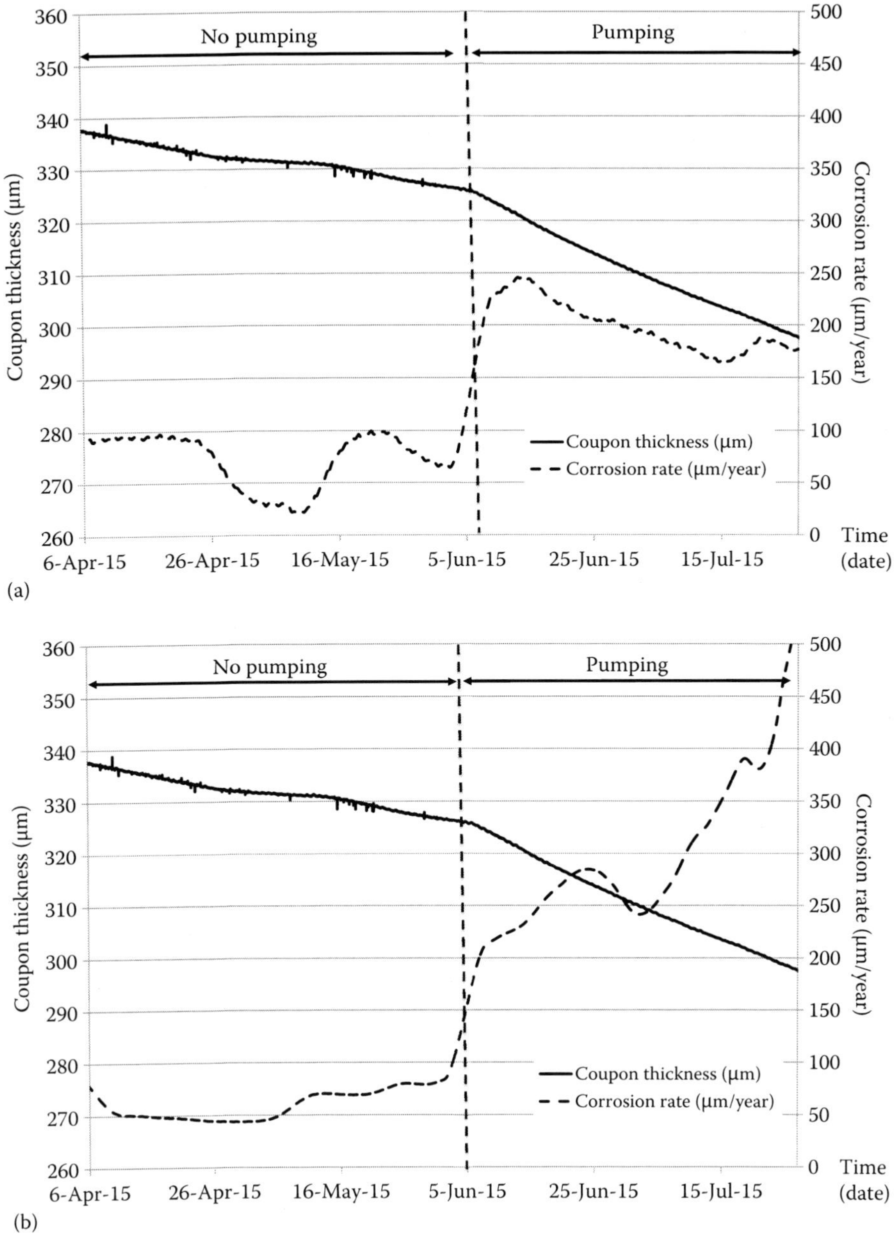

FIGURE 21.9 Effects of change of groundwater flow on the corrosion rate. (a, b) Results for two different MIC ER probes.

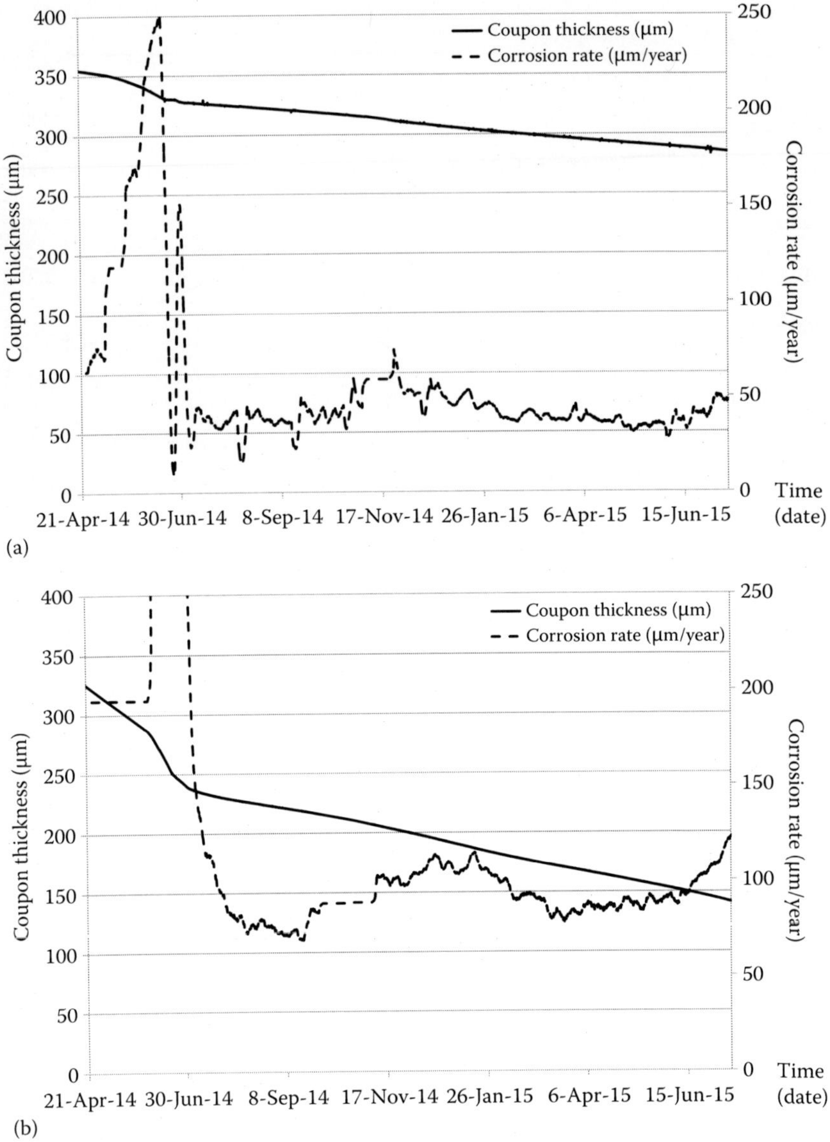

FIGURE 21.10 Effects of transport limitation in the field: MIC rates for ER probes initially exposed to groundwater (before July 1, 2014) and after that exposed to sand. (a) For probe 805; (b) for probe 806 (details: see Table 21.4).

TABLE 21.4
Characteristics and Corrosion Rates of ER Bioprobes Before and After Burial

Probe	Depth of Burial July 1, 2014 (m Below Surface)	Corrosion Rate June 2014 (μm/Year) Before Burial	Corrosion Rate June 2015 (μm/Year) After Burial
805	2.9	250	50
806	2.2	400	100

21.4 CONCLUSIONS

ER bioprobes enabled rapid and controlled assessment of the MIC risk of a specific type of groundwater. MIC on ER probes based on sulfate-reducing microorganisms were quickly and reproducibly started in a laboratory setup by following these steps: (a) stimulation of the sulfate reduction by dosing of sulfate and lactate; (b) when stable sulfate reduction occurs, at fairly modest corrosion rates, administering one or multiple pulses of oxygen; (c) after administration of oxygen, immediately dose lactate again. The key parameters that influence the development of MIC are sulfate reduction (redox potential, organic matter, sulfate), mass transfer, and changing conditions (oxygen, nitrate).

Temperature affected MIC in the tested range from 2°C to 17°C. The effect was particularly large after rapid changes in temperature. At higher temperature levels, the MIC rates increased; at a lower temperature levels, the MIC rates decreased.

Differences in MIC rates were observed for groundwater from different locations. The experiments with groundwater from various locations showed clear and consistent differences. Differences in earlier observations of occurrence of MIC between locations determined in the laboratory were consistent with those measured in the field. At elevated sulfate reduction rates, increased MIC rates were observed. With continuous exposure to high concentrations of nitrate, MIC rates remained low. While exposed to nitrate after active sulfate reduction, the corrosion rate increased.

The MIC ER probes can successfully be applied in the field as well. Due to the larger amount of parameters that may vary in the field compared to the laboratory, the observed variation in results may be larger in the field than in the laboratory. Because of this, a combination of laboratory and field experiments gives the most significant information. Field experiments resemble the environmental conditions of a buried pipeline quite well, and laboratory experiments enable us to quantify the influence of separate parameters. This combined approach indicated the influence of mass transfer as a rate-determining step for MIC: Its importance was consistently shown in the field as well as in the laboratory.

These experiments confirm the applicability of MIC bioprobes for the measurement of the MIC risk in different types of groundwater.

REFERENCES

Davidova, I., Hicks, M. S., Fedorak, P. M., Suflita, J. M. 2001. The influence of nitrate on microbial processes in oil industry production waters. *J Ind Microbiol Biotechnol*, 27: 80–86.

Enning, D., Garrelfs, J. 2014. Corrosion of iron by sulfate-reducing bacteria: New views of an old problem. *Appl Environ Microbiol*, 80: 1226–1236.

Flemming, H. C. 1994. Microbial deterioration of materials–fundamentals: Economical and technical overview. *Mater Corros*, 45: 5–9.

Jack, T. R., Wilmott, M. J., Sutherby, R. L., Worthingham, R. G. 1996. External corrosion of line pipe—A summary of research activities. *Mater Perform*, 35(3): 18.

Jacobs, A. F. G., Heusinkveld, B. G., Holtslag, A. A. M. 2011. Long-term record and analysis of soil temperatures and soil heat fluxes in a grassland area, the Netherlands. *Agric Forest Meteorol*, 151: 774–780.

Jenneman, G. E., McInerney, M. J., Knapp, R. M. 1986. Effect of nitrate on biogenic sulfide production. *Appl Environ Microbiol*, 51: 1205–1211.

Kruger, J. 2011. Cost of metallic corrosion, pp. 15–20. In Revie, R. W. (ed.), *Uhlig's Corrosion Handbook*, 3rd ed. Wiley, Hoboken, NJ.

Von Wolzogen Kühr, C. A. H., van der Vlugt, L. S. 1934. The graphitization of cast iron as an electrobiochemical process in anaerobic soil. *Water*, 18: 147–165.

22 The Use of Multiple Microbial Population Analysis Techniques to Diagnose Microbially Influenced Corrosion Potential in Oil and Gas Systems

Jennifer Fichter and Elizabeth J. Summer

CONTENTS

ABSTRACT

The environments found in oil and gas operations are often ideal for promoting the growth of microbial populations capable of corrosion, fouling, and souring. As a result, microbial monitoring and application of biocides is a routine part of normal operations. A variety of different techniques, both old and new, have become available for monitoring bacterial populations. Each technique varies with respect to mechanism of detection, what component or feature of the bacterial cell is actually being detected, as well as the physiological status of the cells being detected. Since oilfield environments are often quite extreme, the ability of each method to be suitable under different sample physical parameters is also not known. For this study, data are presented from four very distinct O&G field case histories: (1) a sour, highly saline shale gas field in which microbial corrosion was not expected but determined to be present; (2) a natural gas tight sandstone field that was not experiencing MIC in contrast to culture-based analysis that indicated it should be; (3) a light oil shale field where rapid MIC was occurring in the extreme temperature on the surface of fire tubes; and (4) a tight shale gas field experiencing a wide range of microbial-associated problems, including downhole and surface MIC issues. In addition to inquiries about operations, the microbial populations at each location were analyzed using four distinct monitoring methods: (1) culture-based enumeration using bug bottles with indicator media for sulfate-reducing bacteria and acid-producing bacteria, (2) rapid bacterial cell enumeration based on an adenosine triphosphate quantification assay, (3) epifluorescence microscopy, and (4) molecular analysis of the microbial populations by DNA isolation and sequencing of 16S amplicons. Samples from location 4 were also analyzed using a direct bacterial cell counting system and by an assay that quantifies specific enzymatic activity. The usefulness of each method varied from location to location. Considering the variations in physical and chemical characteristics of each field site, as well as the very high bacterial diversity that is present in oil and gas locations, such variations are not unexpected. Microbial population analysis allowed for a diagnosis and course of action to be suggested to the field operators. The relative value of each data set for diagnosing microbial contribution to corrosion issues in the fields is discussed. Ultimately, there is not one method that can be used to accurately monitor microbial populations in all locations. Understanding the strengths and limitations of each method provides the most solid foundation for microbial monitoring and educated decision making in the field.

22.1 INTRODUCTION

Microbial activity negatively impacts the oil and gas industry through the production of biogenic sulfide (H_2S), through the process of microbially influenced corrosion (MIC) of downhole and surface facilities, through fouling and clogging of water treatment facilities, and finally by altering viscosities of drilling and fracturing fluids (Miller and Tiller 1970; Videla 1996). As a result, microbial monitoring and biocidal chemical applications are standard operating procedures for most operations (NACE 1994; Maxwell and Campbell 2006; NACE 2014). In terms of field applications, the ultimate goal for most operators is to predict the impact that microbial activity might have on a given operation and to assist in the decision of how much and what type of biocides to apply. Microbial population monitoring assays available for use include methods that simply determine the presence of bacteria or total bacterial concentrations, as well as those that provide some level of differentiation of bacterial types based on a phenotypic trait or sophisticated genetic identifications of taxonomic affiliations. The phenotypic traits of common interest are sulfidogenisis by SRB, acid production by APB, iron reduction by iron-reducing bacteria (IRB), biofilm formation, and the ability to degrade oilfield chemicals or petroleum products directly.

Microbial population assessment is a critical part of systems operations. Methods in common use by many independent operators include total bacterial quantification by ATP, direct visualization and enumeration by microscopy, culture-based detection and enumeration based on growth in indicator media, and genetic screens for either functional genes or, more commonly, 16S rDNA gene sequencing or detection. The industry standard for field quantification and differentiation of a few phenotypes, such as SRB, GHB, APB, and IRB, is based on growth after serial dilution into indicator culture media, sometimes called "bug bottles" (NACE 2014). Advantages of bug bottles include that they are easy to set up, require minimal specialized equipment or training, and are quite field operator friendly. However, there are several well-documented shortcomings. In many cases, it is difficult to set up culture media and conditions to mimic the system, for example, systems with extreme salt compositions and concentrations, high temperatures, souring, and unique carbon sources or systems where the organisms present are unculturable or very difficult to culture (i.e., methanogens). False negatives or underrepresentation of the system population along with long turnaround time (4 weeks for SRB) means there is a delay between testing and using the results to determine biocide treatment efficacy and biocide selection. In response to increasing awareness of the limitations of bug bottles, there has been an increase in the use of non–culture dependent methods. The most recent revision of the NACE standards includes descriptions of the use of immunological assay for the SRB-specific protein APS reductase, a general microbial concentration determining method based on ATP quantification, as well as a bacterial quantification assay based on measuring hydrogenase activity (NACE 2014). Additionally, new methods have come on the market based on methods such as direct counting via cell staining and flow cytometry as well as biochemical measurements of cellular activity, such as hydrolase activity. An understanding of the strengths and limitations of each method is critical for proper monitoring and control of bacteria in the field. A critical aspect of all methods is proper sample handling as the speed at which a sample is processed after collection greatly affects test results. Field tests that do not require transport prior to analysis have an advantage over methods that require shipment to an off-site laboratory.

Direct visualization and quantification of bacteria in the sample by some form of microscopic or flow cytometry analysis is one method used to enumerate microbial populations in oilfield samples. The technique is comparatively fast and does not rely on growth or metabolic activity. Typically, oilfield samples require filtration to collect the bacteria and differential staining, for example, with acridine orange, 4′,6-diamidino-2-phenylindole (DAPI), fluorescein isothiocyanate (FITC), or rhodamine (Fichter et al. 2012). Microscopic analysis requires a skilled operator and typically does not allow for identification of different types of bacteria without the use of additional expensive and sensitive reagents, thus the method only provides bacterial counts. The greatest source of error associated with this technique is differentiating microbes from suspended solids. To work around the limitations of human error with microscopic analysis, the Rapid B flow cytometer assay was evaluated in one of the case histories compared to the other microbial monitoring methods. This assay uses two different fluorescent probes to differentiate viable from nonviable cells. In addition, the flow cytometer is programmed to be able to differentiate cells from other background sample debris using the particle's shape, size, and translucence, removing the human element from the equation.

Two rapid assays for cellular activity were included in the microbial monitoring programs for the case studies. The first assay is based on quantification of ATP in the sample with the assumption that each picogram of ATP measured corresponds to a defined number of bacterial cells. The method is based on photometric measurements of luminescence generated in an ATP-dependent reaction. The second rapid cellular activity assay detects hydrolase activity via use of a fluorescent-tagged substrate. Hydrolysis of the substrate releases the fluorescent tag, which is then measured by a fluorometer and correlated to relative microbial activity. Both assays allow for rapid (10 to 20 minutes to process a sample) assessment of the relative microbial activity present in a given sample. The assay kits can be taken into the field, allowing the analyzer to trace microbial contamination and associated issues while in the field. As with microscopy, these methods do not differentiate between the various metabolic classes of microorganisms.

The most thoroughly descriptive methodologies are based on analysis of the DNA of organisms in the sample in order to identify them. Genetic tests based on global DNA sequencing technologies have the advantage of allowing for the analysis of thousands of bacteria in each sample, being comparatively unbiased, and offering a great depth of information on exact bacterial types (Summer et al. 2014). Shotgun analysis of 16S sequences yields statistically significant data on both the types and relative abundance of different organisms. The major disadvantages of sequencing methodologies are the comparatively high cost, long turnaround time for results, the higher skill level and training required to conduct and interpret the results, and the expense associated with the equipment required for the testing.

Discrepancies between different methods and field results are not uncommon. Each method has associated biases and limitations. The ideal method for monitoring microbial populations would be fast, accurate, inexpensive, and provide meaningful data that is directly comparable regardless of location and testing service. Six case studies from four field sites are presented in which a comparison of results obtained with the different methods was conducted. These samples were analyzed by some combination of DNA sequence–based population analysis, ATP-based enumeration, enumeration using bug bottle serial dilutions, and enumeration by microscopy, hydrolase activity enumeration, and flow cytometry enumeration. While not every method was used at every site, this provides an insight into the relative value and agreement between the different approaches against a broad cross-section of industrial samples. The results of microbial monitoring at each location and the agreement between each method used is discussed.

22.2 EXPERIMENTAL PROCEDURE

22.2.1 Sample Collection

Water, drilling mud, and pigging envelope samples were collected in clean glass bottles. Each sample point was flushed thoroughly prior to sample collection. All bottles were rinsed with sample fluids prior to collection of the sample used for analyses. Sample bottles were filled to overflowing and capped tightly to maintain sample integrity. The wellhead swab, day tank biofilm, and fire tube sessile samples were obtained by scraping biofilm from the corresponding system surface. A dilution blank was prepared, which contained 9 mL of 0.2 micron filter-sterilized system fresh or produced water collected from the location being sampled. The container used for the dilution blank was a 12 mL glass vial with a Teflon-lined screw-capped lid. Sessile samples were collected from 1 cm^2 of the respective system surface using a sterile cotton swab. The solids and the cotton swab tip were placed into dilution blank fluid. A slurry was created by shaking each sample for 2 to 3 minutes. All analysis was conducted onsite except for DNA extraction and sequencing. For DNA analysis, samples were packed on ice and shipped overnight to an outside lab where they were processed immediately upon arrival.

Shale core discs were split along fracture lines with a surface sterilized chisel and hammer. The shale material used for microbial analysis was scraped from the freshly exposed surface using clean, sterilized metal spatulas.

22.2.2 Culture-Based Bacterial Enumeration

Culture-based bacterial quantification was conducted using serial tenfold dilutions, starting with 1 mL of the sample. Anaerobic indicator media for sulfate-reducing bacteria (SRB), acid-producing bacteria (APB), and general heterotrophic bacteria (GHB) were used for the analysis. Cultures were incubated for 30 days at system temperature and positive growth was recorded to calculate the viable cells per milliliter. For SRB enumeration, dilutions were performed in NACE TMO194-94 Standard recommended modified postgate B broth (MPB) (NACE 1994). For GHB and APB enumeration, dilutions were prepared in phenol red dextrose (PRD). For all samples included in this technical paper, all positive phenol red dextrose bottles had a noticeable color change to orange or yellow, indicating the presence of APB. The results are listed as GHB/APB in the data tables.

22.2.3 ATP Analysis

Microorganisms were enumerated using a standard ATP assay test kit following the manufacturer's protocol (LuminUltra Technologies Inc) (Balch et al. 1977).

22.2.4 Epifluorescence Microscopy

Samples were filtered through black, 25 mm diameter, 0.2 micron, polycarbonate filters. The bacteria trapped on the filters were stained with acridine orange, and the filters visualized at 1000× magnification on a Nikon Eclipse microscope equipped with a fluorescence epi-illuminator. The number of bacteria was determined by averaging the number of bacteria counted over numerous microscopic fields and then calculating bacteria/mL of sample based on the amount of sample filtered, the size of the filter, and the size of the microscopic fields.

22.2.5 Flow Cytometer Analysis

Microorganisms were enumerated by direct counting with a flow cytometry combined with live/dead staining, following the manufacturer's protocol (Petro Chemical Bio-Diagnostics RAPID-B Assay). Briefly, a 0.5 mL sample aliquot is mixed with an equal volume of staining solution containing two independent fluorescent reagent probes. One probe binds to both viable and nonviable cells. The second probe is retained only by cells where the cell wall has been compromised, or the efflux pumping mechanism has been inactivated. This second probe overwhelms the response from the first probe and indicates that the cell is no longer viable. Sample and stains were incubated for 15 min, and then a 100 μL aliquot of sample was analyzed in the flow cytometer. The flow cytometer sorts the microbial cells from suspended debris using the phenotypic properties of the particle (size, shape, translucence, etc.) to electronically ignore the noncellular shaped particles before the software algorithm utilizes the probcs to differentiate the viable from nonviable cells.

22.2.6 Hydrolase Analysis

Microorganisms were enumerated using a hydrolase enzymatic assay test kit following the manufacturer's protocol (The Mycometer, Inc. Bactiquant®-test). The assay utilizes a hydrolase enzyme substrate whose cleavage by endogenous hydrolase results in a fluorescent product. First, bacteria in the samples are concentrated onto a filter, and then the enzyme substrate is added. After a 30-minute incubation, fluorescence is measured. The resulting fluorescence is measure with a fluorometer and the value converted into a total cell count based on an assumption about hydrolase activity per cell. The bacterial enzyme hydrolyzes the enzyme substrate, cleaving the substrate into two molecules. One of these molecules absorbs UV light at 365 nm and re-emits visible light that can be measured by a fluorometer. The amount of fluorescence is correlated to bacterial concentration.

22.2.7 Environmental DNA Extraction

For DNA analysis, raw samples were shipped to an off-site lab for analysis, packed on ice. Bacteria from between 1 mL and 500 mL of liquid of each sample were concentrated either by filtration through a 0.22 μm filter or centrifugation (Fichter et al. 2012; Summer et al. 2014). DNA was extracted from bacteria using a bead-beating extraction method following the manufacturer's recommendations. The DNA yields were used to provide an estimate of bacterial levels in each sample, assuming an average DNA content of 3 fg DNA per cell and that the majority of DNA in the sample is prokaryotic in origin. DNA was isolated directly from ~0.5 g of solid material scraped from the freshly exposed surface of core discs.

22.2.8 Sequence Analysis

Bacterial 16S sequences were determined for each sample by either pyrosequencing or illumumina MISeq. The primers utilized for the illumina MISeq reactions were 926F-AAACTYAAAKGAATTGRCGG and 1392R-ACGGGCGGTGTGTRC. The primers used for pyrosequencing were 939F-TTGACGGGGGCCCGCAC and 1492R-TACCTTGTTACGACTT (Summer et al. 2014). The primer set utilized for pyrosequencing broadly amplifies the 16S gene among the major Eubacterial phyla, including Acidobacteria, Actinobacteria, Aquificae, Bacteroidetes, Caldiserica, Chlamydiae, Chlorobi, Chloroflexi, Chrysiogenetes, Deferribacteres, Deinococcus-Thermus, Dictyoglomi, Fibrobacteres, Firmicutes, Fusobacteria, Gemmatimonadetes, Lentisphaerae, Nitrospira, BRC1, Cyanobacteria, OP10, OP11, TM7, WS3, Proteobacteria, Spirochaetes, Synergistetes, Thermodesulfobacteria, Thermotogae, and Verrucomicrobia. This primer set is not predicted to amplify archaea or members of OD1, SR1, Planctomycetes, or Tenericutes. Resulting sequences were trimmed and quality scored. All sequences passing quality score were compared using Basic Local Alignment Search Tool (BLASTn) to a ribosomal database to make classification. Identity values were used to make assignments to the appropriate taxonomic levels. Metabolic and physiological assignments were made based on the data available for the most closely related organisms.

22.3 RESULTS

22.3.1 Sample Descriptions

Samples were collected from four oil and natural gas fields from the United States and Canada. Table 22.1 outlines a few key characteristics of these hydrocarbon systems. Samples were analyzed by culture-based enumeration, DNA isolation and sequencing, ATP analysis, and epifluorescence microscopy.

22.3.2 Case Study 1: British Columbia, Canada Sour Gas Samples

The first set of samples was collected from a sour gas field in British Columbia, Canada. The system components of interest consist of a single production well

TABLE 22.1
Characteristics of Systems Evaluated in the Study

Field Location	Type of Production	Formation Type	Total Dissolved Solids (mg/L)	Gaseous H_2S (mol%)	Gaseous CO_2 (mol%)	System Temp
British Columbia, CA	Natural gas/some condensate	Shale gas	170,000–300,000	1.0–3.0	5.5	104°F–140°F (40°C–60°C)
Wyoming, U.S.	Natural gas/some condensate	Tight sandstone	300–12,000	0–0.001	1–3	70°F–120°F (21°C–49°C)
Mississippi, U.S.	Light oil/ some natural gas	Oil shale	80,000–180,000	0.0002–0.001	4.5–6.5	80°F–275°F (27°C–135°C)
Texas, U.S.	Light oil/ some natural gas	Oil shale	500–320,000	0–0.500	1–3	65°F–180°F (18°C–82°C)

feeding into a 6 inch, 2.1 mile (3.39 km) pipeline lateral, which ties into a larger gathering system downstream. The production well was brought on line in 2006, and a corrosion inhibition program was implemented to protect the pipeline from the high CO_2 and H_2S concentrations in the natural gas. In September of 2010, the flowline experienced a corrosion pitting failure (Figure 22.1).

An in-line inspection (ILI) was run, revealing an additional 222 metal loss features. Because the system contains high H_2S and CO_2 concentrations, high brine, and total dissolved solids, and because culture media failed to test positive for microbial activity, the team continued to assume that the elevated corrosion rates were abiotic in origin and thus significant modifications were made to the corrosion–inhibition and pigging programs. Four additional ILIs have been run on the line since 2010, each one indicating that, despite the upgrades made to the corrosion-inhibition and pigging program, significant corrosion was continuing to occur, resulting in additional metal loss features and preventative pipeline replacement expenditures. After the 2014 ILI run showed continued elevated corrosion rates, a root cause failure analysis was performed. Based on several key observations, (1) the pipeline superficial gas velocity drop from 3.97 ft/sec (1.21 m/sec) in 2010 to just 0.79 ft/sec (0.24 m/sec) in 2014, (2) the presence of deposition in the pipeline, (3) significant pitting damage observed in low-lying areas of the pipeline segment, (4) the location (5 to 7 o'clock position) of a large portion of the pitting damage in the flowline, and (5) the morphology of a significant amount of the pitting damage ("pit within a pit"), the team decided to look more closely at microbiologically induced corrosion (MIC) as a potential contribution to the corrosion occurring in the pipeline even though the system characteristics are generally considered noncompatible with biotic corrosion.

Three samples were collected from the Canadian sour gas wells (Table 22.2). Collection was performed using extreme care to avoid contamination of the samples as well as to maintain samples as close as possible to in situ conditions. First, a test separator was brought in and connected downstream of the wellhead

FIGURE 22.1 Photograph of the through-wall failure in 2010.

TABLE 22.2
British Columbia Sour Gas Well Microbiological Results

		Microbial Cells per mL/cm²				
Sample ID	**Sample Origin**	**ATP Assay**	**Microscopy**	**GHB/APB Culture**	**SRB Culture**	**DNA**
1.A	CAN Pig Envelope	2.6×10^5	3.1×10^8	$<10^1$	10^3	1.1×10^4
1.B	CAN Wellhead Swab	3.4×10^5	2.0×10^6	10^3	$<10^1$	1.3×10^4
1.C	CAN Separator Water	2.0×10^4	5.0×10^6	10^3	$<10^1$	3.0×10^3

Note: APB = acid-producing bacteria, ATP = adenosine triphosphate, GHB = general heterotrophic bacteria, SRB = sulfate-reducing bacteria.

to collect a sample of the produced water from the production well feeding into the pipeline (1.C, CAN Separator Water). The test separator was cleaned as thoroughly as possible to flush out any prior contamination, and the well fluids were run through the separator for 24 hours prior to collection of the produced water sample for microbial analysis. A sessile biofilm sample (1.B, CAN Wellhead Swab) was collected by opening the wellhead flowline and swabbing a 1 cm^2 surface area that had been constantly exposed to the system produced fluid. The materials from the surface and the swab tip were placed into a dilution blank prepared from filter-sterilized produced water. Finally, a pigging run was made in the flowline feeding from the production well pad to the gathering system trunk line, and a sample of the pigging envelope slurry was collected from the pig receiver (1.A, CAN Pig Envelope).

A detailed microbial survey, including ATP, microscopy, culturing in SRB and GHB/APB indicator media, and 16S DNA sequence analysis, was conducted on the three samples collected from the Canadian sour gas system (Table 22.2). The ATP assay was performed onsite immediately following sample collection and indicated the presence of moderate levels of microorganisms, ranging from 2.0×10^4 to 2.6×10^5 total microbes per milliliter. Quantification of cells in the sample by epifluorescence microscopy provided the highest estimations with values ranging from 2.0×10^6 to 3.1×10^8 cells/mL. Visualization was challenged in part due to the large amount of deposition, making it difficult to differentiate between the bacterial cells and deposit during the counting process. In contrast, culture-based analysis resulted in the lowest microbial activity recorded for any of the different assays used, identifying only 10^3 culturable cells per mL or cm^2 for either GHB/APB or SRB at the most. Historically, culture-based bacterial enumerations from this field typically show no signs of microbial growth (data not shown). In this study, attempts were made to provide culture conditions that mimicked the in situ sample conditions as closely as possible, including incubations at the system temperature of 113°F (45°C) and using media with similar salinity to produced waters from this location (between

TABLE 22.3
British Columbia Sour Gas Well Diversity Analysis

Sample ID	**1.A**	**1.B**	**1.C**	**Total**
# Cells IDed	272,193	198,155	178,732	649,080
# Unique Taxa	61	33	45	73
Metabolism	**% Sample Population; # Taxa**			**Total Taxa**
All Sulfidogens	0.3; 2	18; 2	0; 0	2
– SRB	0.3; 1	18; 1	0; 0	2
– SuRB	0.01; 1	0.01; 1	0; 0	1
Fe(III)RB	0.01; 1	0.01; 1	0; 0	1
APB	9; 5	18; 4	1; 5	8
Biodegradation	39; 10	19; 5	49; 8	11
Methanogen	0.4; 1	5; 1	13; 4	4
Dominant Class	**% Population; # Unique Taxa**			**Total Taxa**
Alphaproteobacteria	46; 10	19; 4	23; 8	11
Gammaproteobacteria	20; 19	15; 12	20; 15	21

Note: APB = acid-producing bacteria; Fe(III)RB = iron-reducing bacteria, SRB = sulfate-reducing bacteria, SuRB = sulfur-reducing bacteria.

150,000 and 200,000 mg/L TDS). Consistent with ATP and microscopy analysis, DNA was successfully isolated from each of the three samples, providing another line of evidence that despite the atypical environment, microorganisms are present in these samples; however, yields were quite low.

The populations of bacteria present in these samples were analyzed by shotgun sequence analysis of 16S genes from between 178,000 and 273,000 bacteria from each sample (Table 22.3).

These sequences were grouped into between 33 and 61 unique taxa. Organisms typically associated with oilfield corrosion or high-temperature formations, such as SRB and IRB, were not found in great abundance in these samples. This model is supported by the observation that the species profiles of the three samples were quite different from one another. Members of the class Alphaproteobacteria numerically dominated the samples. Abundant genera included *Brevundimonas* and *Paracoccus*, which are implicated in biodegradation and acid production (Wrangham and Summer 2013). *Paracoccus* isolates have been found in many diverse environments and are able to derive energy from organic and inorganic compounds, including hydrogen and sulfur (Kryachko et al. 2011). While not classified as such here, it is possible that *Paracoccus* could contribute to sulfidogenisis in these environments.

22.3.3 Case Study 2: Wyoming, U.S. Production System Samples

The second set of samples originated from a tight sandstone natural gas field located in Wyoming (Table 22.4). This system historically has shown minimal bacteria-related operational issues and rarely demonstrates any evidence

TABLE 22.4
Wyoming Tight Sandstone Production System Microbiological Results

		Bacterial Cells per mL				
Sample ID	**Sample Origin**	**ATP Assay**	**Microscopy**	**GHB/APB Culture**	**SRB Culture**	**DNA**
2.A	WY 3-27 Pig Envelope	8.4×10^7	4.4×10^9	10^8	10^8	1.10×10^8
2.A.1	WY 3-27 PRD	nd	nd	10^8	nd	3.33×10^7
2.A.2	WY 3-27 MPB	nd	nd	nd	10^8	1.07×10^8
2.B	WY 2-36 Pig Envelope	6.2×10^7	1.2×10^8	10^7	10^8	1.60×10^7
2.B.1	WY 2-36 PRD	nd	nd	10^7	nd	1.13×10^7
2.B.2	WY 2-36 MPB	nd	nd	nd	10^8	1.83×10^7
2.C	WY Pond 7	2.1×10^7	2.3×10^7	10^6	10^7	5.00×10^7

Note: APB = acid-producing bacteria, ATP = adenosine triphosphate, GHB = general heterotrophic bacteria, SRB = sulfate-reducing bacteria, nd = not determined.

of MIC, plugging, or H_2S production. However, routine culturing of these and many other samples from this site in SRB and GHB/APB bacterial culture media also indicated very high levels of SRB and GHB/APB at this location, up to 10^9 bacteria/mL.

The field risk assessment team wanted to understand more clearly how a site that historically shows no bacterially related issues can conversely provide culture-based evidence of large populations of problematic organisms. Detailed analysis was performed on samples consisting of pigging run envelopes from two gas gathering system pipelines, WY 3-27 Pig Envelope and WY 2-36 Pig Envelope, along with cognate cultures established from each envelope sample in either SRB or GHB/APB indicator media (MPB and PRD, respectively) (Tablc 22.4). Additionally, sample WY Pond 7 was collected from a covered water holding pond for flowback water from wells fractured with fresh and produced water. This water is recycled for use in hydraulic fracturing, and there was interest in understanding how storing the water might affect the population of bacteria.

Each of the three primary samples (3-27 Envelope, 2-36 Envelope, and Pond 7) was subject to ATP analysis, microscopy, and GHB/APB and SRB culture analysis (Table 22.4). Consistent with historical observations from this site, the 3-27 and 2-36 pigging envelope samples, as well as many other samples collected from throughout the field (data not shown), showed very high concentrations of bacteria using either the ATP assay (6.2×10^7 to 8.4×10^7) or by microscopic analysis (1.2×10^8 to 4.4×10^9). Routine culturing of these and many other samples in SRB and GHB/APB bacterial culture media, indicated very high levels of culturable SRB and GHB/APB at this location with culture-based enumeration typically revealing between 10^7 and 10^9 viable bacteria/mL (Table 22.4).

In addition to the standard enumeration methods, amplicon pyrosequencing was completed on the 3-27 and 2-36 raw pigging envelope samples as well as the

MPB and PRD cultures established from each of these, called WY 3-27 SRB, WY 3-27 APB, WY 2-36 SRB, and WY 2-36 APB, respectively (Table 22.5). Between 6,900 and 13,036 sequences were obtained from each population. The populations in the raw envelope samples clustered into 78 and 58 taxa, and between 5 and 19 taxa were obtained from each of the four enrichment cultures. The two raw pig envelope samples were found to be quite different from one another. The bacterial populations from the raw WY 3-27 Envelope sample were dominated by members of the Gammaproteobacteria and Thermotogae (Table 22.5). In contrast, the raw WY 2-36 Envelope sample was dominated by members of class Clostridia and Negativicutes (Table 22.5). However, neither sample was found to contain

TABLE 22.5
Bacterial Populations in Pigging Envelope Raw Samples and MPB and PRD Cultures

Sample ID	PE3-27	PE3-27 PRD	PE3-27 MPB	PE2-36	PE2-36 PRD	PE2-36 MPB	Pond 7	Total
# Bacteria Tested	7,477	13,039	6,975	7,138	10,086	9,164	2,413	56,292
# Unique Taxa	78	19	17	59	9	5	39	144
Traits				**% Population; # Taxa**				**Taxa**
All Sulfidogens	60; 9	11; 2	12; 2	0.3; 4	14; 1	25; 1	20; 7	15
–SRB	0; 0	0; 0	0; 0	0; 0	0; 0	0; 0	19; 3	3
–TRB	60; 9	11; 2	12; 2	0.3; 4	14; 1	25; 1	0.4; 2	10
Citrobacter sp.	33	11	12	0	0	0	0	1
Clostridium beijerinckii	0	0	0	0.01	14	25	0.08	1
Thermotoga sp.	25	0	0	0.2	0	0	0	1
APB	8; 9	69; 3	66; 3	63; 8	52; 2	0; 0	24; 10	18
Fermentative	18; 15	2.6; 2	1.8; 2	13; 18	47; 3	99; 2	26; 8	27
Thermophile	28; 14	0; 0	0; 0	3; 10	14; 1	25; 1	26; 3	17
Biodegradation	0.2; 5	1.6; 2	1.9; 2	0.9; 5	0.3; 1	0; 0	0.2; 5	15
Dominant Class				**% Population; # Taxa**				**Taxa**
Gammaproteobacteria	42; 30	28; 12	30; 10	0; 0	0.01; 1	0; 0	1; 4	39
Thermotogae	25; 3	0; 0	0; 0	0.2; 2	0; 0	0; 0	0; 0	3
Clostridia	8; 28	67; 2	66; 2	77; 41	99; 7	100; 5	23; 13	55
Negativicutes	0.1; 3	0; 0	0; 0	15; 6	0; 0	0; 0	27; 5	11
Deltaproteobacteria	0; 0	0; 0	0; 0	0; 0	0; 0	0; 0	19; 4	4

Note: PE = Pig Envelope. MPB and PRD are the sequence results of cultures growing in the indicated media. # Bacteria Tested = number of sequences from sample. # Unique Taxa = number of sequence clusters. Taxa and physiological groupings values represent the % of total population followed by the number of sequence clusters for that group. Total Taxa is less than the sum of taxa in each sample because each taxa may be present in more than one sample.

organisms classified as true sulfate-reducing bacteria based on 16S gene analysis (Table 22.5).

Considering the strongly positive results obtained from culturing these samples in SRB indicator media, this result was unexpected. To delve into this counter-intuitive result a little further, 16S amplicon pyrosequencing was also performed on the cultures enriched in PRD and MPB from the WY 3-27 and WY 2-36 pigging envelope samples (Table 22.5). The results for the MPB medium indicated that for both originating populations, the sulfidogenic organisms most likely responsible MPB positive results were not true SRB. The population enriched from the WY 3-27 Envelope sample was a subset of the bacteria growing in the PRD media. All 15 species identified in sample WY 3-27 SRB selective media were also present in sample WY 3-27 APB selective media. The three most abundant organisms in each sample were *Acetobacterium*, *Citrobacter*, and *Klebsiella*. *Citrobacter* are sulfidogenic coliforms that generate sulfide via a thiosulfate reduction mechanisms using sulfur-containing amino acids as a substrate (Braunstein and Tomasulo 1978). Both MPB and PRD media types contained significant levels (66%) of the anaerobic, acetogenic genera, *Acetobacterium* (Balch et al. 1977). The population enriched from the WY 2-36 envelope sample were dominated by *Clostridium* species related to *C. beijerinkii*, and *Eubacterium aggregans*, while a third species, *Oscillibacter* sp., was also present in the PRD media culture. Several *Clostridium* species are known to ferment sulfur-containing amino acids and to generate sulfide as a result of sulfur and/or thiosulfate reduction (Hernández-Eugenio et al. 2002). *Eubacterium aggregans* isolates are homoacetogens, capable of acetic acid production via carbon dioxide reduction (Mechichi et al. 1998). The identification of populations of non-SRB sulfidogens in enrichment cultures producing hydrogen sulfide in standard MPB media is an atypical result. NACE standard MPB lacks appreciable thiosulfate to serve as an electron acceptor. One possibility that was investigated was that these results were simply due to incorrect media formulation. This was addressed by culturing in MPB from several independent sources as well as media made in-house according to NACE standards (data not shown). Sulfide generation was observed in all cases (data not shown). There is one report suggesting that there is sulfate reduction in some *Citrobacter*; while this is unexpected, a formal possibility is that this trait is more common than previously appreciated (Qiu et al. 2009).

Regardless of the bacteria involved, identification of a population of bacteria that does not appear to contribute to significant sulfidogenesis in the field but that is capable of sulfidogenisis in the lab demonstrates the need to compare lab results against the field data. The more typical caution when using culture-based enumeration is that the sulfidogens will not grow in the media, resulting in a false negative. While atypical, this demonstrates that it is also possible to have a population of bacteria capable of sulfidogenisis under laboratory conditions but not in the field.

22.3.4 Case Study 3: Mississippi, U.S. Oil Shale Samples

Samples analyzed for case study 3 were collected from an oil shale field in Mississippi in the United States along with samples of the cognate frac fluid source waters (Table 22.6). Sample 4.A originated from the Amite River water and was used to fracture the production well that fed into the heater treater solids collected for Sample 4.B. Sample 4.B was a solid deposit sample scraped from the surface of a heater treater fire tube that experienced an MIC failure after being in service for just 3 months (Figure 22.2).

The surface of the fire tube exhibited a large amount of iron sulfide as well as some calcium carbonate scale. As soon as the failed vessel was opened, a sessile swab sample was collected from the moist fire tube surface right next to the failure.

Due to the high solids content of the sample, microscopic analysis was not performed. The ATP quantification assay was conducted immediately upon sample collection while on location and indicated the presence of high concentrations of bacteria (Table 22.6).

Salinity-adjusted bacterial culture media was also inoculated immediately. No bacterial growth was observed during the incubation period. This may be due to the difficulty in mimicking the high temperature encountered on the fire tube skin

TABLE 22.6
Mississippi, U.S. Sample Microbiological Results

		Bacterial Cells per mL/cm^2				
Sample	**Sample Origin**	**ATP Assay**	**Microscopy**	**GHB/APB**	**SRB**	**DNA**
4.A	Frac Water Source 1	3.2×10^5	2.1×10^6	10^6	10^2	2.8×10^7
4.B	Fire Tube MIC	6.8×10^6	nd	$<10^1$	$<10^1$	6.8×10^8
4.C	Frac Water Source 2	1.4×10^5	3.2×10^6	10^6	10^3	9.2×10^5
4.D	Well Flowback	2.3×10^2	1.1×10^5	10^1	$<10^1$	1.0×10^7

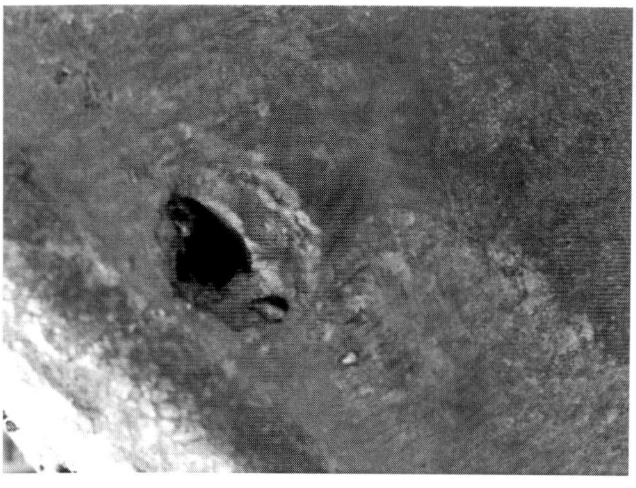

FIGURE 22.2 Photograph of the Mississippi fire tube failure.

surface (estimated 160°F to 180°F or 71°C to 82°C) during the media incubation period; if incubated at this elevated temperature, the medium would have likely degraded. Bacterial levels in sample 4.A, the Amite river water sample, were consistent with levels in a natural water system. Microscopic, ATP quantification, and DNA yield estimations resulted in 10^5, 10^6, and 10^7 cells per mL, respectively. The higher level observed in the DNA yield estimation could be the result of assaying eukaryotic DNA in the sample as this is based on conversion of total DNA yield from the sample and not qPCR. Additionally, both GHB/APB and SRB culture media were positive and gave estimations of 10^6 and 10^2 GHB/APB and SRB per mL, respectively.

The second paired samples from Mississippi, 4.C and 4.D, were collected as part of a larger study to monitor the efficacy of a fracturing fluid biocide program (Table 22.6). Sample 4.D was an aliquot of production well flowback water while sample 4.C was the source water used to fracture the well. Bacterial levels at this location were being monitored by both the ATP quantification assay as well a culture media–based enumeration during the flowback period (Table 22.6). During routine monitoring, both ATP and bug bottle results consistently indicated very good control of the frac fluid bacterial populations. It was decided to spot check a few wells utilizing both microscopic and pyrosequencing analysis. Consistent with earlier field results, ATP quantification and GHB/APB/SRB culture-based enumeration indicated that bacterial levels in sample 4.D were quite low at 10^2 or less. However, both microscopic enumeration and DNA yield quantification revealed moderate to high concentrations (10^5 to 10^7 bacteria/mL) of microbes in the well flowback sample 4.D. In contrast, sample 4.C, the fracturing water source 1 sample, resulted in typical natural water bacterial levels of 10^5 to 10^6 using all three total cell enumeration methods. Culturable GHB/APB and SRB in the fracturing water source 1 sample were 10^6 and 10^3, respectively.

The bacteria diversity of all four Mississippi samples was evaluated by pyrosequencing of 16S amplicons (Table 22.7). Between 30 and 266 taxa were identified in each sample. The fracturing water source 2 sample was enriched in phototrophic organisms, such as *Anabaena* as well many bacterioplankton species of *Limnohabitans*. The profiles of both samples 4.B and 4.D are consistent with anaerobic, possibly corrosive microbial populations. Both samples 4.B and 4.D contained elevated levels of sulfidogenic organisms with 70% and 18% of the bacteria identified in samples 4.B and 4.D, respectively, being SRB. Sulfidogens were not identified in the fracturing water source 2 sample despite culture-based evidence. Further data, for example, sequencing the enriched population, is required to identify the sulfidogenic organisms in this sample.

A direct comparison of the frac source waters and site water/deposit samples was conducted (Figure 22.3). Overall, there was a stronger correlation between the type of sample and the number of shared species; for example, the two river samples and the two well site samples were more similar to each other than either cognate source water/well site pair was to each other. Bacterial populations from the two source waters were most similar to each other, with 21 of the 30 organisms identified in the fracturing water source 2 sample also being present in the fracturing water source 1 sample (data not shown). In contrast,

TABLE 22.7
Mississippi, U.S. Sample Pyrosequencing Results

	Frac Water Source 1	Fire Tube MIC	Frac Water Source 2	Well Flowback	
Sample ID	**4.A**	**4.B**	**4.C**	**4.D**	**Total**
# Bacteria Tested	793	6,013	6,897	6,595	20,298
# Unique Taxa	30	47	266	46	347
Traits		**% Population; # Taxa**			**Taxa**
Sulfidogen	0; 0	71; 19	2.1; 4	32; 14	32
–SRB	0; 0	70; 15	0.1; 3	18; 8	22
–TRB	0; 0	0.26; 2	2; 1	26; 6	8
–SuRB	0; 0	0.4; 3	0; 0	17; 3	6
Fe(III)RB	0; 0	0.4; 4	6; 10	0.5; 2	15
APB	0; 0	0.1; 1	1; 3	0.3; 2	6
Ferm	0.5; 1	2; 12	2; 6	43; 16	29
Phototrophic	23; 2	0; 0	2; 6	0; 0	7
Thermophile	0; 0	2; 6	0.01; 1	30; 12	17
Biodegradation	20; 5	76; 5	9; 32	5; 3	42
Methanogen	0; 0	0; 0	4; 7	0; 0	7

Note: # Bacteria Tested = number of sequences from sample. # Unique Taxa = number of sequence clusters. Taxa and physiological groupings values represent the % of total population followed by the number of sequence clusters for that group. Total Taxa is less than the sum of taxa in each sample because each taxa may be present in more than one sample. SuRB = sulfur reducing bacteria.

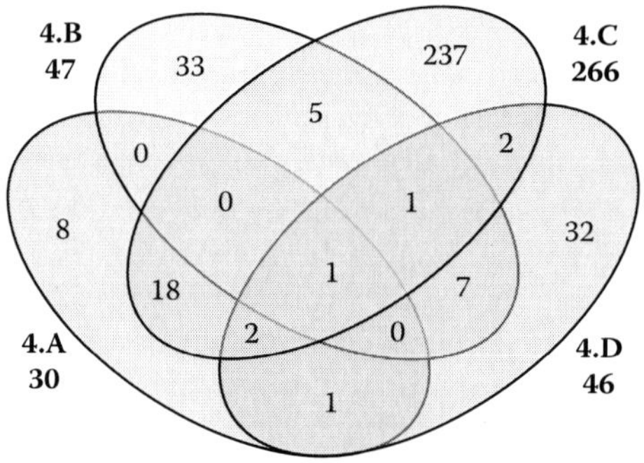

FIGURE 22.3 Venn diagram comparing shared organisms between the four Mississippi samples.

the paired source water–well site samples were not very similar to each other. Samples 4.A and 4.B shared only two organisms, and samples 4.C and 4.D shared only six organisms. Strikingly, the nine species found in samples 4.B and 4.D include those implicated in corrosion and or sulfide generation. These included species of *Desulfovibrio*, *Anaerophaga*, *Sulfurospirillum*, *Desulfonatronovibrio*, *Halanaerobium*, *Flexistipes*, and *Thermovirga* (Mechichi et al. 1998; Dahle and Birkeland 2006; Dahle et al. 2008; Ghosh and Dam 2009; Hubert et al. 2011; Kryachko et al. 2011; Mand et al. 2014). Finally, the low overlap of populations from the frac water source to the well site samples indicates that the frac biocide program is indeed providing a good kill of the frac water microbes. The source of the microbes causing the fire tube MIC failure and in the well flowback water remains to be determined. One source might be the oil shale reservoir itself, a model supported by the ability of these organisms to grown at the high temperature expected on a heater-treater tube. However, the level of coverage for each sample does not provide robust identification of organisms present at less than 0.1% of the population, and so an alternative explanation is that low abundance organisms in the source water proliferated under the extreme local environment.

22.3.5 Case Study 4: Texas, U.S. Oil Shale Samples

A preliminary analysis of the failure history of 1,450 production wells was conducted based on case histories. It was determined that over 4,000 downhole failures had occurred over a 5-year period. The majority of failures (80%) were attributed to rod on tubing wear. Other causes of failures included CO_2 corrosion, MIC, scaling, and debris in the downhole pumps as well as scale and paraffin deposition on tubing and rods, resulting in under-deposit corrosion and rod parts. Evidence implicating bacterial issues throughout the system include drilling mud degradation; drill pipe under-deposit corrosion; downhole and surface equipment under-deposit corrosion; generation of hydrogen sulfide (H_2S); generation of iron sulfide, resulting in “black water,” especially in tank bottoms and interface solids; the formation of emulsions; oil carryover to salt water disposal (SWD) facilities/wells; downhole production well pump plugging; and SWD well plugging issues. Examples of candidate MIC pitting are shown (Figure 22.4)

As part of a total systems project encompassing analysis and remediation of all potential sources of bacterial contamination in a production system, a program was implemented specifically with the purpose of examining the extent and sources of microbial issues. The approach taken was to conduct analysis of microbial populations from 129 samples, including core samples, drilling mud samples, completions source waters (freshwater wells, brackish source waters, and blended waters from frac pits), flowback waters during the first 90 days of production to evaluate the impact of the incumbent frac biocide (ClO_2) program, production wells with a history of MIC failures, and salt water disposal facilities (Table 22.8).

The majority of production well samples were from wells reporting microbial-associated issues, including MIC, H_2S production, and iron sulfide production (black water). Samples were subject to various analysis methods, including water analysis

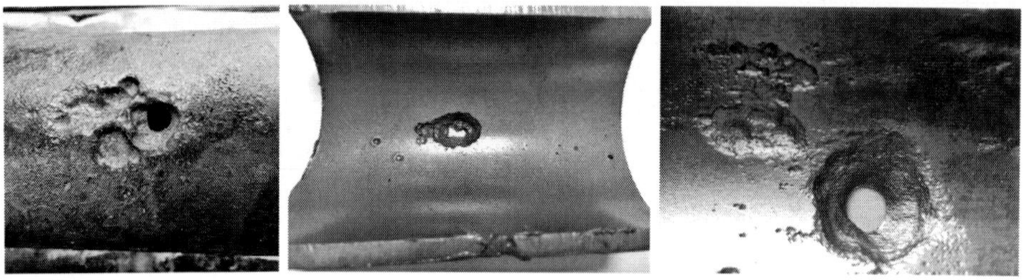

FIGURE 22.4 Representative photographs of MIC-related downhole failures on production tubing.

TABLE 22.8
Summary of Case Study Four Samples

Sample Source	Number of Samples
Shale core	6
Drilling mud	7
Completions water sources	36
Frac biocide flowback samples	9
Production wells, MIC history	51
Salt water disposal wells	17
Total	129

and microbial quantification from each sample using three culture-independent methods as well as traditional indicator media culture based analysis ("bug bottles"). Based on findings, recommendations were made with regards to biocide choices and application protocols.

Chemical analysis, including total dissolved solids (TDS), chlorides, sulfate, calcium, bicarbonates, magnesium, $CaCO_3$, and pH was conducted on completions water and production wells (Table 22.9). Total dissolved solids, chlorides, and calcium levels were significantly higher in the produced waters as compared to completions waters. In contrast, the sulfate levels in frac waters were significantly higher as compared to produced waters. The significant difference in frac versus produced water sulfate concentrations indicates that the sulfate is either getting consumed by sulfate-reducing bacteria or reacting with various cations (barium, strontium, or calcium) to form mineral scale deposits.

Microbial enumeration was conducted on each sample, using four different quantification methods (Table 22.10). One method was traditional bug bottles, using serial tenfold dilutions into indicator media for SRB, APB/GHB, and IRB followed by incubation and scoring for growth. Two methods (ATP assay and the hydrolase assay) were based on biochemical assays rather than direct counting of cells or cell growth. For both the ATP and hydrolase assays, the values are converted into a "cells per mL sample" using a formula that assumes that the per-cell content of

TABLE 22.9
Chemical Analysis of Completions/Frac and Produced Waters

Parameter	Completions Water Avg ppm	Produced Water Avg ppm
TDS	4,475 ± 4,889	116,063 ± 33,844
Chlorides	1,171 ± 2,525	68,571 ± 19,795
Sodium	1,212 ± 1,580	40,220 ± 11,972
Sulfate	925 ± 1,123	371 ± 435
Barium	Below detection limit	7.89 ± 11.86
Strontium	Below detection limit	696 ± 285.66
Calcium	186 ± 136	3443 ± 1814
Bicarbonates	241 ± 117	357 ± 216
Magnesium	136 ± 174	645 ± 878
Iron	<1	78.9 ± 88.78
Manganese	Below detection limit	1.03 ± 11.92
Total hardness (as $CaCO_3$)	1,060 ± 1,254	11,813 ± 6,710
pH	7.92 ± 0.45	6.67 ± 0.54

TABLE 22.10
Average Microbial Levels in 129 Permian Basin Samples

Sample	Bactiquant	ATP	Rapid B
Cores	1.7E + 05 ± 1.6E + 04	3.6E + 05 ± 3.9E + 05	9.6E + 04 ± 3.9E + 05
Drilling muds	2.5E + 06 ± 3.7E + 06	7.9E + 07 ± 1.2E + 08	3.8E + 07 ± 5.1E + 07
Frac water pits	2.6E + 05 ± 3.0E + 05	8.9E + 05 ± 5.5E + 05	3.8E + 05 ± 3.5E + 05
Fresh water wells	1.3E + 05 ± 2.4E + 05	3.6E + 05 ± 2.7E + 05	1.4E + 05 ± 1.7E + 05
Production wells	2.9E + 04 ± 1.5E + 05	1.1E + 06 ± 1.6E + 06	8.2E + 06 ± 1.9E + 07
SWD	6.0E + 04 ± 7.9E + 04	4.2E + 06 ± 4.4E + 06	2.9E + 06 ± 3.1E + 06

the molecule being measured (e.g., ATP and hydrolase) is constant across all cell types. The fourth method used for microbial quantification was a direct cell counting method, using live/dead staining and a flow cytometer. Flow cytometry is prone to several types of artifacts, notably binding of stain to particulate matter and underestimating cell numbers in aggregates. As such, the value of this approach may vary depending on sample type.

The microbial content of each of the 129 samples was analyzed by the four methods. The values for all members within a specific sample type (core samples, drilling mud samples, frac water samples, production well samples, and salt water disposal wells) was averaged and compared against each of the other sample sets (Figure 22.5, Table 22.10). When using the three culture-independent methods (ATP, hydrolase, flow cytometer), bacterial levels in most samples were determined to be in the range of 10^5 to 10^6 cells per mL. The notable exception was drilling muds, where

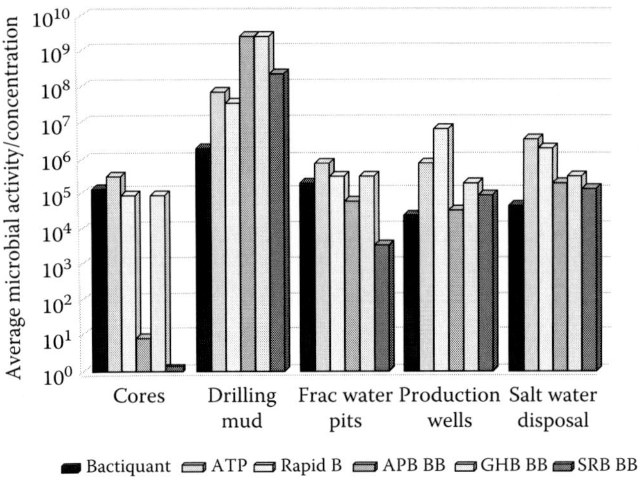

FIGURE 22.5 System profile microbial population averages.

bacterial levels were determined to be in the range of 10^7 to 10^8 using the culture-independent methods.

Regardless of method used, drilling muds were determined to have the highest concentration of bacteria (Figure 22.5, Table 22.10). Drilling muds also contained significantly higher levels of culturable SRB and APB, exceeding over 4 log orders higher in drilling muds than in the other samples (Figure 22.5).

The identities of bacteria in a subset of the samples were also evaluated by 16S amplicon metagenomics, and bacteria identified were given physiological assignments (Table 22.11). The percent of bacteria in each sample (as a function of the total population) for each physiological trait is given (Table 22.11).

The most obvious result from 16S amplicon metagenomic analysis was the low percentage of APB, sulfidogens, and IRB in the core sample. This suggests that the widespread occurrence of problematic microorganisms in this field might be the result of bacteria introduced into the formation during production activities.

An additional component of this system analysis was to evaluate the performance of the completions biocide program. The field was utilizing chlorine dioxide (ClO_2) as the sole biocide for frac fluid treatment. To evaluate the performance

TABLE 22.11
Microbial Population Analysis by Amplicon Metagenomics

Sample	Samples Tested	APB	Sulfidogens	IRB
Cores	6	0% to 3%	0%	0%
Drilling muds	3	<1% to 30%	2% to 35%	0% to 5%
Production wells	5	1% to 23%	0% to 23%	0% to 9%
SWD	2	15% to 18%	23% to 24%	0% to 11%

of the ClO_2 program, produced water flowback samples were collected from 10 production wells that were in early stages of flowback (within first 60 days following frac job). Figure 22.6 shows a comparison of the average microbial population from the 10 flowback samples to the average for the 51 problematic production well samples. The results indicate that the bacterial population numbers in the ClO_2 treated water are essentially identical to population size found in the problematic production wells.

As a result of these findings, a detailed biocide selection study was undertaken with representative frac water sources. In an effort to best mimic the conditions observed in the field, the frac water was treated with the oxidizer (sodium hypochlorite) to the demand of the water, and the samples were allowed to sit for 30 minutes for all chemical reactions to occur. Each bottle was then dosed with various biocides at different concentrations recommended by the manufacturer. An untreated control was run alongside the treated samples to monitor the background bacterial activity and allow for kill assessment by each biocide treatment. The untreated control and all treated samples were incubated at 65°C to mimic reservoir temperature. At 21 days, all samples were rechallenged with 1 mL of a mixture of IRB, APB, and SRB cultures grown previously from the frac water source. Following 24 hours, 7 days, 14 days, 21 days, and 28 days of incubation, the untreated control and all samples were analyzed using the RAPID-B assay, ATP assay, Bactiquant assay, and bacterial culture media. The 28-day results shown in Figure 22.7 are expressed as the percentage reduction of the treated sample from the untreated control.

Interestingly, the data from three rapid assays provided the most insight as to the most effective treating strategies because the three assays are showing the activity of all remaining viable microbial populations, including unculturable organisms. The culture media data was misleading as it indicated that nearly all the treatment strategies provided good microbial control and did not allow for differentiation of the various treatments (data not shown). The RAPID-B, ATP, and Bactiquant assay provided faster results feedback for each treatment strategy at each sampling point so that a rapid decision could be made on what program to recommend for the field.

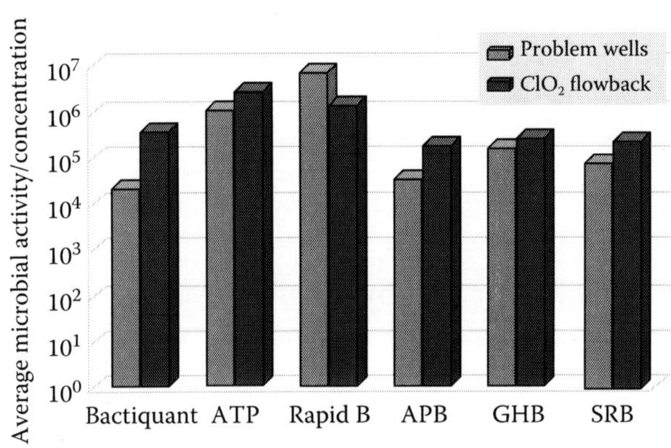

FIGURE 22.6 Comparison of bacterial populations in problematic bacterial issue wells versus ClO_2-treated wells in early flowback.

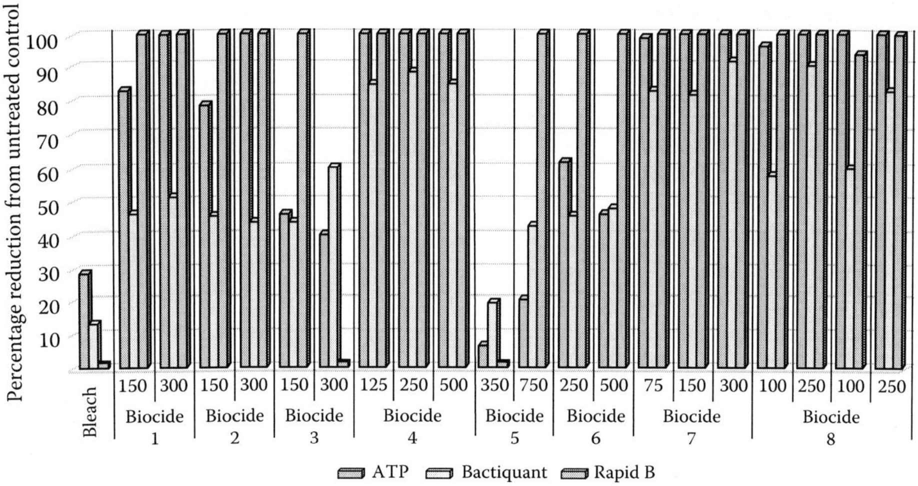

FIGURE 22.7 Planktonic biocide selection testing results, 28-day contact time. Results are expressed as percentage reduction from the untreated control.

A dual treatment strategy was implemented using a sodium hypochlorite oxidizer pretreatment to provide an initial bacterial kill, followed by injection of a preservative biocide chemistry number 7 at the blender for maintenance of control over any surviving microbes in the reservoir. The performance of the frac biocide program was monitored during flowback by measuring biocide residual profiles as well as kill performance using the ATP assay and bacterial culture media. To date, the new dual strategy biocide program has provided quality bacterial kill for up to 120 days post-frac.

In addition to the upgrade in the completions biocide program, a biocide program was implemented in the drilling mud, the water used for the completions plug drill-out process, the water used for well workovers, the flush water used for chemical treatment of the production wells, and the water used for hot-watering wells for paraffin control. In essence, any water source that was introduced to the wells was treated with a microbiocide in an effort to prevent further contamination of the reservoir and wellbore.

22.4 CONCLUSIONS

The bacterial populations in four case studies in which microbial activity affected standard field operations were evaluated using four different methodologies: microscopy, ATP quantification, MPN analysis in "bug bottles," and genetics-based population analysis.

- **Case Study 1:** British Columbia, Canada Sour Gas Samples. High corrosion rates historically attributed to abiotic causes (CO_2, high salt, H_2S) because of lack of growth in bug bottles. Alternative methods (ATP, microscopy, and DNA recovery) all indicated bacteria. *Conclusion*: Extreme salinity and temperatures dictate alternatives to traditional bug bottle enumerations.

- **Case Study 2:** Wyoming, U.S. Production System Samples. Historically, this system did not exhibit MIC or souring even though MPN analysis with bug bottles showed abundant SRB. Sequence analysis of samples did not detect SRB, and organisms growing in the bug bottles were identified as non-SRB sulfidogens. *Conclusion*: This is an atypical situation in which non-SRB sulfidogens gave false positives in bug bottles but do not cause problems in the real system.
- **Case Study 3:** Mississippi, U.S. Oil Shale Samples: MIC of Heater Treater Fire Tube. MIC corrosion affecting a heater-treater tube despite extreme fire tube skin temperature. H_2S, iron sulfide, and elevated corrosion rates were observed. Sequence analysis showed high levels of SRB in samples. *Conclusion*: Extremely high temperature environments can exhibit MIC.
- **Case Study 4:** Texas, U.S. Oil Shale Samples: Downhole MIC Failures. Samples from pristine formation core material and all potential sources of contamination and phases of well lifecycle were collected for microbial evaluation. The core material contained large microbial populations, but the organisms were identified as oil-producing and degrading microbes, not problem-causing microbes. All water sources introduced to the wells were contaminated with high concentrations of problematic bacteria with the drilling mud being the biggest contributor of problematic bacteria. *Conclusion*: The problematic bacteria causing the downhole MIC failures and other operational issues were not present in the indigenous microbial populations isolated from formation core material but rather were introduced to the system during the drilling, completions, and production processes. In addition, treatment of each drop of water introduced to the system is necessary to prevent microbial contamination and subsequent problems in oil and gas systems.

Overall, the rapid microbial monitoring methods evaluated in these case studies correlated well with the more traditional MPN culture media methodology. The rapid feedback gained from these assays can facilitate same-day identification of problematic areas in a system, provide a tool for rapid biocide selection testing, and allow for rapid assessment of biocide performance in the field.

The most important take-home lesson from this project is that for each case study, microbial population analysis data needed to be generated from more than one method and that interpretation required integration of this information within the context of field conditions. There was considerable test-to-test variation when samples from the same location were analyzed in parallel. However, this observation is not unexpected simply because conditions at each location varied considerably, and not every method performs adequately on every type of sample. Some understanding of the strengths and limitations of each method is critical for interpreting results. It is commonly recognized, for example, that to use MPN enumeration, the ability of the organism to grow in the medium provided limits the usefulness of this method for analyzing samples originating from extreme environments. However, methods dependent upon biochemical or genetic-based approaches are also subject to technique-dependent limitations. Calculations of bacterial levels using different methods resulted in orders of magnitude differences in some cases. There are many reasons for the differences. Methods that require cell lysis as part of the sample preparation steps must balance the necessarily harsh lysis conditions with the biochemical assay. Methods that require extraction of the cells from the solids and biomass in a sample are subject to variations due to the difficulty getting to all of the microbes in highly tenacious biofilms or solids matrices. Enzyme activity and ATP levels per cell are not absolute values; there are species level and metabolic state variations. As a rule of thumb, actively growing cells will have higher levels of both. DNA methodologies based on PCR approaches, such as amplicon sequencing, are dependent on primers, which all have inherent bias, some much more than others. It is important to understand which organisms a given primer set detects. Whole genome shotgun sequencing approaches would avoid primer bias; however, it is simply not practical for routine monitoring because of the massive volumes of data that would have to be generated to detect all but the most abundant organisms. Once DNA isolation and primer design is accounted for, the probability of detecting a specific type of bacteria using DNA sequence methodologies is statistically related to the number of unique sequences returned from the sample and the relative abundance in the sample, following a Poisson distribution of likelihood. Together this information provides us "% likelihood that an organism of X abundance in the sample has been detected." Comparing the presence of low abundance organisms between different samples can be misleading because of the likelihood that a given organism has not been detected simply due to sequences read coverage for that sample.

Each method tried here relied on analysis of a different bacterial cell characteristic, including growth characteristics (MPN), whole cell counting (cytometer), quantification of ATP, quantification of enzyme activity, and DNA sequence analysis. These methods generate data that can be categorized as those that provide information on abundance and bacterial type (DNA analysis, MPN growth analysis) and those that provide primarily information on abundance (ATP, flow cytometer, enzyme analysis). The trade-off between abundance and bacterial type is speed of analysis versus time for analysis to be completed. As a rule, the quantification-only methods are much faster with results available in the same day as sampling while the more informative techniques require several weeks. For field operators, several weeks can be too long to wait to make decisions. However, a real understanding of the system is not possible without some in-depth analysis of bacterial types. It seems prudent to utilize a combined approach that generally takes advantage of at least one rapid quantification method and sporadic use of the approaches that provide more in depth investigation of the actual identities of the bacteria present. This could be accomplished simply by adoption of a rapid quantification technique to be used for most samples and subjecting a subset of samples for bacterial identification via DNA sequence analysis.

Taken together, a thorough, global approach to microbial monitoring should ideally incorporate multiple monitoring approaches. Individual bacterial quantification results should not stand on their own but rather be viewed as part of a trend in a consistent, systematic monitoring program using standardized approaches. Any conclusions must be made by thoughtful comparison of results obtained by each method and contrasting these finding with the microbial issues the field operators observe on location.

REFERENCES

Balch, W. E.; Schoberth, S.; Tanner, R. S.; Wolfe, R. S., 1977. *Acetobacterium*, a new genus of hydrogen-oxidizing, carbon dioxide-reducing, anaerobic bacteria. *Int J Syst Bacteriol.* 27, 355–361.

Braunstein, H.; Tomasulo, M., 1978. Hydrogen sulfide-producing *Citrobacter diversus*. A re-emphasis of the potential ability of all Enterobacteriaceae to manifest this quality. *Am J Clin Pathol.* 69 (4), 418–420.

Dahle, H.; Birkeland, N. K., 2006. *Thermovirga lienii* gen. nov., sp. nov., a novel moderately thermophilic, anaerobic, amino-acid-degrading bacterium isolated from a North Sea oil well. *Int J Syst Evol Microbiol.* 56 (Pt 7), 1539–1545.

Dahle, H.; Garshol, F.; Madsen, M.; Birkeland, N. K., 2008. Microbial community structure analysis of produced water from a high-temperature North Sea oil-field. *Antonie Van Leeuwenhoek.* 93 (1–2), 37–49.

Fichter, J.; Wunch, K.; Moore, R.; Summer, E. J.; Braman, S.; Holmes, P., 2012. How hot is too hot for bacteria? A technical study assessing bacterial establishment in downhole drilling, fracturing, and stimulation operations. Paper no. 01310. Proc. CORROSION 2012 (Houston, TX: NACE International 2012).

Ghosh, W.; Dam, B., 2009. Biochemistry and molecular biology of lithotrophic sulfur oxidation by taxonomically and ecologically diverse bacteria and archaea. *FEMS Microbiol Rev.* 33 (6), 999–1043.

Greene, A. C.; Patel, B. K.; Sheehy, A. J., 1997. *Deferribacter thermophilus* gen. nov., sp. nov., a novel thermophilic manganese- and iron-reducing bacterium isolated from a petroleum reservoir. *Int J Syst Bacteriol.* 47 (2), 505–509.

Hernández-Eugenio, G.; Fardeau, M. L.; Cayol, J. L.; Patel, B. K.; Thomas, P.; Macarie, H.; Garcia, J. L., 2002. *Clostridium thiosulfatireducens* sp. nov., a proteolytic, thiosulfate- and sulfur-reducing bacterium isolated from an upflow anaerobic sludge blanket (UASB) reactor. *Int J Syst Evol Microbiol.* 52 (Pt5), 1461–1468.

Hubert, C. R.; Oldenburg, T. B.; Fustic, M.; Gray, N. D.; Larter, S. R.; Penn, K.; Rowan, A. K.; Seshadri, R.; Sherry, A.; Swainsbury, R.; Voordouw, G.; Voordouw, J. K.; Head, I. M., 2011. Massive dominance of Epsilonproteobacteria in formation waters from a Canadian oil sands reservoir containing severely biodegraded oil. *Environ Microbiol.* 14 (2): 387–404.

Kryachko, Y.; Dong, X.; Sensen, C. W.; Voordouw, G., 2011. Compositions of microbial communities associated with oil and water in a mesothermic oil field. *Antonie Van Leeuwenhoek.* 101 (3), 493–506.

Mand, J.; Park, H. S.; Jack, T. R.; Voordouw, G., 2014. The role of acetogens in microbially influenced corrosion of steel. *Front Microbiol.* 5, 268.

Maxwell, S.; Campbell, S., 2006. Monitoring the mitigation of MIC risk in pipelines. Paper 06662. CORROSION NACE 2006 (San Diego, CA: NACE International).

Mechichi, T.; Labat, M.; Woo, T. H.; Thomas, P.; Garcia, J. L.; Patel, B. K., 1998. *Eubacterium aggregans* sp. nov., a new homoacetogenic bacterium from olive mill wastewater treatment digestor. *Anaerobe.* 4 (6), 283–291.

Miller, J.; Tiller, A., 1970. *Microbial Aspect of Metallurgy*, Elsevier, New York (61–105).

NACE Standard TM0194-94, 1994. Field monitoring of bacterial growth in oilfield systems, 1994. Houston, TX: NACE International, item no. 21224.

Qiu, R.; Zhao, B.; Liu, J.; Huang, X.; Li, Q.; Brewer, E.; Wang, S.; Shi, N., 2009. Sulfate reduction and copper precipitation by a *Citrobacter* sp. isolated from a mining area. *J Hazard Mater.* May 30;164 (2–3), 1310–1315.

Summer, E.; Duggleby, S.; Janes, C.; Liu, M., 2014. Microbial populations in the O&G: Application of the knowledge. Paper no. 4376. Proc. CORROSION 2014 (San Antonio, TX: NACE International 2014).

TM0194-2014, 2014. Field monitoring of bacterial growth in oil and gas systems. NACE International, item no. 21224.

Videla, H. A., 1996. *Manual of Biocorrosion.* Boca Raton, FL: CRC Lewis Publishers.

Wrangham, J.; Summer, E. J., 2013. Planktonic microbial population profiles do not accurately represent same location sessile population profiles. Paper no. 02780. Proc CORROSION 2013 (Houston, TX: NACE International 2013).

23 Multidisciplinary Approach to Evaluate MIC of Two Stainless Steels in a Hypersaline Environment Using a Dynamic System

Viviane de Oliveira, Diogo Coutinho, Mariana Galvão, and Márcia Lutterbach

CONTENTS

ABSTRACT

The oil and gas industrial systems have different characteristics that may favor the colonization of metal surfaces by microorganisms, which can lead to biocorrosion problems. The diversity of environments found in these industries is very large and poses a growing challenge to the industry since deeper and more complex environments are now being explored. Microorganisms can cause problems of biocorrosion in underground tanks and pipes buried in different types of soil, in metal structures submerged in water of different salinities and depths, and so on. Therefore, the search for new materials capable of withstanding and/or slowing down the microbiological corrosion process is common in the oil and gas industries. Similarly, laboratory tests that simulate the environmental conditions found in industrial systems are fundamental to predicting the performance of these materials.

The aim of this work was to evaluate the performance of two stainless steel alloys (duplex and superduplex) in a hypersaline environment with the presence of microorganisms. The experiment was conducted in a dynamic system able to simulate the environmental conditions found in a pipeline. In this system, coupons of the two types of steels were inserted. The process fluid used was hypersaline water collected directly from an oil reservoir and previously characterized chemically and microbiologically. The total duration of the test was 360 days with periodic withdrawals of specimens to evaluate the corrosion process. Several analyses were performed on the coupons, such as metal surface analysis using SEM and quantification of the weight loss. The coupons were also used to study the biofilm formation through the quantification of sessile microorganisms using traditional culture techniques and molecular biology, and visualization of cell adhesion by SEM.

The two alloys showed high corrosion resistance during the period tested. However, a large number of bacteria, especially of the iron bacteria group, were found adhered to both types of coupons. Even though there was no corrosive process detected in these materials, the presence of these bacteria can still cause other problems for the oil and gas industry. Iron bacteria are recognized for their ability to precipitate iron, which may lead to the formation of tubercles on the inside of pipelines. These tubercles can cause a decrease in the flow rate inside the pipe until they completely clog the line. This study shows that a multidisciplinary approach, employing different types of analyses, is critical to evaluating the use of new materials in the oil and gas industry.

23.1 INTRODUCTION

The oil and gas industrial facilities have different characteristics that may favor the colonization of surfaces by microorganisms. This colonization can cause serious problems for these installations, such as clogged pipelines, fuel contamination, and biocorrosion. Various sectors of the oil production chain can suffer from the damage caused by biocorrosion, including exploration, refining, distribution, and storage.

Biocorrosion or microbiologically influenced corrosion (MIC) is a phenomenon already reported and studied decades ago by researchers from different fields. However, a multidisciplinary approach, including specialists in oil exploration and production (geologists, geochemists, chemists, engineers) and microbiologists, is essential to better understand the environment and conditions in which MIC can occur. Although the presence and activity of bacteria in oil reservoirs and industrial facilities have been recognized for a long time, our knowledge of the nature and diversity of bacteria growing in these ecosystems is still not fully understood and their metabolic activities in situ largely ignored.

In this chapter, we describe an experimental procedure specially designed to simulate the conditions of a pipeline in laboratory. Different materials were tested against biocorrosion and different types of analyses in this device: Microbiological quantification, surface analyses, and weight loss measurements were employed.

23.2 NEW ENVIRONMENTS AND NEW CHALLENGES

The growing concern over depletion of world oil reserves, the political instability in major oil-exporting regions, and the development of new technologies have encouraged the industry to explore deeper and harder-to-reach fields. However, these new frontiers pose new challenges for the oil and gas industry. Among the challenges are high temperature and pressure and high concentration of salt and CO_2. Therefore, due to the many challenges in oil extraction from these harsh and remote environments, the oil industry has been constantly investing in solutions and new technologies.

The diversity of environments in which oil and gas industries operate creates a challenge for the materials used during drilling and production. Many new technologies require materials with combinations of properties not usually met by conventional stainless steel. The use of corrosion-resistant alloys (CRA) is necessary to provide long-term resistance to corrosion for different components exposed to oil and gas production environments (Craig and Smith 2011). Among the CRAs, duplex and superduplex stainless steel are being used especially in offshore environments due to their high resistance to pitting corrosion. However, the selection of the right CRA for producing and transporting corrosive oil and gas can be a complex procedure, and if improperly carried out, can lead to mistakes in application and misunderstanding about the performance of a CRA in a specific environment. One of the best ways to select the right metal alloy is to do a test that simulates the particular part of the field environment under study with a possible range of materials.

The limited number of publications regarding the quantity and diversity of microorganisms found in harsh and remote environments also presents a challenge for the oil and gas industry. It is more difficult to choose the right material and to predict its performance without knowing the microbiological conditions of the environment.

If there is a consensus among researchers it is that there is still a lot to investigate about the deep seas. Research should focus not only on oil extraction itself but also on the physical, chemical, and biological environments found in the reservoirs.

23.3 TRADITIONAL CULTURE TECHNIQUE VERSUS DNA-BASED TECHNIQUES

The use of traditional culture techniques in combination with molecular biological techniques for the study of microbial communities is essential for the characterization of the microorganisms in an environment. Since the discovery of aerobic and anaerobic microorganisms associated with oil/water systems, many species have been identified with the aid of molecular techniques. Many new microorganisms have been isolated from these environments, belonging to various groups, such as sulfate reducers, fermenting, iron reducers, methanogens, acetogens, and even archaea (Magot 2005).

The standard culture technique used to characterize microbial communities includes the isolation and characterization of microorganisms through the use of culture media. The cultivation of microorganisms in the laboratory is still very important for the study of the physiological potential as well as for understanding the role of these microorganisms in the studied ecosystem. However, culture-based methods have several drawbacks, mainly when used to determine the number of bacteria. These methods are known for their lack of reliability since bacteria can only be cultivated in the laboratory if their metabolic and physiological requirements can be reproduced in vitro (Nadkarni et al. 2009). The problems with using culture methods for community analysis arise from the fact that an artificial homogeneous medium typically allows the growth of only a small fraction of microorganisms. Culture techniques usually cannot simulate the ecological niches and symbiotic relationships found in complex natural environments needed to support the entire spectrum of microbial diversity. In addition to selectively allowing the growth of some species and suppressing the growth of others, the composition of a cultivable fraction community is distorted during cultivation. The generation time of microorganisms vary with fast-growing species being more efficient and growing faster than others (Nocker et al. 2007). To try to overcome these disadvantages and attempt to isolate the largest number of microorganisms in a sample, many researchers are developing and using alternative culture media. Such media are formulated with different carbon sources and can be incubated under different conditions of temperature, salinity, pressure, and oxygen, aiming to simulate in the laboratory the environmental conditions in which the microorganisms were found.

The approach based on culture techniques, while it is extremely important for understanding the metabolism of microorganisms, does not necessarily provide complete information on the composition of the microbial communities (Van Hamme et al. 2003). A more reliable characterization of microbial communities became possible with the introduction of molecular methods in the study of microbial ecology. These methods are effective for the assessment of microbial communities in various environments. Furthermore, culture-independent methods allow phylogenetic detection and identification of fastidious or noncultivable microorganisms (Juck et al. 2000). In the petroleum microbiology, the use of molecular techniques in oilfield community analysis is still relatively recent (Orphan et al. 2000; Bonch-Osmolovskaya et al. 2003) despite its huge potential to help culture-based techniques to gather important and complementary data. These methods include both direct cloning and sequencing DNA fragments as well as previous amplification of target sequences by polymerase chain reaction (PCR). The PCR products can be cloned and sequenced or subjected to a variety of genetic characterization methods (Nocker et al. 2007). Thus, the ideal approach to best characterize the microbial community of an environment is by using both methods. Culture-based methods help with a better understanding of the microorganism's metabolism, and DNA-based methods help with a better assessment of the diversity of microorganisms present in the sample.

23.4 DYNAMIC TEST

The study of biofilm formation is often carried out in static systems (Merritt et al. 2005). This type of system is widely used to study the early stages of cell adhesion and is very easy to use since these systems require common equipment in research laboratories. However, in static systems, the biofilm may not receive the necessary flow of nutrients and oxygen; thus, a mature biofilm may not form. To circumvent these problems, the use of dynamic systems becomes increasingly common (Stepanović et al. 2003). The use of dynamic systems can simulate actual field conditions such as flow rate and temperature.

23.4.1 System

A dynamic system was designed in order to simulate in the laboratory the conditions found in a pipeline. This system was used to test the performance of two stainless steels and to study biofilm formation (Figure 23.1). The two main parts of the system, where the metal coupons were inserted, were composed of acrylic material and were connected through PVC pipes. The whole system had valves and an electric pump to control the flow rate. The flow rate was kept at 17 L/min, which represents a turbulent flow (Re = 4.224). Two types of stainless steels with different amount of chromium were tested: 22Cr duplex steel (with 22% of Cr) and 25Cr superduplex steel (25% Cr). Small metal coupons (4 cm × 1 cm) of both alloys were inserted in the system.

23.4.2 Hypersaline Fluid

The hypersaline fluid used to fill the system was collected from an onshore hypersaline well and transported refrigerated to the laboratory in a 20 L container. It took 1 week for the laboratory to receive the sample. The physicochemical characterization of the fluid is shown in Table 23.1.

Hypersaline systems are harsh environments that have salt concentrations much greater than that of seawater—often close to or exceeding salt saturation. Hypersaline environments usually have low biological diversity because of the challenging environment.

The microbiological quantification of the main microbial groups of interest for the oil and gas industry was performed through the most probable number (MPN)

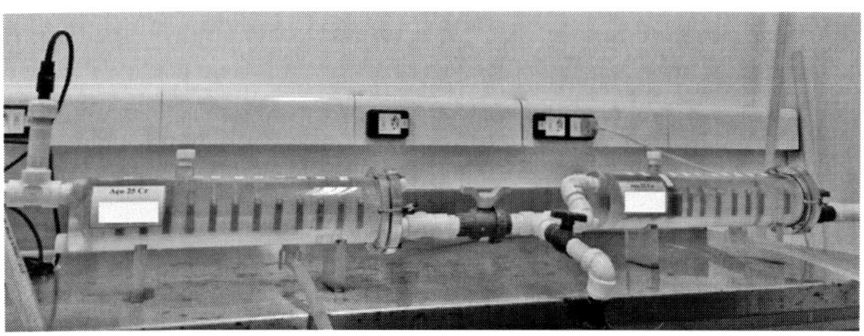

FIGURE 23.1 Dynamic system.

TABLE 23.1
Physicochemical Characteristics of the Hypersaline Water

Analyses	Results (mg/L)
pH	6.8
Electrical conductivity (dS/m)	164.18
K (Potassium)	270.9
Na (Sodium)	7597.59
Ca (Calcium)	29599.57
Mg (Magnesium)	45.30
Fe (Iron)	3.24
Cu (Copper)	0.23
Zn (Zinc)	0.60
Mn (Manganese)	4.86
CO_3 (Carbonate)	–
HCO_3 (Bicarbonate)	76.25
S (Sulfur)	10.90
Cl (Chlorine)	80256.00
Pb (Lead)	3.60
Cd (Cadmium)	0.38

technique. The fluid contained a low number of aerobic heterotrophic bacteria (aproximately 10^1 MPN/mL), but a high number of iron bacteria were detected (10^9 MPN/mL). No sulfate-reducing bacteria (SRB) were detected in the fluid, and approximately 10^6 MPN/mL of anaerobic heterotrophic bacteria were detected.

23.4.3 Experimental Procedure

The dynamic test lasted 360 days with monthly analyses. Every month, quantifications of planktonic bacteria were performed, and coupons of both alloys were withdrawn for sessile microorganism quantification and surface analyses with scanning electron microscopy (SEM) FEI model Inspect S50 and stereoscopic observation with Zeiss stereoscope model Stereo Discovery.V8. Another set of coupons was used for the determination of the weight loss and corrosion rate. The biofilm was observed with SEM after the coupons were fixed, dehydrated, and dried.

The coupons used for the sessile microorganism quantification were removed from the system using aseptic conditions, and the biofilm was scraped using a sterile spatula into an appropriate preserving liquid (sterile saline solution for the quantification of aerobic bacteria and sterile reducing solution for the quantification of anaerobic bacteria, NACE Standard TM0 194-2004).

23.5 MICROBIOLOGICAL INVESTIGATION

The microbiological diversity of the hypersaline fluid was evaluated before the actual dynamic system assay to identify the main microorganisms in the water. Bacterial adhesion on the metal surfaces was also assessed using the dynamic system.

23.5.1 Microbiological Diversity

Two methodologies were used for the microbiological study of the hypersaline water: one based on culture and a DNA-based technique. These two methodologies were used in order to increase the number of microorganisms identified in the sample.

The culture-dependent method consisted of the direct inoculation of the sample in different culture media followed by the isolation of the bacteria in pure cultures (Figure 23.2): nutrient broth for aerobic bacteria, broth fluid to thioglycolate for anaerobic, broth citrate ammonium green for iron bacteria, and modified Postgate E for the sulfate-reducing bacteria. An alternative culture media, using the sample itself, was used in the culture medium in place of distilled water.

The DNA-based technique used involved the collection of the cells through the filtration of the sample in Sartorius 0.45 μm regenerated cellulose membranes (Figure 23.3). The total DNA was extracted straight from these membranes containing the cells. Clone libraries for the 16S rRNA genes were then generated using the TOPO TA cloning kit (Invitrogen).

The same bacteria identified through the DNA-based technique were also isolated in culture media. Three bacterial strains were isolated, identified, and stored in our culture collection for further studies. *Chromohalobacter salexigens*, *Marinobacter gudaonensis*, and *Halomonas elongata* are halophiles that have been previously

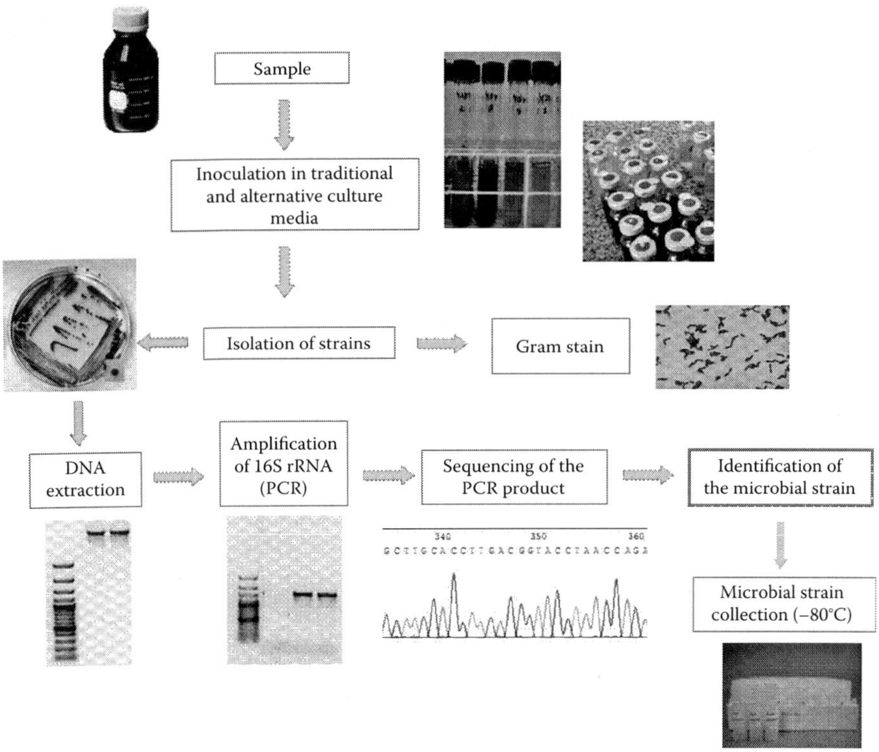

FIGURE 23.2 Culture-dependent method used for the identification of microorganisms.

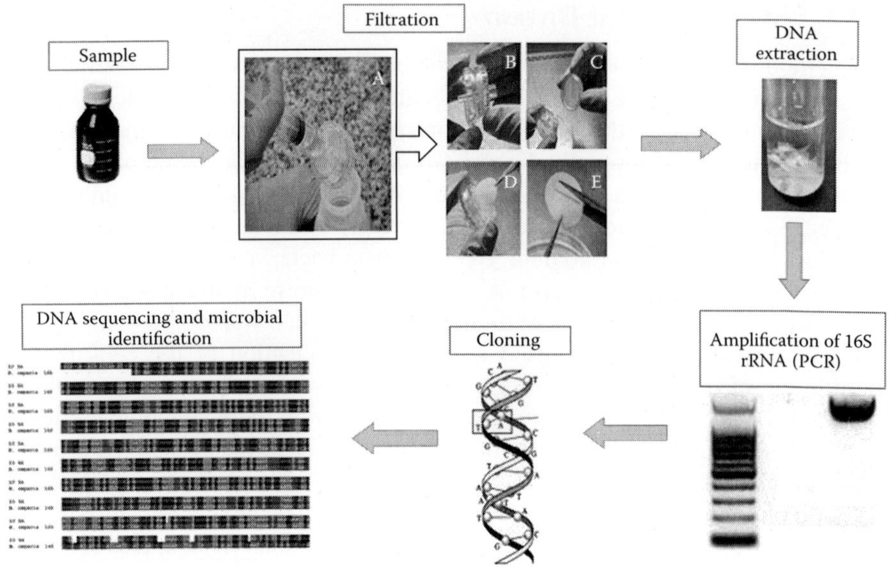

FIGURE 23.3 DNA-based technique used for the identification of microorganisms.

described as able to grow in hypersaline environments (Oren et al. 2005; Kurz et al. 2006; Gu et al. 2007). The use of alternative culture media proved to be a good method for the isolation of microorganisms. Among all the clone sequences obtained, 21 were identified as belonging to the *Chromohalobacter* genre. *C. salexigens* was first isolated from an oil-polluted saline soil in a Chinese oilfield (Gu et al. 2007). The other clone sequences were also affiliated with sequences previously obtained from other hypersaline environments.

23.5.2 Biofilm Formation on the Stainless Steel Coupons

Biofilm formation on the two metal alloys was studied using the dynamic system for 360 days. Every month, three coupons were removed from the system (where they had been exposed to the hypersaline fluid), and the number of sessile microorganisms was assessed. Figure 23.4 shows the bacteria adhesion on the coupons of 22Cr and 25Cr stainless steels through time. After 30 days of experiment, a biofilm composed of iron bacteria and heterotrophic anaerobic bacteria was already formed on the surface of both metal alloys. A large number of bacteria, especially of the iron bacteria group, were found adhered to both types of coupons. This particular group is of great concern to the oil and gas industry. Iron bacteria are recognized by their ability to precipitate iron, which may lead to the formation of tubercles on the inside of pipelines. These tubercles can cause a decrease in the flow rate inside the pipe until they completely clog the line (Tuovinen et al. 1980).

Anaerobic heterotrophic bacteria were present in a slightly higher number in 22Cr coupons, and iron bacteria and aerobic heterotrophic bacteria were predominant in 25Cr coupons. However, according to the results, there was not a significant

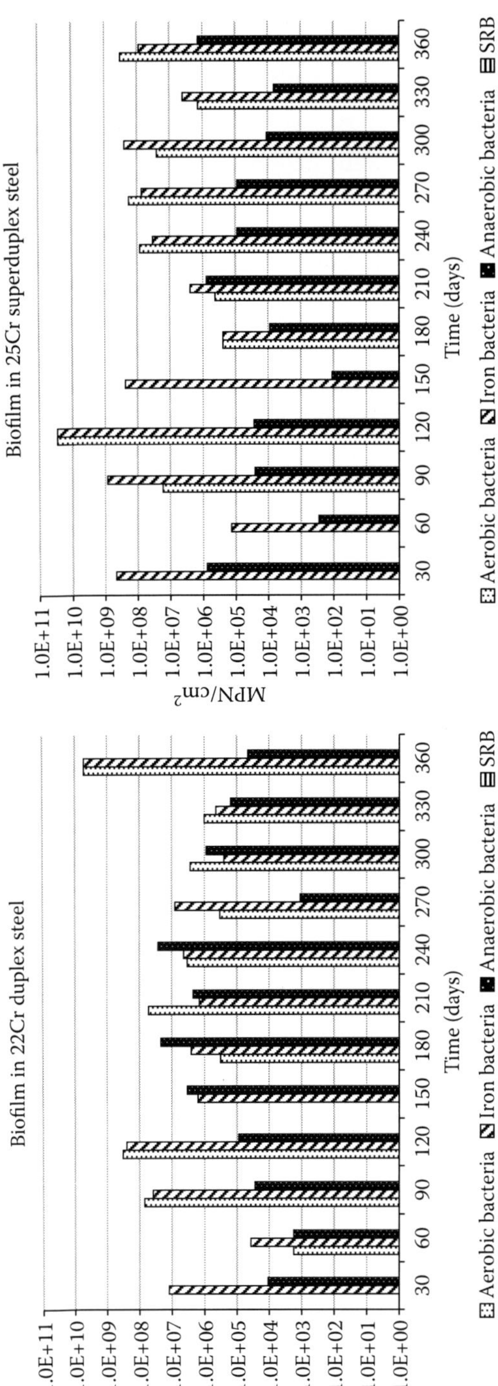

FIGURE 23.4 Bacterial adhesion on 22Cr and 25Cr stainless steel coupons through time.

difference between biofilm formation on the two alloys. No SRB were detected in the experiment.

23.6 CORROSION RATE

The weight losses of the coupons were analyzed in order to quantify the actual corrosion process. Three coupons of each metal alloy were removed each month for this determination. The coupons were treated with an acid solution to remove corrosion products adhered to its surface according to the standard methods ASTM G1-03 (ASTM 2011). After this treatment, the coupons were weighed and the final weight subtracted from the initial weight to obtain the weight loss. The weight loss is used to calculate the corrosion rate.

Figure 23.5 shows the corrosion rate for the two types of stainless steels during the whole experiment in the dynamic system. The corrosion rate of the 22Cr duplex steel was significantly different in the first 2 months. This fact can be due to the initial exposure of the metal surface to the aggressive fluid, which has a high content of chloride ions. After the initial months, the corrosion rates were very low probably because of the formation of a passive film, which prevented further surface corrosion by blocking oxygen and chloride diffusion to the steel surface. The amount of chromium in the steel plays an important role in the formation of the passive film (Craig and Anderson 1995).

Even though the initial corrosion rate of the duplex steel was higher than the others, the general corrosion rates of both alloys were very low throughout the experiment. According to the National Association of Corrosion Engineers (NACE) Standard RP-07-75, corrosion rates lower than 0.025 are considered low (ASTM 2011). The low corrosion rates of these types of stainless steels show the high corrosion resistance of these materials.

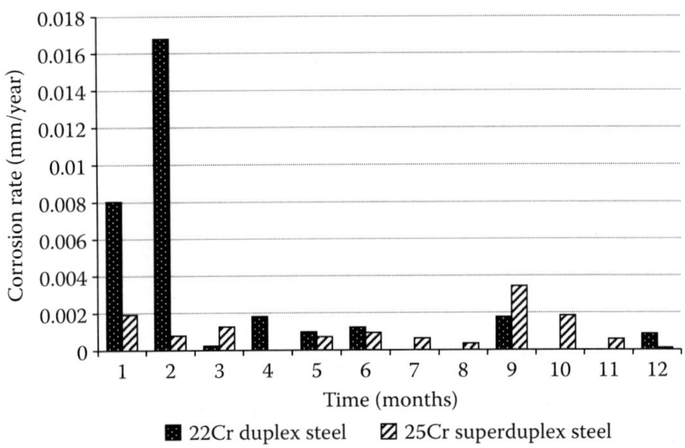

FIGURE 23.5 Corrosion rate of 22Cr and 25Cr stainless steels during the dynamic system assay.

23.7 SURFACE CHARACTERIZATION

The surfaces of the coupons were also analyzed with the aid of stereoscope and SEM to check for localized corrosion and also to visualize the biofilm. No localized corrosion was identified on the surfaces of the coupons. The two stainless steels had a similar appearance at the end of the experiment, which once again shows their high corrosion resistance.

The gradual biofilm formation was followed with SEM. The increase in the production of polymeric material was especially evident after 90 days. Production of polymeric substances is usually related to the formation of mature biofilms. The exopolysaccharides (EPS) produced by some bacteria can create physical and nutritional conditions for the development of anaerobic bacteria (Flemming et al. 2007). Moreover, EPS protects the cells against the flow and hinders the diffusion of biocides, which can help the biofilm formation (Beech and Sunner 2004). There was no significant difference between the biofilms formed on the surfaces of 22Cr and 25Cr steel coupons.

23.8 EFFECT OF BIOCIDE TREATMENT

Biocide treatment is a common practice to mitigate biocorrosion within oilfield facilities. The types of biocides that have been used historically by the oil and gas industries include tetrakis hydroxymethyl phosphonium sulfate (THPS), glutaraldehyde, and formaldehyde among others (Cheung and Beech 1996). Since the number of bacteria (especially of the iron bacteria group) found in the hypersaline fluid and, consequently, on the metal surfaces was high, a biocide was tested in order to check its efficiency and possible use against these microbial groups.

A glutaraldehyde-based biocide (Buckman) was tested in a final concentration of 250 and 360 ppm (concentrations recommended by the manufacturer). Duplex 22Cr and superduplex 25Cr metal coupons with 360-day-old biofilms adhered to them were treated for 3 hours with the biocide. After the treatment, the biofilms were removed from the surface, and the bacteria were quantified.

Figure 23.6 shows the results of the biocide treatment experiment. The biocide was very effective against the biofilms formed on both metal alloys coupons. Even in the lowest biocide concentration (250 ppm), the chemical product was able to eliminate all three bacterial groups in the biofilms. The use of this kind of biocide would be a good alternative to control the formation of biofilms in pipelines of these metal alloys.

23.9 DISCUSSION

Due to the great challenge involved in oil extraction from deeper and more remote areas, the oil and gas industries are heavily investing in new technologies and in the development of new materials. However, the few numbers of publications regarding the microbiological diversity of these environments contribute to the difficulties that those industries face.

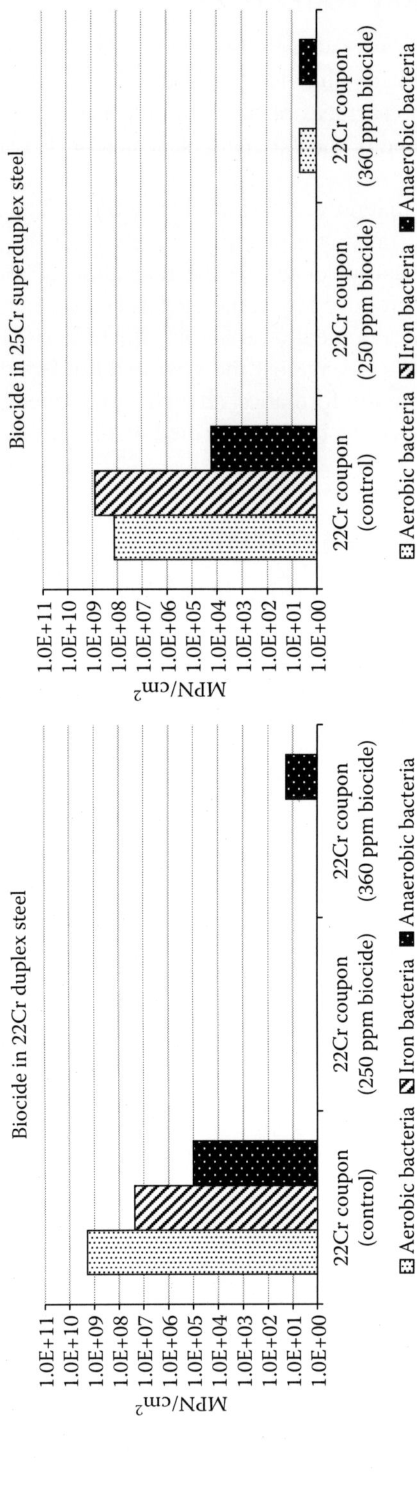

FIGURE 23.6 Biocide treatment in 22Cr and 25Cr stainless steel coupons.

The use of dynamic systems to simulate in the laboratory the conditions found in the field is an excellent strategy to study the MIC process and to test different materials. In this work, two types of stainless steel (22Cr duplex and 25Cr superduplex) showed corrosion resistance against a hypersaline fluid. No localized corrosion was observed on the metal surfaces after 360 days of experiment, and the corrosion rates were very low. However, a very high number of bacteria, especially iron bacteria, were detected on both metals. This is particularly concerning for the oil and gas industries since iron bacteria are recognized for their ability to precipitate iron. These iron compounds may lead to the formation of tubercles on the inside of pipelines causing a decrease in the flow rate and even completely clog the line. According to the SEM images, the biofilms were mature with considerable amounts of EPS. Since the number of bacteria present on the biofilms was high, a glutaraldehyde-based biocide was tested as a method to control the biofilm formation. The biocide was effective against the biofilms on both coupons even at the lower recommended concentration (250 ppm).

Most ecological studies in hypersaline environments have been based on the traditional methods of isolation and identification of pure cultures. However, recent molecular techniques are improving our knowledge of the microbial diversity of these habitats. All bacteria that were either isolated from the hypersaline fluid or identified through a DNA-based technique in this study were related to hypersaline environments, proving the survival of microorganisms to survive in those harsh conditions.

This work showed how a multidisciplinary approach, involving different science areas (such as chemical engineering and microbiology), can help in a better understanding of microbiologically influenced corrosion. The combination of different analyses, such as weight loss measurement, surface analyses, and microbiological quantification, gave a much broader view of the susceptibility of the metal alloys.

REFERENCES

ASTM. 2011. ASTM Standard G1-03: Standard Practice for Preparing, Cleaning and Evaluating Corrosion Test Specimens, ASTM—American Society for Testing Materials.

Beech, I. B., and Sunner, J. 2004. Biocorrosion: Towards understanding interactions between biofilms and metals. *Current Opinion in Biotechnology* 15: 181–186.

Bonch-Osmolovskaya, E. A., Miroshnichenko, M. L., Lebedinsky, A. V., Chernyh, N. A., Nazina, T. N., Ivoilov, V. S., Belyaev, S. S. et al. 2003. Radioisotopic, culture-based, and oligonucleotide microchip analyses of thermophilic microbial communities in a continental high-temperature petroleum reservoir. *Applied and Environmental Microbiology* 69: 6143–6151.

Cheung, C. W. S., and Beech, I. B. 1996. The use of biocides to control sulphate-reducing bacteria in biofilms on mild steel surfaces. *Biofouling* 9: 231–249.

Craig, B. D., and Anderson, D. B. 1995. *Handbook of Corrosion Data.* 2nd Edition. ASM International, Ohio.

Craig, B. D., and Smith, L. 2011. *Corrosion Resistant Alloys (CRAs) in the oil and gas industry—Selection guidelines update*, 3rd Edition. Nickel Institute Technical series no. 10073, Toronto, Canada.

Flemming, H.-C., Neu, T. R., and Wozniak, D. J. 2007. The EPS matrix: The "house of biofilm cells." *Journal of Bacteriology* 189 (22): 7945–7947.

Gu, J., Cai H., Yu, S. L., Qu, R., Yin, B., Guo, Y. F., Zhao, J. Y., and Wu, X. L. 2007. *Marinobacter gudaonensis* sp. nov., isolated from an oil-polluted saline soil in a Chinese oilfield. *International Journal of Systematic and Evolutionary Microbiology* 57 (Pt 2): 250–254.

Juck, D., Charles, T., Whyte, L. G., and Greer, C. W. 2000. Polyphasic microbial community analysis of petroleum hydrocarbon-contaminated oils from two northern Canadian communities. *FEMS Microbiology Ecology* 33: 241–249.

Kurz, M., Brünig N. S., and Galinski, E. A. 2006. NhaD type sodium/proton-antiporter of *Halomonas elongata*: A salt stress response mechanism in marine habitats? *Saline Systems* 2: 10.

Magot, M. 2005. Indigenous microbial communities in oil fields. In: *Petroleum Microbiology*, ed. ASM Press, Washington, DC, 365 p.

Merritt, J. H., Kadouri, D. E., and O'Toole, G. A. 2005. Growing and analyzing static biofilms. *Current Protocols in Microbiology*. 0 1: Unit–1B.1. doi: 10.1002/9780471729259.mc01b01s00.

NACE RP0775-2005. 2005. Preparation, Installation, Analysis, and Interpretation of Corrosion Coupons in Oilfield Operations.

NACE TM0194-2004. 2004. Field Monitoring of Bacterial Growth in Oil and Gas Systems.

Nadkarni, M. A., Martin, F. E., Hunter, N., and Jacques, N. A. 2009. Methods for optimizing DNA extraction before quantifying oral bacterial numbers by real-time PCR. *FEMS Microbiology Letters* 296: 45–51.

Nocker, A., Burr, M., and Camper, K. 2007. Genotypic microbial community profiling: A critical technical review. *Microbial Ecology* 54: 276–289.

Oren, A., Larimer, F., Richardson, P. et al. 2005. How to be moderately halophilic with broad salt tolerance. *Extremophiles* 9: 275–279.

Orphan, V. J., Taylor, L. T., Hafenbradl, D., and Delong, E. F. 2000. Culture-dependent and culture-independent characterization of microbial assemblage associated with high temperature petroleum reservoirs. *Applied and Environmental Microbiology* 66 (2): 700–711.

Stepanovic, S., Cirkovic, I., Mijac, V., and Svabic-Vlahovic, M. 2003. Influence of the incubation temperature, atmosphere and dynamic conditions on biofilm formation by *Salmonella* spp. *Food Microbiology* 20 (3): 339–343.

Tuovinen, O. H. et al. 1980. Bacterial, chemical, and mineralogical characteristics of tubercles in distribution pipelines. *American Water Works Association*, 72 (11): 626–635.

Van Hamme, J. D., Singh, A., and Ward, O. P. 2003. Recent advances in petroleum microbiology. *Microbiology and Molecular Biology Reviews* 67: 503–549.

24 Identifying, Treating, and Monitoring for MIC in a Heavily Waterflooded Oilfield

Jodi B. Wrangham

CONTENTS

ABSTRACT

Widespread microbial contamination was suspected as a major contributor to the frequent corrosion failures occurring in a waterflooded oilfield. Biocide applications were initiated in order to decrease the failure rate; however, these applications appeared to be ineffective in controlling the microbial load. Thorough examination, utilizing multiple microbial detection and identification methods, confirmed the presence of organisms implicated in MIC and the formation of endospores throughout the field. Several potential sources of microbial ingress were also identified, including the injection and workover waters. Planktonic monitoring was found to be an inaccurate means by which to determine biocide treatment efficacy in this contaminated reservoir environment. In addition to an improved microbial monitoring protocol, changes were implemented to the biocide program, which considered the presence of sessile and endospore bacteria in the chemistry selection and application processes.

24.1 FIELD CHARACTERIZATION AND CHALLENGES

An oil- and gas-producing field had been experiencing a high number of corrosion failures with the incidents ever increasing in severity and frequency. Microbial contamination was suspected as playing a prominent role in the majority of these failures as associated culture analyses revealed high bacteria counts, and most wells pulled during workovers displayed significant pitting on the tubing and rods. Numerous operational problems directly associated with microbiologically influenced corrosion (MIC) were noted in failure analyses, including pitted pumps, rod parting, and tubing failures.

The field contains approximately 500 wells with just over 75% of those being production wells. All production wells utilize a rod pump lift mechanism, are vertically completed to a depth of 5000–7000 feet, and have a maximum downhole temperature of 120°F. The remaining wells are water injection wells utilized for pressure maintenance and enhanced oil recovery. This asset has been in production for approximately 30 years and is divided into two sides, designated as A and B. Side A is the older side of the field and contains the majority of the wells. All wells in A and B produce from three possible production zones with most wells producing from at least two zones. While there is currently no new drilling or completion activity, select existing production wells are undergoing refracturing in order to open up and produce from additional production zones.

In addition to MIC, other operational challenges include widespread calcium carbonate and iron sulfide scaling, and paraffin deposition. Current chemical means of control include the majority of production wells receiving corrosion and scale inhibitor, paraffin inhibitor, and hot watering on select wells as needed for paraffin and biocide on wells that have exhibited MIC pitting. All challenges are deemed under control with the exception of MIC.

Of particular concern is the observation that the microbes in this field appear to be unaffected by biocides, regardless of type or concentration. Biocide chemistries that have been utilized include glutaraldehyde, tetrakis(hydroxymethyl) phosphonium sulfate (THPS), quaternary amine compounds, diamine acetate, acrolein, and various combinations thereof. Administered concentrations have exceeded 5000 ppm of active biocide with up to a 48-hour well shut-in time. Despite aggressive treatment, it is reported that microbial counts remain high post-treatment or are temporarily lowered before quickly returning to pretreatment levels of contamination.

24.2 MICROBIAL TEST METHODS

An extensive microbiological study was conducted in order to characterize and enumerate the microbial inhabitants of the field in an effort to confirm if MIC is the major failure mechanism and to proactively protect the facilities against future MIC attack and biotic sulfide generation. Microbial tests performed included culture analysis via bug bottles for the enumeration of problematic metabolic classes of bacteria, adenosine triphosphate (ATP) testing to determine microbial activity levels, quantitative polymerase chain reaction (qPCR) to accurately quantify bacteria and archaea of interest, and 16S ribosomal RNA (rRNA) sequencing for microbial population identification.

Both planktonic (free-floating microbes in liquid) and sessile (microbes attached to surfaces or contained within solids) samples were analyzed. Planktonic samples were obtained from wellheads and pump discharges and sessile samples were obtained by collecting solids off of corrosion and paraffin coupons.

Bug bottles, a culture analysis method, are bacteria detection vials, which contain media specific and selective for the growth of a class or group of bacteria of interest based upon metabolism. In typical oilfield settings, the classes of most concern and therefore most often tested for are APB (acid-producing bacteria), IRB (iron-reducing bacteria), and SRB (sulfate-reducing bacteria). After the inoculation of the bug bottles with the water of interest, the bottles will yield a count of the class of bacteria within 2–4 weeks. To provide context for interpretation of the results obtained in this study, Table 24.1 shows the concentrations considered to be a low, moderate, and high levels of contamination, which allows for analysis of risk. This table is to serve as a reference point only as no range is considered typical, and each system is different.

ATP analysis is a rapid field analytic technique, which measures microbial activity. This method relies on the reaction that occurs when luciferase, an enzyme produced by a firefly, comes into contact with ATP (the primary energy carrier for all life forms), causing a light emission. The amount of light emitted serves as a function of the amount of ATP present. A luminometer is utilized to measure the amount of light emitted and, therefore, the amount of biological activity present. To provide context for interpretation of the results obtained in this study, Table 24.2 shows the concentrations considered to be a low, moderate, and high levels of activity, which allows for analysis of risk. This table is to serve as a reference point only as no range is considered typical, and each system is different.

TABLE 24.1
Risk Levels Associated with Cultured Microbial Counts

Sample Type	Low	Moderate	High
Planktonic[a]	$\leq 10^2$	10^3–10^4	$\geq 10^5$
Sessile[b]	$\leq 10^3$	10^4–10^6	$\geq 10^7$

[a] Expressed as bacteria per mL of fluid.
[b] Expressed as bacteria per in^2 of surface or per gram of solid.

TABLE 24.2
Risk Levels Associated with Microbial Activities

Sample Type	Low	Moderate	High
ATP[a]	≤100	100–999	≥1000

[a] Expressed as pg of ATP per mL of fluid, in^2 of surface, or gram of solid.

qPCR is a quantitative technique in which a DNA sequence of interest is amplified in a thermocycler utilizing sequence-specific DNA probes. This technique is typically used for quantifying either a gene or organism of interest. Here, qPCR is targeting a conserved (universal) bacterial region of the 16S rRNA gene in order to quantify the total bacterial load of a sample. This technique more accurately quantifies microbes as compared to culture analyses, assuming an accurate DNA extraction of the sample has occurred. The risk analysis of a sample is not tied to the quantification number alone, but also to the inhabitants of that population. Therefore, qPCR is paired with 16S rRNA sequencing for a complete risk evaluation. This is accomplished by utilizing a singular DNA extraction for use in the coupled qPCR analysis and 16S speciation.

16S rRNA sequencing is an analysis that allows for exact identification and speciation of all microbial inhabitants of a sample of interest. Specifically, this technique allows for direct identification of those species known to be implicated in MIC and sulfide genesis without the bias and inaccuracies of other frequently used detection methods. This microbial population identification utilizes PCR to amplify conserved regions of the 16S rRNA, which are later sequenced and identified. These speciation results are analyzed in conjunction with the qPCR quantification, which allows for exact identification and enumeration of problematic organisms.

Dominant species of the population, defined as comprising at least 1% of a sample's microbial consortium, are assigned a metabolic classification relevant to the petroleum microbiology industry. Metabolic assignments of taxons, which are phylogenetically inferred, are defined as the following: APB (acid-producing bacteria), HDB (hydrocarbon-degrading bacteria), IRB (iron-reducing bacteria), NRB (nitrate-reducing bacteria), TRB (thiosulfate-reducing bacteria), SOB (sulfur-oxidizing bacteria), SRB (sulfate-reducing bacteria), and UNK (microbes of unknown significance).

24.3 WATER SOURCE WELLS AND INJECTION WATER

The goal of this study was to both identify areas of high microbial contamination and to identify any potential source(s) of microbial ingress into the system. Any waters put downhole, whether for drilling and completions, workovers, chemical flush, or pressure maintenance, are potential sources of microbial contamination and thus must be investigated.

The field is set up such that groups consisting of approximately 20 wells feed into a satellite, and those satellites flow into two central tank batteries (CTB). Both A and B contain a CTB, which separates the produced gas/oil/water. CTB-A and CTB-B also receive makeup water from water source wells (A-SW and B-SW), and the comingled produced water and makeup water is injected into the three production zones via the approximately 100 injection wells that spread out geographically across the field. Approximately 35,000 barrels of water per day (BWPD) are injected with approximately 10,000 BWPD of that being recycled produced water. The total dissolved solids (TDS) of the produced water averages 10%, and the water from the two source wells averages 1%.

Each CTB contains the typical separation vessels and storage tanks. Previous to this study, high total suspended solids (TSS) were an ongoing issue as measured by

Millipore analysis at the water injection pump discharges (A-WIPD and B-WIPD), and all vessels were mechanically cleaned to remove solids that had accumulated at the tank bottoms. This mechanical cleaning was followed by a high concentration biocide treatment with the aim to kill any residual biofilm present on the tank or vessel walls and bottoms. Thermal imaging confirmed that the tank bottoms were indeed removed and that a good cleaning was achieved. This has implications on the microbial load of the system as equipment that contains a large amount of solids often functions as a microbial incubator and is a source for further contamination downstream, which here includes the water utilized for injection.

All of the vessels and tanks at CTB-A and CTB-B were tested for APB, IRB, and SRB via culture analysis and for microbial activity via ATP. All exhibited high SRB counts and microbial activity levels (results not shown). No one tank had a significantly higher count or activity than another or than that of the WIPD water. This indicates that there is not a specific localized area of heavy contamination at CTB-A or CTB-B that is contributing to the high microbial load, but rather that the produced fluids received into the CTBs are responsible for the contamination.

The injection water quality and the role that the source wells play, if any, as a source of microbial ingress, is investigated next. CTB-A and CTB-B each receive a twice weekly treatment of a diamine acetate/alkyl dimethyl benzyl ammonium chloride (ADAC) quat combo at 50 ppm (active concentration) for a 24-hour period, injected immediately downstream of the free-water knockout (FWKO) vessels, which are the first vessels to receive incoming fluids from the satellites. Waters from the source wells flow into produced water tanks ahead of the WIPs and receive no additional biocide.

Water source well and injection water microbial results for both CTBs are displayed in Table 24.3. 16S rRNA speciation results of the dominant and metabolically significant microbial inhabitants isolated from the water source wells and injection waters are displayed in Table 24.4. Table 24.5 contains shared microbial inhabitants between each source well and the injection water from its respective CTB as well as between the injection waters of the two CTBs.

TABLE 24.3
SW and WIPD Microbial Results

	ATP	Culture Analysis			qPCR
Sample ID	Activity (pg/mL)	APB (per mL)	IRB (per mL)	SRB (per mL)	Total Bacteria (per mL)
A-SW: 1% TDS bug bottles	2.3	10	10	10^2	5.2×10^2
A-SW: 10% TDS bug bottles	2.3	0	10	0	5.2×10^2
A-WIPD: pre-biocide	702.4	10	10^2	10^5	ND
A-WIPD: post-biocide	243.8	10	10	10^2	ND
B-SW: 1% TDS bug bottles	17.76	10	10	10^2	5.6×10^2
B-SW: 10% TDS bug bottles	17.76	0	10	10	5.6×10^2
B-WIPD: pre-biocide	221.8	10	10	10^5	ND
B-WIPD: post-biocide	216.5	10	10	10	ND

TABLE 24.4
SW and WIPD Microbial Isolates

Metabolism	Isolate	A-SW (%)	A-WIPD (%)	B-SW (%)	B-WIPD (%)
		Bacteria			
NRB/SOB	*Arcobacter* sp.	0	72.29	0	0
HDB	*Azospira* sp.	0	0	1.36	0
HDB	Burkholderia; unknown	0	0	13.29	0
SRB	*Desulfomicrobium* sp.	0	3.62	0	42.57
SRB	Desulfuromonas; unknown	0	8.41	0	27.48
APB	Halothiobacillus; unknown	0	6.95	0	0
NRB	*Ochrobactrum* sp.	0	0	2.81	0
SRB	*Pelobacter* sp.	0	0	0	13.82
UNK	Phaeospirillum; unknown	0	0	1.39	0
APB/NRB	Pseudomonas; unknown	0	0	11.11	0
APB/NRB	*Pseudomonas* sp.	0	0	4.21	0
SRB	*Soehngenia* sp.	0	0	3.41	0
HDB	*Sphingomonas* sp.	0	0	17.29	0
NRB/SOB	Sulfurospirillum; unknown	99.10	0	0	0
APB/SRB	*Thermovirga* sp.	0	0	0	7.66
SRB	Vibrio; unknown	0	3.71	0	0

A-SW exhibited a low microbial load and activity. Two sets of bug bottles were inoculated from a single sampling, one to represent the TDS of the source water and one that is representative of the TDS that the microbes will encounter downhole. Not surprisingly, the bug bottles that more closely mimic the TDS of the microbes' native environment cultured up more bacteria than those at the elevated TDS. The implication is that the higher TDS that the microbes encounter in the reservoir may slow or prevent their reproduction and thus would aid to minimize contamination from A-SW.

Speciation results revealed that more than 99% of the microbial inhabitants of A-SW are of the *Sulfurospirillum* genus, which roughly correlates to 5.15×10^2 *Sulfurospirillum* per mL of fluid. These bacteria have the dual metabolism of NRB and SOB (NRSOB).

A-WIPD exhibited a high SRB load and moderate microbial activity. Samples were obtained just prior to and just after biocide injection into the FWKO outlet area. qPCR quantification was not determined (ND). Post-biocide, qPCR enumeration is not performed as DNA is a very stable molecule and would be detected in a dead organism as it would in a viable cell when using many biocide chemistries, and thus no significant change would be anticipated. A modest decrease is observed post-biocide in the microbial activity level and in the IRB and SRB culture analyses. These planktonic samples illustrate that CTB-A is receiving low levels of microbes from A-SW and significant SRB from the incoming wells. Once the biocide clears the system, no change will be observed due to the fact that there is not a contaminated

TABLE 24.5
SW and WIPD Shared Microbial Isolates

Isolate	A-SW (%)	A-WIPD (%)
Bacteria		
Desulfoglaeba; unknown	0.01	0.03
Desulfomicrobium sp.	0.70	3.62
Desulfovibrio sp.	0.01	0.04
Desulfuromonas; unknown	0.01	8.41
Order Campylobacterales; unknown	0.03	0.75
Phylum Firmicutes; unknown	0.02	0.01
Phylum Proteobacteria; unknown	0.02	2.68
Isolate	**B-SW (%)**	**B-WIPD (%)**
Bacteria		
Azospira sp.	1.36	0.01
Burkholderia; unknown	13.29	0.04
Desulfomicrobium sp.	0.12	42.57
Desulfovibrio; unknown	0.03	0.44
Family Enterobacteriaceae; unknown	2.93	0.01
Order Anaerolineales; unclassified	2.80	0.00
Order Burkholderiales; unknown	0.04	0.03
Order Clostridiales; unknown	0.10	0.01
Phylum Proteobacteria; unknown	0.12	2.57
Pseudomonas; unknown	11.11	0.06
Isolate	**A-WIPD (%)**	**B-WIPD (%)**
Bacteria		
Desulfoglaeba; unknown	0.03	0.02
Desulfomicrobium sp.	3.62	42.57
Desulfovibrio; unknown	0.06	0.44
Desulfuromonas; unknown	8.41	27.48
Dethiosulfovibrio; unknown	0.48	0.16
Halothiobacillus; unknown	6.95	0.88
Order Desulfovibrionales; unclassified	0.01	0.10
Order Desulfuromonadales; unknown	0.04	0.36
Pelobacter sp.	0.13	13.82
Phylum Proteobacteria; unknown	2.68	2.57

vessel present where an impact could be seen. Additionally, 50 ppm of biocide is a very low dose and is likely not sufficient to achieve a sustained sanitization.

Over 72% of the microbial inhabitants of A-WIPD are of the *Arcobacter* genus, which have the dual metabolism of NRB and SOB (NRSOB). SRB also dominate with over 17% of the microbes being capable of this metabolism.

A-SW and A-WIPD share seven possible isolates. The injection water from A-WIPD shares approximately 15% similarity of microorganisms with A-SW. A-SW and A-WIPD were each dominated by a single, different genus that was not present in the other sample. The lack of similarity between the samples is unexpected given that the majority of the water passing through the A-WIP is from A-SW.

B-SW exhibited a low microbial load and activity. As was done with A-SW, two sets of bug bottles were inoculated from a single sampling: one to represent the TDS of the source water and one that is representative of the TDS that the microbes will encounter downhole. Again, the bug bottles that more closely mimic the TDS of the microbes' native environment cultured up more bacteria than those at the elevated TDS. This effect, if replicated in the field, would serve to minimize contamination from B-SW.

The HDB of genuses *Burkholderia* and *Sphingomonas* comprised over 30% of the microbial inhabitants with the APB/NRB *Pseudomonas* genus also dominant at approximately 15%. Interestingly, the *Arcobacter* that dominated at A-WIPD is also present in B-SW albeit at less than 1% of the population (result not shown).

B-WIPD exhibited a high SRB load and a moderate activity level. As was done with A-WIPD, samples were obtained just prior to and just after biocide injection into the FWKO outlet area. The addition of biocide did not change the microbial activity but did decrease the SRB counts at B-WIPD, which is similar to the scenario observed at CTB-A.

Over 92% of the microbial inhabitants of B-WIPD are classified as SRB. Dominating the SRB are the *Desulfomicrobium* at 42% of the population.

B-SW and B-WIPD share 10 possible isolates. The injection water from B-WIPD shares approximately 45% similarity of microorganisms with B-SW. This similarity is expected as the majority of the water passing through the B-WIP is from B-SW.

A-WIPD and B-WIPD share 10 possible isolates. Approximately 22% of the isolates found in A-WIPD are found in B-WIPD, and approximately 88% of the isolates in B-WIPD are found in A-WIPD. Both injection streams contain problematic species, especially the strong presence of SRB. Because CTB-A and CTB-B operate in the same production zones and are in close proximity to one another, waters are considered comingled and thus forward analyses will utilize the isolates found in both WIPs.

In summary, A-SW and B-SW do contribute organisms that cause MIC and souring to the production zones, however, not at the magnitude as does the recycled produced water that is reinjected. While the overall microbial loads are low at the source wells, a high level of contamination at these locations is not necessary in order to contaminate their corresponding production zones and wells, causing significant MIC. The lack of an effective biocide control program at the CTBs and corresponding injection water as well as the low reservoir temperature are very conducive to microbial growth and widespread contamination.

24.4 WELL REFRACTURING

No new drilling or completion activity is occurring in this field. However, multiple production wells are undergoing refracturing in order to open up and produce from additional production zones. This process involves introducing foreign water downhole and thus could be a possible source of microbial contamination.

Municipal (city) water is most commonly used throughout the area for oilfield processes. Typically, municipal water is chlorinated and thus is considered to be an excellent source of quality water. This particular field uses municipal water for drilling, fracturing, workovers, and chemical treatments. A designated water station is located within the field parameters for collection of this water. Water-hauling trucks move water as needed and are instructed to add biocide into their tanks prior to filling. This has the dual purpose of sanitizing the truck tank and treating the water that is pulled at the station.

The water station and B-31 microbial results are displayed in Table 24.6. Table 24.7 displays the dominant and metabolically significant microbial inhabitants isolated from the water station and B-31.

Water station municipal water, which is used for refracturing, exhibited a low microbial activity and bug bottle result. Two sets of bug bottles were inoculated from a single sampling: one to represent the TDS of the water station and one that is representative of the TDS that the microbes will encounter downhole. The TDS of the culture bottles did not illustrate a difference as all reads were negative for growth. This correlates with the very low activity level present within the samples. In contrast, the qPCR detected a moderate amount of bacteria. This could signify three possible scenarios or combinations thereof: (1) The bacteria were not viable due to chlorination; (2) the bacteria present are not capable of an APB, IRB, or SRB metabolism; and/or (3) bacteria of those metabolisms are present and viable, but

TABLE 24.6
Water Station and B-31 Microbial Results

	ATP	Culture Analysis			qPCR
Sample ID	**Activity (pg/mL)**	**APB (per mL)**	**IRB (per mL)**	**SRB (per mL)**	**Total Bacteria (per mL)**
Water station: 1% TDS bug bottles	0.1	0	0	0	1.8×10^4
Water station: 10% TDS bug bottles	0.1	0	0	0	1.8×10^4
B-31: pre-frac	124.6	10^3	0	0	5.2×10^2
B-31: post-frac	6.1	10	10	10^4	ND
B-31: post-frac (1 week)	65.6	10	0	0	ND
B-31: post-frac (after swabbing; 3 weeks)	41.8	0	0	10	ND
B-31: post-frac (1 month)	5332.8	10^4	10	10^4	ND
B-31: post-frac (6 weeks)	3163.41	10^6	10^2	10^6	ND
B-31: post-frac (2 months)	3474.82	10^4	10	10^3	ND

TABLE 24.7
Water Station and B-31 Microbial Isolates

Metabolism	Isolate	Water Station (%)	B-31 (%)
	Bacteria		
APB	*Actinomyces* sp.	5.07	0
APB	Bacteroides; unknown	1.62	0
APB	Bacteroides uniformis	1.08	0
HDB	Burkholderia; unknown	1.89	0
APB/HDB	Corynebacterium; unknown	3.99	0
SRB	*Desulfohalobium retbaense*	0	10.85
APB	*Flexistipes sinusarabici*	0	8.92
TRB	Halanaerobium; unknown	0	5.98
NRB/SOB	Methylobacterium; unknown	7.17	0
SRB	Order Desulfovibrionales; unclassified	0	15.48
SRB	Order Desulfovibrionales; unknown	0	41.41
Varies	Phylum Firmicutes; unknown	36.98	0
APB/NRB	Pseudomonas; unknown	7.30	0
APB	Roseburia; unknown	1.42	0

are not culturable. Culture-based detection methods do not facilitate the growth of all organisms of their respective metabolisms and thus often grossly underestimate what is actually present (Wrangham et al. 2010).

The speciation results reveal multiple organisms capable of APB or SRB metabolism and, therefore, were not detected via culture analysis due to either chlorination (this is the likely scenario as the microbial activity read was very low as well), or the isolates are not culturable.

The microbial population was dominated by members of the phylum Firmicutes, whose members vary in metabolic capabilities. The significance of the Firmicutes is the ability of many of the members of this taxon to produce endospores, which are resistant to desiccation and can survive extreme conditions (Filippidoua et al. 2015, p. 299). It is possible for these microbes to form dormant endospores in reaction to low levels of chlorine, and once the chlorine is diluted or reacted out (as it would be downhole) the cells could become vegetative and metabolically active again.

A water-hauling truck used in the refracturing of B-31 was sampled and tested (results not shown). No bacteria were cultured, and the microbial activity was very low. It was indicated that an unspecified oxidizer-type biocide was applied to the truck tank prior to filling.

B-31 was evaluated during the refracturing process. Pre-fracturing, the microbial activity and microbial load levels were low. Post-process, the produced water became black and contained high levels of solids. Throughout the months of evaluation post-frac, the well was repeatedly shut in for various operational issues. Approximately 3 weeks after refracturing, the well was swabbed. The following weeks exhibited high microbial activities and high APB and SRB counts. Speciation confirmed that the microbial population of this well is dominated by SRB with the majority belonging to the order Desulfovibrionales.

The well became very contaminated post-fracturing, especially after the 1-month mark. Because municipal water was the only water received by the well, it is not likely the primary culprit behind this fouling. Rather the frequent shut-ins due to operational issues, causing the fluids to sit stagnant, is thought to be the reason for the elevated microbial contamination.

24.5 TREATMENT OUTCOMES OF PROBLEM WELLS

A large number of wells are designated as problem wells, defined as those that fail multiple times in a year. Problem wells are not geographically clustered to any area of the field, but rather are spread throughout. Most problem wells exhibit MIC.

Dozens of wells were treated with biocide and evaluated; however, only a few are characterized here, each of which represents a specific type of post-biocide trend observed throughout the field. Those post-biocide treatment outcomes are (1) wells that showed an immediate kill; however, microbial levels returned to pretreatment levels very quickly; (2) wells that showed little to no change; and (3) wells that showed a significant kill that was maintained. Two biocide chemistry options are utilized for problem well treatment: a diamine acetate/ADAC quat combo product or acrolein.

Pre- and post-biocide microbial results from B-91, A-63, A-85, and A-172 are displayed in Table 24.8. Table 24.9 displays the dominant and metabolically significant microbial inhabitants isolated from B-91, A-63, and A-172. B-91 exhibited MIC and hydrogen cracking associated with SRB colonization as shown in Figure 24.1.

B-91 experienced a mechanical failure, and the well was pulled in order to replace the damaged equipment. One week post-failure, the microbial activity and load was low with the exception of the APB count. However, each subsequent week saw an increase in microbial contamination. At 6 weeks post-failure, a biocide treatment was initiated. A high-concentration diamine acetate/ADAC quat combo product was injected down the casing (backside), circulated, and the well was shut-in for 24 hours. Post-biocide, the microbial activity and load is similar to that of the pre-biocide read. However, that pre-biocide read is much lower than that of the previous weeks for unknown reasons.

16S rRNA identification revealed that more than 90% of the isolates from B-91 are classified as SRB. This is expected, given the SRB colonization that is present on the rods and tubing.

A-63, A-85, and A-172 are problem wells that were proactively treated with acrolein. The goal of these applications is not exclusively for microbial control, but to also scavenge dissolved hydrogen sulfide (H_2S) and to dissolve iron sulfide solids from the casing and immediate wellbore area. All wells receive 6000 ppm of acrolein slipstreamed down the backside of the well followed by a 48 hour shut-in.

A-63 is a problem well that typically presents with black water (iron sulfide) and low levels of H_2S, both dissolved and gaseous. Figure 24.2 is an example of the extensive pitting found within A-63. Despite low culture analysis results, A-63

TABLE 24.8
Pre- and Post-Biocide Microbial Results

	ATP	Culture Analysis			qPCR
Sample ID	**Activity (pg/mL)**	**APB (per mL)**	**IRB (per mL)**	**SRB (per mL)**	**Total Bacteria (per mL)**
B-91: post-failure (1 week)	146	10^5	10	10	1.6×10^2/mL
B-91: post-failure (2 weeks)	606.5	10^5	10	10^3	ND
B-91: post-failure (5 weeks)	1872.07	10^4	10	10^6	ND
B-91: post-failure (6 weeks); pre-biocide	358.9	10^2	10	0	ND
B-91: post-biocide (2 weeks)	86.7	10^3	10	10	ND
B-91: post-biocide (3 weeks)	93.2	10^3	10	10^2	ND
A-63: pre-biocide	1081	10^2	10	10	7.2×10^5
A-63: post-biocide	3124.2	10^2	10	10	ND
A-63: post-biocide (1 week)	275.6	10^2	10^2	10^4	ND
A-63: post-biocide (2 weeks)	1860.8	10^2	10	10	ND
A-63: post-biocide (1 month)	1648	10^4	10^2	10^2	ND
A-63: post-biocide (2 month)	1570.4	10	10	10	ND
A-85: pre-biocide	1629.3	10^2	10	10	ND
A-85: post-biocide	1.9	0	0	0	ND
A-85: post-biocide (1 week)	15.5	0	0	0	ND
A-85: post-biocide (2 weeks)	4202.1	10^3	10	10^2	ND
A-85: post-biocide (1 month)	2468.2	10^2	10	10	ND
A-172: pre-biocide	185.2	10^4	10^2	10^3	1.6×10^2
A-172: post-biocide	3.2	0	0	0	ND
A-172: post-biocide (1 week)	1.8	0	0	0	ND
A-172: post-biocide (1.5 week)	1.6	0	0	0	ND
A-172: post-biocide (2 weeks)	1.65	0	0	0	ND
A-172: post-biocide (3 weeks)	10.47	10	0	0	ND
A-172: post-biocide (1 month)	1	0	0	0	ND

exhibited a high microbial activity level and a high total bacteria count via qPCR. Immediately post-acrolein, the bug bottle counts remained steady while the microbial level increased significantly. In the following weeks, the activity remained high, and the bug bottle counts increased as well. Increasing microbial activities and culture counts from planktonic samples immediately following biocide administration often signifies an effective downhole cleaning and subsequent release of biofilm. This effect may have been delayed in this particular well as it is a low producer, and the biofilm would not flush out immediately. Speciation revealed that while a significant portion of the microbial consortium of A-63 are SRB, the majority of the population consisted of the *Pseudomonas*, which are capable of both APB and NRB metabolisms.

A-85 is a problem well that displays generalized corrosion, which includes possible MIC pitting. Figure 24.3 displays corrosion on a parted rod from this well.

TABLE 24.9
B-91, A-63, and A-172 Microbial Isolates

Metabolism	Isolate	B-91 (%)	A-63 (%)	A-172 (%)
	Bacteria			
APB	*Anaerophaga* sp.	0	0	7.99
SRB	*Desulfomicrobium* sp.	0	6.69	5.90
SRB	Desulfovibrio; unknown	0	0	2.04
SRB	Desulfuromonas; unknown	0	26.34	3.98
SRB	Geotoga; unknown	0	2.22	0
TRB	Halanaerobium; unknown	3.92	0	0
TRB	Halanaerobium; unknown	0	0	13.08
NRB	*Marinilabilia salmonicolor*	0	0	1.40
SRB	Order Desulfovibrionales; unclassified	85.71	0	5.59
SRB	Order Desulfovibrionales; unknown	5.67	0	1.85
SRB	*Pelobacter* sp.	0	0	17.89
SRB	Petrotoga; unknown	0	0	2.39
APB/NRB	Pseudomonas; unknown	0	51.28	0
APB/SRB	*Thermovirga* sp.	0	0	2.01

FIGURE 24.1 B-91 MIC.

Despite low culture analysis results, A-85 exhibited a high microbial activity level. Post-acrolein, the microbial activity and bug bottle counts dropped significantly to very low levels. However, these results do not persist, as the 2-week read exhibits a very high level of microbial contamination. qPCR and 16S speciation analyses were not completed on this well.

A-172 exhibited a moderate microbial load and a low level of activity. Post-acrolein, the microbial activity and bacterial counts drop significantly to very low

FIGURE 24.2 A-63 pitting corrosion.

FIGURE 24.3 A-85 corrosion.

levels, and this kill persists a month later. This well is dominated by SRB with a strong APB and TRB presence.

In an effort to understand why this well responded favorably to biocide when most have not, all parameters of this well were investigated and compared to other problem wells that did not respond favorably to biocide. The only observable difference is the fact that this well makes very little water (<1 BWPD). Because very little water is being produced, it is thought that perhaps this well was producing connate water, and the contaminated waterflood water was not feeding through to this well as of yet. To test this hypothesis, the microbial isolates from the water produced from A-172 were compared to the microbial isolates from the WIPDs to evaluate similarity. Those results are displayed in Table 24.10. A-172 shares approximately 77% of its bacteria isolates with the injection water population. Because this field has been heavily waterflooded, it is not possible to predict what the connate water would look like in terms of microbial inhabitants. However, the high level of similarity observed between A-172 and the waterflood water, many of whose isolates originated from make-up (source water), would not be predicted if this were connate water. This is one of the very few wells that has responded favorably to biocide as determined by typical planktonic monitoring.

In summary, three different outcomes were seen in problem wells following biocide administration regardless of chemistry or concentration. Those post-biocide treatment outcomes are (1) wells that showed an immediate kill; however, microbial levels returned to pretreatment levels very quickly (A-85); (2) wells that showed little to no change (B-91 and A-63); and (3) wells that showed a significant kill that was

TABLE 24.10
WIPDs and A-172 Shared Microbial Isolates

Isolate	A-WIPD (%)	B-WIPD (%)	A-172 (%)
	Bacteria		
Aminomonas sp.	0	0.03	0.21
Anaerophaga sp.	0	0	7.99
Bacillus; unknown	0	0.02	0.19
Burkholderia; unknown	0	0.04	0.19
Citrobacter; unknown	0	0.02	0.24
Class Bacilli; unknown	0	0.15	1.66
Desulfohalobium retbaense	0	0	0.47
Desulfomicrobium sp.	3.62	42.57	5.90
Desulfotignum; unknown	0.30	0	0.14
Desulfovibrio; unknown	0.06	0.44	2.04
Desulfuromonas; unknown	8.41	27.48	3.98
Family Clostridiaceae; unknown	0	0	11.82
Halanaerobium; unknown	0	0.02	13.08
Marinilabilia salmonicolor	0	0.46	1.40
Order Burkholderiales; unknown	0	0.03	0.26
Order Clostridiales; unclassified	0	0.36	0.05
Order Clostridiales; unknown	0	0.01	0.02
Order Desulfobacterales; unclassified	0	0.10	0.69
Order Desulfovibrionales; unclassified	0.01	0.10	5.59
Order Desulfovibrionales; unknown	0.01	0	1.85
Pelobacter sp.	0.13	13.82	17.89
Phylum Cyanobacteria; unclassified	0	0.25	2.77
Thermovirga sp.	0	7.66	2.01

maintained (A-172). The differences in outcomes between the wells may, at least in part, be explained by the differences in water production and the waterflooding of the field. In general, those wells that saw little change or returned to prior levels quickly are high water producers, and that waterflood may be providing the influx of microbes detected soon after treatment.

It is often difficult to ascertain success or failure of a downhole biocide treatment based upon wellhead planktonic sampling alone in fields that are heavily waterflooded and where the injected water has broken through to producers. This is because it is not practical or economically feasible to sterilize that injection water, which will lead to some level of eventual contamination of the reservoir/production zones and their associated production wells. This is especially true when produced water is reinjected. In addition, this injection water will be producing back at the production wells, and planktonic samples from wellheads will test as contaminated even if a well was cleaned up effectively by biocide. A more accurate means of monitoring post-biocide in these scenarios could be achieved with sessile sampling, which can demonstrate sessile kill and biofilm removal. However, this analysis may be difficult to accurately execute as a corrosion coupon in a wellhead may not be representative of the downhole equipment in either biofilm accumulation, biocide contact, or kill rate. The true test of a biocide program's effectiveness is the lack of or prolonging of failures due to MIC on a treated well.

24.6 PLANKTONIC AND SESSILE POPULATIONS ARE HIGHLY SIMILAR IN ISOLATES

Because planktonic samples are those that are routinely used to base important decisions on, it is useful to determine if planktonic samples are similar to the sessile population (those that cause the problems). Previous studies have found that planktonic organisms are not necessarily representative of sessile organisms from the same location (Wrangham and Summer 2013).

Table 24.11 contains the planktonic and sessile (coupon) microbial results for B-31, B-82, and B-91. The wellhead coupons were swabbed with a sterile cotton swab, removing all of the solids. Thus, the swab result accounts for total bacteria colonized on the coupon surface. All coupons remain in the system for approximately 1 month before retrieval. ATP and culture analyses were not conducted on the coupon sample as this would alter the qPCR total bacteria result and potentially contaminate the sample prior to 16S identification.

The microbial isolates from the planktonic and sessile samples of the three wells were compared to evaluate similarity. Those results are displayed in Table 24.12. For all three wellhead planktonic samples and their corresponding coupons, the total microbial counts are relatively low. The sessile bacteria colonized on the coupons are found in the produced fluids at approximately 90% in B-31, 87% in B-82, and 77% in B-91.

This field exhibits a high level of similarity between planktonic and sessile isolates although they are present as different percentages of the population. The microbial loads, however, would be and are different, depending upon multiple factors, such as length of time the coupon is in the system and chemical treatments. For these

TABLE 24.11
Planktonic and Sessile Microbial Results

	ATP	Culture Analysis			qPCR
Sample ID	Activity (pg/mL)	APB (per mL)	IRB (per mL)	SRB (per mL)	Total Bacteria (per mL or swab)
B-31	124.6	10^3	0	0	5.2×10^2/mL
B-31 coupon	ND	ND	ND	ND	5.8×10^3/swab
B-82	760.37	10^2	10^2	10^3	1.3×10^3/mL
B-82 coupon	ND	ND	ND	ND	2.2×10^5/swab
B-91	146	10^5	10	10	1.6×10^2/mL
B-91 coupon	ND	ND	ND	ND	1×10^3/swab

TABLE 24.12
Planktonic and Sessile Shared Microbial Isolates

Isolate	B-31 (%)	B-31 Coupon (%)
Bacteria		
Desulfohalobium retbaense	10.85	3.88
Flexistipes sinusarabici	8.92	1.51
Halanaerobium; unknown	5.98	0.20
Halanaerobium sp.	0.21	35.14
Order Desulfovibrionales; unclassified	15.48	22.95
Order Desulfovibrionales; unknown	41.41	7.29
Order Halanaerobiales; unknown	0.18	18.83
Isolate	**B-82 (%)**	**B-82 Coupon (%)**
Archaea		
Family Methanosarcinaceae; unknown	0.01	0.09
Bacteria		
Anaerophaga; unknown	0.11	0.01
Anaerophaga sp.	0.62	0.02
Citrobacter; unknown	0.01	0.06
Class Deltaproteobacteria; unknown	3.08	0.29
Class Planctomycetia; unclassified	2.03	0.01
Desulfocaldus sp.	6.71	22.08
Desulfoglaeba; unknown	8.13	0.27
Desulfomicrobium sp.	21.27	55.33
Desulfovibrio sp.	0.02	0.10
Family Comamonadaceae; unknown	0.02	0.20
Family Synergistaceae; unknown	18.97	1.80
Geotoga; unknown	0.51	0.73
Halanaerobium sp.	1.67	0.00
Halospirulina sp.	0.06	0.11
Kingdom Bacteria; unclassified	0.18	0.21
Kingdom Bacteria; unknown	0.93	1.35
Methylobacterium; unknown	0.22	0.21
Nocardioides sp.	0.01	0.54
Order Anaerolineales; unclassified	1.51	2.40
Order Clostridiales; unclassified	0.05	0.03
Order Desulfobacterales; unclassified	1.06	0.03
Order Desulfovibrionales; unclassified	0.80	0.10
Order Desulfovibrionales; unknown	3.03	0.07
Order Halanaerobiales; unknown	0.95	0.02
Order Thermotogales; unclassified	2.32	0.11

(Continued)

TABLE 24.12 (CONTINUED)
Planktonic and Sessile Shared Microbial Isolates

Isolate	B-82 (%)	B-82 Coupon (%)
Pelobacter sp.	9.28	0.83
Petrotoga; unknown	2.02	0.22
Phylum Actinobacteria; unclassified	0.06	0.06
Phylum Chloroflexi; unknown	0.02	0.12
Phylum Firmicutes; unknown	1.05	0.13
Phylum Proteobacteria; unknown	0.07	0.16
Phylum Spirochaetes; unclassified	3.40	0.11
Phylum Spirochaetes; unknown	0.10	0.12
Phylum Thermotogae; unknown	0.23	0.06
Sphingomonas sp.	0.01	0.64
Thermovirga sp.	0.13	0.01

Isolate	B-9C1 (%)	B-91 Coupon (%)
Archaea		
Methanohalophilus sp.	0.20	0.14
Bacteria		
Geotoga; unknown	0.02	0.05
Halanaerobium; unknown	3.92	4.94
Order Desulfovibrionales; unclassified	85.71	69.70
Order Desulfovibrionales; unknown	5.67	0.88
Phylum Bacteroidetes; unknown	0.01	0.02
Phylum Proteobacteria; unknown	0.08	1.34
Pseudomonas; unknown	0.01	0.14

reasons, it is still best practice to utilize as many sessile samples as possible as part of the microbial monitoring program despite the similarity between the planktonic and sessile isolates.

24.7 BIOCIDE EFFICACY TESTING

Biocide chemistry selection testing, often referred to as a time-kill test, was conducted on multiple waters sampled from problem wells and the WIPDs according to the NACE standard test method TM0194-2014 (NACE 2014).

Biofilm and associated solids are a concern throughout the field, and therefore, the biocide chemistry must be capable of penetrating and removing these solids. Because this selection testing method is planktonic, it is important to eliminate chemistries that would perform well in a planktonic test but not translate well in the field against sessile build-up. Three microbiocide products (glut/ADBAC quat combo, diamine acetate/ADAC quat combo, and acrolein) were tested at three concentrations based on activity (50, 100, and 250 ppm) over a 2 hour contact time.

The biocide selection testing results for B-31, B-91, A-WIPD, and B-WIPD are displayed in Table 24.13.

APB and IRB results are not shown as all products and concentrations effectively eliminated all organisms of those metabolisms and therefore did not allow for differentiation between the chemistries. This is not detrimental to the biocide selection process as this field is dominated by SRB rather than APB or IRB. This is especially true of the sessile samples obtained from the coupons, which were almost entirely SRB.

B-31 and B-91 saw acrolein and the glut quat combo offer a full kill at low concentrations. The diamine acetate quat combo was not as effective against this population of SRB.

A-WIPD and B-WIPD exhibited an SRB population that was much more resistant to the biocide chemistries as compared to the problem wells. The diamine acetate quat combo did not perform well here even at the highest concentration of 250 ppm. Likewise, the glut quat combo could not inhibit completely.

The WIPDs receive some biocide as the CTBs receive 50 ppm of the diamine acetate quat combo twice a week. However, as can be seen from the biocide selection testing, this chemistry at such a low concentration could not be sufficient in order to effectively treat the injection water or to carry any residual biocide into the reservoir, where the contamination problem intensifies. It is important to note that the contact time of the test was 2 hours, which may have placed the diamine acetate combo at a disadvantage as it may require 4 hours or more for its full effect to be observed. Both glutaraldehyde and acrolein are fast-acting chemistries. Therefore, in problem well treatments in which there is a 24–48 hour shut-in time, the diamine acetate may perform better than observed here.

TABLE 24.13
Kill Test Results

Biocide Chemistry	Active Biocide Concentration (ppm)	B-31 SRB (per mL)	B-91 SRB (per mL)	A-WIPD SRB (per mL)	B-WIPD SRB (per mL)
Control	–	10^5	10^6	10^7	10^6
Glutaraldehyde/ADBAC quat	50	0	10^2	10	10^5
Glutaraldehyde/ADBAC quat	100	0	0	10	10^5
Glutaraldehyde/ADBAC quat	250	0	0	10	10
Diamine acetate/ADAC quat	50	10	10^6	10^6	10^5
Diamine acetate/ADAC quat	100	10	10^6	10^5	10^5
Diamine acetate/ADAC quat	250	0	10^3	10^4	10
Acrolein	50	0	0	10	10
Acrolein	100	0	0	0	0
Acrolein	250	0	0	0	0

Acrolein is the ideal chemistry for this field's particular population of bacteria as was illustrated in all kill tests. While 100 ppm inhibited all bacterial growth in all tests, a higher concentration may be required in the field. Many biocides work best on planktonic organisms while fewer biocides are able to reduce sessile populations at the same dosage (Kelland 2009, p. 337). The planktonic time-kill test is exclusively a chemistry selection test and cannot accurately predict the optimized field dosage. Only monitoring, once the chemistry is applied in the field, can determine what the needed dosage will be.

24.8 PRESENCE OF ENDOSPORE-FORMING BACTERIA

Endospore-producing bacteria were detected throughout the system via 16S rRNA identification. As previously mentioned, members of the phylum Firmicutes were detected in the water station sample. Members of this taxon, including *Geobacillus stearothermophilus*, were also detected at the water source wells, injection water, and problem wells in both planktonic and sessile samples (results not shown). This signifies a widespread contamination of species capable of endospore-formation.

The strong presence of endospore-forming bacteria is significant in that this should be factored into the biocide selection process. Typically, a vegetative cell (metabolically active) will not form an endospore (dormant cell) unless presented with adverse conditions, including chemical presence. Once a spore is formed, it is relatively impervious and resistant to chemicals, such as biocides, largely due to the structure and chemical composition of bacterial spores differing considerably from those of vegetative cells (Leggett et al. 2012, p. 485). These spores can remain viable for thousands of years. When spores are exposed to favorable conditions, such as biocide-free fluids, they can germinate into a vegetative cell within 90 minutes.

Many of these bacteria can form endospores in approximately 6 to 8 hours after being exposed to adverse conditions. Therefore, a fast-acting chemistry, such as an oxidizer (bleach, chlorine dioxide), glutaraldehyde, THPS, or acrolein should comprise at least some portion of the biocide regimen. This will likely eliminate the possibility that these cells will have the time necessary to form chemically resistant endospores. The slower acting biocide that is currently being administered at the CTBs at a low treating dose may be forcing these bacteria into endospores, which rapidly become vegetative cells again once the biocide has either passed or is diluted in the reservoir. This scenario, over time, would favor and select for a population of bacteria that can form endospores when biocide is administered, rendering the biocide program ineffective.

24.9 SUMMARY

An oil- and gas-producing field was experiencing a high number of corrosion failures, many of which involved MIC. This field is an ideal environment for microbial contamination as the reservoir temperature does not exceed 120°F, it is heavily waterflooded with recycled produced water, and the injection water is not effectively treated with biocide. It is unlikely that the low dose of biocide could sufficiently treat the injection water or to carry any residual biocide into the reservoir, where the contamination problem intensifies.

Microbes that are implicated in MIC were identified as present throughout the field. Because the field has been in production for several decades with multiple waters being introduced during that time, it is not possible to determine which, if any, problematic microbes are indigenous or pre-existing in the production zones versus what has been introduced. However, waters that are currently being utilized in the field were investigated and identified as sources of contamination. While the overall microbial loads are low at some sources, a high level of contamination is not necessary in order to contaminate the production zones and wells, causing significant MIC. The produced water, which is subsequently reinjected into the reservoir, was identified as the most significant source of microbial contamination.

This system exhibits a high level of similarity between planktonic and sessile isolates although they are present as different percentages of the population. The microbial loads, however, are different. Despite a similarity between planktonic and sessile microbial isolates, it is still best practice to utilize as many sessile samples as possible as part of the microbial monitoring program.

The major concern in this field is the cleanup and maintenance of problem wells experiencing MIC. Particularly challenging to the operator was the observation that the microbes appear to be largely unaffected by biocides regardless of type or concentration. This impression is due to planktonic-only monitoring coupled with the waterflood, providing the influx of planktonic microbes detected soon after treatment. A more accurate means of monitoring post-biocide is sessile monitoring, which can demonstrate sessile kill and biofilm removal although the best test of the biocide program's effectiveness is the lack of or prolonging of failures due to MIC on a treated well.

Biocide selection testing identified acrolein as the most effective chemistry against this population of bacteria. This is a fast-acting chemistry, which is ideal given the strong presence of endospore-forming bacteria throughout this field.

REFERENCES

Filippidoua, S., T. Juniera, T. Wunderlina, C.-C. Loc, P.-E. Lic, P. S. Chainc, and P. Juniera. 2015. Under-Detection of Endospore-Forming Firmicutes in Metagenomic Data. *Computational and Structural Biotechnology Journal* 13: 299–306.

Kelland, M. 2009. *Production Chemicals for the Oil and Gas Industry.* Boca Raton, FL: CRC Press.

Leggett, M. J., G. McDonnell, S. P. Denyer, P. Setlow, and J. Y. Maillard. 2012. Bacterial Spore Structures and Their Protective Role in Biocide Resistance. *Journal of Applied Microbiology* 113: 485–498.

NACE Standard TM0194-2014. Field Monitoring of Bacterial Growth in Oil and Gas Systems.

Wrangham, J. B., J. E. Penkala, S. D'Imperio, B. M. Peyton, and K. G. Wunch. 2010. Utilization of a 16S rRNA Gene Microarray to Analyze the Efficacy of Oil and Gas Industry Bacteria Culture Media. NACE *Corrosion* 2010, Paper 10408.

Wrangham, J. B., and E. J. Summer. 2013. Planktonic Microbial Population Profiles Do Not Accurately Represent Same Location Sessile Population Profiles. NACE *Corrosion* 2013, Paper 2780.

25 MIC Case Histories in Oil, Gas, and Associated Operations

Faisal M. Alabbas

CONTENTS

ABSTRACT

Since the commissioning of oil and gas systems, scientists, engineers, and microbiologists have explored the microbial communities and their subsequent effects on the integrity of operations and structures to ultimately design and operate effective monitoring and mitigation strategies. Microbiologically influenced corrosion (MIC) is one of several considerable concerns to the oil and gas industry. MIC is considered one of the most dominant failure modes in gas and oil operations, which has a high impact on integrity, operations, and maintenance costs. MIC failures result from accelerated deterioration driven by different indigenous microorganisms that naturally reside in the hydrocarbon and associated water systems. Different microorganisms thrive in oil, gas, and associated water injection systems because the essential elements for their lives are present in these environments. Microorganisms need four elements to thrive: carbon source, water, electron donor, and electron acceptor. Hydrocarbons and other organic compounds are an excellent carbon (food) source for a wide variety of microorganisms. The presence of microbial activities accelerates the rate of anodic and/or cathodic reactions. MIC is not a distinct type of corrosion, but rather is the synergistic interaction of microorganisms with resulting biofilms and metabolic products that enhance corrosion processes. In most cases, MIC morphologies are localized types of corrosion that manifest in pitting, crevice corrosion, under deposit corrosion, cracking, enhanced erosion–corrosion, and dealloying. MIC investigations are complex and involve interdisciplinary science that includes microbiology, metallurgy, and corrosion. This chapter presents several real-field cases related to MIC failures that were recorded and reported from oil and gas systems and associated operations. The cases present failures of various engineering materials under the effects of different biological and chemical factors. The data offered by this chapter provide a wealth of information that might guide field engineers in how to deploy systematic approach by utilizing the advanced microbiological and analytical methods to understand the nature of failure and to design appropriate mitigation measures.

25.1 INTRODUCTION

Corrosion is a significant problem in the oil and gas industry. In 1996, the estimated annual cost of pipeline corrosion for the gas in the U.S. industry was about $840 million (Buk et al. 1996). In 2001, the estimated annual cost of corrosion to the U.S. oil and gas industries was $13.4 billion (Koch et al. 2001). Moreover, reports by the U.S. Pipeline and Hazardous Materials Safety Administration (PHMSA) revealed that corrosion accounted for 30% of the accidents occurring in hazardous liquid pipeline, which is greater than any other failure mode investigated (Al-Abbas et al. 2012). Microbiologically influenced corrosion is one of several considerable concerns to the oil and gas industry. MIC has been reported in oil and gas treating facilities, such as refineries and gas fractionating plants, pipeline systems, and exporting terminals (Magot et al. 2000). MIC is one of the common failure modes for pipeline corrosion (Graves and Sullivan 1992). It has been estimated that MIC is responsible for 20% to 40% of serious corrosion cases (Walch 1992). Also, 70% to 95% of pipeline internal leaks are reportedly due to localized corrosion mainly caused by MIC (Little and Lee 2007). Microbial activities are thought to be responsible for greater than 75% of corrosion in productive oil wells and for greater than 50% of the failure of pipeline systems (Jacobson 2007). In 2006, MIC was suspected to be one of the two major factors that shut down the major Alaska Prudhoe Bay oilfield pipeline

(Jacobson 2007). The leak caused turmoil in the global oil market. This chapter presents failures of various engineering materials under the effects of different biological and chemical factors. The data offered in the chapter provide wealthy information to guide field engineers and microbiologists in how to deploy a systematic approach by utilizing the advanced microbiological and analytical methods to understand the nature of failure and to design appropriate mitigation measures.

25.2 CASE 1: FAILURE IN OUTLET PIPING IN A HIGH-PRESSURE WET CRUDE PRODUCTION TRAP

25.2.1 Background

The plant operator reported a massive leak due to internal corrosion at the outlet of one of the high-pressure production traps at the Gas Oil Separation Plant (GOSP). The leaks were found on the outlet side and occurred on 20-inch standby piping and 24-inch bypass line. The failure resulted in shutdown of production facility, which led to significant production lost (for about 70% of the operation capacity). The system specifications are as follows:

- Service life: 2 years
- Material: Carbon steel (API 5L)
- Initial thickness: 0.375 inch
- Operating temperature: 95°F–115°F
- Operating pressure: 50 psi
- Service: Crude oil with 1.2% H_2S and 3.5% water cut

25.2.2 Diagnosis and Analysis

The parts as received are shown in Figures 25.1 through 25.3. Close visual examination showed very deep localized corrosion and pitting and a significant amount of corrosion deposits. The morphology and shape of the corrosion pits were deep, localized, circular, and cup-like in appearance with striation (Figures 25.1b, 25.3b, c).

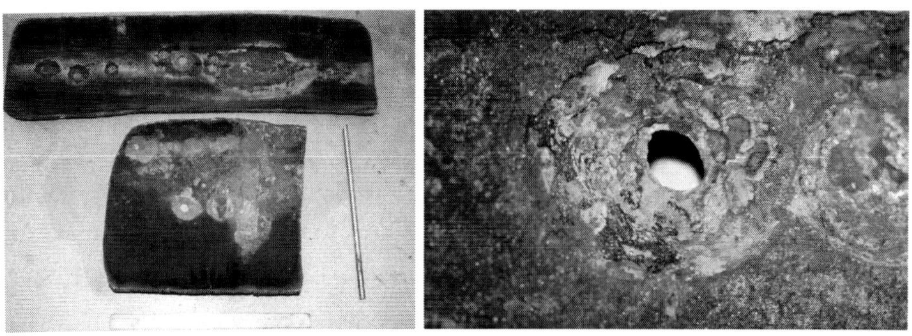

(a) (b)

FIGURE 25.1 (a) Failed outlet pipe; (b) localized hole through pitting.

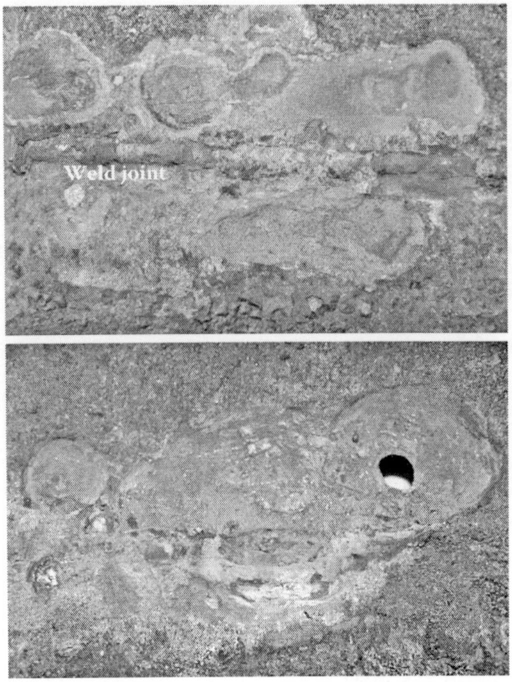

FIGURE 25.2 Localized elongated pitting along the HAZ welds.

In general, the smaller pits were discrete and hemispherical in shape. There was also some corrosion near the heat-affected zone welds as shown Figure 25.2. The heat-affected zone (HAZ) refers to the area of base material that is not melted and has had its microstructure and properties altered due to intensive heat introduced by welding operations.

There is a trend to establish a relationship between the pit shape and MIC. Previous investigations have found that initial pit formations caused by microbes have special identifying characteristics. Pope et al. (1990) investigated gas pipelines that contained large numbers of acid-producing bacteria (APB) and organic acids and reported specific metallurgical characteristics: large craters ranged from 5 to 8 cm or greater in diameter surrounded by uncorroded metal, cup-shaped or hemispherical pits on the pipe surface, striations, and tunnels at the ends of the craters. In this failure, the morphology and shape of the corrosion pits was identical to those formed by microbes. The pits were localized and cup-shaped in appearance (Figure 25.1b).

The X-ray diffraction (XRD) analysis indicated that the corrosion deposits consisted of iron oxide and iron hydroxide products. Inside the pits, the deposits also contained significant quantities of iron hydroxide and chloride corrosion products. Furthermore, iron sulfide and iron carbonate corrosion products and elemental sulfur were present all samples.

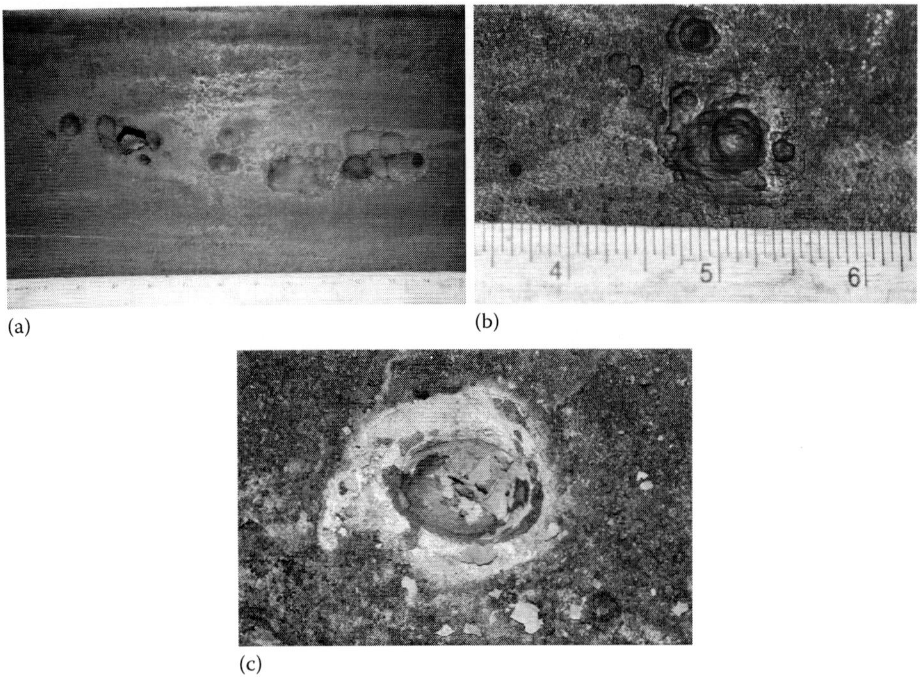

FIGURE 25.3 (a) Failed bypass line; (b) striations in the direction of rolling in the bottom of the pit; (c) localized corrosion and corrosion deposits inside.

Sterile tools and equipment were used to collect three solid surface film samples from different locations along the failed parts. The dry surface film samples were analyzed for specific corrosion-causing microorganisms using quantitative polymerase chain reaction (qPCR). The qPCR technique is considered more suitable than the traditional most probable number (MPN) technique since the only available samples were dry and have been exposed to air (Little and Lee 2007). Moreover, the MPN is a culture-based technique requiring the cultivation of microbes from samples by using selective growth media, which can detect only a fraction of microbes that can be detected by using qPCR. The qPCR was carried out as described by Zhu et al. (2006). Nucleic acids were extracted from the solid samples and used as a template to determine the number of eight different types of microorganisms in each sample (Table 25.1).

The microbial testing showed the presence of a wide range of microbial flora in relatively high numbers. As shown in Table 25.1, the identified microbial groups includes sulfate-reducing bacteria (SRB), methanogens, and acid-producing bacteria.

The diagnosis suggests that the main damage mechanism is under-deposit corrosion accelerated by MIC. The low flow or stagnation conditions promoted deposition on the indigenous scales in the crude oil and associated formation water. This accumulation creates a suitable environment for microbial growth as evident by microbial assay.

TABLE 25.1
qPCR Results of the Three Dry Samples Collected from the Failed Part

Sample	**Sample Type**	**Dry Sample**
	Description	Solid from Larger Pits
	Sample quantity (gram)	0.571
Microbial Count/Measuring Unit		**Gene Copies/Gram**
Microbial Type	Total bacteria	1.38×10^8
	Total archaea	6.31×10^8
	Sulfate-reducing bacteria (SRB)	8.10×10^4
	Sulfate-reducing archaea (SRA)	ND
	Acid-producing bacteria (APB)	1.37×10^3
	Iron-oxidizing bacteria (IOB)	ND
	Nitrate-reducing bacteria (NRB)	4.69×10^5
	Methanogens	4.54×10^7

Note: ND: none detected.

25.2.3 Summary

The root cause of the failure was attributed to under-deposit corrosion accelerated by MIC. The water accumulation created a suitable environment for microbial growth. The low or stagnant (bypass line) flow in the system prompted water to drop out and solids to build up. This creates a suitable environment for a diverse indigenous microbial population to flourish under the deposited scales.

25.2.4 Remedial Actions

A periodic flushing program and extensive cleaning was necessary to avoid scale buildup and biofilm development. Batch biocide treatment of the flowline system and internal coating of the standby spools were recommended. Rigorous inspection and microbial monitoring programs were implemented. The implementation of these measures has reduced the leaks and failures for this system significantly.

25.3 CASE 2: REFINERY SEAWATER COOLER TUBE LEAKS

25.3.1 Background

Unexpected failures and leaks were observed from damaged tubes from a refinery's seawater cooling system, which resulted in shutdowns and significant production losses. The affected tubes came from four heat exchangers at different process units: the secondary cooling water cooler, the naphtha rundown cooler, the drier overhead precondenser for kerosene hydrodesulfurization process, and the lube oil cooler.

For each heat exchange shown in Table 25.2, cooling water systems are designed to control the process temperatures and transfer waste heat from the refinery process

TABLE 25.2
Affected Heat Exchanger Information

Description	Secondary Cooling Water	Naphtha Rundown Cooler	Drier Overhead Precondenser	Lube Oil Cooler
Shell side medium	Water	Naphtha	Kerosene	Lube Oil
Tube side medium	Seawater	Seawater	Seawater	Seawater
Shell operating Temp. (°C)	85	57	96	58
Tube operating Temp. (°C)	40	44	38	40
Tube materials	Brass (B111-C68700)	Brass (B111-C68700)	Cu-Ni (B111-C71500)	Brass (B111-C68700)

units. In the oil and gas industry close to the sea, the favorable medium for cooling is seawater because it is readily available, inexpensive, has an ability to absorb large amounts of heat per square area, and can be safely discharged.

25.3.2 Diagnosis and Analysis

The corroded tube sections can be seen as received in Figure 25.4. Visual inspection after cleaning typically showed tubes were found without ferrules and minor tube end erosion in the first pass (Figure 25.5a). A subsequent shell side test with the test

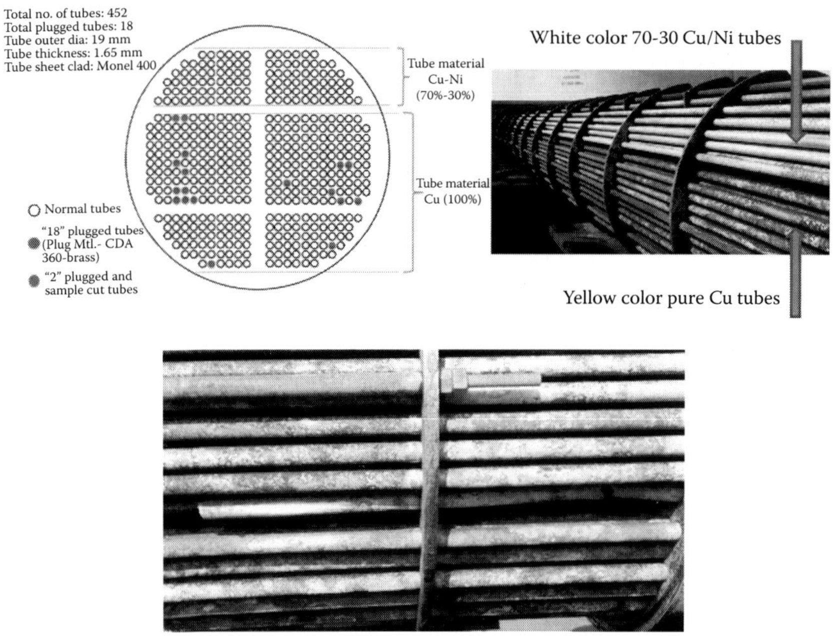

FIGURE 25.4 Location of damaged heat exchanger tubes.

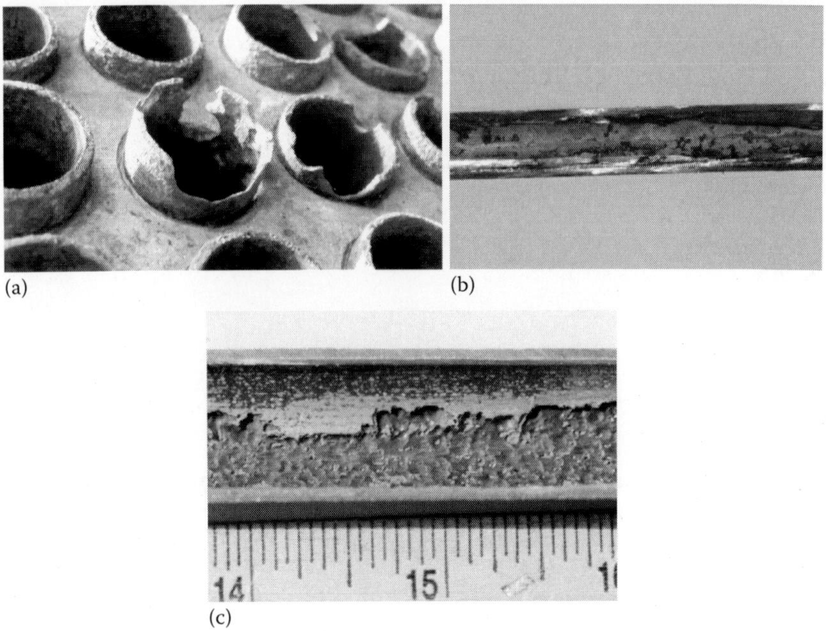

FIGURE 25.5 (a) Tube end corrosion; (b) internal localized pitting corrosion; (c) extensive metal removal.

ring reveals tubes were leaking from the units that already have ferrules while tube ends were in good condition located in the second pass. Splitting the tube revealed the tube suffered from scattered, isolated pitting, and the hole was located about 2 meters away from the floating head tube sheet (Figure 25.5b).

The macroscopic examination of the corrosion damages on tube internal diameter are shown in Figures 25.5c and 25.6. They revealed localized corrosion and a significant amount of deposits. Some of the observed pits were deep, localized, circular, and cup-like in appearance. Morphology and shape of evaluated corrosion pits were typical of those formed by MIC. As observed, corrosion pits may be characteristically rounded with etched sides and containing pits within pits as can be seen in microscopic images in Figure 25.6.

Four different solid samples collected from the internal side of the failed tube sections were analyzed by a PCR-based molecular microbiological method to determine the abundance and specific groups of microorganisms. The characterization and quantification of microorganisms on the internal surface of the samples is shown in Table 25.3. As shown in Table 25.3, medium to high numbers of acid-producing bacteria ranged from 5.19×10^3 to 1.16×10^5 cells/gram were detected in all samples. Furthermore, high numbers of iron-oxidizing bacteria (IOB) (5.28×10^5 to 4.49×10^6 cell/gram) were detected in all samples.

It is worth mentioning that a number of limited microbial analyses were conducted by the operator prior to the failures. The analyses focused on only SRB detection and have failed to detect SRB in an open system. The results appear to have

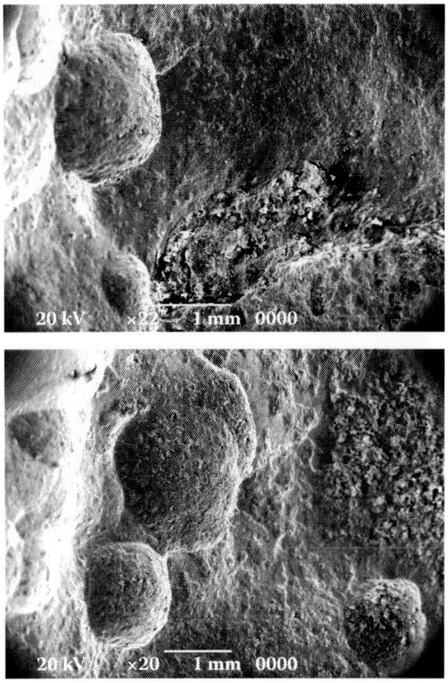

FIGURE 25.6 SEM images showing small pits within larger pits with deposit.

TABLE 25.3
qPCR Results of Four Samples from the Area Surrounding the Failed Tube Sections

Sample	ID	1	2	3	4
	Description	Solid	Solid	Solid	Solid
Microbial Count/Measuring Unit		**#/gram**	**#/gram**	**#/gram**	**#/gram**
Microbial Type	Total bacteria	5.67E+06	9.56E+07	1.99E+07	4.91E+07
	Total archaea	ND	ND	ND	ND
	Sulfate-reducing bacteria (SRB)	ND	ND	ND	ND
	Acid-producing bacteria (APB)	5.19E+03	1.16E+05	1.28E+04	4.46E+04
	Acetic acid–producing bacteria	5.19E+03	5.49E+04	1.28E+04	1.62E+04
	Butyric acid–produing bacteria	BDL	6.11E+04	BDL	2.84E+04
	Iron-oxidizing bacteria (IOB)	8.05E+05	4.49E+06	5.28E+05	9.06E+05
	Leptothrix and *Sphaerotilus* species	4.58E+04	4.94E+04	4.13E+04	4.24E+04
	Gallionella species	7.59E+05	4.44E+06	4.87E+05	8.63E+05
	Denitrifying bacteria (DNF)	ND	ND	ND	6.57E+04
	Methanogens	ND	ND	ND	ND

Note: ND: none detected; BDL: below detection limit.

led the operator to underestimate the role of other bacteria and have significantly contributed to overlooking the chlorine treatment.

Apparently, the complementary action of APB and IOB has driven tube corrosion. The numbers and diversity of detected bacteria are a cause of concern, and it is more likely that uncontrolled microbial growth has contributed to the rapid failure of the system.

To obtain a holistic understanding of the failures, the treatment strategies and operation variables were also investigated. The reviews revealed that the chlorination system was not functioning properly. The lack of chlorination and presence of hydrocarbon and other organic compounds might lead to a flourishing microbial population, including corrosion-causing species and subsequent rapid MIC.

Seawater velocity was set at an integrity operating window (IOW) for these heat exchangers. the IOW is the process limit variable that can affect the integrity of the equipment if the process operation deviates from it for a predetermined length of time. The available history records showed that the velocity was within acceptable limits, which confirmed that stagnation was not a causal factor that derives MIC failures.

25.3.3 Summary

Microscopic analysis conducted on all four leaking tubes with different metallurgies (70–30 Cu–Ni and Al–Brass) revealed pits that were deep, localized, circular, and cup-like in appearance typical of those formed by MIC. Specialized testing done showed that all deposit samples were colonized by various types of acid-producing bacteria and iron-oxidizing bacteria. However, no sulfate-reducing bacteria were detected.

25.3.4 Remedial Actions

A comprehensive control and monitoring plan was developed to safeguard the integrity of the system as follows:

1. Critical exchangers
   a. Replace in-kind (with similar materials and design) at the next available opportunity
   b. Perform high-velocity cleaning followed by shock chlorination (5–7 ppm)
2. Noncritical exchangers (can resolve online)
   a. Isolate, hydro jet clean, and inspect
   b. Alternatively, replace if dictated by economics
3. Chlorination
   a. Switch to continuous dosing without shocking
   b. Maintain a 0.1 ppm minimum return water residual
   c. Research use of outsourced chlorine dioxide replacement system
   d. Ensure availability by sparing and perform timely preventive maintenance
4. Nutritional contamination
   a. Take samples of the inlet and return water to check for hydrocarbon contamination

b. Check chlorine demand/shocking needs
c. Check other carbon sources that can be used by bacteria
d. Adjust chlorination treatment accordingly

25.4 CASE 3: PREMATURE FAILURES OF MULTIPLE FIRE HYDRANTS

25.4.1 Background

Fire water protection is a critical system for oil and gas facilities. In most cases, the majority of the fire protection systems service life is spent in a static, no flow condition. However, when required to operate in an emergency, the fire protection systems can be critically important to the safety of the assets and the continued operation of a facility. Therefore, the fire protection system components (piping, connections, valves, and hydrants) must be designed to provide reliable operation over an extended period of time. In general, groundwater or untreated seawater are used for these systems. In this case, unexpected and premature failures of fire hydrants occurred after 15 months in service. These hydrants are part of fire protection system that protects an oil storage terminal facility.

25.4.2 Diagnosis and Analysis

Different samples were prepared from the received couplant (Figure 25.7) for different mechanical, metallurgical, and microscopic evaluations. A cross-sectional sample from the fractured coupling was mounted, polished, and etched for detailed examination. The chemical and mechanical composition of the failed part is shown in Table 25.4. Positive material identification (PMI) of the submitted coupling was performed, indicating that the material was brass with 24% Zn content. Examining the coupling fracture surface revealed the effects of two distinctive failure mechanisms, namely intergranular stress corrosion cracking (SCC) and selective leaching (dezincification). Metallographic examination of a cross-sectional sample taken from the fractured coupling, Figure 25.8, showed multiple branching intergranular cracks propagating in all directions. These findings indicated material susceptibility to environmental cracking.

The qPCR results indicated the microbial structures include both bacteria and archaea (Table 25.5). This community structure (total bacteria and archaea) indicates the overall environmental conditions are suitable for prokaryotic growth. The main reason for the detection of SRB in only one of the three samples is most likely due to oxygen inhibition in the used raw seawater. SRB predominantly live in oxygen-free environments; however, they can tolerate oxygen for a long period of time (Cypionka 2000). The results of the molecular microbiology analysis confirmed the presence of two types of corrosion causing bacteria: sulfate-reducing bacteria and nitrate-reducing bacteria (Table 25.5). The sample coded "Attached Nozzle Ring" contained moderate concentrations of total bacteria of 3.46×10^5 cell/gram, archaea of 2.60×10^4 cell/gram, and denitrifying bacteria of 1.30×10^4 cell/gram whereas the sample coded "Detached Nozzle Ring" enumerated a

FIGURE 25.7 Images showing the damaged fire hydrant and collected couplant sample.

TABLE 25.4
Chemical (Weight%) and Mechanical Properties of the As-Received Materials

Mg	Cu	Al	Fe	Zn	Sn	Pb
3.67	62.94	5.62	2.69	24.69	0.04	0.10

Mechanical Properties	
Tensile strength (PSI)	94,500
Yield strength (PSI)	66,000
Elongation (%)	3.0
Hardness (HBW)	229

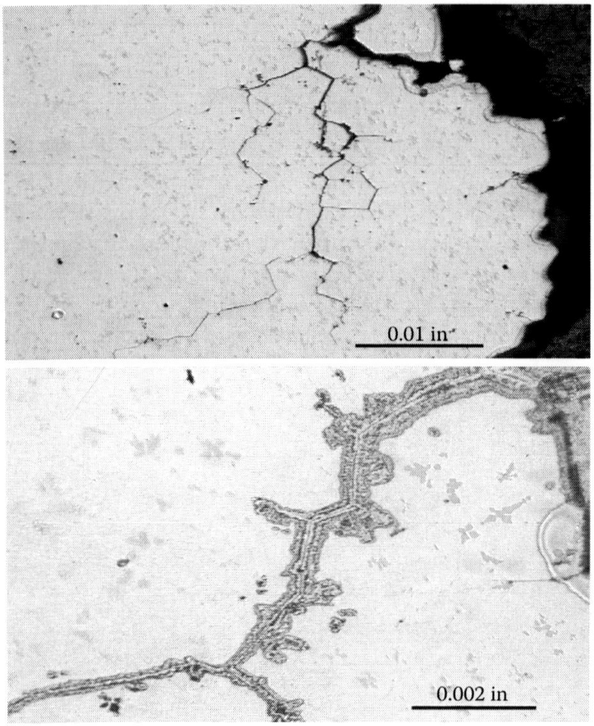

FIGURE 25.8 Metallography showing intergranular stress corrosion cracking and selective leaching (dezincification).

TABLE 25.5
qPCR Results of the Three Dry Samples Collected from the Failed Nozzle Zone

	Sample Code	**Attached Nozzle Ring**	**Detached Nozzle Ring**	**Cap Scale**	**Detection Limit/Gram**
Sample	Description	Solid	Solid	Solid	
	Sample quantity (gram)	0.04301	0.22389	0.0841	
Microbial Count/Measuring Unit		**Cell/Gram**	**Cell/Gram**	**Cell/Gram**	
Microbial Type	Total bacteria	3.46×10^5	1.72×10^5	2.38×10^5	1.32×10^3
	Total archaea	2.60×10^4	BDL	BDL	1.32×10^2
	Sulfate-reducing bacteria (SRB)	ND	7.30×10^3	ND	1.32×10^1
	Sulfate-reducing archaea (SRA)	ND	ND	ND	1.32×10^1
	Acid-producing bacteria (APB)	BDL	BDL	ND	1.32×10^1
	Iron-oxidizing bacteria (IOB)	BDL	BDL	BDL	1.32×10^2
	Nitrate-reducing bacteria (NRB)	1.30×10^4	ND	1.09×10^4	1.32×10^1
	Methanogens	BD	BDL	BDL	1.32×10^2

Note: ND: none detected; BDL: below detection limit.

moderate concentration of total bacteria of 1.72×10^5 cell/gram and a low concentration of sulfate-reducing bacteria of 7.30×10^3 cell/gram. The sample coded "Cap Scale" contained moderate concentrations of total bacteria at 2.38×10^5 cell/gram and nitrate-reducing bacteria at 1.09×10^4 cell/gram. The detection of a low level of nitrate (1 mg/L) in the collected firewater sample might be attributed to microbial consumption.

In many industrial applications, it is well established that copper alloys are susceptible to SCC when exposed to ammonia, sulfates, nitrites, and nitrates in the presence of residual or applied stress (Rao and Nair 1998). In our case, one possible source of ammonia might be bacterial activities in the used seawater. The generation of ammonia might be attributed to the presence of nitrate-reducing bacteria, which was confirmed by qPCR results.

25.4.3 Summary

The prematurely failed nozzle is constructed of a copper base alloy, and the detection of denitrifying bacteria in the surface film samples is of particular concern. This group of microorganisms is capable of reducing available oxidized nitrogen compounds through different pathways. The specific pathway for the reduction depends on a number of factors, including colonizing strains, carbon availability, and temperature and operating conditions. Depending on the specific pathway, ammonia may be intermediately produced, which is of particular concern in the case of copper.

25.4.4 Remedial Actions

Material upgrading and periodic flushing of installations with a potent antibacterial agent are among the options to be considered.

25.5 CASE 4: FAILURE OF A DIESEL SHIPPING LINE

25.5.1 Background

A 10-inch carbon steel pipe of a diesel shipper pump recycle line failed due to severe corrosion, which caused a leak. A pinhole leak was observed on the 10-inch pipe. The location of the leak was at the 6 o'clock position of a horizontal carbon steel pipe. The system specifications are as follows:

- Service life: 14 years
- Pipe size: 10 inches
- Thickness: 0.365 inch (standard schedule)
- Fabrication material: Carbon steel API 5L
- Service environment: Diesel
- Operating temperature: 100°F–150°F
- Operating pressure: 330–390 psig

25.5.2 Diagnosis and Analysis

The failed part as received can be seen in Figure 25.9. Visual examination showed a significant amount of corrosion deposits. Upon removal of the deposits, very deep localized corrosion and pitting were found (Figure 25.10). The morphology and shape of the corrosion pits was deep.

A sample of the deposit was collected from the internal pipe surface to determine the chemical composition using the X-ray powder diffraction technique. XRD results (Table 25.6) showed that the samples consisted mainly of iron oxide, iron hydroxide, and iron carbonate. Minor quantities of sodium chloride (NaCl) were also present.

The morphology and shape of the corrosion pits was typical of those formed by MIC. The pits were localized, circular, and ring-like in appearance. In addition, the presence of deposits and the 6 o'clock position point to MIC as the culprit. The presence of iron carbonate (siderite) in a system that shouldn't have any carbonate or carbon dioxide was also indicative of MIC.

Looking to the corrosion product color and compositions, it was postulated the MIC was driven by the presence of a microbial consortium the might include metal oxidizing. Rajasekar et al. (2007) investigated the role of *Bacillus cereus ACE4* on degradation of diesel and its influence on corrosion of API 5LX steel. This study revealed that *B. cereus* degraded the aliphatic and aromatic in diesel and is capable of oxidizing ferrous/manganese into oxides.

The use of biocides to control microbes is not a cost-effective solution due to the high volumes that pass through the system each day and the difficulty in protecting the steel under deposits and in stagnant lines.

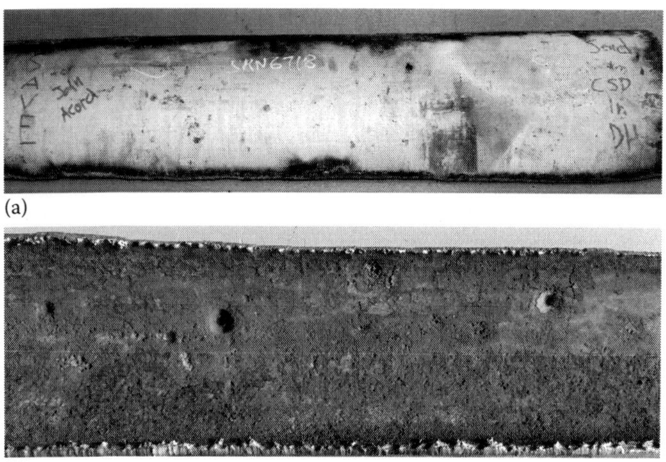

FIGURE 25.9 (a) The external pipe surface in the as-received condition; (b) close-up on the internal surface showing perforation (after cleaning).

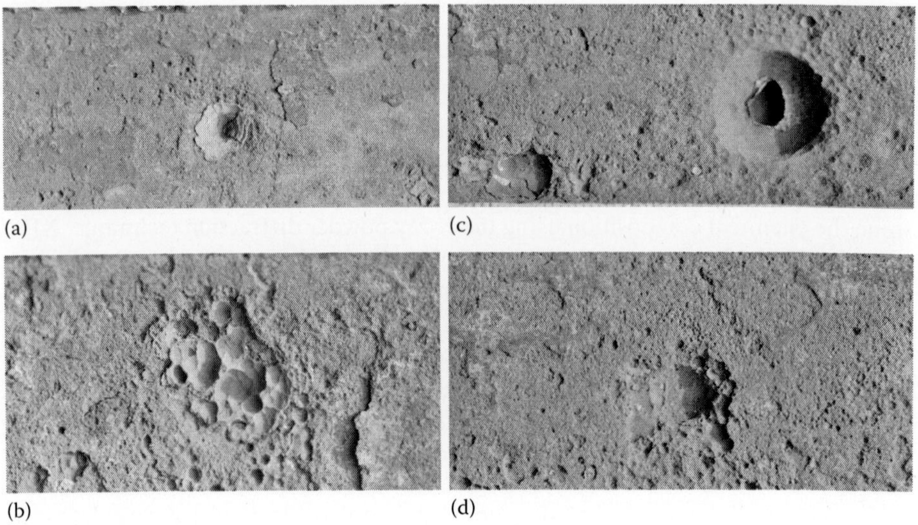

FIGURE 25.10 (a) Localized corrosion and deposits; (b) circular and cup-like in appearance pit; (c) circular shape of the hole through the pit; (d) smaller pits were discrete and hemispherical in shape.

TABLE 25.6
Crystalline Phases Identified by XRD

Compound	wt%
Magnetite–Fe_3O_4	37
Akaganeite–FeOOH	25
Goethite–FeOOH	32
Siderite–$FeCO_3$	4
Halite–NaCl	2

25.5.3 Remedial Actions

It is important to note that stagnant or low-flow conditions may allow and even promote the growth of microorganisms and biofilm development. Routinely flushing stagnant lines and increasing the velocities in low flowlines, corrosion products, and solids are less likely to deposit on the piping thus minimizing the possibility of MIC. It is also more effective to apply internal coatings to all dead-leg portions of the system to provide more protection.

25.6 CONCLUSION

During oil and gas operations, the system networks are subjected to different corrosion deterioration mechanisms that result from the interaction between the fluid process and the substrates. Among these mechanisms is microbiologically influenced corrosion that results from accelerated deterioration caused by different indigenous microorganisms that naturally reside in the hydrocarbon and associated water systems. This chapter discussed several real field cases related to MIC failures that were recorded and reported from oil and gas facilities and associated operations.

It has been demonstrated that MIC-related failure investigations require a combination of microbiological, metallurgical, and chemical analyses and encompasses three different tasks: (1) identify causative microorganisms, either in the bulk medium or associated with the corrosion products; (2) correlate pit morphology with an MIC mechanism; and (3) identify a corrosion product chemistry that is consistent with the causative organisms.

Historically, the identification and accurate enumeration of potentially problematic microbes in oil and gas systems has proven to be a challenge for the oil and gas industry. Cultivation methods are considered the most often used practice in the industry. However, it has been recognized that culture-based techniques reflect only up to 15%. This figure is not surprising as biodiversity, predominantly microbial diversity, is dominated in most ecosystems by uncultured microorganisms. These limitations of culture-based methods have inspired the application of molecular microbiological methods (MMM) such as qPCR. The utilization of molecular microbiological analyses has recently increased in the oil and gas industries.

The data offered by this chapter provide a wealth of information and might guide field engineers in how to deploy a systematic approach by utilizing advanced microbiological and analytical methods to understand the nature of failure and to design appropriate mitigation measures.

REFERENCES

Al-Abbas F. M., J. R. Spear, A. Kakpovbia, N. M. Balhareth, D. L. Olson and B. Mishra. 2012. Bacterial attachment to metal substrate and its effects on microbiologically influenced in transporting hydrocarbon pipelines. *Journal Pipeline Engineering*, vol. 11, no. 3, pp. 63–72.

Al Nabulsi K. M., T. Y. Rizk, F. M. Al-Abbas and O. C. Dias. 2016. Sea water cooler tubes corrosion and leaks due to microbiologically induced corrosion. NACE Corrosion Conf. & Expo, Vancouver, Canada.

Al Nabulsi K. M., F. M. Al-Abbas, T. Y. Rizk and A. M. Salameh. 2015. Microbiologically assisted stress corrosion cracking in the presence of nitrate reducing bacteria. *Eng. Fail. Anal.*, vol. 58, pp. 165–172.

Buck E., G. C. Maddux and R. L. Sullivan. 1996. Internal corrosion cost impact study—United States natural gas exploration and production industry. Gas Res. Inst., Des Plaines, IL, Rep. GRI-96/0056 (96–1466).

Cypionka H. 2000. Oxygen respiration by *Desulfovibrio* species. *Annu. Rev. Microbiol.*, vol. 54, no. 1, pp. 827–848.

Graves J. W. and E. H. Sullivan. 1992. Internal corrosion in gas gathering systems and transmission lines. *Materials Protection*, vol. 5, pp. 33–37.

Jacobson G. 2007. Corrosion at Prudhoe Bay—A lesson on the line. *Materials Performance*, vol. 46, pp. 27–34.

Koch G. H., M. P. H. Brongers, N. G. Thompson, Y. P. Virmani and J. H. Payer. 2001. Corrosion costs and preventive strategies in the U.S. Dept. of Transp. Dublin, OH, Rep. FHWA-RD-01-156, 2001.

Little B. J. and J. S. Lee. 2007. *Microbiologically Influenced Corrosion.* Hoboken, NJ: Wiley & Sons.

Magot M., B. Ollivier and B. K. C. Patel. 2000. Microbiology of petroleum reservoirs. *Antonie Van Leeuwenhoek*, vol. 77, no. 2, pp. 103–116.

Pope D. 1990. *GRI Field Guide: Microbiologically Influenced Corrosion (MIC): Methods of Detection in the Field.* Chicago: Gas Research Institute.

Rajasekar A., T. G. Babu, S. K. Pandian, S. Maruthamuthu, N. Palaniswamy and A. Rajendran. 2007. Biodegradation and corrosion behavior of manganese oxidizer *Bacillus cereus* ACE4 in diesel transporting pipeline. *Corrosion Science*, vol. 49, no. 6, pp. 2694–2710.

Rao T. S. and K. Nair. 1998. Microbiologically influenced stress corrosion cracking failure of admiralty brass condenser tubes in a nuclear power plant cooled by fresh water. *Corrosion Science*, vol. 40, no. 11, pp. 1821–1836.

Walch M. 1992. Corrosion, Microbial. In *Encyclopaedia of Microbiology*, vol. 1, J. Lederberg, Ed. New York: Academic, pp. 585–591.

Zhu X. Y., H. Modi, A. Ayala and J. J. Kilbane II. 2006. Rapid detection and quantification of microbes related to microbiologically influenced corrosion using quantitative polymerase chain reaction. *Corrosion*, vol. 62, pp. 950–955.

Index

Page numbers followed by f and t indicate figures and tables, respectively.

C

D

G

H

N

O

Q

R

S

T

U